Quantum Measurements in Optics

NATO ASI Series

Advanced Science Institutes Series

A series presenting the results of activities sponsored by the NATO Science Committee, which aims at the dissemination of advanced scientific and technological knowledge, with a view to strengthening links between scientific communities.

The series is published by an international board of publishers in conjunction with the NATO Scientific Affairs Division

A	**Life Sciences**	Plenum Publishing Corporation
B	**Physics**	New York and London
C	**Mathematical and Physical Sciences**	Kluwer Academic Publishers
D	**Behavioral and Social Sciences**	Dordrecht, Boston, and London
E	**Applied Sciences**	
F	**Computer and Systems Sciences**	Springer-Verlag
G	**Ecological Sciences**	Berlin, Heidelberg, New York, London,
H	**Cell Biology**	Paris, Tokyo, Hong Kong, and Barcelona
I	**Global Environmental Change**	

Recent Volumes in this Series

Volume 278—Painlevé Transcendents
edited by D. Levi and P. Winternitz

Volume 279—Fundamental Aspects of Inert Gases in Solids
edited by S. E. Donnelly and J. H. Evans

Volume 280—Chaos, Order, and Patterns
edited by Roberto Artuso, Predrag Cvitanović, and Giulio Casati

Volume 281—Low-Dimensional Structures in Semiconductors: From Basic Physics to Applications
edited by A. R. Peaker and H. G. Grimmeiss

Volume 282—Quantum Measurements in Optics
edited by Paolo Tombesi and Daniel F. Walls

Volume 283—Cluster Models for Surface and Bulk Phenomena
edited by Gianfranco Pacchioni, Paul S. Bagus, and Fulvio Parmigiani

Volume 284—Asymptotics beyond All Orders
edited by Harvey Segur, Saleh Tanveer, and Herbert Levine

Volume 285—Highlights in Condensed Matter Physics and Future Prospects
edited by Leo Esaki

Series B: Physics

Quantum Measurements in Optics

Edited by

Paolo Tombesi

University of Camerino
Camerino, Italy

and

Daniel F. Walls

University of Auckland
Auckland, New Zealand

Springer Science+Business Media, LLC

Proceedings of a NATO Advanced Research Workshop
on Quantum Measurements in Optics,
held January 21-25, 1991,
in Cortina d'Ampezzo, Italy

Library of Congress Cataloging-in-Publication Data

Quantum measurements in optics / edited by Paolo Tombesi and Daniel F.
Walls.
p. cm. -- (NATO ASI series. Series B, Physics ; v. 282)
"Proceedings of a NATO Advanced Research Workshop on Quantum
Measurements in Optics, held January 21-25, 1991, in Cortina
d'Ampezzo, Italy"--T.p. verso.
"Published in cooperation with NATO Scientific Affairs Division."
Includes bibliographical references and index.
ISBN 978-0-306-44101-1 ISBN 978-1-4615-3386-3 (eBook)
DOI 10.1007/978-1-4615-3386-3
1. Quantum optics--Measurement--Congresses. 2. Quantum
electronics--Congresses. I. Tombesi, P. (Paolo) II. Walls, D. F.
III. NATO Advanced Research Workshop on Quantum Measurements in
Optics (1991 : Cortina d'Ampezzo, Italy) IV. North Atlantic Treaty
Organization. Scientific Affairs Division. V. Series.
QC446.15.Q33 1992
535--dc20 91-43646
CIP

ISBN 978-0-306-44101-1

Originally published by Plenum Press, New York in 1992

ORGANIZING COMMITTEE

P. Tombesi	University of Rome "La Sapienza", Italy
D.F. Walls	University of Auckland, New Zealand
E. Giacobino	Lab.Spectr. Hertzienne de l'ENS, France
P. Meystre	University of Arizona, USA.
R.E. Slusher	AT & T Bell Lab., USA

INVITED SPEAKERS

I.Abram	Centre Nationale d'Etudes des Télécommunications, France
L.Accardi	Centro Matematico "V.Volterra", Università Roma II, Italy
R.J.Brecha	Sektion Physik, Universität München, Germany
M.Brune	Lab. Spectroscopie Hertzienne de l'ENS, France
R.K.Bullough	The University of Manchester, U.K.
F.De Martini	University of Rome "La Sapienza", Italy
P.D.Drummond	University of Queensland, Australia
C.Fabre	Lab. Spectroscopie Hertzienne de l'ENS, France
K.M.Gheri	University of Auckland, New Zealand
E.Giacobino	Lab. Spectroscopie Hertzienne de l'ENS, France
R.J.Glauber	Harvard University, U.S.A.
P.Grangier	Institut d'Optique, France
F.Haake	Universität Essen, Germany
N.Imoto	University of Essex, U.K.
P.Knight	Imperial College of Science, U.K.
L.A.Lugiato	Università di Milano, Italy
V.I.Man'ko	Lebedev Institute of Physics, USSR
M.B.Mensky	Lebedev Institute of Physics, USSR
J.Mlynek	University of Konstanz, Germany
E.R.Pike	King's College London, U.K.
J.G.Rarity	Royal Signals and Radar Establishment, U.K.

W.Schempp	Universität Siegen, Germany
W.P.Schleich	Max-Planck Institut fur Quantenoptik, Germany
M.Scully	Max-Planck Institut fur Quantenoptik, Germany
S.Stenholm	University of Helsinki, Finland
P.Tombesi	University of Rome “La Sapienza”, Italy
E.Yablonovitch	Bellcore Lab., USA
Y.Yamamoto	Basic Research Lab., NTT, Japan
A.Zeilinger	Atominstitut, Austria
P.Zoller	Universität Innsbruck, Austria

PREFACE

The NATO Advanced Research Workshop on Quantum Measurements in Optics was held in Cortina d'Ampezzo, Italy, January 21-25, 1991. This workshop was attended by 70 participants from 16 different countries.

The subjects discussed at this workshop concentrated on quantum measurements in optics made possible by the recent advances in the generation and detection of light with low quantum noise. These advances have occurred simultaneously with the development of atomic traps capable of trapping a single atom for a considerable period of time. The interaction of a single two level atom with the single mode of the electromagnetic field is now possible in high Q microcavities. A new field of cavity QED has developed studying the properties of Rydberg atoms in microwave cavities.

At this meeting we heard the first report of an atomic interferometer where a single atom passing through the two slits exhibits wave like interference phenomena. This new field involving the transfer of momentum from photons to atoms has lead to new possibilities for quantum nondemolition measurements on an optical field. We heard suggestions for such measurements at this meeting.

With the new light sources available the possibility of using low quantum noise light in optical communications becomes close to reality. The problem of the propagation of quantum light field in optical fibres was actively discussed at this meeting.

New results in the development of low noise light sources were reported particularly in the area of twin beam generation where the detection of one beam enables one to accurately deduce the noise properties of its twin. Such twin photon generation has also been used in intensity interference experiments which demonstrate the nonlocality of the photon field via a violation of Bell's inequalities.

The workshop successfully covered a wide range of topics in quantum measurements in optics from fundamental problems in quantum mechanics to applications in optical technology.

The special flavour of the meeting was due to the beautiful environment in which it was held. Its success owed much to the cooperation and assistance we received from Mrs. Flavia Bertozi, the hotelier who generously allowed the evening sessions to be held in the very comfortable Hotel Ancora and provided us with such splendid meals and convivial atmosphere. We also wish to thank the Consiglio di Istituto of the Istituto d'Arte of Cortina, in particular the Dean Prof. Emilio Bassanin, who provided the venue for the lectures. We are also grateful to the local Consorzio per lo Sviluppo e il Turismo for their assistance in many ways.

Finally we wish to thank our sponsors NATO, Consiglio Nazionale delle Ricerche and Istituto Nazionale di Fisica Nucleare without whose generous support the workshop could never have been held.

We have had full collaboration from Plenum Press in the publication of this volume, and the success of the meeting and the not inconsiderable work before and afterwards owes much to our uncomplaining secretaries Angela Di Silvestro and Marcella Mastrofini.

Paolo Tombesi

Daniel F. Walls

June 1991

CONTENTS

I. ATOMIC INFERFEROMETRY AND TRAPPING

II. QUANTUM MEASUREMENTS

III. QUANTUM NOISE REDUCTION AND SQUEEZING

IV. OPTICAL PROPAGATION, COMMUNICATION AND IMAGING

V. PHOTON INTERFERENCE AND BELL'S THEOREM

I. ATOMIC INTERFEROMETRY AND TRAPPING

QUANTUM THEORY OF PARTICLE TRAPPING BY OSCILLATING FIELDS

Roy J. Glauber

Lyman Laboratory of Physics
Harvard University, Cambridge, MA 02138

1. INTRODUCTION

The technique of trapping atomic particles in electromagnetic fields has opened the possibility of dramatic advances in high-precision spectroscopy. Most theoretical analyses of trapping have thus far been carried out in classical terms, but since the very purpose of such techniques is to minimize the energy of particle motion, it is clear that their success implies a need for quantum mechanical analysis. Only by treating such systems quantum mechanically can we hope to understand the influence of residual particle motions on radiative interactions, or any fundamental limits on the lowest attainable particle temperatures.

Since there is no static field configuration in which charged particles can be permanently trapped, we must deal with their motion in explicitly time-dependent fields. The quantum states we must find are therefore not stationary states, or energy eigenstates of any kind. Finding them thus presents somewhat novel problems in the context of quantum mechanics, and developing simple means for doing that possesses a certain methodological interest in its own right.

The Paul trap[1] offers a particularly simple example of a trapping problem since its basic equations of motion, even though they contain a time-dependent coefficient, remain fundamentally linear in structure. Those equations can easily be solved by a simple technique[2] for generating a constant of the motion. The result, we shall show, is that even though the particle motion has no stationary states, we can construct for it a complete set of non-stationary states that are in an elementary one-to-one correspondence with the energy eigenstates of ordinary (static-field) harmonic oscillators. We can combine these states, furthermore, to form analogues of the coherent states of the static-field oscillators. These are correspondingly useful wave-packet states that follow, in their mean behavior, all the details of the classical motion of a particle in the time-dependent field.

Many years ago P. Kapitza[3] suggested an interesting approximation for treating classically the motion of a charged particle in a rapidly oscillating field. He assumed that the motion can be separated into two components, one of large amplitude that varies more slowly than the applied field, and the other much smaller in amplitude but varying with the same frequency as the applied field. The approximation suggested

Quantum Measurements in Optics, Edited by P. Tombesi and D.F. Walls, Plenum Press, New York, 1992

that the slow component of the motion can be regarded as taking place in a certain effective static potential. We shall show, by formulating an exact solution for the Paul trap, that the Kapitza approximation can hold quite accurately in quantum mechanical as well as classical terms.

2. FORMULATION OF CONSTANTS OF THE MOTION

The Paul trap is based on an oscillating quadrupole field, that is to say, an oscillating potential quadratic in each of the three cartesian coordinates of a charged particle. Since the problem is thus separable, it suffices to consider the motions of the three cartesian coordinates individually. We shall just consider one of them for the present, and call it $q(t)$. The time dependence of the quadratic potential, on the other hand, need not be restricted to simple oscillations. We shall see that the problem of finding the quantum mechanical motion can be solved in large measure without specifying the time dependence of the field at all. Let us consider therefore the motion that follows from the Hamiltonian

$$H_0(q,p,t) = \frac{p^2}{2m} + \frac{m}{2}W(t)q^2, \tag{2.1}$$

in which p and q are momentum and coordinate operators respectively and $W(t)$ expresses the time dependence of the externally applied field.

The Hamiltonian equations of motion

$$\dot{q} = \frac{p}{m} \tag{2.2}$$

and

$$\dot{p} = -mW(t)q \tag{2.3}$$

can be combined into the second order equation

$$\ddot{q} + W(t)q = 0. \tag{2.4}$$

If $q_1(t)$ and $q_2(t)$ are any two solutions of this equation, they must obey the Wronskian relation

$$q_1\dot{q}_2 - \dot{q}_1q_2 = \text{constant}. \tag{2.5}$$

Our strategy will be to use this Wronskian relation to generate a particularly useful constant of the motion of the quantum mechanical system. We note, for this purpose, that q_1 and q_2 may be *any* solutions of the differential equation (2.4). We are free, for example, to let $q_1(t)$ be an ordinary c–number solution of Eq. (2.4), and to let $q_2(t)$ be the general operator solution $q(t)$ to Eqs. (2.2) and (2.3), regarded as Heisenberg equations of motion.

We will find it useful therefore to introduce the complex c–number function $u(t)$, which we define to satisfy the differential equation

$$\ddot{u} + W(t)u = 0, \tag{2.6}$$

and the initial condition

$$u(0) = 1. \tag{2.7}$$

A second initial condition is necessary for the specification of $u(t)$ and we shall write that as

$$\dot{u}(0) = i\omega, \tag{2.8}$$

where ω is a real positive parameter that we are free to choose quite arbitrarily, at least for the present. A useful property of $u(t)$ is that it too obeys the Wronskian identity

$$u^*(t)\dot{u}(t) - u(t)\dot{u}^*(t) = 2i\omega. \tag{2.9}$$

When this function $u(t)$ and the coordinate $q(t)$ are substituted in Eq. (2.5), we see that the combination $u(t)\dot{q}(t) - \dot{u}(t)q(t)$ must be constant in time. By including a factor of the mass m in this expression, and giving it a convenient normalizing factor we can define the function

$$C(t) = \frac{i}{\sqrt{2m\hbar\omega}}\left\{u(t)p(t) - m\dot{u}(t)q(t)\right\}, \tag{2.10}$$

which is evidently an operator constant of the motion. It is therefore equal at all times to its initial value

$$C(t) = C(0) = \frac{i}{\sqrt{2m\hbar\omega}}(p - im\omega q), \tag{2.11}$$

an expression that we can recognize as the annihilation operator for excitations of an ordinary harmonic oscillator of frequency ω. It will be convenient then to use the quantum states of just such a static–field oscillator as a basis for describing our time–dependent system.

The operators p and q, which in Eq. (2.11) stand for the initial values of $p(t)$ and $q(t)$, can also be taken to be the corresponding time–independent operators characteristic of the Schrödinger picture. If we define the annihilation operator

$$a = \frac{i}{\sqrt{2m\hbar\omega}}(p - im\omega q) \tag{2.12}$$

for the excitations of a static–field oscillator of frequency ω, its ground state $|n{=}0\rangle_\omega$ must obey the relation

$$a|n{=}0\rangle_\omega = 0. \tag{2.13}$$

The constancy of $C(t)$, expressed by Eq. (2.11), then assures us that

$$C(t)|n{=}0\rangle_\omega = 0 \tag{2.14}$$

at all times. This fact furnishes us a simple means of determining the state at any time for our time–dependent system. The operator $C(t)$ is defined in Eq. (2.10) within the Heisenberg picture; that is, it is a linear combination of the Heisenberg operators $p(t)$ and $q(t)$. It is the time dependence of the linear combination coefficients that compensates for the time dependence of $p(t)$ and $q(t)$ and reduces $C(t)$ to constancy.

The unitary transformation from the Schrödinger to the Heisenberg picture takes the characteristic form

$$q(t) = U^{-1}(t)qU(t), \tag{2.15}$$

in which the operator U obeys the time-dependent Schrödinger equation and has the initial value $U(0) = 1$. We shall not, however, need to find $U(t)$ at all. It will suffice simply to observe that $C(t)$ is related to its Schrödinger-picture counterpart $C_s(t)$ by the relation

$$C(t) = U^{-1}(t)C_s(t)U(t), \tag{2.16}$$

and that $C_s(t)$ is given by

$$C_s(t) = \frac{i}{\sqrt{2m\hbar\omega}}\left\{u(t)p - m\dot{u}(t)q\right\}, \tag{2.17}$$

which is a time-dependent combination of time-independent operators.

3. SOLUTION FOR THE WAVE FUNCTIONS

To find the time-dependent states we note that we can rewrite Eq. (2.13) or (2.14) in the form

$$C_s(t)U(t)|n{=}0\rangle_\omega = 0, \tag{3.1}$$

that follows from Eq. (2.16). The unitarily transformed state vector $U(t)|n{=}0\rangle_\omega$ is the Schrödinger state of the time-dependent system that evolves from the initial ($t = 0$) static-field oscillator state $|n{=}0\rangle_\omega$. If we write it as

$$|n{=}0, t\rangle \equiv U(t)|n{=}0\rangle_\omega, \tag{3.2}$$

then we see that it obeys the equation

$$C_s(t)|n{=}0, t\rangle = 0, \tag{3.3}$$

or

$$\left\{u(t)p - m\dot{u}(t)q\right\}|n{=}0, t\rangle = 0. \tag{3.4}$$

Expressed in coordinate space, this is the first-order differential equation

$$\left\{u(t)\frac{\hbar}{i}\frac{\partial}{\partial q'} - m\dot{u}(t)q'\right\}\langle q'|n{=}0, t\rangle = 0, \tag{3.5}$$

which has the normalized solution

$$\langle q'|n{=}0, t\rangle = \left(\frac{m\omega}{\pi\hbar}\right)^{1/4}\frac{1}{\{u(t)\}^{1/2}}\exp\left\{\frac{im}{2\hbar}\frac{\dot{u}(t)}{u(t)}q'^2\right\}. \tag{3.6}$$

This is at all times a Gaussian wave function similar in several ways to the ground state wave function for the static-field oscillator, and it reduces to the latter wave function for the constant field specified by $W(t) = \omega^2$ and $u(t) = \exp(i\omega t)$. In the more general case it differs from the static wave function only by a time-dependent complex scale factor.

It is worth emphasizing that, because we have not needed to specify the time dependence of the field, the wave function (3.6) applies as well to the motion of unbound particles as it does to ones that are trapped by the time-varying field. The rapid decrease of the wave function as $q'^2 \to \infty$, and its square integrability are, in fact, not related to the question of trapping. They only require that $\mathcal{I}m\{\dot{u}(t)/u(t)\}$ be non-negative, and that condition is assured, for arbitrary time dependence of the field, by writing the Wronskian condition (2.9) in the form

$$\mathcal{I}m\left\{\frac{\dot{u}(t)}{u(t)}\right\} = \frac{\omega}{|u(t)|^2} \geq 0. \tag{3.7}$$

The density function that characterizes this state simplifies, when use is made of the Wronskian relation (2.9), to the form

$$|\langle q'|n{=}0,t\rangle|^2 = \left\{\frac{m\omega}{\pi\hbar|u(t)|^2}\right\}^{1/2} \exp\left\{-\frac{m\omega}{\hbar}\frac{q'^2}{|u(t)|^2}\right\}. \tag{3.8}$$

It shows that the expectation value of $q^2(t)$ is given by

$$\langle q^2(t)\rangle_{n=0} = \frac{\hbar}{2m\omega}|u(t)|^2. \tag{3.9}$$

We could, alternatively, have started our time-dependent system out in any of the n-quantum states of the static-field oscillator

$$|n\rangle_\omega = \frac{a^{\dagger n}}{\sqrt{n!}}|0\rangle_\omega. \tag{3.10}$$

The time-dependent Schrödinger state vector would then be

$$|n,t\rangle = U(t)|n\rangle_\omega = \frac{[C_s^\dagger(t)]^n}{\sqrt{n!}}|0,t\rangle. \tag{3.11}$$

When this relation is recast in the coordinate representation, it generates the Hermite polynomials much as the corresponding relation (3.10) does for the static-field oscillator. In this way we find the entire sequence of normalized coordinate-space wave functions

$$\begin{aligned}\langle q'|nt\rangle = &\frac{1}{\sqrt{n!}}\left(\frac{m\omega}{\pi\hbar}\right)^{1/4}\frac{1}{[u(t)]^{1/2}}\left\{\frac{u^*(t)}{2u(t)}\right\}^{n/2} H_n\left(\left\{\frac{m\omega}{\hbar|u(t)|^2}\right\}^{1/2} q'\right)\times \\ &\times \exp\left\{\frac{im}{2\hbar}\frac{\dot{u}(t)}{u(t)}q'^2\right\}.\end{aligned} \tag{3.12}$$

Wave functions that fall into this form for the case in which the function $W(t)$ has a cosine dependence on the time have been derived by M. Combescure[4]. The present approach leaves the time-dependent spring constant $W(t)$ unrestricted.

It is equally easy to return to Eq. (3.4) and use it, by similar steps, to solve for the entire sequence of wave functions in momentum space. They are found to be

$$\langle p'|nt\rangle = \frac{1}{\sqrt{n!}}\left(\frac{\omega}{\pi m\hbar}\right)^{1/4}\frac{1}{[\dot{u}(t)]^{1/2}}\left\{\frac{\dot{u}^*(t)}{2\dot{u}(t)}\right\}^{n/2} H_n\left(\left\{\frac{\omega}{m\hbar|\dot{u}(t)|^2}\right\}^{1/2} p'\right)\times$$
$$\times \exp\left\{\frac{-i}{2m\hbar}\frac{u(t)}{\dot{u}(t)}p'^2\right\}. \tag{3.13}$$

In particular the momentum space density for the $n = 0$ state is

$$|\langle p'|0t\rangle|^2 = \left\{\frac{\omega}{\pi m\hbar|\dot{u}(t)|^2}\right\}^{1/2} \exp\left\{-\frac{\omega}{m\hbar}\frac{p'^2}{|\dot{u}(t)|^2}\right\}. \tag{3.14}$$

and the expectation value of $p^2(t)$ is

$$\langle p^2(t)\rangle_{n=0} = \frac{m\hbar}{2\omega}|\dot{u}(t)|^2. \tag{3.15}$$

4. THE COHERENT STATES

The n–quantum states for the ordinary harmonic oscillator of frequency ω form a complete set. We have found the time–dependent wave function that evolves from each of them when it is chosen as the initial state. The wave function that evolves from an arbitrary choice of the initial state may therefore be written as a linear combination of these wave functions.

A particularly interesting motion of the time–dependent system is one that follows from an initially coherent state. We take the Schrödinger state vector at time $t = 0$, in this case, to be the coherent state $|\alpha\rangle_\omega$, which satisfies the relation

$$C(t)|\alpha\rangle_\omega = a|\alpha\rangle_\omega = \alpha|\alpha\rangle_\omega \tag{4.1}$$

for an arbitrary complex value of α. The time dependent state $|\alpha, t\rangle$ that evolves from this initial state then obeys the relation

$$C_s(t)|\alpha, t\rangle = \alpha|\alpha, t\rangle. \tag{4.2}$$

To find the corresponding configuration space wave function we can either solve this equation, as we solved the previous ones, or form the appropriate linear combination of the solutions given by Eq. (3.12). The result in either case may be written as the Gaussian wave function

$$\langle q'|\alpha, t\rangle = \left(\frac{m\omega}{\pi\hbar}\right)^{1/4}\frac{1}{[u(t)]^{1/2}}\exp\left\{\frac{im}{2\hbar}\frac{\dot{u}(t)}{u(t)}q'^2 + \frac{\alpha q'}{u(t)}\sqrt{\frac{2m\omega}{\hbar}} - \frac{\alpha^2 u^*(t)}{2u(t)} - \frac{1}{2}|\alpha|^2\right\}. \tag{4.3}$$

It is a form that shares and considerably generalizes some of the remarkable properties of the coherent state wave function for an ordinary harmonic oscillator. The configuration space density function, for example, takes the form

$$|\langle q'|\alpha, t\rangle|^2 = \left\{\frac{m\omega}{\pi\hbar|u(t)|^2}\right\}^{1/2} \exp\left\{\frac{-m\omega}{\hbar|u(t)|^2}\left[q' - \left(\frac{2\hbar}{m\omega}\right)^{1/2}\mathcal{R}e\{\alpha u^*(t)\}\right]^2\right\}. \tag{4.4}$$

It is a Gaussian function of the same form as the ground state density given by Eq. (3.8), but is centered on the mean coordinate

$$\langle q(t)\rangle_\alpha = \sqrt{\frac{2\hbar}{m\omega}}\mathcal{R}e\{\alpha u^*(t)\}. \tag{4.5}$$

The width of the Gaussian density distribution, in other words, varies with time precisely as the ground state distribution does, while its center is carried along an essentially classical trajectory.

5. UNCERTAINTY PRODUCTS

The coherent states of an ordinary harmonic oscillator are well known to yield the minimal uncertainty product

$$\langle(\delta q)^2\rangle\langle(\delta p)^2\rangle = \frac{\hbar^2}{4}, \tag{5.1}$$

where $\delta q = q - \langle q\rangle$, and $\delta p = p - \langle p\rangle$. For the generalized coherent states defined by Eqs. (4.2) and (4.3), on the other hand, the uncertainty product is the same as that in the $n = 0$ state. It is given, in other words, by the product of the expressions in Eqs. (3.9) and (3.15).

$$\langle(\delta q)^2\rangle\langle(\delta p)^2\rangle = \frac{\hbar^2}{4\omega}|u(t)|^2|\dot{u}(t)|^2. \tag{5.2}$$

This expression indeed takes the value $\hbar^2/4$ at time $t = 0$, but at later times, as we shall see, it is in general larger.

There is nonetheless a well-defined sense in which the uncertainty product (5.2) is minimal. The combination of the Schwarz inequality and the canonical commutation rule provides that

$$\langle(\delta q)^2\rangle\langle(\delta p)^2\rangle \geq \frac{1}{4}\langle\delta q\delta p + \delta p\delta q\rangle^2 + \frac{1}{4}\hbar^2. \tag{5.3}$$

The Hamiltonian of an ordinary harmonic oscillator possesses a kind of rotational symmetry in the variables p and $m\omega q$ that makes the expectation value of the anticommutator $\{\delta q, \delta p\}$ vanish in any stationary state. That symmetry is in general broken, however, by the time dependence of the function $W(t)$ in the Hamiltonian stated in Eq. (2.1). The expectation value of the anticommutator need not vanish, and in fact we find

$$\langle\{\delta q(t), \delta p(t)\}\rangle^2 = \hbar^2\{|u(t)|^2|\frac{\dot{u}(t)}{\omega}|^2 - 1\}, \tag{5.4}$$

an expression that can never be negative, and so it follows that the uncertainty product, though minimal, must in general exceed $(1/4)\hbar^2$. In more physical terms the uncertainty relation (5.2) differs from that for an ordinary static–field oscillator by the inevitable presence of correlations between position and momentum in the generalized coherent states.

6. EFFECT OF EXTERNALLY APPLIED FIELDS

The simplest way to alter the state of a trapped particle is to apply an external forcing field. Let us assume that the particle described by the Hamiltonian H_0 of Eq. (2.1) is subjected to an additional uniform time–dependent force field of magnitude $mF(t)$. Its Hamiltonian is then

$$H = H_0 - mF(t)q, \tag{6.1}$$

and the coordinate $q(t)$ must satisfy the inhomogeneous differential equation

$$\ddot{q}(t) + W(t)q(t) = F(t). \tag{6.2}$$

The presence of the inhomogeneous term $F(t)$ means that the expression $C(t)$ that we constructed in Eq. (2.10) is no longer a constant of the motion. In fact we find, from the Heisenberg equations of motion, that

$$\frac{d}{dt}C(t) = i\sqrt{\frac{m}{2\hbar\omega}}u(t)F(t), \tag{6.3}$$

but that relation shows us that we can construct a new constant of the motion in the form

$$C'(t) = C(t) - i\sqrt{\frac{m}{2\hbar\omega}}\int_0^t u(t')F(t')dt'. \tag{6.4}$$

This new constant of the motion, when evaluated at the initial time again equals the annihilation operator a, and so must retain that value at all times,

$$C'(t) = a. \tag{6.5}$$

If we assume, for example, that the particle is initially in the state $|n{=}0\rangle_\omega$, then we find that its state $|t\rangle$ at time t obeys the relation

$$C'_S(t)|t\rangle = 0, \tag{6.6}$$

where C'_S is the Schrödinger–picture version of $C'(t)$. This relation can also be written in the form

$$C_S(t)|t\rangle = \alpha(t)|t\rangle, \tag{6.7}$$

with

$$\alpha(t) = i\sqrt{\frac{m}{2\hbar\omega}}\int_0^t u(t')F(t')dt'. \tag{6.8}$$

By comparing this equation with Eq. (4.2), that we solved earlier, we see that the new state generated by the forcing field is the coherent state

$$|t\rangle = |\alpha(t), t\rangle. \tag{6.9}$$

The effect of such a forcing field can always be represented by a unitary displacement transformation in the complex α–plane. It therefore always carries an initially coherent state, such as the $n = 0$ state, into another coherent state.

7. TRAPPING BY PERIODIC FIELDS

It is clear from the structure of the wave functions (3.6) and (3.12) and from their corresponding densities, such as that of Eq. (3.8), that the condition for trapping is the essentially classical requirement that $|u(t)|^2$ remain bounded. The most obvious example of trapping is the static–field oscillator for which W is a positive constant. If we then choose $\omega = \sqrt{W}$, we have $u(t) = e^{i\omega t}$ and $|u(t)|^2 = 1$. An equally obvious example of unstable behavior is one in which the potential is constant in time and repulsive, *i.e.*, W is a negative constant. Then if we choose $\omega = \sqrt{-W}$, we find $|u(t)|^2 = 1 + 2\sinh^2 \omega t$, and the density distributions spread out explosively and without bound.

The quadrupole field in the Paul traps is made to vary periodically in time. If its period is T, then we have

$$W(t+T) = W(t), \tag{7.1}$$

and the differential equation (2.6) for the function $u(t)$ becomes the well-known equation of Hill, a second order equation with a periodic coefficient[5]. It is always possible to find two linearly independent solutions u_1 and u_2 to this equation with the property that increasing t by one period multiplies the solutions by two constants, say σ_1 and σ_2 respectively,

$$u_j(t+T) = \sigma_j u_j(t) \quad j = 1, 2. \tag{7.2}$$

If either of the constants σ_1 or σ_2 should have absolute value greater than unity, the corresponding solution $u_j(t)$ will clearly increase in modulus without bound over many periods.

The two constants σ_1 and σ_2 are in fact linked together in value by the constancy of the Wronskian. If we evaluate that expression both at t and at $t+T$, we find

$$\sigma_1 \sigma_2 = 1, \tag{7.3}$$

so if one solution increases in modulus with time, the other must decrease. Now the initial condition on the derivative of $u(t)$ contains the parameter ω, which we are still free to choose arbitrarily; its value only specifies a set of basis functions. If the particle we are describing is truly trapped, $|u(t)|^2$ must remain bounded for any and all real, positive values of ω. Since both solutions $u_1(t)$ and $u_2(t)$ must be combined in general to meet this initial condition, neither of them can be allowed to grow in modulus without bound. It follows then that we must have

$$|\sigma_1| = |\sigma_2| = 1. \tag{7.4}$$

If we define the characteristic exponent μ by the relations

$$\sigma_1 = e^{i\mu T}, \qquad \sigma_2 = e^{-i\mu T}, \tag{7.5}$$

then we see that μ must be real if trapping is to occur, and that we may take it to be positive as well. We can furthermore define the functions

$$\varphi_1(t) = e^{-i\mu t} u_1(t) \tag{7.6}$$

$$\varphi_2(t) = e^{i\mu t} u_2(t), \tag{7.7}$$

which are both periodic with period T

$$\varphi_j(t+T) = \varphi_j(t) \qquad j = 1, 2. \tag{7.8}$$

The solutions $u_j(t)$ may therefore be expressed in the characteristic Floquet form,

$$u_1(t) = e^{i\mu t}\varphi_1(t), \tag{7.9}$$

$$u_2(t) = e^{-i\mu t}\varphi_2(t), \tag{7.10}$$

in which the exponent μ is real and the periodic functions φ_j can be taken to have the initial values

$$\varphi_j(0) = 1 \qquad j = 1, 2. \tag{7.11}$$

Let us assume, for simplicity, that the function $W(t)$ is even–valued

$$W(-t) = W(t). \tag{7.12}$$

Then the solution $u_2(t)$ can be taken simply to be $u_1(-t)$. The most general solution of the differential equation (2.6) can thus be written in the form

$$u(t) = Au_1(t) + Bu_1(-t), \tag{7.13}$$

where the constants A and B are to be determined by the initial conditions

$$u(0) = 1 = A + B, \tag{7.14}$$

$$\dot{u}(0) = i\omega = (A - B)\dot{u}_1(0). \tag{7.15}$$

Whatever values these equations give to A and B, the value of $|u(t)|^2$ remains bounded for the solution (7.13).

It is worth remembering once more at this point that the parameter ω can be chosen arbitrarily and that there may be considerable simplification to be gained in the description of the motion by making an appropriate choice. If we give to ω, for example, the value

$$\omega = -i\dot{u}_1(0), \tag{7.16}$$

then we see that the solution of the equations (7.14) and (7.15) reduce to

$$A = 1, \qquad B = 0, \tag{7.17}$$

and the solution for $u(t)$ becomes just

$$u(t) = u_1(t) = e^{i\mu t}\varphi_1(t). \tag{7.18}$$

Our special choice of the basis states, in other words, has greatly simplified the way we see the motion.

What this isolated Floquet solution for $u(t)$ does is to separate the motion of the particle, in effect, into two components, very much as Kapitza suggested[3]. The exponential $e^{i\mu t}$ best describes the relatively slow part of the motion while the periodic function $\varphi_1(t)$ describes the rapid oscillations at the frequency of the driving field (and its harmonics) that make up what is often called the "micromotion". Indeed, when the Floquet solution (7.18) is substituted into the wave functions we have constructed, they all simplify considerably in their behavior. That is evident because the function $|u(t)|^2$ which figures in them becomes periodic, as does the logarithmic derivative

$$\frac{\dot{u}(t)}{u(t)} = i\mu + \frac{\dot{\varphi}_1(t)}{\varphi_1(t)}, \qquad \text{etc.} \tag{7.19}$$

The density function (3.8) for the n=0 state, for example, can be seen to pulsate in size about the normal value for an oscillator of the frequency specified by Eq. (7.16). The coherent state wave packets undergo the same periodic pulsations in size while they follow the classical trajectories of trapped particles. In this sense they are rather like the coherent states of ordinary oscillators. They undergo no secular spread or diffusion. Their new feature is the throbbing that represents "micromotion", in addition to their following a more intricate classical path.

To be a bit more specific about the correspondence of our results with the Kapitza approximation, let us take the function $W(t)$ to have the form

$$W(t) = \omega_1^2 \cos \Omega t, \tag{7.20}$$

where Ω is the angular frequency of the driving field and ω_1^2 is a measure of its strength. Then, according to the Kapitza approximation, as long as we have $\Omega >> \omega_1$, the slow component of the motion oscillates with frequency $\omega_1^2/\Omega\sqrt{2}$ in an effectively static harmonic oscillator potential[3].

When the function $W(t)$ has the cosine time dependence indicated in Eq. (7.19), the solution $u_1(t)$ formulated in Eq. (7.9) becomes a Mathieu function[5]. For the case in which $\omega_1/\Omega << 1$, it is possible to expand both the characteristic exponent μ and the logarithmic derivative $\dot{u}_1(0)/u_1(0)$ in powers of the ratio ω_1/Ω [6]. We then find for the characteristic exponent the value

$$\mu = \frac{\omega_1^2}{\Omega\sqrt{2}}\left\{1 + \frac{25}{32}\left(\frac{\omega_1}{\Omega}\right)^4 + \cdots\right\}, \tag{7.21}$$

while our parameter ω, determined by Eq. (7.16), is

$$\omega = \frac{\omega_1^2}{\Omega\sqrt{2}}\left\{1 + 2\left(\frac{\omega_1}{\Omega}\right)^2 + \cdots\right\}. \tag{7.22}$$

In the lowest approximation then we have $\omega \approx \mu$, and both of these values agree with the Kapitza frequency. The analysis we have carried out clearly extends the Kapitza approximation to higher orders in ω_1/Ω.

I am greatly indebted to Professor W. Paul for demonstrating a model of his trap for me, to Jene Golovchenko for asking many questions about trapping, and to Vladimir Man'ko for telling me many new things about harmonic oscillators. This work was supported in part by a grant from the Department of Energy.

Paper delivered at International Conference on Quantum Optics, Hyderabad, India, January 5–10, 1991, and at NATO Advanced Research Workshop on Quantum Measurements in Optics, Cortina d'Ampezzo, Italy, January 21–25, 1991.

REFERENCES

1. For a review see: W. Paul, *Rev. Mod. Phys.* 62, 531 (1990).
2. V. V. Dodonov and V. I. Man'ko, in *Invariants and the Evolution of Nonstationary Quantum Systems*, ed. M. A. Markov (Nova Science Publishers, Commack, N.Y., 1989) p. 103.
3. Summarized in L. D. Landau and E. M. Lifschitz, *Mechanics*, (Pergamon Press, Oxford, 1969).
4. M. Combescure, *Ann. Inst. Henri Poincaré* 44, 293 (1986).
5. E. T. Whittaker and G. N. Watson, *Modern Analysis* (Macmillan, New York 1943).
6. J. Meixner and F. W. Schäfke, *Mathieusche Funktionen und Sphäroidfunktionen* (Springer Verlag, Berlin 1954) p. 119.

QUANTUM MEASUREMENTS ON LOCALIZED PARTICLES

Stig Stenholm

Research Institute for Theoretical Physics
Siltavuorenpenger 20 C
SF-00170 Helsinki, Finland

1. INTRODUCTION

During recent years many areas of physics have taken an interest in the dynamics of localized quantum states, i.e. wave packets. The reasons for this are manifold. In molecular physics the femtosecond pulses have offered a tool to interrogate molecular motion in real time[1]. In our own work, we have addressed some questions in this field[2]. It has become possible to prepare Rydberg atoms in planetary states and follow their evolution[3]. Ballistic phenomena in consensed substances and at their surfaces have also attracted attention lately. Last but not least, the availability of large and fast computers makes it feasible to approach the theoretical questions with some confidence.

The conceptual use of wave packets is, however, old. Originally Schrödinger[4] thought it to be an essential part of the interpretation of quantum physics, but such hopes were thwarted. On the other hand, the wave packets have occupied a central role as a tool for discussing basic questions of measurements. In this lecture, I plan to follow through the conceptual and historical development of quantum wave packets with the aim to place them in the framework of measurements in quantum optics.

First I want to stress that I adhere to the consensus theory of measurement as presented e.g. by Bohm in his book on quantum theory[5]. No novel aspects or reinterpretations are introduced here. An outline of the standard argument is as follows:

The state of the system under investigation is expanded in a preselected basis

$$\psi = \sum_i c_i u_i \ . \tag{1}$$

After this the chosen basis is in a favoured position; it defines the quantity to be measured and the unitary equivalence between bases is broaken by our choice. The measurement is performed by a pointer device[6], which is originally in the state f_0 which belongs to a set of states

$$f_0 \in \{f_i\} \ . \tag{2}$$

Quantum Measurements in Optics, Edited by P. Tombesi and
D.F. Walls, Plenum Press, New York, 1992

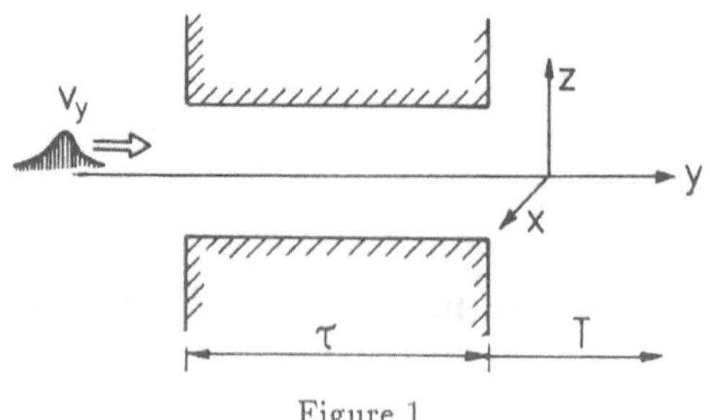

Figure 1

The measurement interaction transforms the uncorrelated initial state into the correlated one according to

$$f_0\psi \Rightarrow \sum_i c_i f_i u_i \ . \tag{3}$$

This formulation was to the best of my knowledge first published by Landau and Peierls[7]. If the pointer device is a macroscopic system in the sense intended by Bohr[8], we can determine the frequency of the occurrence of each f_i in (3), and this gives the coefficient $|c_i|$ in the original expansion (1). The distinguishability of the states f_i determines the resolution of the measurement.

The internal states of a particle are often separated by forcing them to follow different trajectories for the different states. This was, in fact, taken to be the starting point for quantum measurement theory by Pauli[9]:

> The simplest way of investigating the state of a system consists in placing it in an external force field, with which the different states of the system can react differently.

The prototype for such processes has become the Stern-Gerlach device, which has been elevated to the role of a paradigm for measurements. In the following text, we are going to analyze the distinguishability of wave packets by experimental setups designed to separate the different internal states spatially.

2. THE STERN-GERLACH EXPERIMENT

We consider a particle with the magnetic moment μ passing through an inhomogenous field according to the arrangement in Fig. 1. The dipole moment gives a force on the particle in the z-direction according to

$$F_z^M = \mu \frac{\partial B_z}{\partial z} \ . \tag{4}$$

For an electron we have the following relation between its charge, mass and dipole moment

$$\mu_0 = \frac{e\hbar}{2m_0} \ . \tag{5}$$

For spin 1/2 the force (4) takes two different values only. If the force is allowed to act for a period τ it gives a momentum kick to the two states in opposite directions. After an additional drift time T, we have two wave packets separated by the distance

$$\Delta z = \frac{2F_z^M \tau}{M} T , \tag{6}$$

where M is the mass of the particle. When the initial width of the wave packet is δz, we have a spreading in the momentum given by

$$\delta p_z \geq \frac{\hbar}{\delta z} . \tag{7}$$

If the drift time is long enough to make the spreading dominate the initial width, we have to require the relations

$$\delta z \ll \frac{\hbar}{M \delta z} T \ll \frac{2F_z^M \tau}{M} T , \tag{8}$$

in order to be able to distinguish the states. This implies that

$$\frac{\hbar}{\delta z} \ll F_z^M \tau , \tag{9}$$

which says that the momentum kick achieving the separation has to be larger than the initial spread in momentum.

The analysis above is, however, incomplete. Because the magnetic field has to be divergencefree in the interaction region we must have

$$\frac{\partial B_z}{\partial z} + \frac{\partial B_x}{\partial x} = 0 . \tag{10}$$

Even if we take the magnetic field to be zero at $x = 0$, for an incoming wave packet with the initial width δx, we have an uncertainty in the magnitude of the field of the order

$$\delta B_z \simeq \frac{\partial B_z}{\partial z} \delta x . \tag{11}$$

If the particle has charge Q, this gives rise to a Lorentz force of the magnitude

$$F_z^L \simeq Q v_y \frac{\partial B_z}{\partial z} \delta x . \tag{12}$$

The velocity v_y gives the speed of the system through the magnet, and it determines the de Broglie wavelength of the wave function according to

$$\lambda = \frac{2\pi\hbar}{M v_y} . \tag{13}$$

If we want to be able to separate the two spin states spatially, we must require the magnetic force (4) to dominate the Lorentz force uncertainty (12). Thus a large ratio

$$\frac{F_z^M}{F_z^L} = \frac{\mu}{Q v_y \delta x} = \left(\frac{M}{m_0}\right)\left(\frac{\mu}{\mu_0}\right)\left(\frac{e}{Q}\right)\frac{1}{4\pi}\left(\frac{\lambda}{\delta x}\right) \tag{14}$$

is required. But if the first three factors are unity, the requirement is violated by the last factor which by quantum mechanical necessity is small. The atomic Stern-Gerlach experiment is possible thanks to the first factor, but for an electron no separation of the spin states seems possible. Considerations like these were behind Bohr's insistance that the spin of a free electron is not observable. The spin force is seen only as a small quantum correction to the classical force acting on the charge of the electron. The detailed argument above is due to Pauli[10] and Mott[11]. In a different context, I hope to be able to analyze the possible loopholes in the argument, here I present it as a mere historical account only.

3. OBSERVATION OF LOCALIZED STATES

Let us assume that the initial stage of a measurement consists in transferring the desired information to two initial components $|\psi_\pm>$, see Eq.(3). If we let these develop according to the unitary time evolution operator

$$U(t) = e^{-iHt/\hbar} , \tag{15}$$

how is the possibility to separate them evolving in time? The answer can be obtained from the overlap between the states

$$<\psi_+(t)|\psi_-(t)>=<\psi_+(0)|U^\dagger(t)U(t)|\psi_-(0)>=<\psi_+(0)|\psi_-(0)> , \tag{16}$$

which shows that the wave packet overlap cannot be made to decrease in time under the action of a single Hamiltonian[12]. How does this come about?

Let us look at a minimum uncertainty packet. For macroscopic measuring devices, no doubt the uncertainty $\delta p \delta z$ much exceeds its minimum value, but we can still gain useful understanding from the simple state

$$\psi_0(z) = (2\pi b^2)^{-1/4} \exp\left[-\frac{(z-\bar{z})^2}{4b^2}\right] . \tag{17}$$

The measurement is performed by kicking this into different directions by the momentum $\pm\bar{p}$, and the information thus resides in the two states

$$\psi_\pm(z) = e^{\pm i\bar{p}z/\hbar}\psi_0(z) . \tag{18}$$

The initial overlap between these two states is given by

$$<\psi_+|\psi_->=\int e^{-i2\bar{p}z/\hbar}|\psi_0(z)|^2 dz . \tag{19}$$

If the momentum kick $\bar{p}$ is large enough to overcome the momentum uncertainty, see Eq.(9), already the initial integral (19) is very small. It is thus possible to extract the information from the states at once, e.g. by measuring the Doppler shift of light scattered from the components. In time, however, the lack of substantial overlap will manifest itself as a spatial separation.

For a free particle, we can compute the time evolution of the wave packets (18) to be

$$\begin{aligned}\psi_\pm(z,t) = (2\pi)^{-1/4}\left(b + i\frac{\hbar t}{2mb}\right)^{-1/2} \exp\left(-\frac{ip^2}{2m\hbar}t\right) e^{\pm i\bar{p}z/\hbar} \\ \times \exp\left[-\frac{(z-\bar{z}\mp\bar{p}t/m)^2}{2b^2+2i\hbar t/m}\right] .\end{aligned} \tag{20}$$

If we take the state

$$\psi(z,t) = \left[c_+\psi_+(z,t) + c_-\psi_-(z,t)\right] , \tag{21}$$

its spatial evolution will, after an initial time, show the well separated wave packets (see Fig. 2 where $c_+ = c_-$), but for short initial times the rapidly varying phase difference manifests itself as rapid oscillations. These interference fringes show that the difference between the wave packet components is accessible at any time. We can thus formulate the following:

Thesis. The information content in a quantum state cannot be changed by a unitary time evolution.

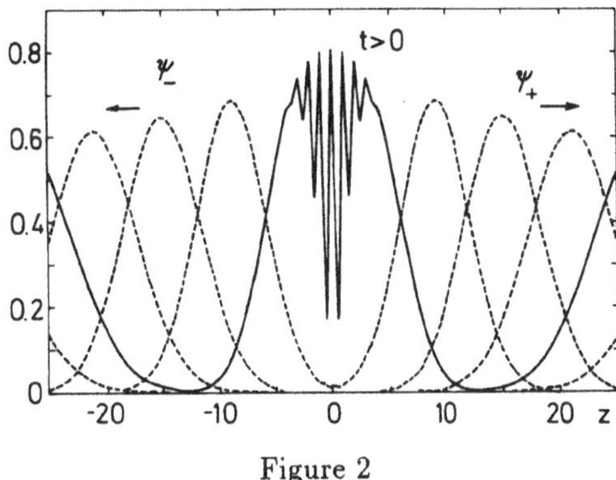

Figure 2

We are here talking about information in a loose sense, not referring to any formal definition. Assuming one Hamiltonian, we have the following:

Corollary. If two states can be distinguished at one time by a measurement, there is a measurement that can separate them at any other time.

We do not, however, observe wave packet overlaps, but only the response of localized detectors. The special role of position measurements is stressed by Landau and Lifshitz[13]. A particle detector localized at Z is described by a local operator

$A(z-Z)$. If we calculate its expectation value in the state (21) we find

$$
\begin{aligned}
<A> &= <\psi(t)|A(z-Z)|\psi(t)> \\
&= |c_+|^2 \int \psi_+^*(z) A(z-Z) \psi_+(z) dz + |c_-|^2 \int \psi_-^*(z) A(z-Z) \psi_-(z) dz \\
&+ 2Re\left[c_+^* c_- \int \psi_+^*(z) A(z-Z) \psi_-(z) dz\right] .
\end{aligned}
\tag{22}
$$

The first two terms give the probability of finding the system at position Z, and by moving the detector we can read off the probabilities $|c_\pm|^2$. The inaccuracy is given by the interference term. To illustrate the physical contents of Eq.(22), we assume the ideal limit of a point detector. This gives the result

$$
\begin{aligned}
<A> = |c_+|^2 \exp\left[-\frac{(Z-z_+(t))^2}{\sigma^2(t)}\right] + |c_-|^2 \exp\left[-\frac{(Z-z_-(t))^2}{\sigma^2(t)}\right] \\
+ 2Re\left[c_+^* c_- e^{i2\bar{p}z/\hbar}\right] \exp\left[-\frac{(Z-\bar{z})^2}{\sigma^2(t)}\right] \exp\left[-\left(\frac{\bar{p}}{\delta p}\right)^2\right] .
\end{aligned}
\tag{23}
$$

Here the trajectories are the center of mass motion for the two states and the spreading $\sigma^2(t)$ is determined by Eq.(20). We see that the interfence term is small if the initial momentum kick is large compared with the momentum uncertainty and/or the detector position Z is far removed from the initial position $\bar{z}$; precisely the conditions we formulated in Sec. 2. The overlap is independent of time expect for the slow spreading in the width $\sigma^2(t)$. From this example we formulate a further:

Thesis. With localized detectors, wave packets can be separated even with a unitary time evolution.

4. WIGNER-FUNCTION DESCRIPTION

The motion of a quantum system in phase space can be described well by the Wigner function derived from the density matrix ρ through

$$W(R,P) = (2\pi\hbar)^{-1} \int e^{-iPz/\hbar} < R + \frac{z}{2}|\rho|R - \frac{z}{2} > dz . \tag{24}$$

Its equation of motion in the potential $U(R)$ is of the form

$$\frac{\partial W}{\partial t} + \frac{P}{M}\frac{\partial W}{\partial R} - \frac{\partial U}{\partial R}\frac{\partial W}{\partial P} = O\left(\hbar^2 \frac{\partial^3 U}{\partial R^3}\right) , \tag{25}$$

where right-hand term denotes quantum corrençtions to the classical motion. For potentials of the form $U(R) \propto R^n$ $(n = 0,1,2)$, these corrections vanish and quantum mechanics enters the problem only through the initial conditions.

We define a set $\{A(t),\ B(t),\ C(t)\}$ of fundamental solutions of the classical motion in the given potential $U(R)$ by setting

$$
\begin{aligned}
A(0) = \dot{B}(0) = 1 \ ; \ \dot{A}(0) = B(0) = 0 \\
C(0) = \dot{C}(0) = 0 ,
\end{aligned}
\tag{26}
$$

and we also define the functions

$$D(t) = A\dot{B} - \dot{A}B \, ,$$
$$F_1(t) = \frac{(\dot{B}C - B\dot{C})}{D} \; ; \; F_2(t) = M\frac{(\dot{A}C - A\dot{C})}{D} \, . \tag{27}$$

The Wigner function for the minimum uncertainty wave packet (17) is given by

$$W_0(R,P) = C \exp\left[-\frac{(R-\bar{z})^2}{2b^2}\right] \exp\left[-\frac{2b^2(P-\bar{p})^2}{\hbar^2}\right] , \tag{28}$$

and taking this as the initial condition, we obtain the general solution of Eq.(25) in the form

$$W(R,P,t) = C \exp\left[-\frac{1}{2b^2}\left(\frac{\dot{B}}{D}R - \frac{B}{DM}P + F_1 - \bar{z}\right)^2\right]$$
$$\times \exp\left[-\frac{2b^2}{\hbar^2}\left(\frac{A}{D}P - M\frac{\dot{A}}{D}R + F_2 - \bar{p}\right)^2\right] . \tag{29}$$

The marginal distribution in configuration space can be obtained by integrating out the momentum variable giving

$$W_P(R,t) = N \exp\left[-\frac{(R - R(t))^2}{2\sigma^2(t)}\right] , \tag{30}$$

where $R(t)$ is the classical trajectory in the potential, and the spreading $\sigma^2(t)$ is calculated in a straightforward way. For the three potential cases with $n = 0, 1$ and 2, the parameters of the Wigner function are given in TABLE I.

TABLE I

$U(R)$	A	B	C	$R(t)$	$\sigma^2(t)$
0	1	t	0	$\bar{z} + \frac{\bar{p}t}{M}$	$b^2 + \frac{\hbar^2}{4b^2M^2}t^2$
$-MgR$	1	t	$\frac{gt^2}{2}$	$\bar{z} + \frac{\bar{p}t}{M} + \frac{gt^2}{2}$	$b^2 + \frac{\hbar^2}{4b^2M^2}t^2$
$\frac{M\Omega^2R^2}{2}$	$\cos\Omega t$	$\frac{\sin\Omega t}{\Omega}$	0	$\bar{z}\cos\Omega t + \frac{\bar{p}}{M\Omega}\sin\Omega t$	$\cos^2\Omega t \, b^2 + \frac{\sin^2\Omega t}{\Omega^2}\frac{\hbar^2}{4b^2M^2}$

If we choose the initial width of the wave packet according to

$$b = \sqrt{\frac{\hbar}{M\Omega^2}} \tag{31}$$

the oscillator wave packet retains its width constant; this is the Schrödinger wave packet corresponding to the coherent state.

If we have two different quantum states, it is easy to show from the definition (24) that the phase space overlap between their Wigner functions gives the square of the overlap of the states

$$2\pi\hbar \int dR \int dP \; W_1(R,P)W_2(R,P) = |<\psi_1|\psi_2>|^2 \,. \tag{32}$$

Using the various cases in Table I, we can easily evaluate the wave function overlap from Eq.(32). If the two states develop with the same Hamiltonian, no change in the overlap will emerge as found in Eq.(16). This leads e.g. to the amusing situation shown in Fig. 3, where two broad harmonic oscillator states, resting at $t = 0$, are given momentum kicks in different directions. At their respective turning points they are sharp and well resolved, but their overlap integral has not changed.

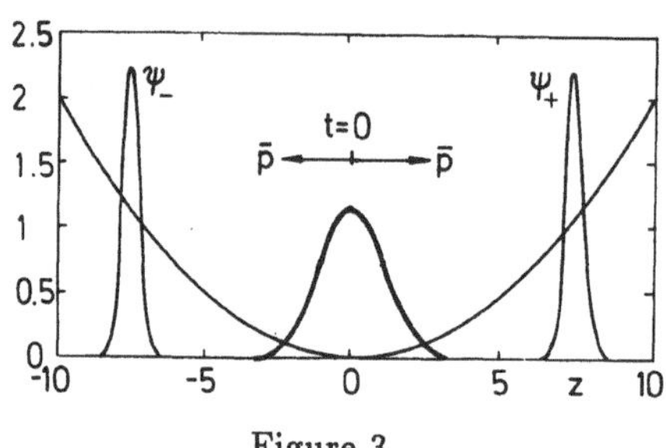

Figure 3

If the states to be separated evolve under different Hamiltonians the situation is different. When the two states to be separated are falling freely in oppositely directed constant fields, we can use Table I with $\pm g$. The overlap between them is found to evolve according to

$$|<\psi_-|\psi_+>|^2 \propto \exp\left[-\left(\frac{gt^2/2}{b}\right)^2 - \left(\frac{Mgt}{\delta p}\right)^2\right] , \tag{33}$$

where $\delta p = (\hbar/2b)$. Thus freely falling particles separate at rates measured by their widths in position and momentum.

Another example is offered by two broad states resting in different harmonic potential wells. If they are given identical momentum kicks, they climb their respective potential edges, and sharpen up at different turning points. The evaluation of the overlap can be carried out using the data in Table I, but the main features of the situation are easily seen from Fig. 4. If we let these packets oscillate in their potentials continuously, we are approaching the case termed the "continuous Stern-Gerlach effect" by Dehmelt. This interesting special case will be the topic of the next section.

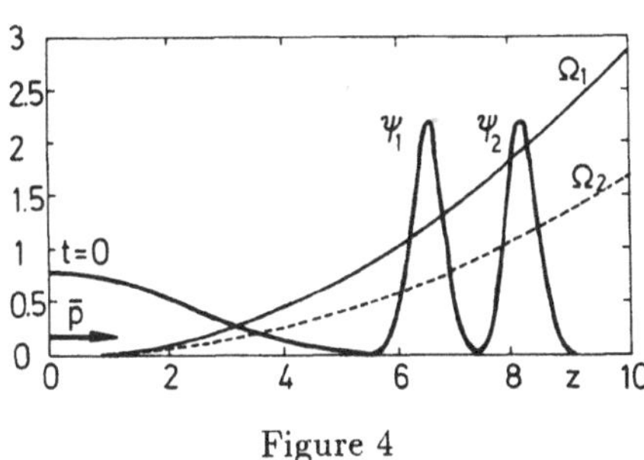

Figure 4

5. CONTINUOUS STERN-GERLACH EFFECT

During his long and successful program to measure the magnetic moments of elementary particles, Hans Dehmelt has developed a method called the continuous Stern-Gerlach experiment[14]. In this he is able to observe the force acting on an electron continuously by introducing a spin dependent component in the trapping potential. Thus the different spins see different harmonic potentials as indicated in Fig. 5. The frequency difference is small, $\omega_+ - \omega_- \simeq 1Hz$, but it can be monitored by picking up the radiofrequency signal emitted by the vibrating electron. A sensitive detection scheme allows one to monitor the frequancy of the signal, and when the spin state changes by a quantum jump, the frequency is seen to change as indicated in the right-hand part of Fig. 5. The actual experiment is much more complex, but this simplified description suffices for our purposes.

From the frequency change in the field gradient the value of the electronic magnetic moment can be extracted, and Dehmelt[14] concludes:

> [It] does measure the spin magnetic moment of the free electron and also the Bohr magneton. Therefore the general theorem formulated by Pauli, which claims that such experiments are impossible in principle ..... seems to require correction.

One may argue that the electron is not free but trapped. Rudolf Peierls, however, concedes that the electron is free in the sense intended by Bohr, who was wrong on this point[15].

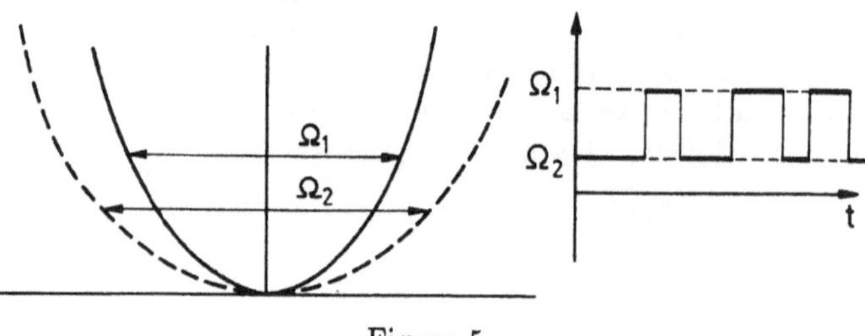

Figure 5

The simple model of the continuous Stern-Gerlach experiment which we outline in Fig. 5 could easily be analyzed in some detail. Here I only make a few comments concerning the physics of the device:

1. The readout arrangement provides a damping mechanism so that the harmonic motion has to be described by the equation of motion

$$\ddot{x} + \Omega_{\pm}^2 x = -\beta \dot{x} \ . \tag{34}$$

2. This damping prevents any coherence of the type (21) to develop between the spin states. The coefficients $c_{\pm}$ acquire random phases, and the spin states alternate through a sequence of random quantum jumps.
3. The corresponding noise broadens the spin resonance lines. This manifests itself so that the coupling to the dipole weakens. This effect is described by Dehmelt[16] as "the toothless ghost of the formidable Zeno no-go theorem".
4. The observed spectrum can be calculated using a Random Telegraph model for the jumping process.

6. THE OPTICAL STERN-GERLACH EFFECT

Recently there has been considerable interest in the possibilities to steer atomic motion by the influence of laser light. This opens up the field of atomic optics, where atomic inference, reflection and focusing are investigated. The possibilities to construct atomic interferometers has been discussed, but the simplest configuration is that of Bragg scattering from a standing wave[17]. The setup is indicated in Fig. 6, where a beam of two-level atoms inpinges on a standing laser wave nearly in resonance with the atoms. If the quantum state is written in the form

$$|\psi> = a|\psi_a> + b|\psi_b> \ , \tag{35}$$

the Schrödinger equation becomes

$$\begin{aligned} i\dot{a} &= (T + \Delta)a + Vb \\ i\dot{b} &= Tb + Va \ . \end{aligned} \tag{36}$$

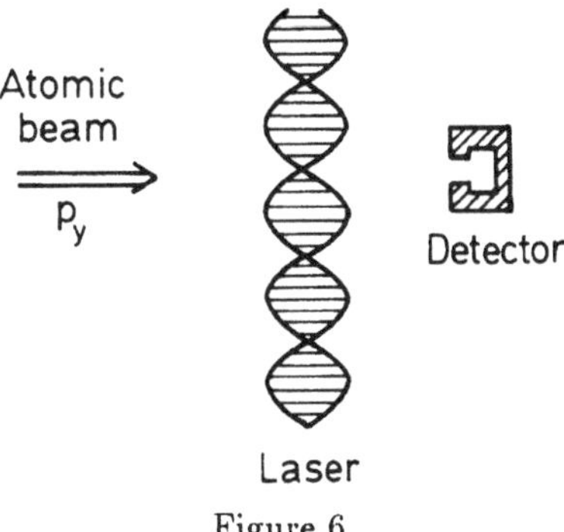

Figure 6

T is the kinetic energy operator, Δ is the laser detuning and the coupling V is spatially varying. For a simple standing wave we have

$$V(z) = \wp E_0 \cos kz \; . \tag{37}$$

The two coupled equations (36) could be solved for wave packets just like in the ordinary Stern-Gerlach case. Here we look at two simple special cases only:

Resonance $\Delta = 0$:
By adding and subtracting the two equations in (36) we separate the variables and find

$$i\frac{d}{dt}(a+b) = (T+V)(a+b) \tag{38}$$

$$i\frac{d}{dt}(a-b) = (T-V)(a-b) \; .$$

In this representation the two states see the potentials $\pm V$, which gives a situation very similar to the separation of spin states which we have discussed above. This phenomen was named Optical Birefringence by Kazantsev in 1974[18].

Far detuned $\Delta \gg 0$:
In this case the excited state can be eliminated adiabatically by solving

$$a = -\frac{V}{\Delta} \; . \tag{39}$$

For the lower level, we obtain the equation of motion

$$i\frac{d}{dt}b = \left(T - \frac{V(z)^2}{\Delta}\right)b \; . \tag{40}$$

This clearly describes Bragg scattering from the intensity distribution of the laser light. The theoretical analysis of the experiments by Pritchard et al.[19] has been carried out in this limit.

The considerations presented above for Stern-Gerlach experiments with spin can all be applied to the optical cases as well. In addition, we need to consider the effects of atomic dissipation, which have been discussed in some detail by Arimondo et al.[20] and Tanguy et al.[21].

REFERENCES

1. H. Metiu and V. Engel, *JOSA*, **B7**, 1709, (1990).
2. B. Garraway, K.-A. Suominen and S. Stenholm, *Proceedings of the Workshop: "Light Induced Kinetic Effects on Atoms, Ions and Molecules"*, May 2-5, 1990, Elba, Italy. To appear.
3. G. Alber and P. Zoller, "*Laser excitation of electronic wave packets in Rydberg atoms*", Physics Report, to appear.
4. E. Schrödinger, *Naturwiss.*, **14**, 664, (1926).
5. D. Bohm, *Quantum Theory*, (Dover Publ. Inc., New York, 1989).
6. W.H. Zurek, *Phys. Rev.*, **D24**, 1516, (1981).
7. L.D. Landau and R. Peierls, *Z. Phys.*, **69**, 56, (1931).
8. N. Bohr, *Phys. Rev.*, **48**, 696, (1935). See in particular p. 701.
9. W. Pauli, *General Principles of Quantum Mechanics*, (Springer-Verlag, Heidelberg, 1980).
10. W. Pauli in Le Magnetisme, *Proceedings of the VI Solvay Conference*, (Gauthier-Villars, Paris, 1932).
11. N.F. Mott, *Proc. Roy. Soc.*, **A124**, 440, (1929).
12. This argument was presented to me by Paul Berman many years ago.
13. L.D. Landau and E.M. Lifshitz, *Quantum Mechanics*, §7, (Pergamon Press, Oxford, 1958).
14. H. Dehmelt, *Z. Phys.*, **D10**, 1278, (1988).
15. R. Peierls in private discusion with the author.
16. H. Dehmelt, *Am. J. Phys.*, **58**, 17, (1990).
17. P.J. Martin, B.O. Oldaker, A.H. Miklich and D.E. Pritchard, *Phys. Rev. Lett.*, **60**, 515, (1988).
18. A.P. Kazantsev, *Sov. Phys. -JETP*, **40**, 825, (1975).
19. P.J. Martin, P.L. Gould, B.G. Oldaker, A.H. Miklich and D.E. Pritchard, *Phys. Rev.*, **A36**, 2495, (1987).
20. E. Arimondo, A. Bambini and S. Stenholm, *Phys. Rev.*, **A24**, 898, (1981).
21. C. Tanguy, S. Reynaud and C. Cohen-Tannoudji, *J. Phys. B:At. Mol. Phys.*, **17**, 4623, (1984).

INTERFEROMETRY WITH METASTABLE RARE GAS ATOMS

T. Sleator, O. Carnal, A. Faulstich, and J. Mlynek

Universität Konstanz, Fakultät für Physik
D-7750 Konstanz

1. INTRODUCTION

Matter wave interferometry is a well established field in physics. Interferometers with de Broglie waves have been demonstrated for electrons[1] and neutrons,[2] and recently also for atoms.[3,4]

The main interest in interferometry with massive particles is due to the fact that, in contrast to light interferometry, gravitational effects can be studied and that the low particle velocity increases the interaction time in the interferometer, vastly increasing the sensitivity. Therefore, neutron and electron interferometers have been extensively used for fundamental tests of quantum mechanics and general relativity.[5] Moreover, interferometers with atoms offer additional possibilities to study effects involving the atomic internal structure. Examples are the effect of electric fields on the atomic polarizability and the Casimir effect. Atoms can also be prepared in many different internal states, and effects due to resonant light-atom interactions and to spontaneous decay processes can be investigated.

The realization of an atom interferometer, however, has only been successful in the last year.[3] The construction of an atom interferometer is rendered difficult by the fact that atoms carry no charge and do not penetrate through condensed matter. Therefore, beam splitters different from those used in neutron or electron interferometry had to be developed and demonstrated.[6]

We report on the first experimental realization of an atom interferometer, which is based on a Young's double slit arrangement.[3] This interferometer belongs to a class of interferometers based on the principle of wavefront splitting. We used a beam of metastable helium atoms together with microfabricated transmission structures with dimensions in the micrometer range. With the present setup it is possible to measure differential phase shifts on the order of 0.3 radian in less than 10 minutes.

This paper is organized as follows. In the Section 2 we present the experimental results obtained with the double slit interferometer and analyze its performance. Section 3 discusses potential applications of this device that make use of induced or spontaneous atomic transitions in a laser field and their effect on the phase of the atomic wavefunction. Of great importance for precise phase measurements is a highly coherent atomic beam. In this respect we present in Section 4 the first experiments

Quantum Measurements in Optics, Edited by P. Tombesi and
D.F. Walls, Plenum Press, New York, 1992

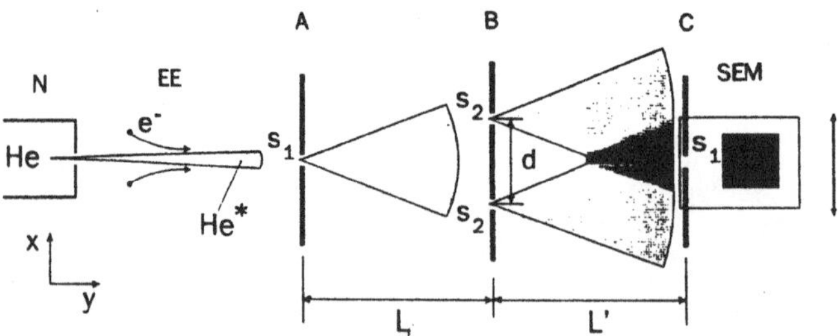

Figure 1. Schematic representation of the experimental setup: Nozzle system and gas reservoir, N; electron impact excitation, EE; entrance slit A, double slit B, and detector screen C; secondary electron multiplier, SEM. Dimensions $d = 8\ \mu m$, $L = L' = 64$ cm; slit widths: $s_1 = 2\ \mu m$, $s_2 = 1\ \mu m$.

on a beam of metastable argon atoms demonstrating an efficient and stong velocity compression by laser light forces. The obtained velocity ratio $v/\Delta v > 500$ represents the hitherto most monochromatic c.w. atomic beam.

2. THE ATOM INTERFEROMETER

In our experiments we have probably chosen the easiest configuration to realize an atom interferometer. The interferometer consists of an entrance slit, a double slit, and a detector slit. The double slit defines the two possible paths between entrance and detector slit. We have used metastable helium atoms as atomic species: helium is a light atom, which leads to a large de Broglie wavelength, and the production of a very bright atomic beam source by supersonic expansion is a standard technique. Moreover, metastable helium atoms can be detected very efficiently on almost zero background with a secondary electron multiplier, and they have dipole transitions in the near infrared so that they can be manipulated by laser light. This offers the possibility to improve the coherence properties by laser light pressure as well as to carry out experiments with the interferometer where phase changes can be induced by atom-laser interactions.

The scheme of our experimental setup is shown in Fig. 1; details are presented elsewhere.[3] A supersonic gas expansion followed by a collinear electron beam excitation creates an intense and fairly monochromatic beam of metastable helium atoms in the two states 2^1S_0 and 2^3S_1. The variable temperature of the gas reservoir and the nozzle system defines the mean velocity of the atoms in the beam and thus the de Broglie wavelength. At present, the de Broglie wavelength can be set to either $\lambda_{dB} = 56$ pm or to $\lambda_{dB} = 103$ pm.

The first defining slit following the beam source is 2 μm wide. After passing through this slit the coherence length in the transverse direction is sufficiently large that the double slit, which is located 64 cm downstream, is irradiated coherently. The double slit consists of two 1 μm wide slits separated by 8 μm. The atomic waves emerging from this double slit are superimposed coherently and create an interference fringe pattern in the atomic density distribution. This fringe system is monitored another 64 cm behind the double slit with either a single 2 μm wide slit or a grating whose periodicity of 8 μm is matched to the periodicity of the interference pattern. The complete detector system consists of a secondary electron multiplier (SEM) behind a gold foil with the two transmission structures described above. The system can be moved perpendicular to the atomic beam axis in steps of 1.9 μm. The pulses coming out of the SEM are preamplified and discriminated to eliminate unwanted detector noise. The microfabricated transmission structures imprinted in thin gold foils are approximately 2-4 mm high and have been manufactured by Heidenhain Inc., Traunreut, Germany by a special photolithographic technique.

The interference pattern obtained at the two different de Broglie wavelengths were first scanned with the single 2 μm detector slit. The obtained fringe spacings of 4.5 ± 0.6 μm ($\lambda_{dB} = 56$ pm) and 8.4 ± 0.8 μm ($\lambda_{dB} = 103$ pm) are in good agreement with the predictions given by the Fraunhofer diffraction theory. The extremely low signal at the detector (approximately 5-15 counts/minute) has the disadvantage of requiring long integration times, leading to a high sensitivity to slow thermal drifts in the beam machine.

This problem can be overcome by using a grating instead of a single slit to monitor the intensity distribution. The integration over many interference maxima or minima increased the count rate by a factor of 10 and therefore allowed us to decrease the integration time. The 8 μm periodicity of the grating is matched to the interference pattern at $\lambda_{dB} = 103$ pm. Scanning the interference fringes with the grating is of interest in experiments where only the phase change between the two paths and not the exact shape of the interference pattern is of importance. A scan over the atomic inteference structure is shown in Fig. 2. The periodicity was determined to be 7.7 ± 0.5 μm, which is within 5% of the theoretical predictions. The visibility amounts to 30%, whereas 50% can be expected under ideal conditions. The mean relative error at each detector position is less than 10% when integrating 5 minutes per point, and is due to the stochastic nature of the arrivals at the detector. This error corresponds to an accurracy of our device on phase changes of 0.3 radian in less than 10 minutes.

One main inconvenience with the present setup is the very low count rate at the detector and therefore long integration times to monitor the interference pattern. There are two different approaches to increase the signal: one can either improve the detection efficiency (approximately 15 % at present) or the intensity of our metastable atomic beam beam. The former is very difficult to achieve, but the beam intensity can be very efficiently increased (at least by a factor of 10) by transverse laser cooling.

A further improvement would be a compression of the velocity distribution in longitudinal direction. The higher monochromaticity in the beam would allow more precise studies of atom interactions with the applied potentials. Section 4 will discuss first experiments on metastable argon atoms to decrease the velocity spread in the atomic beam.

3. ATOM INTERFEROMETRY AND ATOM-LIGHT INTERACTIONS

An important difference between atom interferometers and other matter interferometers such as those using neutrons or electrons, is that atoms have a complex internal structure that can be conveniently manipulated with lasers. A second important difference is that atoms exhibit spontaneous emission. In the case of an interferometer, it is important to know what influence spontaneous emission will have on the interference pattern. Besides, the influence of spontaneous emission is also of fundamental interest.

This section is divided into two parts. The first part discusses proposed experiments in which a phase shift is introduced into the atomic wave function by optical means. The influence of spontaneous emission on atomic coherence is treated in the second part. These two parts contain only the main ideas of the proposed experiments; more details will be given elsewhere.[7]

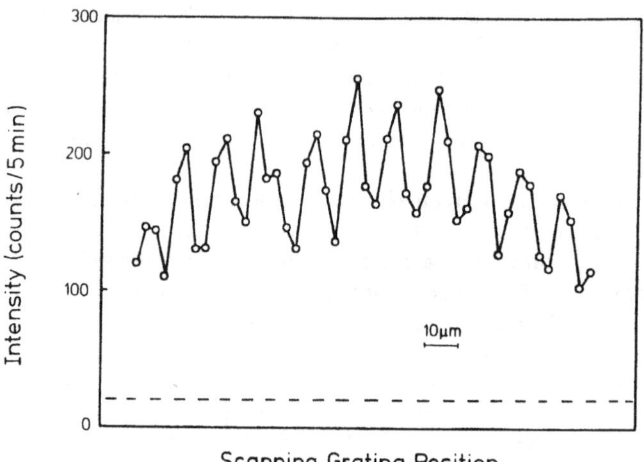

Figure 2. Atomic density profile monitored with the 8 μm grating in the detector plane, as a function of the lateral grating displacement. The dashed line connecting the experimental points is a guide to the eye.

3.1. *Coherent Atom-Light Interactions*

Here, we discuss some potential experiments in which a phase shift in the atomic wave function is introduced as a result of the coherent interaction of a laser beam with the atoms in one path of the interferometer (see Fig. 3). This phase shift leads to a displacement of the atomic fringe pattern on the screen. The experimental configuration is interesting because it represents a relatively simple way of introducing a phase shift into an atom interferometer, and thus demonstrates the device as an interferometer, and not just as a means of observing diffraction of atoms. In addition, this experiment represents a way of probing the properties of a light field in which no energy is exchanged with a given photon mode. This has the possibility of leading to "quantum non demolition" measurements of the photon number.[8]

We now calculate the phase shift induced by the interaction of atoms with a laser field. We assume that the atoms pass through the entrance slit in the ground state (For He, this would be a metastable state, either $^1S_1$ or $^3S_1$.) In addition, we restrict our discussion for the moment to atom-light interactions that a) return the atoms to the ground state, and b) produce no spontaneous emission.

Since one is ultimately measuring the center of mass of the atoms, one must calculate the influence of the interaction on the center of mass. The interaction of the laser with the internal degrees of freedom produces an effective potential that is seen by the atomic center of mass. We assume in this calculation that the atom can be described by a 2-level system with an energy difference in frequency units of ω_0. The laser is tuned to frequency ω with a detuning $\Delta = \omega - \omega_0$, and the laser field amplitude is expressed in terms of the Rabi frequency ω_1.

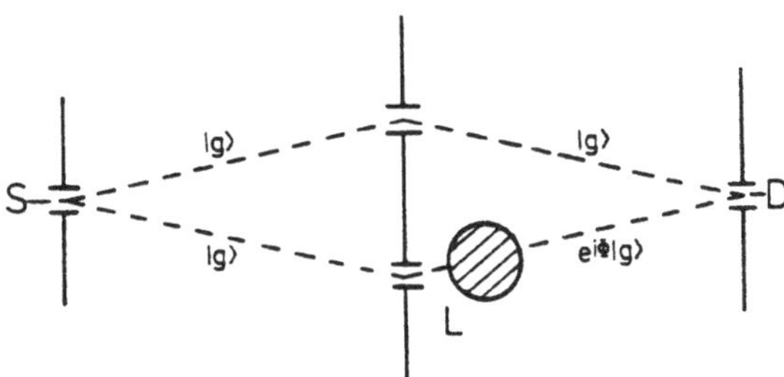

Figure 3. Schematic diagram of a phase shift experiment with an atom interferometer: Atom source, S; Detector, D; Laser beam, L.: $|g>$ represents an atom in the ground state, and Φ the phase shift produced by the interaction with the laser.

We consider two cases. In the first case, we assume that the laser is detuned from the atomic resonance ($\Delta \neq 0$), and that the laser intensity increases slowly enough that the internal degrees of freedom evolve adiabatically. i.e., we assume there are no transitions between the eigenstates of the "dressed-atom"[9] Hamiltonian. In this case the center of mass sees a well defined potential given by[9] $U = (\hbar\Delta/2)[1+\omega_1^2(x)/\Delta^2]^{1/2}$. For a particle with initial velocity v_0, and initial wave vector k_0, the phase shift between the two paths is given by the integral over the path of the atoms:

$$\begin{aligned} \Phi &= \int [k(x) - k_0]dx \\ &= (\Delta/2v_0) \int \{[1+\omega_1^2(x)/\Delta^2]^{1/2} - 1\}dx, \end{aligned} \tag{1}$$

where $\hbar^2k^2(x)/2m$ is the "instantaneous" kinetic energy of the particle. Note that the factor $(\Delta/2)\{[1+\omega_1^2(x)/\Delta^2]^{1/2} - 1\}$ in Eq. (1) is just the dynamical frequency shift[10] of the atom in the electromagnetic field. We see that the phase shift depends both on the laser intensity and detuning.

During the interaction, there is a possibility of spontaneous emission, which, as will be discussed later, can destroy the fringe pattern. To evaluate whether spontaneous emission can be neglected, we calculate the number of photons spontaneously emitted during the interaction time. This is given by $N = A\int P_e(t)dt$, where A is the spontaneous emission rate (equal to twice the natural linewidth) and $P_e(t) = \frac{1}{2}\{1 - [1+\omega_1^2(t)/\Delta^2]^{-1/2}\}$ is the probability that the atom is in the excited state. If we require that $N < 0.1$, and $\Phi = 2\pi$, then we find that the detuning must be greater than 100 natural linewidths. For the $\lambda = 1.083$ μm transition from the metastable 2^3S_1 triplet state of He, and an atomic velocity v_0 of 1000 m/s, the power required to produce this phase shift with a laser beam with 25 μm waist is 0.5 mW, which is quite easy to obtain.

We now consider another case in which the laser is tuned exactly to resonance ($\Delta = 0$) and the internal degrees of freedom flop between the ground and excited states. Let us assume that the intensity and spatial distribution of the laser beam are such that the atom is put into the excited state and immediately de-excited by stimulated emission back to the ground state (2π pulse). In this case, the motion of the center of mass is not determined by a unique potential as in the adiabatic case described earlier. Here, the light beam acts as a "birefringent" medium for the atoms.[11] Each of the two eigenstates of the dressed-atom Hamiltonian sees a potential given by $\pm\hbar\omega_1(x)/2$ respectively.[9] The ground state of the atom can be expressed as a linear combination of these two states, which accumulate an equal and opposite phase shift $\Phi_\pm = \pm(1/2v_0)\int \omega_1(x)dx$ during the interaction with the laser beam. Here we consider the case where the atom-light interaction has produced a phase shift $\Phi_+ = \pi$ and $\Phi_- = -\pi$. In this case the phase difference $\Phi_+ - \Phi_- = 2\pi$ and the atom has returned to the ground state. Since both dressed-atom eigenstates have accumulated a phase shift of π (or $-\pi$) then the total wave function has accumulated a phase shift of π. This means that on the detector screen the positions of the intensity maxima and minima will be exchanged.

Again one has to account for spontaneous emission. If, however, one can arrange the experiment so that the interaction time of the atoms with the field is much less than the spontaneous emission time, then one can neglect spontaneous emission. To carry out this experiment with metastable He atoms traveling at 1000 m/s using the $\lambda = 1.083$ μm transition would require a light field with diameter less than 25 μm with a power of about 1 μW.

3.2. *Influence of Spontaneous Emission on Atomic Coherence*

We present two problems relating to the effect of spontaneous emission on atomic coherence. The first problem is as follows: Imagine that by some means, perhaps by the coherent excitation by a laser beam, the atoms in **only one** of the two paths are put into the excited state and shortly afterwards spontaneously decay to the ground state by emitting a photon [see Fig. 4(a)]. We assume that both excitation and spontaneous emission occur in the vicinity of the slits. The question is, will interference fringes be observed. In this case the answer is no. A simple argument for this is that one could in principle determine which slit the atom went through by observing (or not observing) the emitted photon.

The second problem is: does one see interference when the atoms in **both** paths are put into the excited state and subsequently decay to the ground state [Fig. 4(b)]. To answer this question, the effect of spontaneous emission on the atomic wave packet must be considered, which is a result of the atomic recoil from the emitted photon. Assume that the atom is initially in the excited state $|e>$ with total wavefunction $|\psi_i> = \psi_e(\mathbf{r})|e>$, where $\psi_e(\mathbf{r})$ is the center-of-mass wave function. After emitting a photon with momentum $\mathbf{q}$, the atom will be in the ground state $|g>$ with the wavefunction $|\psi_f> = \psi_g(\mathbf{r})|g>$, where $\psi_g(\mathbf{r}) = e^{i\mathbf{q}\cdot\mathbf{r}}\psi_e(\mathbf{r})$.[12] Hence, the ground state atomic wavefunction is correlated to the direction of the emitted photon, and the total

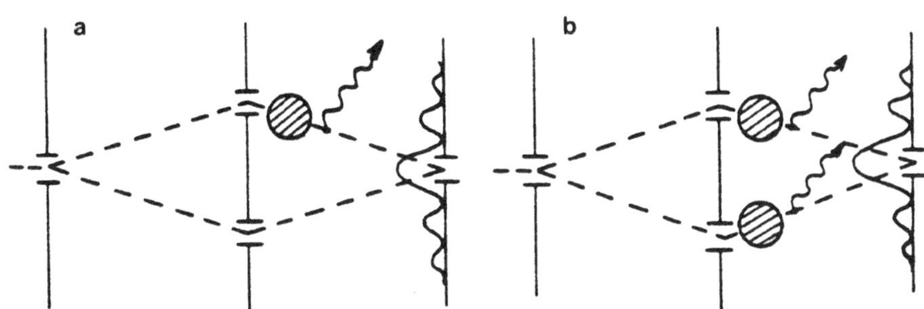

Figure 4. Schematic diagram of interferometer experiments involving spontaneous emission. (a) Spontaneous emission only from atoms on one path. (b) Spontaneous emission from atoms on both paths.

wave function must be expressed as a sum over $\mathbf{q}$, with appropriate weights, of the product states $e^{i\mathbf{q}\cdot\mathbf{r}}\psi_e(\mathbf{r}) \otimes |\mathbf{q}>$, where $|\mathbf{q}>$ represents a photon state with momentum $\mathbf{q}$. The atomic "state" is obtained by tracing the total density matrix over the photon states.[12]

We now return to the two slit problem by considering the situation immediately after the atom has passed through the double slit and has been put into the excited state by the laser beams [see Fig. 4(b) and the inset to Fig. 5]. We model the situation by assuming that the wave function of the excited state atom ψ_e is a linear combination of two localized wave packets (i.e. with spatial extent much less than the photon wavelength) labeled $|1>$ and $|2>$. State $|1>$ ($|2>$) represents the wave packet of an atom that has just passed through slit 1 (slit 2), as shown in the inset of Fig. 5. We take these wave packets $|1>$ and $|2>$ to be located at positions $\pm(d/2)\hat{\mathbf{z}}$ on the z-axis, where d is the spacing between the slits and $\hat{\mathbf{z}}$ is perpendicular to the propagation direction of the atoms. We also assume that the atom has equal amplitudes to pass through slits 1 and 2 and write $\psi_e = (|1> + |2>)/\sqrt{2}$. After the spontaneous emission, the ground state wave-function will be

$$\begin{aligned}\psi_g &= e^{i\mathbf{q}\cdot\mathbf{R}}(|1> + |2>)/\sqrt{2} \\ &= [e^{iqd(\cos\theta)/2}|1> + e^{-iqd(\cos\theta)/2}|2>]/\sqrt{2} \qquad (2)\end{aligned}$$

where $\mathbf{R}$ is the position operator and θ is the angle between $\mathbf{q}$ and $\hat{\mathbf{z}}$.

If the photon wavelength is much larger than the slit spacing ($qd << 1$) there will be no relative phase shift produced between the wavepackets $|1>$ and $|2>$. In this case $\psi_g = \psi_e$, and the fringe pattern will be unaffected. This is equivalent to saying that the recoil momentum of the atom is too small to deflect the atomic trajectory a significant fraction of an interference fringe.

The situation where the photon wavelength λ is on the order of, or smaller than the slit spacing d is more complicated. In this case a typical deflection of the atom due to recoil will be on the order of, or greater than the fringe spacing. To solve the problem, one must, in effect, calculate the fringe pattern for each direction of the emitted photon, and then integrate this fringe pattern over all directions of the emitted photon, weighted by the probability of emitting a photon in a given direction. A convenient way to do this is to use the density matrix formalism to describe the atomic state (which is not a pure state).

We now give an example of what the fringe visibility is expected to look like in the case of an electric dipole transition. We choose the case of a $\Delta m = \pm 1$ transition and a quantization axis along $\hat{\mathbf{z}}$, where the emission probability is proportional to $1 + \cos^2\theta$. This could be achieved with metastable He atoms, for example with the 1.083 μm transition between the $J = 2$ excited state and the $J = 1$ metastable state. To give the result in a convienient way, we generalize the concept of visibility. If the positions of the maxima and minima in the fringe pattern are unchanged by the spontaneous emission process, then we take the usual definition: $V = V_{usual} \equiv (I_{max} - I_{min})/(I_{max} + I_{min})$. If the positions of the maxima and minima are exchanged, we define $V = -V_{usual}$. Figure 5 shows the visibility of the fringes V *vs.* d/λ. The case of $V = 1$ corresponds to 100% modulation of the atomic intensity distribution, $V = 0$ corresponds to no fringe pattern at all, and $V < 0$ refers to the situation in which the positions of the maxima and minima are exchanged.

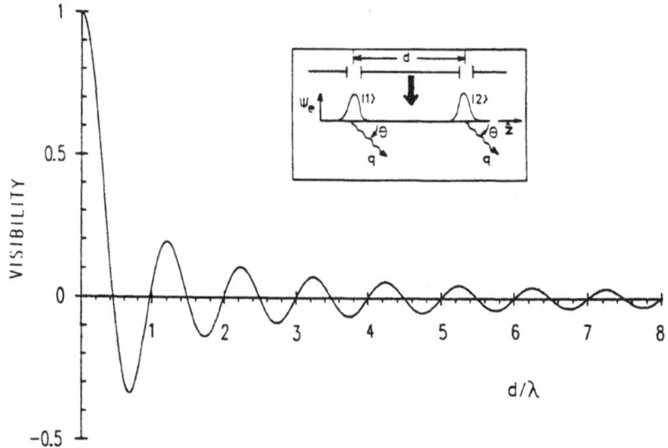

Figure 5. Coherence after spontaneous emission. Visibility of the fringes *vs.* d/λ. The result is shown for a $\Delta m = \pm 1$ transition with quantization axis along $\hat{\mathbf{z}}$. The probability of emitting a photon an angle θ from the z-axis is proportional to $1 + \cos^2\theta$. Inset: Geometrical configuration used in calculation of the fringe visibility. θ is the angle between the direction of the emitted photon and the z-axis. The fat arrow indicates the propagation direction of the atoms.

The calculation shows that when $d << \lambda$, the visibility is 100%, in agreement with our earlier discussion. In contrast, when the slit spacing becomes much larger than the wavelength ($d >> \lambda$), the fringe pattern tends to disappear. This we expect, because by looking at the emitted photon, one could in principle determine from where it was emitted (to a precision of roughly λ), and therefore, which slit the atom passed through. It is very interesting, however, to notice that even for slit spacings several times the wavelength there is still significant visibility, and for some values, this visibility is negative.

This calculation represents an idealization of the real situation. In it, we have assumed that the slit width is infinitely narrow, whereas the width of the slits used in our present interferometer is on the order of a wavelength. Such wide slits would tend to reduce the visibility below that shown in Fig. 5.

An example of an experiment that could be done to test the above ideas would be to construct a pair of slits, each with about 0.2μm width, and a 0.7μm spacing between them (for metastable He, the transition wavelength is about 1μm). Then, by putting the atoms into the excited state with a π pulse from a laser beam, one should be able to see an inversion of the fringe pattern (see Fig 5). As the laser power is increased further, the original fringe pattern should return when a 2π pulse is reached.

Another possible experiment would be to measure the correlation of the direction of the emitted photon with the fringe pattern of the atoms. One would find for large slit spacings that spontaneous emission would wipe out the fringe pattern. However, if one looked only at those atomic events in which a photon was emitted in a certain direction, then one should recover the fringe pattern.[13] Correspondingly, if one looked only at those events in which the atom strikes the detector at a particular position, then one would expect to see a fringe pattern in the angular distribution of the emitted photons.

4. VELOCITY COMPRESSION OF A SUPERSONIC ATOMIC BEAM

A highly monochromatic atomic beam is of great importance for precise phase measurements in atomic interferometry. It is also a prerequisite for other investigations in atom optics, such as the realization of mirrors and lenses for atoms. Since atoms can be cooled by the spontaneous light scattering force, it is possible to increase the monochromaticity of an atomic beam by longitudinal laser cooling, as well as to increase its intensity and brightness by applying the same mechanism in transverse dimensions.[14] In this section we report on our recent results on longitudinal laser cooling of a beam of metastable argon atoms, using optical molasses in a co-moving frame of reference.

Argon, as a noble gas, is in comparison to alkali atoms easy to handle. Due to the fact that it is impossible to excite noble gas atoms from the ground state by laser radiation, we excite them by electron impact, populating the metastable states $1s_3$ and $1s_5$. With lifetimes of several seconds, these metastables serve as "ground state atoms" for the cooling process.

The absence of a hyperfine splitting in the argon level scheme makes it easy to find a closed two level system, namely the transition $1s_5(J=2) \rightarrow 2p_9(J=3)$, that can be excited without causing problems due to unwanted effects by optical pumping. This transition has a wavelength of 811 nm and a natural lifetime of 25 ns, corresponding to a natural linewidth of 6.3 MHz (FWHM). Since we are interested in a fast atomic beam of high mean velocity and narrow velocity spread, we generate the beam by supersonic nozzle expansion (see Section 2), where the atoms experience cooling during the adiabatic expansion. In our beam the measured mean velocity is 560 m/s, while the FWHM-velocity spread Δv is approximately 50 m/s.

To compress the velocity distribution of the atomic beam, we overlapped two laser beams, one co- and one counterpropagating, as described in Ref. 15 and shown in Fig. 6. Each laser is tuned to about one natural linewidth below the Doppler shifted atomic resonance frequency, thus creating an optical molasses in the atom's moving frame.

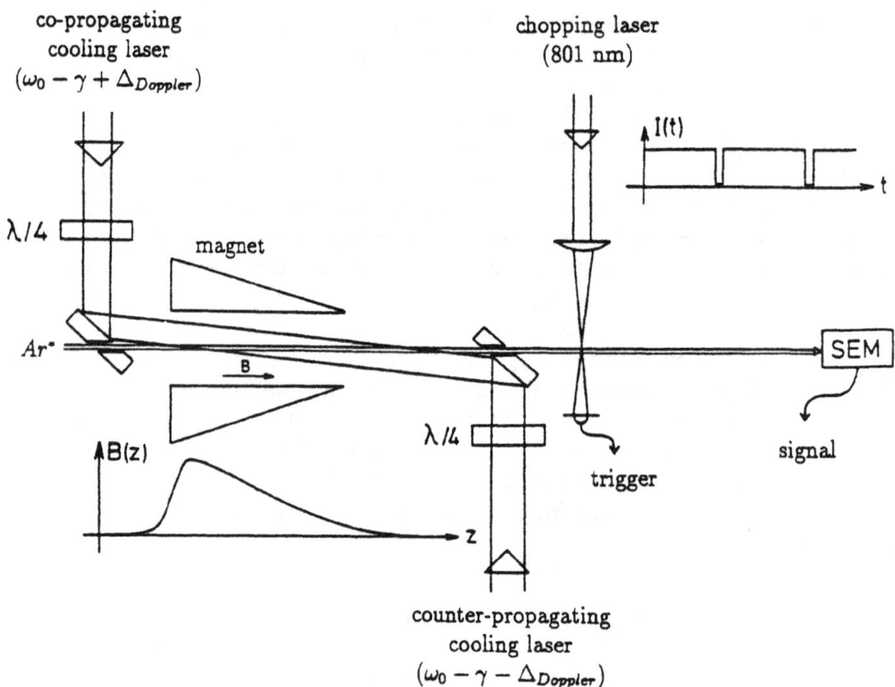

Figure 6. Scheme of the setup used for the laser cooling experiment. The source of the metastable argon atoms is on the left side of the figure, and the secondary electron multiplier (SEM) is shown on the right side. ω_0, γ, and $\Delta_{Doppler}$ denote the atomic resonance, natural linewidth (FWHM), and Doppler shift respectively. The magnet with field profile $B(z)$ is used to tune the atomic resonance by the Zeeman effect. Also shown is the chopping laser, which is focused onto the atomic beam and has the time dependent intensity I(t).

To create these light fields, the laser had to be locked with a well defined frequency shift with respect to the atomic resonance line. The lowest temperature, called the Doppler limit, is expected when the red-detuning is exactly half of the natural linewidth of the transition. In our experiment, a fraction of the laser light was frequency shifted by modulators and then locked to an argon dc discharge. To compensate for the Doppler shift of the atomic beam, the rest of the laser light was shifted in frequency to ±720 MHz by a high frequency, high bandwidth electro-optic modulator. The laser intensities used in this experiment ranged from 1 to 10 mW/cm^2.

The optical molasses alone does not suffice to cool the entire atomic velocity distribution, since the Doppler width corresponding to 50 m/s is still ten times broader than the natural linewidth of 6.3 MHz, which determines the capture range of the optical molasses. To cool all atoms of the atomic velocity distribution, special arrangements have to be used.

One possibility is to increase the power density of the cooling light field. This leads to a broadening of the atomic resonance line due to saturation effects, thus increasing the capture range. On the other hand optical saturation increases the temperature limit of the cooling molasses. The second way to capture all atoms is to sweep the cooling laser in frequency,[16] thus creating cooled bunches of atoms within the atomic beam. Since we are aiming for a cold and continuous atomic beam, this method is unpractical. A third possibility is to Zeeman tune the atomic resonance over a well defined area of space.[17] This has the advantage to create a cooled cw atomic beam, but due to the Zeeman splitting the transition is no longer a two level system. However, a two level system can be maintained if the cooling laser is purely σ^+ polarized, exciting the transition $1s_5(m_J = +2) \rightarrow 2p_9(m_J = +3)$ only. Of course, all atoms within the velocity distribution have to be pumped into the $(m_J = +2)$ ground state sublevel, to participate in the cooling process.

We have chosen the last alternative and inserted a 300 mm long tapered magnet into the beam apparatus, which created an axial magnetic field of 50 G at its maximum. The gradient of the field is designed in such a way, that the Zeeman detuning of the transition exactly compensates the change in Doppler shift due to the change in velocity, when maximum radiation force is applied. A Monte Carlo simulation of this cooling process that takes into account the exact dimensions of our experimental setup has shown[15] that $v/\Delta v$ figures of more than 1300 can be achived, as predicted by the Doppler limit.

A problem that arises here is the detection of such narrow velocity distributions. The detection of resonance flourescence in our case is limited to a $v/\Delta v$ of about 100, as can be easily calculated from the natural linewidth and the Doppler shift of 720 MHz. To improve the velocity resolution we have chosen a time of flight method (TOF) to measure the velocity spread of the cooled beam. The atomic beam is chopped by a laser beam that is tuned to the $1s_5(m_J = +2) \rightarrow 2p_8(m_J = +2)$ transition at 801 nm, pumping the atoms into the $1s_4(m_J = +1)$ state, which has a decay channel to the atomic ground state. Since our detector is not sensitive to ground state atoms, it is possible to modulate the atomic beam in intensity by a modulated laser beam. This laser beam is intensity modulated by an accusto-optic modulator and is transversely focussed onto the atomic beam by a cylindrical lens (see Fig. 6). We measured the total time resolution of our chopping and detection system to be 1 μs. With this time resolution and with a flight distance of 78 cm, corresponding to a time of flight of 1.4 ms, the limit of the TOF measurement is 1400 and thus sufficient to resolve a velocity spread down to the Doppler limit.

A typical result of our cooling measurements is shown in figure 7. Plot a) of the figure shows the initial velocity distribution without cooling laser. In b) the cooling laser and the Zeeman magnet are turned on. The FWHM of the time of flight signal of the cooled atomic beam is 3.6 μs, which corresponds to a $v/\Delta v$ of 500.[18] The intensity of this cooled beam of metastable atoms was roughly 10^8 s^{-1} striking a 2 mm diameter area of the detector. To our knowledge this is the most narrow velocity distribution ever obtained in a cw atomic beam. The reason why we did not observe cooling down to the Doppler limit is still subject to discussion. One possibility could be a heating effect by one of the two laser beams, when the forces exerted on the atoms do not cancel exactly due to a mismatch in the alignment of the counterpropagating light beams.

5. CONCLUSION

Recent experimental progress in the area of neutral atom interferometry[3,4] has given birth to a whole new subfield of atom and laser physics. We expect that in the next few years, tremendous progress will be made both experimentally and theoretically, and that many experiments of a fundamentally new type will be possible, some of which have been mentioned in this paper.

We have reported on our recently developed atom interferometer, which is based on wavefront splitting and is one of the conceptually easiest realizations of an atom interferometer. Despite its simple design, we expect to carry out some novel experiments with this device.

As an example of possible applications of an atom interferometer, we have discussed experiments in which atom-laser interactions induce phase shifts or changes in the visibility of the fringes by induced or spontaneous emission processes. These experiments would be a new class of applications of matter wave interferometers unknown to neutron or electron interferometry.

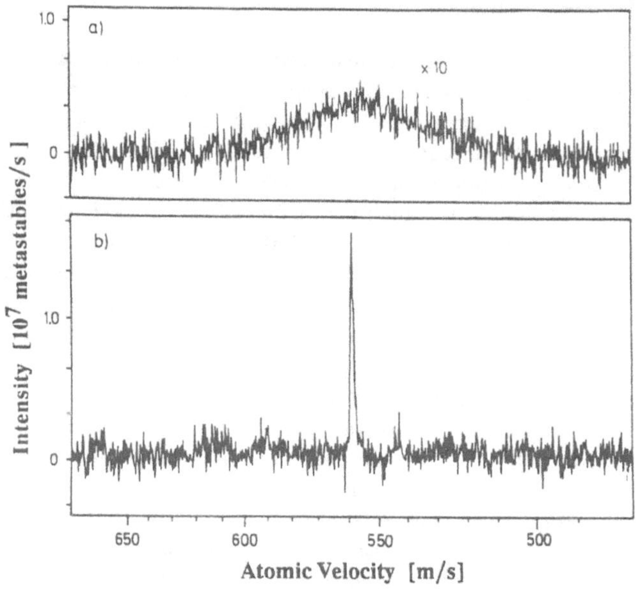

Figure 7. A time of flight spectrum of the uncooled (a) and the cooled (b) atomic velocity distribution. For better visibility the intensity in (a) is enhanced by a factor of 10.

After the first successful demonstations of atom interferometers, much work needs to be done on the technological development of interferometers with better performance. Important goals are to improve the precision and the signal to noise ratio. Such goals could be achieved by improving the spectral quality and intensity of the atomic beam used in the interferometer. As an example of such beam conditioning, we have presented a cooling scheme where a velocity compression in the centre-of-mass frame is achieved by means of light forces. The obtained velocity ratio of $v/\Delta v = 500$ would make it possible to carry out precision measurements in the interferometer or other experiments in the field of atom optics.

ACKNOWLEDGMENTS

We gratefully acknowledge financial support by the Deutsche Forschungsgemeinshaft. One of us (T.S.) thanks the Alexander von Humboldt foundation for a fellowship. We also thank S. Hahn for technical assistance and A. Schnetz, D. Leipold, and M. Sigel for their help in the experiments. Finally we are grateful to H. Takuma and V. Balykin for stimulating discussions.

REFERENCES

1. L. Marton, J. A. Simpson, and J. A. Suddeth, Electron beam interferometer, *Phys. Rev.* 90:490 (1953).

2. H. Maier-Leibnitz and T. Springer, Ein Interferometer für langsame Neutronen, *Z. Physik* 167:386 (1962).

3. O. Carnal A. Faulstich, T. Sleator, and J. Mlynek, Demonstration eines Atominterferometers mit einer Young'shen Doppelspaltanordnung, *Verhand. DPG (VI)*, 26:867 (1991); O. Carnal and J. Mlynek, Young's double slit experiment with atoms: a simple atom interferometer, to appear in *Phys. Rev. Lett.*

4. D.W. Keith, C.R. Ekstrom, Q.A. Turchette, and D.E. Pritchard, An interferometer with atoms, submitted to *Phys. Rev. Lett.* ; F. Riehle, Th. Kisters, A. Witte, J. Helmcke, and Ch.J. Bordé, Optical Ramsey spectroscopy in a rotating frame: Sagnac effect in a matter wave interferometer', submitted to *Phys. Rev. Lett.*

5. *Proc. of the Int. Workshop on Matter Wave Interferometry*, G. Badurek, H. Rauch and A. Zeilinger, editors, *Physica B* 151 (1988), and references therein.

6. P.L. Gould, G.A. Ruff and D.E. Pritchard, *Phys. Rev. Lett.* 56:827 (1986); D.W. Keith, M.L. Schattenburg, H.I. Smith, and D.E. Pritchard *Phys. Rev. Lett.* 61:1580 (1988); O. Carnal, A. Faulstich and J. Mlynek, submitted to *Appl. Phys. B.*

7. T. Sleator *et al.*, to be published .

8. See, e.g., M. Brune, S. Haroche, V. Lefevre, J.M. Raimond, and N. Zagury, Quantum nondemolition measurement of small photon numbers by Rydberg-atom phase-sensitive detection, *Phys. Rev. Lett.* 65:976 (1990), and references therein.

9. See, e.g. J. Dalibard and C. Cohen-Tannoudji, Dressed-atom approach to atomic motion in laser light: the dipole force revisited, *J. Opt. Soc. Am. B* 2:1707 (1985) .

10. S. Stenholm, "*Foundations of laser spectroscopy*", Wiley, New York (1984) p. 163.

11. A.P.Kastinov, Recoil effect in a strong resonant field, *Sov. Phys.-JETP* 40:825 (1975) .

12. S. Stenholm, Redistribution of molecular velocities by optical processes, *Appl. Phys.* 15:287 (1978); J. Dalibard and C. Cohen-Tannoudji, Atomic motion in laser light: connection between semiclassical and quantum descriptions, *J. Phys. B* 18:1661 (1985).

13. See, e.g., M.O. Scully and K. Drühl, Quantum eraser: a proposed photon correlation experiment concerning observation and "delayed choice" in quantum mechanics, *Phys. Rev. A* 25:2208 (1982).

14. *J. Opt. Soc. Am.* B6, (1989) Special issue on laser cooling and trapping of atoms,

15. T. Haslwanter and J. Mlynek, Laser cooling in the center of mass system: A proposal for the creation of a monoenergetic atomic beam, *Ann. Phys.* 47:583, (1990)

16. W. Ertmer, R. Blatt, J.L. Hall and M. Zhu, Laser manipulation of atomic beam velocities: demonstration of stopped atoms and velocity reversal, *Phys. Rev. Lett.* 54, 996, (1985)

17. W.D. Philips and H. Metcalf, Laser deceleration of an atomic beam *Phys. Rev. Lett.* 48, 596, (1982)

18. A. Faulstich *et al.*, to be published

QUANTUM MEASUREMENTS IN ATOMIC INTERFEROMETRY

M. A. M. Marte[1], P. Zoller[2], M. J. Holland[3],
R. Graham[4] and D. F. Walls[4]

1. Institut für Theoretische Physik, Universität Innsbruck
A-6020 Innsbruck (Austria)
2. Joint Institute for Laboratory Astrophysics
Department of Physics and
National Institute for Standards and Technology
University of Colorado, Boulder, Co 80309 (USA)
3. Clarendon Laboratory, Parks Road, Oxford,
OX1 3PU (England)
4. Department of Physics, University of Auckland
Auckland (New Zealand)

1. INTRODUCTION

Deflection of atoms by laser light has been the subject of various theoretical and experimental investigations during the last couple of years.[1-6] For sufficiently short interaction times of the atom with the radiation field spontaneous emission can be neglected during excitation and the scattering process becomes coherent.[2] Thus the wavepacket of a single atom incident on a light wave will be transformed by the radiative interaction into a *superposition* of product states of internal atomic levels and center of mass wavepackets; according to the momentum transfer by the light field these wavepackets will propagate into different directions and finally separate, forming a macroscopic superposition of quantum states. Apart from the intrinsic interest in the study of effects of macroscopic atomic superposition states, coherent scattering of atomic beams by travelling and standing laser light waves is one of the key elements in the optical realization of a beam splitter and mirror in wave matter interferometry with atoms[7-13] (see also Refs. 14 - 16).

Here we summarize some aspects of our recent work on quantum measurements in atomic interferometry. In Section 2 we will study the electromagnetic field emitted by atoms prepared in linear superpositions of several internal states each of which is attached to a different center of mass wavepacket by the preparation process.[17] It is shown that the motion and mutual separation of the wavepackets can be monitored either by observing the coherent spontaneous emission in a heterodyne experiment or by measuring the energy absorption rate from a weak probe-laser beam.

Quantum Measurements in Optics, Edited by P. Tombesi and
D.F. Walls, Plenum Press, New York, 1992

As an example in Section 3 we will discuss an arrangement of three laser interaction zones consisting of counterpropagating waves of opposite circular polarization, which acts as an interferometer for atoms which have a $J = 1$ groundstate and an appropriate $J = 1$ excited state (such as metastable He* for example).[11]

Finally, we show in Sec.4 that the exchange of momentum between atoms and photons can be used as a sensitive probe of the photon statistics of an optical cavity field.[18] We present a new quantum nondemolition (QND) measurement scheme[19–21] using this interaction. In particular, we show that monitoring the position of deflected atoms at appreciable distance from the interaction region allows observation of quantum jumps[22] in the photon number state.

2. EMISSION FROM ATOMS IN LINEAR SUPERPOSITION OF CENTER OF MASS WAVEPACKETS

We consider the scattering of a stationary atomic beam of two-level atoms with ground state $|g\rangle$ and excited state $|e\rangle$ from a running light wave. Initially the atoms are prepared in the ground state, and propagating in the z-direction of our coordinate system. In the plane $z = 0$ the beam passes through a hole of transverse dimension L, and interacts with a running wave of frequency $\omega = ck$ with wavevector in the x-direction. The atomic dipole moment and the polarization of the light are assumed to be oriented along the z-axis. After passing the laser field the atom is in a coherent superposition of the ground and excited states, each of which is attached to its own wavepacket travelling with its own mean velocity fixed by the energy and momentum (and eventually angular momentum) conservation of the excitation process. The wavepackets propagate into the region $z > 0$, overlapping at first, but eventually separating. We are interested in the radiation emitted by the atom when they spontaneously return to their ground state.

The field emitted from the wavepackets in the region where they still overlap contains interferences which are absent in the radiation emitted from regions where the wavepackets have no overlap. These interferences can be used to monitor the separation of the wavepackets. Here we summarize some of the essential features for the case of two-level systems in travelling waves. Details of the calculation and a discussion of three-level systems in standing wave fields can be found in Ref. 17.

The Hamiltonian of the atom interacting with a light wave is

$$H(t) = \frac{\hat{p}^2}{2m} + H_A + H_F + H_I(t). \tag{1}$$

Here $\hat{p}^2/2m$ is the kinetic energy operator and $H_A = \hbar\omega_0|e\rangle\langle e|$ is the atomic Hamiltonian for the internal degree of freedom for the two-level system with transition frequency ω_0. H_F is the Hamiltonian for the vacuum modes of the radiation field. The interaction term of the two-level system with the laser field is

$$H_I(t) = -\int d^3x \mathbf{D}^{(-)}(\mathbf{x}) \cdot (\mathcal{E}_C^{(+)}(\mathbf{x},t) + \mathcal{E}_R^{(+)}(\mathbf{x},t)) + \text{h.c.} \tag{2}$$

with $\mathbf{D}^{(-)}(\mathbf{x}) = d^*\epsilon^*|e\rangle\langle g| \otimes |\mathbf{x}\rangle\langle\mathbf{x}|$ the negative frequency part of the dipole operator density, $d\epsilon$ the dipole matrix element, and $|\mathbf{x}\rangle$ the atomic center of mass position eigenstate. $\mathcal{E}_R^{(+)}(\mathbf{x})$ is the positive frequency part of the quantized electric field at position $\mathbf{x}$ and $\mathcal{E}_C^{(+)}(\mathbf{x},t) = \epsilon_y \mathcal{E}_G(\mathbf{x}) e^{ik_L x - i\omega_L t}$ is the positive frequency part of the travelling wave with slowly varying amplitude (in the z–direction).

The state vector of the system $|\psi_E(t)\rangle$ can be decomposed into with respect to Fock states of the radiation field describing a situation where no spontaneous photon, one photon, etc. has been emitted. We are interested the case where the interaction time with the laser light is short in comparison with the life time of the excited state. For the vacuum part we make the ansatz of stationary oscillations with the laser frequency ω_L,

$$|\psi_{E_0}(t)\rangle = e^{-\frac{i}{\hbar}Et}|0\rangle \otimes \int d^3x e^{imv_E z/\hbar}(\tilde{\phi}_g(\mathbf{x})|g\rangle + e^{-i\omega_L t}\tilde{\phi}_e(\mathbf{x})|e\rangle) \otimes |\mathbf{x}\rangle, \qquad (3)$$

where $|0\rangle$ denotes the vacuum of the electromagnetic field. In writing (3) we have assumed that the kinetic energy $\frac{1}{2}mv_E^2$ along z is large compared with the kinetic energy of the transverse motion, so that $\tilde{\phi}_{g,e}(\mathbf{x})$ are slowly varying amplitudes which satisfy the *time–independent* Schrödinger equation

$$\begin{aligned} i\hbar v_E \frac{\partial}{\partial z}\tilde{\phi}_g &= -\frac{\hbar^2}{2m}\left(\frac{\partial^2}{\partial x^2}+\frac{\partial^2}{\partial y^2}\right)\tilde{\phi}_g - d\mathcal{E}_G^{(-)}(\mathbf{x})e^{-ik_L x}\tilde{\phi}_e \\ i\hbar v_E \frac{\partial}{\partial z}\tilde{\phi}_e &= \left[-\frac{\hbar^2}{2m}\left(\frac{\partial^2}{\partial x^2}+\frac{\partial^2}{\partial y^2}\right) - \hbar\left(\delta_L + i\frac{\gamma}{2}\right)\right]\tilde{\phi}_e \\ &\quad -d^*\mathcal{E}_G^{(+)}(\mathbf{x})e^{ik_L x}\tilde{\phi}_g \end{aligned} \qquad (4)$$

where $\delta_L = \omega_L - \omega_0$ is the detuning and γ is the spontaneous decay rate of the excited level. The statistical operator of the atomic beam associated with (3) is of the form

$$\rho^{(0)}(t) = \int dE\, W(E)|\psi_{E_0}(t)\rangle\langle\psi_{E_0}(t)|, \qquad (5)$$

where $W(E)$ is the energy distribution of the atomic beam.

Further simplification of (4) is possible if we assume that the dispersion of the wavepacket can be neglected during the interaction (Raman–Nath approximation), which allows us to drop the second order derivatives with respect to x and z. The resulting equation is analogous to the Schrödinger equation for the Rabi problem and can be solved accordingly.

After passing the laser beam the atoms are left in linear superpositions of excited states and the ground state, travelling as free wavepackets with differing mean center of mass velocities and spreading in the directions transverse to the atomic beam. The solution of the Schrödinger equation is

$$\begin{aligned} \tilde{\phi}_g(\mathbf{x}) &= C_g f_E(x,y,z) \\ \tilde{\phi}_e(\mathbf{x}) &= C_e e^{i(\delta_L' + i\gamma)z/v_E + ik_L x} f_E(x - \frac{\hbar k_L z}{m v_E}, y, z) \end{aligned} \qquad (6)$$

where $C_{g,e}$ are excitation amplitudes, $\delta_L' = \omega_L - \omega_0 - \frac{\hbar k_L^2}{2m}$ and $f_E(x,y,z) \simeq f(x,y)$ for $z \ll L^2 m v_E/\hbar$ with $f(x,y)$ the initial distribution given by the slit.

We now wish to evaluate the coherent electromagnetic radiation spontaneously emitted by the atoms after their coherent excitation. To this end we consider the expectation value of the positive frequency part of the emitted quantized electromagnetic field

$$\langle \mathcal{E}_R^{(\pm)}(\mathbf{x},t)\rangle = \tag{7}$$

$$-\frac{1}{c^2}\int dE\, W(E)\int \frac{d^3x'}{|\mathbf{x}-\mathbf{x}'|}\frac{\partial^2}{\partial t^2}\langle \psi_{E_0}(t-\frac{|\mathbf{x}-\mathbf{x}'|}{c})|\mathbf{D}_\perp^{(\pm)}(\mathbf{x}')|\psi_{E_0}(t-\frac{|\mathbf{x}-\mathbf{x}'|}{c})\rangle$$

where $\mathbf{D}_\perp^{(\pm)}(\mathbf{x})$ is the transverse projection of $\mathbf{D}^{(\pm)}(\mathbf{x})$. The coherent emission from the atomic beam may be detected via homodyning the emitted field with the field of a local classical oscillator of the same frequency which has a well-defined phase ϕ with respect to the driving laser field. For the positive frequency part of the electric field we find

$$\langle \mathcal{E}_R^{(+)}(\mathbf{x},t)\rangle = \tag{8}$$

$$k_L^2 d e^{-i\omega_L t}\int dE\, W(E) C_g^* C_e \int \frac{d^3x'}{|\mathbf{x}-\mathbf{x}'|}$$

$$\left(\epsilon_y e^{ik_L(|\mathbf{x}-\mathbf{x}'|+x')+i(\delta_L'+i\gamma/2)z'/v_E} f_E^*(\mathbf{x}') f_E(\mathbf{x}'-\frac{\hbar k_L z'}{m v_E}\mathbf{e}_x)\right)_\perp .$$

It is clear from this result that only the spatially overlapping parts of the wavepackets $f_E(\mathbf{x})$ and $f_E(\mathbf{x}-\frac{\hbar k_L z}{m v_E}\mathbf{e}_x)$ of the ground state and the excited state contribute to this coherently radiating dipole density. As soon as the two components of the center of mass wavefunction are completely separated spatially, the coherent emission stops and the excited state can only decay via incoherent emission. Indeed, after complete spatial separation, the two linearly superposed components of the atomic beam are non-overlapping both in center of mass coordinate and momentum and behave indistinguishably from two mutually independent and incoherent atomic beams of corresponding intensity. However, if the two separated components of the beam are

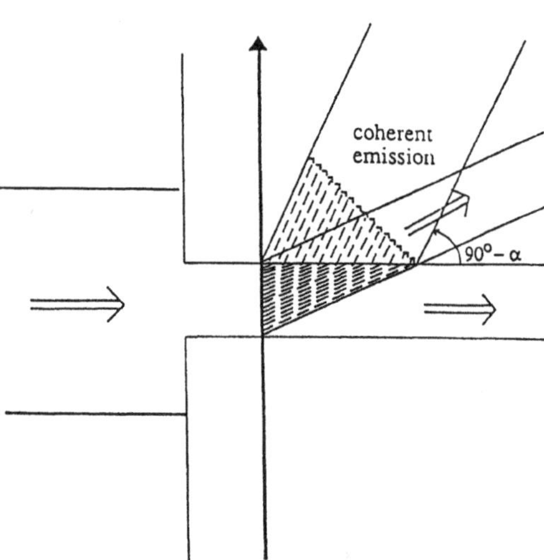

Fig. 1. The coherent radiation emitted from a two–level atom is depicted. The lengths of the arrows represent the amplitude of the radiation.

somehow brought to overlap again, as is done in an atomic interferometer, interference effects and coherent emission will reappear, provided the decay due to spontaneous emission is not yet completed.

Analytical evaluation of the integral (8) is simplified by the following conditions: first we assume that the area of the hole is large compared to the laser wavelength; furthermore we neglect atomic beam diffraction and consider only the the far field of the radiation field. The resulting analytical expression is, however, quite complicated and will not be reproduced here in its general form.[17] Instead we discuss some limiting cases which allow a simple physical interpretation.

The dependence of the coherent emission on the detuning δ'_L shows some interesting features: the coherent emission from a *stationary* beam of atoms must occur with the frequency ω_L of the driving laser frequency. On the other hand, the excited state wavepacket of the atom in the laboratory frame differs in energy from the ground state wave-packet by $\hbar\left(\omega_0 + \frac{\hbar k_L^2}{2m}\right)$ which is different from $\hbar\omega_L$ if $\delta'_L \neq 0$. To compensate this difference the atoms emit at an angle α defined by $\sin\alpha = \frac{\delta'_L}{k_L v_E}$, i.e. the emitted light is Doppler–shifted (see Fig. 1). Here $\alpha > 0$ (blue shift) if $\omega_L > \omega_0 + \frac{\hbar k_L^2}{2m}$ or $\alpha < 0$ (red shift) if $\delta'_L < 0$. Thus the field is essentially emitted along a cone of angle $(\pi/2 - \alpha)$ around the z-axis.

A major feature of the coherent radiation is the fact that it can only be observed in a certain region of space. It cannot be observed in the region $z > \sqrt{x^2+y^2}\tan\alpha + \frac{2mv_EL}{\hbar k_L}$, i.e. in the cone of angle $(\pi/2 - \alpha)$ around the z-axis whose vertex lies at the point where the wave-packets separate.

Having discussed the effects of detuning $\delta'_L \neq 0$, we may now restrict our attention to the case $\delta'_L = 0$, $\alpha = 0$. The dominant radiation is emitted in the (x, z)-plane, $y = 0$. We find

$$\begin{aligned}
&\langle \mathcal{E}_R^{(+)}(x,0,z)\rangle = \qquad\qquad (9)\\
&\quad \epsilon_y \frac{(2\pi)^{1/2}}{|\,x\,|^{1/2}} d k_L^{3/2} e^{-i\omega_L t + i\pi/4} \int dE\, W(E) C_g^* C_e \Theta\left(1 - \frac{\hbar k_L z}{2mv_E L}\right) e^{-\frac{\gamma}{2v_E}z}\\
&\quad \left[\left(1 - \frac{\hbar k_L z}{2mv_E L}\right) e^{ik_L x}\Theta(x) + e^{-ik_L\left(x - \frac{\hbar k_L}{mv_E}z\right)} \frac{\sin\left[2k_L L\left(1 - \frac{\hbar k_L z}{2mv_E L}\right)\right]}{2k_L L}\Theta(-x)\right].
\end{aligned}$$

This shows that the amplitude of the radiation goes to zero continuously as one approaches the border of the region where there is no wavepacket overlap. It also shows a pronounced asymmetry between the radiation emitted in the upward x direction (the direction of propagation of the exciting laser field) and the radiation emitted in $(-x)$ direction which tends to be suppressed.

Instead of observing the coherent spontaneous emission one can observe the overlap of the wavepackets and its disappearance by measuring the absorption rate of energy from a weak probe beam with frequency $\omega_L + \Delta\omega$ and phase which is fixed with respect to the exciting laser.[17] The energy absorption rate is proportional to the probing field strength, oscillates with the difference of the probing and exciting laser frequencies $\Delta\omega$, and, for a sufficiently mono-energetic atomic beam, exhibits a pronounced directional characteristic. The directional characteristic arises because of the spatial extension of the atomic beam in all directions is assumed large compared to the laser wavelength, and because of the spatial coherence of the atomic beam over its cross-sectiono.

Related signatures are also found in incoherent spontaneous emission.[17]

3. ATOMIC TRIPLE LAUE INTERFEROMETER

We study a beam of effective three-level atoms, for example the two Zeeman sublevels $m = \pm 1$ of the 2^3S_1 groundstate and the excited $m = 0$ sublevel of the 2^3P_1 excited state of metastable helium,[11,23] propagating along the z-axis. The atoms interact with two waves of circular polarization and equal amplitude: a σ^+ wave propagating in the $+x$ direction and a σ^- wave propagating along $-x$, as depicted in Fig.2.

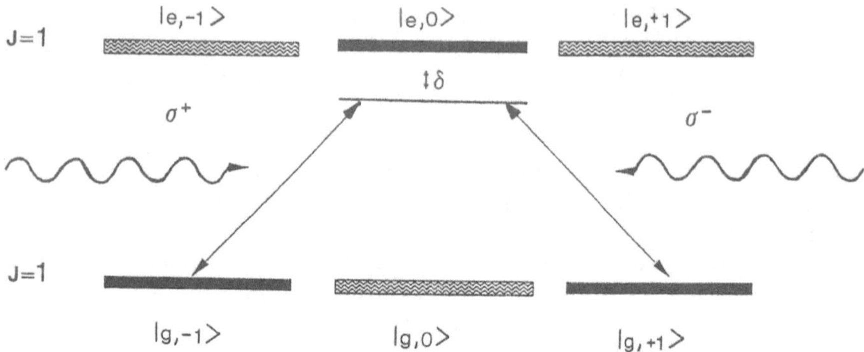

Fig. 2. Energy level diagram with Raman transition induced by counter-propagating σ^+ and σ^- waves.

In this case one has to insert

$$\mathbf{D}^{(-)}(\mathbf{x}) = \left[d_+^* \epsilon_+^* |e\rangle\langle g_+| + d_-^* \epsilon_-^* |e\rangle\langle g_-|\right] \otimes |\mathbf{x}\rangle\langle\mathbf{x}| \tag{10}$$

for the negative frequency part of the dipole operator density (where $d_\pm$ are the dipole matrix elements of the two transitions, $\epsilon_\pm$ are unit vectors), and

$$\mathcal{E}_C^{(+)}(\mathbf{x}, t) = \boldsymbol{\epsilon}_+ \mathcal{E}_+(\mathbf{x}) e^{ik_L x - i\omega_L t} + \boldsymbol{\epsilon}_- \mathcal{E}_-(\mathbf{x}) e^{-ik_L x - i\omega_L t} \tag{11}$$

for the sum of the positive frequency parts of the two counterpropagating σ^+ and σ^- electric fields with slowly varying amplitudes $\mathcal{E}_\pm(\mathbf{x})$ and polarization vectors $\boldsymbol{\epsilon}_\pm = \mp\frac{1}{\sqrt{2}}(\mathbf{e}_x \pm i\mathbf{e}_z)$ into the rotating wave interaction Hamiltonian (2). (Note that here we have made use of the fact that the problem can be reduced to two spatial dimensions; thus in this sectiono $(\mathbf{x})$ denotes (x, z).)

Due to conservation of energy and angular momentum an atomic state with arbitrary momentum p_x and with the internal degree of freedom prepared e.g. in the $m = 1$ groundstate, denoted by $|g_-, p_x\rangle$, is coupled to $|e, p_x + \hbar k_L\rangle$, similarly $|g_+, p_x\rangle$ couples only to $|e, p_x - \hbar k_L\rangle$. Thus (2) reduces to

$$H_I(t) = -\int d^2x \left[\frac{\hbar\Omega_+^*}{2}|e\rangle\langle g_+|e^{-ik_L x - i\omega_L t} + \frac{\hbar\Omega_-^*}{2}|e\rangle\langle g_-|e^{ik_L x - i\omega_L t}\right] \otimes |\mathbf{x}\rangle\langle\mathbf{x}| + \text{h.c.}, \tag{12}$$

where $\Omega_\pm = 2d_\pm \epsilon_\pm \cdot \varepsilon^*_\mp \mathcal{E}^*_\mp$ are the Rabi frequencies, which are assumed constant *inside* the interaction zone. Here we have left out the quantized radiation field used in order to describe spontaneous emission into empty modes, since in the limit of large detuning, which we will be interested in, the population of the excited state and its spontaneous decay is negligible.

Analogous to (3) we make the ansatz

$$|\psi_E(t)\rangle = e^{-iEt/\hbar} \int d^2x \left[\psi_+(\mathbf{x})|g_+\rangle + \psi_-(\mathbf{x})|g_-\rangle + e^{-i\omega_L t}\psi_e(\mathbf{x})|e\rangle\right] \otimes |\mathbf{x}\rangle. \tag{13}$$

For simplicity we restrict ourselves to the case of an atomic beam with a *large quasi-monochromatic* z-component of the velocity v_z, i.e. the general expression (5) reduces to the projector to one state $|\psi_E(t)\rangle$ of energy $E \approx E_z = p_z^2/2m$, but allow for a small Gaussian spread in transverse momenta p_x. Such a beam could be achieved by apertures or transverse cooling.[1]

Choosing the detuning $\delta = \omega_L - \omega_A$ very large (i.e. $\hbar\delta \gg E$), the population of the excited level becomes negligible and can be eliminated. In this case a simple picture of induced Raman-type transitions between the two subgroups of atomic states $|g_-, p_x - \hbar k_L\rangle$ and $|g_+, p_x + \hbar k_L\rangle$, imposing a recoil of two photon momenta $G = 2\hbar k_L$, becomes appropriate. Along the same lines as in Section 2 we make the slowly varying amplitude approximation, that is drop the second order derivatives with respect to z in the equation analogous to (4) for the remaining two slowly varying amplitudes $\tilde{\psi}_\pm(\mathbf{x}) = e^{-imv_z z/\hbar}\psi_\pm(\mathbf{x})$. With these simplifications the problem is reduced to solving the 'Schrödinger equation'

$$i\hbar v_z \frac{\partial}{\partial z}\begin{pmatrix} u_-(p_x;z) \\ u_+(p_x;z) \end{pmatrix} = \begin{pmatrix} V_0 & \hbar\Omega \\ \hbar\Omega^* & V_0 + \frac{(p_x+G)^2}{2m} - \frac{p_x^2}{2m} \end{pmatrix}\begin{pmatrix} u_-(p_x;z) \\ u_+(p_x;z) \end{pmatrix} \tag{14}$$

with fictitious time z/v_z for the amplitudes $u_\pm(x;z)$ of $|\psi_E(t)\rangle$ in the momentum representation

$$\begin{aligned}\langle p_x; z|\psi_E(t)\rangle &= e^{-iEt/\hbar}e^{imv_z z/\hbar}\left[u_-(p_x;z)|g_-\rangle + u_+(p_x;z)|g_+\rangle\right] \qquad (15)\\ &= e^{-iEt/\hbar}e^{imv_z z/\hbar}\int \frac{dp_x}{\sqrt{2\pi\hbar}} e^{ip_x x/\hbar}\left[\tilde{u}_-(x;z)|g_-\rangle + e^{iGx/\hbar}\tilde{u}_+(x;z)|g_+\rangle\right]\end{aligned}$$

with 'initial condition'

$$\begin{aligned} u_+(p_x; z=0) &= 0 \qquad (16)\\ u_-(p_x; z=0) &= f(p_x) \equiv \frac{1}{\sqrt[4]{2\pi(\Delta p_x)^2}} e^{-\frac{(p_x-p_0)^2}{4(\Delta p_x)^2}}. \end{aligned}$$

In (14) G again stands for the recoil momentum $2\hbar k_L$ conveyed by a Raman transition from $|g_-\rangle$ to $|g_+\rangle$, $\Omega = e^{i\phi}|\Omega_+\Omega_-/4\delta|$ is the effective Rabi frequency. For the sake of simplicity it has been assumed that the Rabi frequencies $\Omega_\pm$ of the σ^+ and σ^- transitions have the same magnitude. V_0 denotes the Stark shift $V_0 = \hbar\frac{|\Omega_+|^2}{4\delta} = \hbar\frac{|\Omega_-|^2}{4\delta}$.

Eq.(14) can be solved exactly:

$$\begin{pmatrix} u_-(p_x;z) \\ u_+(p_x;z) \end{pmatrix} = \mathcal{U}(z)\begin{pmatrix} u_-(p_x;0) \\ u_+(p_x;0) \end{pmatrix}, \tag{17}$$

with the unitary matrix $\mathcal{U}(y)$ describing the interaction with the light fields given by

$$\mathcal{U}(z) = e^{i(-\frac{V_0}{\hbar v_z} - \pi\frac{Y}{\Delta_0})z}\begin{pmatrix} cos(\Theta) + \frac{iY}{\sqrt{1+Y^2}}sin(\Theta) & \frac{-ie^{i\phi}}{\sqrt{1+Y^2}}sin(\Theta) \\ \frac{-ie^{-i\phi}}{\sqrt{1+Y^2}}sin(\Theta) & cos(\Theta) - \frac{iY}{\sqrt{1+Y^2}}sin(\Theta) \end{pmatrix} \tag{18}$$

with $\Delta_0 = \pi\frac{v_z}{|\Omega|}$, $\alpha = \frac{2p_xG+G^2}{2mE}$, $Y = \frac{\alpha E}{2|\Omega|\hbar}$ and $\Theta = \frac{\pi}{\Delta_0}z\sqrt{1+Y^2}$. Free evolution for a distance z_F after a laser interaction zone of length z is easily taken into account by multiplying $\mathcal{U}(z)$ with the free space 'propagator' $\mathcal{F}(z_F)$ given by

$$\mathcal{F}(z_F) = \lim_{|\Omega_\pm|\to 0} \mathcal{U}(z_F) = \begin{pmatrix} 1 & 0 \\ 0 & e^{-i\alpha E\frac{z_F}{\hbar v_z}} \end{pmatrix}. \tag{19}$$

For small Y the unitary evolution (18) resembles the Rabi problem of a two-level system : For example, by choosing the length z of the interaction zone to be one quarter of the characteristic length Δ_0, one can prepare a coherent superposition of $|g_-, p_x - \hbar k_L\rangle$ and $|g_+, p_x + \hbar k_L\rangle$, analogous to the application of a $\frac{\pi}{2}$ pulse on a two-level system. Changing the thickness to $\Delta_0/2$, a wavepacket initially peaked around $p_0 = -G/2$ is shifted by an amount G to $+G/2$. Enlarging the interaction zone to Δ_0 brings us back to the original position of the peak - apart from the spreading of the wavepacket which is ignored inside the interaction zone in the Kapitza-Dirac regime of short interaction times.[10] Δ_0 is called the 'Pendel-length' in analogy to a phenomenon encountered in neutron interferometry.[15]

Thus choosing the thickness z of the interaction zone with the light field equal to $\Delta_0/4$ the laser light zone acts as a *coherent beamsplitter*: an initially single-peaked momentum distribution around $-G/2$ splits up into two peaks separated by the momentum transfer G. Subsequent free evolution can lead to a spatial separation into two localized peaks of the atomic wavepacket - provided that the spread of p_x in the initial wavepacket does not violate the necessary condition $\Delta p_x < G/2$ (for detail see Ref.11), a requirement which is of the order of magnitude achievable e.g. by cooling the transverse velocity down to the recoil limit.[1] With the above matrix formalism the generalization of the theory to the more complex geometrical set-up of a 'triple Laue interferometer', consisting of three interaction zones, a splitter of length $z_S = \Delta_0/4$, a mirror of $z_M = \Delta_0/2$ and an analyser of $z_A = \Delta_0/4$ and free propagation in between is easily carried out, as it reduces to simple matrix multiplication: The total unitary operator is given by the product

$$\mathcal{U}(z) = \mathcal{F}(z_{F_3})\mathcal{U}(z_A)\mathcal{F}(z_{F_2})\mathcal{U}(z_M)\mathcal{F}(z_{F_1})\mathcal{U}(z_S) \tag{20}$$

with $z = \sum_{i\in I} z_i$. Such a device is called a 'Laue' interferometer, because the momentum transfer from the light fields to the atom is along the transverse direction x, which resembles the Laue (as opposed to Bragg) case in neutron diffraction theory with the **k**-vectors in the diffracting crystal being parallel to the surface plane of the crystal.

From solution (18) one can immediately calculate the momentum and position distribution:

$$F(p_x; z) \equiv |\langle p_x; z|\psi_E(t)\rangle|^2 = |u_+(p_x; z)|^2 + |u_-(p_x; z)|^2 \tag{21}$$

and

$$\tilde{F}(x; z) \equiv |\langle x; z|\psi_E(t)\rangle|^2 = |\tilde{u}_+(x; z)|^2 + |\tilde{u}_-(x; z)|^2, \tag{22}$$

where $u_+(p_x; z)$ is the Fourier transform of $e^{iGx/\hbar}\tilde{u}_+(x; z)$ and $u_-(p_x; z)$ of $\tilde{u}_-(x; z)$. Fig.3 displays a 3D-plot of the position distribution function of a single atom, as it passes through the interferometer.

There is an important qualitative difference to a triple Laue *neutron* interferometer: For example let us consider a symmetrical set-up consisting of 'splitter' and 'mirror' with $z_S = \Delta_0/4$, $z_M = \Delta_0/2$ and $z_{F_1} = z_{F_2}$ and let us place a detector at the spot where the two branches recombine. There the atom is in a superposition of

$|g_-, -G/2\rangle$ and $|g_+, +G/2\rangle$, but one will *not* get any oscillating interference term, when measuring the center of mass position or momentum. This is so, because the atomic state is a superposition of two states, which have different internal quantum numbers and are thus orthogonal. This is the reason for the absence of interference terms in the expressions for the position and momentum distribution (cf. (21) and Eq.(22)). However, if one measures other observables, which 'rotate' the pseudo-spin J, e.g. turning a $|g_+, +G/2\rangle$ into a $|g_-, +G/2\rangle$, one will see interference. However, this does not imply that such a device cannot be used as sensor for relative phase shifts in the spatially separated branches. One simply has to add a third interaction zone of length $z_A = \Delta_0/4$, aboved referred to as 'analyser', which recombines the two branches. Then the set-up is a sensitive interferometer in the following sense: a nondispersive phase shift in one of

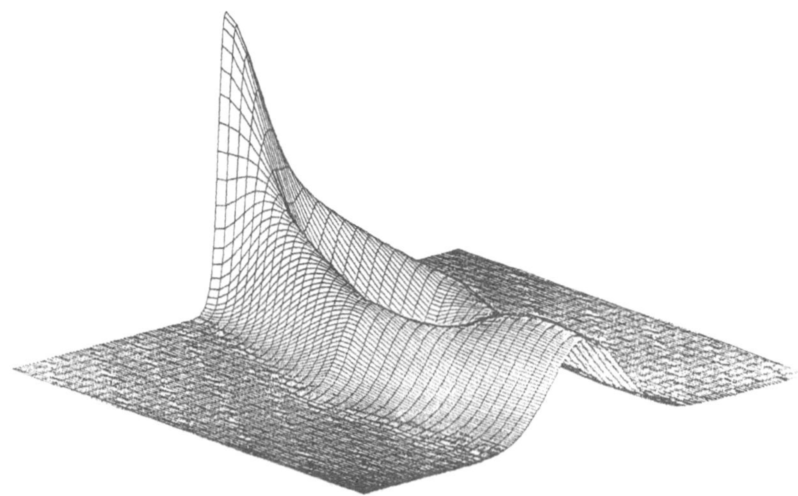

Fig. 3. 3D-plot of the position distribution function of a single atom passing through the interferometer (arbitrary units).

the two branches shifts the location of the peak of the atomic wavefunction behind the analyser. Such a nondispersive phase shifter can be realized by scanning the phase ϕ_M of the effective Rabi frequency Ω of the 'mirror'. The numerical calculation[11] shows that the peak of the position distribution oscillates with a period of π between two points as a function of ϕ_M.

4. QUANTUM NONDEMOLITION OBSERVATION OF QUANTUM JUMPS BY ATOMIC BEAM DEFLECTION

We now wish to discuss a new type of QND coupling[19–21] generated by the exchange of momentum between atoms and photons.[1] Meystre *et al.*[4,5] have shown that the atomic beam deflection by a light field is a sensitive function of the field photon statistics. In our proposal we monitor the time evolution of an optical standing wave field containing a small number of quanta in a high quality cavity. The QND observable is the photon number of the field. The measurement probe is a beam of two-level atoms crossing perpendicularly through the center of the cavity. The transverse momentum transfer of the atom-photon interaction is such that at an appreciable distance from the interaction region the position of the atom carries a significant amount of information about the intensity of the field.

The Hamiltonian for the combined atom and radiation field system in the rotating wave approximation can be written as

$$H = \hbar\omega_a a^\dagger a + \hbar\omega_o \sigma_z + \frac{p^2}{2m} + \hbar\left(g\sigma_- a^\dagger + g^* \sigma_+ a\right)\cos kx. \tag{23}$$

Here a is the boson annihilation operator for the field, the Pauli operators refer to the internal atomic degrees of freedom and $p^2/2m$ denotes the kinetic energy operator for the transverse atomic motion. $\cos kx$ refers to the mode function fo the cavity field with x the atomic position operator and k is the wavenumber of the cavity mode. g is the single photon Rabi frequency. We consider the Raman-Nath regime in which the transverse kinetic energy absorbed by the atom during the interaction can be neglected. Spontaneous emission can be neglected since we consider only the solution in the region of large detuning corresponding to undisturbed photon statistics $|g|\sqrt{n}/\Delta \ll 1$, where $\Delta = \omega_o - \omega_a$ now denotes the detuning, and n is the characteristic photon number. In this perturbation limit, the resulting equations of motion can be derived from the *effective* interaction Hamiltonian

$$V_i = \hbar\frac{|g|^2}{\Delta}\sigma_z a^\dagger a(\cos 2kx + 1) \tag{24}$$

in which neither the atomic inversion σ_z nor the photon number $a^\dagger a$ are altered by the evolution of the atom in the cavity. The strength of the perturbation is proportional to the intensity of the field. Note that the $\cos 2kx$ dependence of the Hamiltonian describes the momentum transfers $\pm 2\hbar k$. The output momentum distribution is given by

$$\begin{aligned} P(p,t) &= P(p,0) * \sum_r Q_r(t)\,\delta(p - 2r\hbar k), \\ Q_r(t) &= \sum_n J_r^2\left(\frac{|g|^2 nt}{2\Delta}\right) P(n). \end{aligned} \tag{25}$$

where $*$ denotes momentum convolution, and $P(n)$ describes the photon statistics of the cavity. The amplitudes Q_r represent the probability that a momentum of $2r\hbar k$ will be transferred to the atom in the cavity.

Since the position wavefunction at large distances from the interaction region is dependent on the statistics of the intracavity field, each atomic position measurement reduces the density operator corresponding to the field state. If the dynamical scale of the cavity is large compared to the atom flow, Eq.(25) can be inverted and a sufficient number of repeated measurements will eventually completely determine the photon statistics. However, if the cavity is not significantly coupled to any external reservoir, continuous probing of the cavity will eventually result in the complete collapse of the field state to that of an exact number state corresponding to the absolute energy of the radiation within the normalization volume. We illustrate this effect by simulating repeated atomic position measurements and examining the residual density of states. The procedure adopted is a generalization of the technique given by Brune *et al.*.[24]

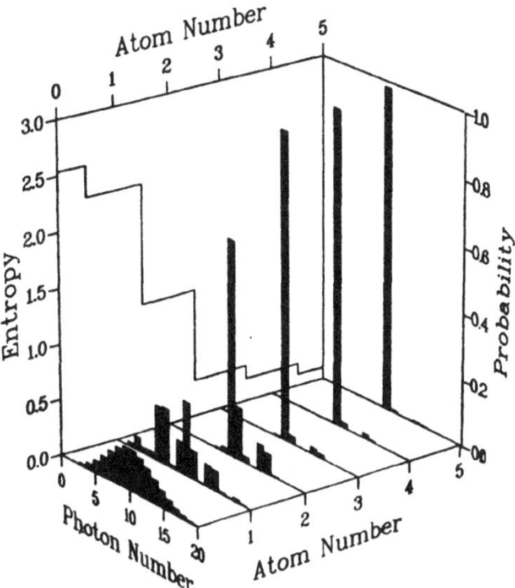

Fig. 4. Simulation of the collapse of the field density of states to a single photon number. The non-relaxing cavity was initially described by a coherent state of mean 10 photons with Poissonian photon statistics. The atomic interaction parameters were $(|g|^2 t)/(2\Delta) = 50$. The vertical *probability* scale on the right corresponds to all 6 of the bar graphs. Projected on to the back wall is the entropy of the field state with scale denoted by the *entropy* vertical axis on the left. As information about the field is accumulated, the entropy is reduced.

Fig.4 illustrates a simulation of 5 probe atoms with a non-relaxing field initially described by a coherent state with a mean photon number of 10. Each such simulation collapses the field to a single photon number, which then does not change any more. The number of times that each number is selected is completely determined by the initial photon statistics. Note that the position measurement of each individual atom extracts partial information from the field, and it is only the cumulative information contained in the full measurement sequence which contracts the field to a well defined number state.

The final entropy can be used as an indicator of the quality of measurement. In the case of an appreciable longitudinal velocity distribution of atoms in the probe beam, it is necessary to select a velocity as well as exit momentum for each atom (see Ref. 23). Such a simulation contains the same characteristics of projection to a single photon number state after a number of probe atoms, but the velocity spread reduces the measurement quality and consequently retards the rate of entropy collapse.

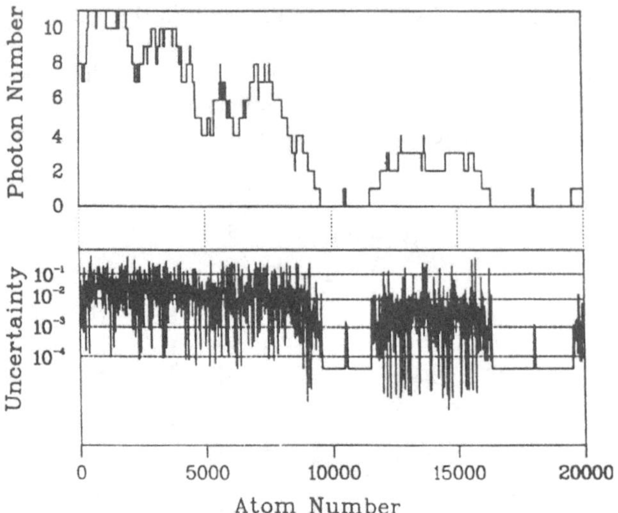

Fig. 5. Simulation of a cavity weakly coupled to a thermal reservoir of mean $\overline{n} = 3$. The field was initially in a coherent state with a mean of 10 photons. The interaction was described by $(|g|^2 t)/(2\Delta) = 50$, and the decay rate of the cavity corresponded to 5000 atoms. The top graph illustrates the most likely photon number showing quantum jumps in the state, and the relaxation of the cavity towards stationary occupation times for each photon number distributed by the photon statistics of the reservoir. The bottom graph displays the entropy measure of field uncertainty on a logarithmic scale. The vertical axis of this graph is labelled with threshold values of 10^{-1}, 10^{-2}, 10^{-3}, and 10^{-4}, corresponding to the total possible residual probability not contained in the photon number state quoted in the top graph.

Provided that the cavity is only weakly relaxing, the repeated atomic measurements can still project the field into a well defined photon number and it is then possible to observe quantum jumps in the state. Fig.5 illustrates this for a field weakly coupled to a thermal bath of mean $\overline{n} = 3$. The field was initially described by a coherent state of mean 10, and the decay rate of the cavity γ corresponded to 5000 atoms. As an indicator of the quality of the measurement as the simulation evolves, we have illustrated the entropy of the field displayed on a logarithmic scale. The vertical axis is labelled by four threshold values, 10^{-1}, 10^{-2}, 10^{-3}, and 10^{-4}, corresponding to residual uncertainty in the photon state.

ACKNOWLEDGMENTS

M.M. and P.Z. were supported by the Österreichische Fonds zur Förderung der wissenschaftlichen Forschung. R. G., who is on leave from the Fachbereich Physik, Universität GHS Essen, D-4300 Essen, Germany, wishes to thank the 'Deutsche Forschungsgemeinschaft' for support through the 'Sonderforschungsbereich 237, Unordnung und große Fluktuationen.'

REFERENCES

1. For articles on mechanical effects of light and laser cooling and trapping see special issues in: *J. Opt. Soc. Am. B* 2, No.11 (1985) and *J. Opt. Soc. Am. B* 6, No.11 (1989)

2. B. W. Shore, *The Theory of Atomic Excitation*, Vol. 2, (J. Wiley and Sons, New York, 1990), p. 1009

3. See the contribution by W. Schleich in these Proceedings

4. P. Meystre, E. Schumacher and S. Stenholm, *Optics Commun.* 73, 443 (1989);

5. B. W. Shore, P. Meystre, and S. Stenholm, preprint

6. For experiments see for example P. J. Martin, P. L. Gould, B. Oldaker, A. H. Miklich and D. E. Pritchard, *Phys. Rev. A* 36, 2495 (1987); P. J. Martin, B. Oldaker, A. H. Miklich and D. E. Pritchard, *Phys. Rev. Lett.* 60, 515 (1988)

7. V. P. Chebotayev, B. Y. Dubetskiǐ, A.P. Kazantsev and V. P. Yakovlev, *J. Opt. Soc. Am. B* 2, 1791 (1985)

8. J. F. Clauser, *Physica B* 151, 262 (1988)

9. Ch. J. Bordé, *Phys. Lett. A* 140, 10 (1989)

10. E. M. Wright and P. Meystre, *Opt. Commun.* 75, 388 (1990)

11. M. A. M. Marte, I. J. Cirac and P. Zoller, unpublished

12. For scattering from gratings see D. W. Keith, M. L. Schattenburg, H. I. Smith and D. E. Pritchard, *Phys. Rev. Lett.* 61, 1580 (1988)

13. See the contribution by J. Mlynek in these Proceedings

14. H. Rauch and D. Petrascheck, in: *Neutron Diffraction* , ed. H. Dachs. (Springer, Berlin, 1978), p. 303

15. *Neutron Interferometry*, ed. U. Bonse and H. Rauch, (Clarendon Press, Oxford, 1979)

16. See contributions in *Proceedings of the International Workshop on Matter Wave Interferometry, Vienna, Austria, 14-16 September 1987*, eds. G. Badurek, H. Rauch and A. Zeilinger, *Physics B+C*, 151, Nos. 1 and 2 (1988)

17. R. Graham, D. F. Walls and P. Zoller, unpublished

18. M. Holland, D. F. Walls and P. Zoller, unpublished

19. V. B. Braginsky, Y. I. Vorontsov, and F. Y. Khalili, *Zh. Eksp. Teor. Fiz.* 73, 1340 (1977) [*Sov. Phys. JETP* 46, 705 (1977)]; W. G. Unruh, *Phys. Rev. D* 18, 1764 (1978); C. M. Caves, K. S. Thorne, R. W. P. Drever, V. D. Sandberg, and M. Zimmerman, *Rev. Mod. Phys.* 52, 341 (1980)

20. G. J. Milburn and D. F. Walls, *Phys. Rev. A* 28, 2065 (1983); N. Imoto, H. A. Haus, and Y. Yamamoto, *Phys. Rev. A* 32, 2287 (1985)

21. M. D. Levenson, R. M. Shelby, M. Reid, and D.F. Walls, *Phys. Rev. Lett.* 57, 2473 (1986). N. Imoto, S. Watkins, and Y. Sasaki, *Opt. Commun.* 61, 159 (1987); A. La Porta, R. E. Slusher, and B. Yurke, *Phys. Rev. Lett.* 6, 28 (1989)

22. R. Blatt and P. Zoller, *Eur. J. Phys.* 9, 250 (1988)

23. A. Aspect, E. Arimondo, R. Kaiser, N. Vansteenkiste and C. Cohen-Tannoudji, *Phys. Rev. Lett.* bf 61, 826 (1988)

24. M. Brune, S. Haroche, V. Lefevre, J. M. Raimond, and N. Zagury, *Phys. Rev. Lett.* 65, 976 (1990)

AN ATOMIC-BEAM QUANTUM-ERASER GEDANKEN EXPERIMENT

Berthold-Georg Englert,[2] Heidi Fearn,[1] Marlan O. Scully,[1,2] and Herbert Walther[2]

1. Center for Advanced Studies and Department of Physics and Astronomy University of New Mexico, Albuquerque, New Mexico 87131
2. Max-Planck Institut für Quantenoptik Sektion Physik, Universität München D-8046 Garching, Germany

1. INTRODUCTION

In recent work[1] we have analyzed a which path (German: welcher Weg) interference experiment, involving an atomic beam incident on two slits. Behind each slit is a micromaser cavity and beyond them a screen to observe interference fringes of the outgoing atomic beams. This scheme is one in which the principle of complementarity is manifest but the position-momentum uncertainty relation plays no role. Essentially, we have found a way to obtain welcher Weg information, concerning which slit the atom passed through, without exposing the interfering beams to uncontrollable scattering processes. The interference fringes vanish once we have welcher Weg information but their disappearance is due to correlations between the field inside the micromaser detectors and the atomic beam wavefunctions. The "erasure" of these correlations was shown to restore the interference pattern. However in the example given, "shutters" were introduced which could not be treated via a simple Hamiltonian dynamics. In the present note we improve upon this model and introduce a Gedanken experiment which is completely calculable.

As in our previous paper, we begin by discussing an old Gedanken experiment by Einstein which illustrates the kind of set up used to examine complementarity in quantum mechanics. We refer of course to the famous recoiling-slit arrangement[2] depicted in Fig. 1. Plane light waves are incident on two slits which are allowed to recoil via mechanical rollers when light passes through either slit. Photons arriving at the distant screen, at a point Δx from the central maximum, possess different momenta $\hbar k_1$ or $\hbar k_2$ depending on which slit they passed through to reach the screen. (Δx is taken to be the fringe spacing). The x-component of the momenta is approximated by

$$\Delta k_x \cong \frac{2\pi}{\Delta x}. \tag{1}$$

Quantum Measurements in Optics, Edited by P. Tombesi and D.F. Walls, Plenum Press, New York, 1992

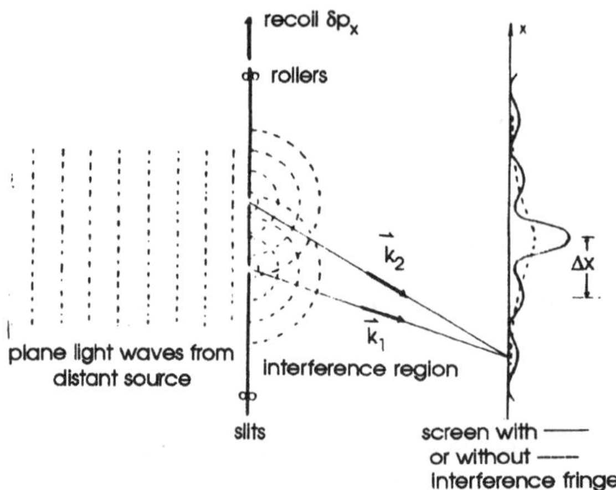

Fig. 1. Einstein's recoiling-slit experiment. The slits can recoil and potentially reveal through which slit the photon reached the screen, since only one of the wavevectors $\vec{k}_1$ or $\vec{k}_2$ is consistent with a known amount of recoil momentum.

The recoil momentum of the slits must therefore be determined with a precision $\delta p_x \leq \hbar \Delta k_x$, to determine through which slit the photon passed. If we take δx to be the uncertainty in position of the slits then Heisenberg's uncertainty relation

$$\delta p_x \cong \frac{\hbar}{\delta x}, \tag{2}$$

implies

$$\Delta k_x \geq \frac{1}{\delta x}. \tag{3}$$

From Eq. 1, we see that $\delta x \geq \Delta x$, which means that the uncertainty in locating the slits (and therefore the fringes) is larger than the spacing between the fringes, and so the fringe pattern is washed out. Therefore, if we have information which allows us to determine through which slit the photon came, we no longer have an interference pattern on the screen. Welcher Weg information and wave-like interference phenomenon are therefore incompatible due to Heisenberg's uncertainty relation. We may now ask the question, is it always the case that an instrument for determining which-path information will inevitably disturb the spatial wavefunction of the photon (or particle) in such a way as to prevent the observation of interference fringes? In the standard examples including Heisenberg's famous microscope,[3] complementarity is always enforced by the position-momentum uncertainty relation. However, in the following experimental arrangement we outline a procedure to obtain welcher Weg information without exposing the interfering beams to uncontrollable scattering processes. Complementarity is enforced in this case, not by Heisenberg's uncertainty relation, but simply by correlations between the measuring apparatus and the system being observed.

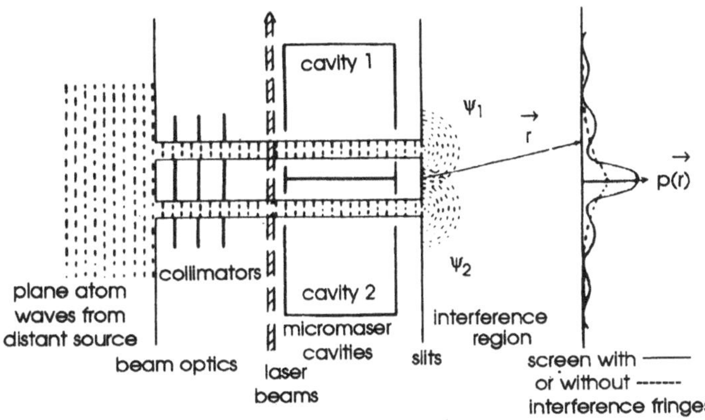

Fig. 2. Two slit experiment with atoms. A set of wider slits collimates two atomic beams that illuminate the narrow slits where the interference pattern originates. Two high-quality micromaser cavities provide which-path information.

2. THE MICROMASER WELCHER WEG DETECTOR

Let us consider an atomic beam instead of photons, incident on two slits as depicted in Fig. 2. The center-of-mass wavefunction of the atoms, near the screen, is given by the sum of two terms referring to the two slits,

$$\psi(\vec{r}) = \frac{2}{\sqrt{2}}[\psi_1(\vec{r}) + \psi_2(\vec{r})]. \tag{4}$$

The probability density of the atoms falling on the screen, denoted by $P(\vec{r})$, is given by the squared modulus of $\psi(\vec{r})$,

$$P(\vec{r}) = \frac{1}{2}[|\psi_1(\vec{r})|^2 + |\psi_2(\vec{r})|^2 + \psi_1^*(\vec{r})\psi_2(\vec{r}) + \psi_2^*(\vec{r})\psi_1(\vec{r})]. \tag{5}$$

The interference phenomenon is a result of the cross terms $\psi_1^*\psi_2 + \psi_2^*\psi_1$. The atomic beam in this Gedanken experiment allows for more internal degrees of freedom than the photons of Einstein's original recoil slit experiment. One can imagine preparing the atoms in an excited state and allowing them to pass through the micromaser cavities. The atoms would then interact with the microwave field present inside the masers and spontaneously emit a photon, thereby leaving welcher Weg information inside the cavity.

The micromasers will only serve as detectors if it is possible to detect the one extra photon deposited by the atom as it passes through. As an ideal example, we may have no photons initially inside the cavities and then an atom leaves one behind. By detecting the single photon we can tell through which micromaser (and hence slit) the atom passed through on its way to the screen. Clearly, we require a quantum

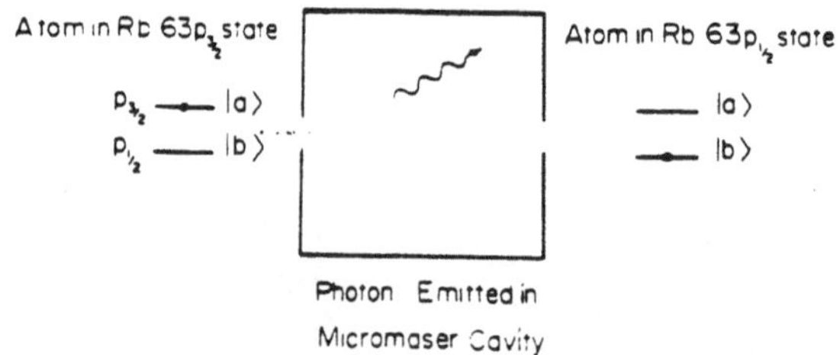

Fig. 3. Rubidium Rydberg atoms in the 63 $P_{3/2}$ state pass through a micromaser cavity, spontaneously emitting a microwave photon and make a transition to the 63 $P_{1/2}$ state.

radiation field, such as a number state field, since a classical radiation field would have so many photons present that an extra photon would not be detectable.

Careful analysis shows that this type of atom-field interaction would not disturb the center-of-mass wavefunctions $\psi_i(\vec{r}), i = 1, 2$ of the atoms.[4] This kind of welcher Weg detector does not suffer from the position-momentum uncertainty relation arguments but complementarity is still upheld by examination of the correlations between the photon degrees of freedom in the micromaser cavities and the atomic beam wavefunctions. These correlations are responsible for the loss of interference fringes (coherence) in the present Gedanken experiment.

The key ingredient in the micromaser welcher Weg detector scheme is an excited atom which emits a photon inside one of the micromasers (not outside the cavity) on its way to the screen. A likely candidate for such an atom is a long lived Rydberg state, such as the 63 $P_{3/2}$ state of Rubidium. On passing through one of the micromaser cavities, this atom couples strongly to the state 63 $P_{1/2}$ at 396 MHz, as indicated in Fig. 3. Also, when such an excited atom is placed in a resonant cavity it couples much more strongly to the microwave field and decays rapidly from the 63 $P_{3/2}$ (state a) to 63 $P_{1/2}$ (state b) because the mode spacing in the cavity is much larger than in free space. It can be realized in practice that a Rydberg atom in a state a spontaneously emits a microwave photon and makes a transition to state b with unit probability when passing through an empty cavity.

Let us place ultracold (vacuum) micromaser cavities behind each slit as in Fig. 2. The atomic beam of Rubidium atoms is excited to the long lived Rydberg state a, by application of a laser beam. After passing through the cavities and making the transition $a \to b$ via spontaneous emission, the state of the correlated atomic beam and maser cavity system is given by

$$\psi(\vec{r}) = \frac{1}{\sqrt{2}}[\psi_1(\vec{r})|1_1 0_2\rangle + \psi_2(\vec{r})|0_1 1_2\rangle]|b\rangle, \tag{6}$$

where $|1_1 0_2\rangle$ denotes the micromaser state in which one photon is in cavity 1 and none are in cavity 2. The atomic probability density at the screen now becomes,

$$P(\vec{r}) = \frac{1}{2}[|\psi_1|^2 + |\psi_1|^2 + \psi_1^* \psi_2 \langle 1_1 0_2 | 0_1 1_2 \rangle + \psi_2^* \psi_1 \langle 0_1 1_2 | 1_1 0_2 \rangle]\langle b|b\rangle. \tag{7}$$

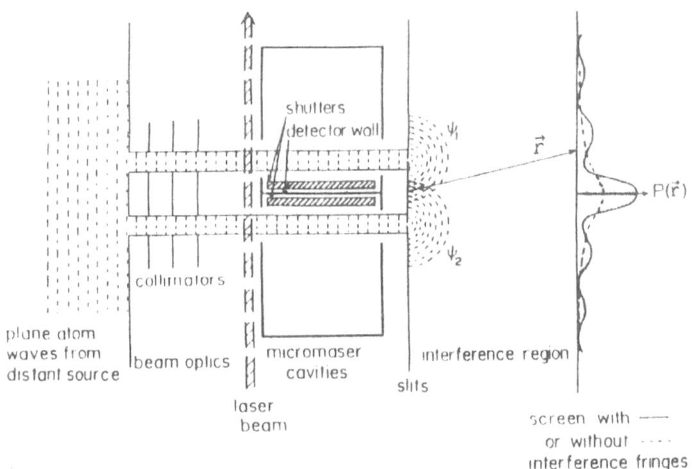

Fig. 4. Quantum-erasure configuration in which electro-optic shutters separate microwave photons in two cavities from the detector wall which absorbs the photons and acts as a photodetector.

But since $\langle 1_1 0_2 | 0_1 1_2 \rangle$ vanishes, the interference terms now disappear. This implies that when we know which cavity contains a photon, we know through which slit the atom passed. When we have welcher Weg information we no longer have interference fringes. In this discussion, it should be noted that no uncontrollable scattering events were involved in destroying the interference (wave-like) behavior.

We would like to consider what happens when one erases the which-path information (by absorbing the microwave photon) contained in the welcher Weg micromaser cavities. Could the interference fringes reappear? This problem has been considered previously in Refs. 5-9. We shall see that the interference fringes can be restored by manipulating the welcher Weg detectors long after the atoms have passed through.

Previously[1], our welcher Weg detectors had been separated by a shutter-detector system, see Fig. 4. This allows for a configuration in which the quantum eraser becomes possible. In Fig. 4 we see two shutters arranged so that radiation can be constrained to either the upper or the lower cavity when the shutters are closed. However, when the shutters are opened, light would be allowed to interact with the photodetector wall. In this way the photon, which is left behind in either the upper or lower cavity, depending on which path the atom took, will be absorbed and the welcher Weg information would be erased. This scheme suffers from the fact that the shutters cannot be treated simply. In the next section we outline a new and improved scheme for calculating the effects of a quantum eraser.

3. ATOMIC-BEAM QUANTUM ERASER

In this section we shall enhance the micromaser welcher Weg detector scheme by the addition of two atomic eraser beams (also Rubidium atoms) which pass through cavities 1 and 2 and absorb any photon present, see Fig. 5.

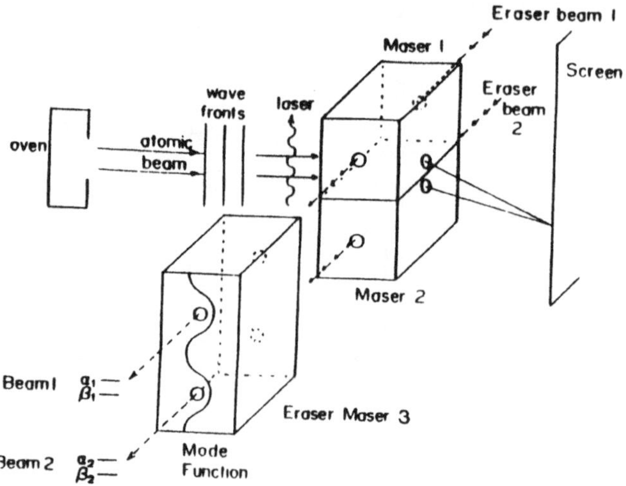

Fig. 5. Atomic-beam quantum eraser. After a Rb atom passes through the cavities and makes a transition from 63 $P_{3/2}$ to 63 $P_{1/2}$ leaving behind a photon in either cavity 1 or 2, we pass two beams of eraser atoms (also Rubidium) in the 63 $P_{1/2}$ configuration through the cavities. These atoms absorb a photon when they find one. The eraser atoms pass into the eraser maser. If a symmetric atomic state is excited, corresponding to fringes on the screen, atoms leave a photon in the eraser maser; if an antisymmetric combination of eraser atoms is produced, no photon will be left and this corresponds to antifringes on the screen.

Let us denote the upper (lower) state of the eraser atom in beam 1 by $\alpha_1 (\beta_1)$, with a similar notation for the atoms in beam 2. Before the two eraser beams enter the maser cavities, the state in which we have one photon in maser 1 and none in 2 is given by $|1_1 0_2\rangle|\beta_1 \beta_2\rangle$. After passing through the maser cavities this state changes to $|0_1 0_2\rangle|\alpha_1 \beta_2\rangle$. Similarly, if to begin with there is one photon present in maser 2, the initial state is given by $|0_1 1_2\rangle|\beta_1 \beta_2\rangle$ and the final state becomes $|0_1 0_2\rangle|\beta_1 \alpha_2\rangle$. The entire atomic -beam micromaser-eraser-beam state vector is now given by

$$\Psi = \frac{1}{\sqrt{2}}[\psi_1(\vec{r})|\alpha_1 \beta_2\rangle + \psi_2(\vec{r})|\beta_1 \alpha_2\rangle]|0_1 0_2\rangle. \tag{8}$$

Next, we pass the eraser beams into microwave cavity 3 in such a way that both beams are at antinodes of the microwave field. Each beam then couples to the field in the same manner. We may now write the states $|\alpha_1 \beta_2\rangle$ and $|\beta_1 \alpha_2\rangle$ in terms of the symmetric $|S_+\rangle$ and antisymmetric $|S_-\rangle$ state as

$$|\alpha_1 \beta_2\rangle = \frac{1}{\sqrt{2}}(|S_+\rangle + |S_-\rangle), \tag{9a}$$

$$|\beta_1\alpha_2\rangle = \frac{1}{\sqrt{2}}(|S_+\rangle - |S_-\rangle), \tag{9b}$$

where

$$|S_\pm\rangle = \frac{1}{\sqrt{2}}(|\alpha_1\beta_2\rangle \pm |\beta_1\alpha_2\rangle). \tag{9c}$$

Inserting Eqs. (9) into (8) we obtain,

$$\Psi = \frac{1}{\sqrt{2}}[(\psi_1(\vec{r}) + \psi_2(\vec{r}))|S_+\rangle + (\psi_1(\vec{r}) - \psi_2(\vec{r}))|S_-\rangle]|0_1 0_2\rangle \tag{10}$$

Inside the eraser maser 3, the symmetric state of the eraser atoms $|S_+\rangle$ will decay to the state $|\beta_1\beta_2\rangle$ by emitting a photon into the cavity, whereas the antisymmetric state $|S_-\rangle$ will not decay. It is useful to introduce the symmetric and antisymmetric atomic states defined by

$$\Psi_\pm(\vec{r}) = \frac{1}{\sqrt{2}}[\psi_1(\vec{r}) \pm \psi_2(\vec{r})] \tag{11}$$

The total state vector and the atomic probability density at the screen then become

$$\Psi = [\psi_+(\vec{r})|0_1 0_2 1_3\rangle|\beta_1\beta_2\rangle + \psi_-(\vec{r})|0_1 0_2 0_3\rangle|S_-\rangle]\,, \tag{12}$$

where $|1_3\rangle$ and $|0_3\rangle$ refer to one and zero photons in maser 3 respectively, and

$$P(\vec{r}) = [\psi_+^*(\vec{r})\psi_+(\vec{r})\langle 1_3|1_3\rangle + \psi_-^*(\vec{r})\psi_-(\vec{r})\langle 0_3|0_3\rangle]\,. \tag{13}$$

In this scheme we correlate the eraser-maser photon with a symmetric fringe pattern and the absence of an eraser-maser photon with atoms producing antifringes on the screen. Without this correlation between the fringe pattern and the eraser maser field, the fringes and antifringes would add destructively to wash out the fringe pattern. The eraser-maser actively sorts the information held by cavities 1 and 2. Without the eraser-maser we would wave which-path information and so no fringes. With the eraser-maser and using the correlation between the eraser-maser field and the state of the atomic wavefunction, the fringe pattern reappears.

This scheme provides a relatively simple Gedanken experiment which is fully calculable and clearly demonstrates the effects of a quantum erasure measurement.

4. ACKNOWLEDGMENTS

This work was supported by the Office of Naval Research.

REFERENCES

1. M. O. Scully, B.-G. Englert and H. Walther, Nature **351**, 111 (1991).
2. A detailed analysis of A. Einstein's "recoiling-slit experiment" is given by W. Wooters and W. Zurek, Phys. Rev. D **19**, 473 (1979). See also M. Jammer, *The Philosophy of Quantum Mechanics*, Wiley, New York (1974).

3. W. Heisenberg, Z. Phys. **43**, 172 (1927).
4. M. O. Scully, B.-G. Englert and J. Schwinger, Phys. Rev. A **40**, 1775 (1989); M. O. Scully and H. Walther, Phys. Rev. A **39**, 5229 (1989); and B.-G. Englert, J. Schwinger and M. O. Scully, in *New Frontiers in Quantum Electrodynamics and Quantum Optics*, ed. A. O. Barut, Plenum Press (1990).
5. M. O. Scully and K. Drühl, Phys. Rev. **25**, 2208 (1982).
6. D. Greenberger, private communication. See, for example, D. Greenberger and A. Ya'sin in *New Techniques and Ideas in Quantum Measurement Theory*, Annal. of N. Y. Acad. of Sci. **480**, 449 (1986), ed. D. Greenberger.
7. E. Jaynes in *Foundations of Radiation Theory and Quantum Electrodynamics*, Plenum Press 1980, ed. A. Barut.
8. A. Peres, Phys. Rev. D **22**, 879 (1980).
9. A. Zajac in *Coherence and Quantum Optics*, ed. L. Mandel and E. Wolf, Plenum Press 1984.

II. QUANTUM MEASUREMENTS

QUANTUM MECHANICAL WATCH-DOG EFFECT AND MEASUREMENT-INDUCED STATE REDUCTION IN A SEMICONDUCTOR LASER

Yoshihisa Yamamoto[1], Wayne H.Richardson[2] and Susumu Machida[1]

1. NTT Basic Research Laboratories
   Musashino-shi, Tokyo 180 (Japan)
2. Jet Propulsion Laboratories, California Institute of Technology
   Pasadena, CA 91109 (U.S.A.)

1. INTRODUCTION

Quantum nondemolition (QND) measurements are the measurements performed on a quantum system, in which the interaction between the measured system and the measuring instrument does not disturb or demolish the system's quantum state. QND measurement is one example of Pauli's first-kind measurement,[1] for which the quantum state immediately after the measurement can be described by von Neumann's projection postulate.[2] Suppose the density operator of the system is $\hat{\rho}_s$ immediately before the first-kind measurement of an observable $\hat{y}$; then the quantum state immediately after the first-kind measurement jumps into one of the eigenstates $|y\rangle$ of the observable $\hat{y}$ with a probability $\langle y|\hat{\rho}_s|y\rangle$. If subsequent measurements of the same observable $\hat{y}$ yield the same eigenvalue y, the measurement is considered to be a QND measurement.

Landau and Peierls were the first to point out the possibility of such a QND measurement, specifically QND measurement of a free particle's momentum.[3] The interest has been rekindled recently in connection with a study on the sensitivity of a gravitational wave antenna. It has been shown that the standard quantum limit (SQL) is imposed by Heisenberg's uncertainty principle on the measurement accuracy of a free mass position and that the SQL is surpassed by the QND measurement for the energy of a mechanical harmonic oscillator.[4-6] In order to achieve a QND measurement, the observable $\hat{y}$ to be measured must commute with the system Hamiltonian $\hat{H}_S$ itself, i.e. $[\hat{y}, \hat{H}_S] = 0$, and the observable $\hat{y}$ must also commute with the interaction Hamiltonian $\hat{H}_I$ imposed between the system and the measuring instrument, i.e. $[\hat{y}, \hat{H}_I] = 0$. The observable $\hat{y}$ and the interaction Hamiltonian $\hat{H}_I$ satisfying these conditions are respectively called a QND observable and back action evading interaction Hamiltonian.

Photon number is a QND observable when the system Hamiltonian is described by $\hat{H}_S = \hbar\omega_s\left(\hat{n}_s + \frac{1}{2}\right)$. The measurement back action noise is imposed on the conjugate observable, i.e. the phase, but the phase noise does not couple to the photon

Quantum Measurements in Optics, Edited by P. Tombesi and D.F. Walls, Plenum Press, New York, 1992

number. The QND measurement of photon number is achieved by utilizing the inter-phase modulation between the signal wave and probe wave in a Kerr nonlinear medium.[7-9] The interaction Hamiltonian $\hat{H}_I = \hbar\chi\hat{n}_s\hat{n}_p$ for the inter-phase modulation between the signal and probe waves satisfies the above back action evading condition. Such a QND measurement of photon number has been experimentally demonstrated using an optical fiber as a nonlinear medium.[10]

The measurement error can be decreased arbitrarily by decreasing the phase noise of the probe wave, i.e. by simply increasing the photon number of the coherent probe wave.[9] It has been theoretically demonstrated that the measurement error $\langle\Delta\hat{n}^2\rangle_{meas}$ determined by the probe phase noise and the back action noise $\langle\Delta\hat{\phi}^2\rangle_{back}$ determined by the probe photon number noise satisfy the minimum uncertainty product if the probe wave is in a minimum uncertainty state.[9] It has also been theoretically demonstrated that the intrinsic absorption loss associated with a Kerr nonlinearity does not place any fundamental limit on the achievable measurement accuracy or signal photon number dissipation.[11]

On the other hand, the QND measurement for an electromagnetic wave trapped inside a cavity ("standing photon") has an inherent problem of a finite dissipation rate of a cavity. Krause, Scully and Walther proposed the preparation scheme of a photon number state in a high-Q micromaser.[12] The decay time for microwave photons in a high-Q superconductor cavity is very long, typically longer than 1 second. Excited two-level atoms are injected into the cavity to interact with the cavity internal field. After they leave the cavity, the number of atoms in the lower state is counted by the field ionization scheme. The number of the lower state atoms is equal to the total number of photons in the maser field, if the cavity internal field is initially in a vacuum state $|0\rangle$ and the photon dissipation is negligible during the whole process. This is another example of Pauli's first kind measurement-induced preparation of a photon number state. However, if a photon number state $|n\rangle$ is thus prepared at a time $t = 0$, the field decays with the rate ω/Q and the field at a later time $t > 0$ is no longer a pure number state. Strictly speaking, however, the photon dissipation proceeds even during the preparation (atom-field interaction and subsequent measurement) process.[13] This is also the inherent problem for the proposed QND measurement of standing photons.[14,15]

Atomic energy (level) is also a QND observable when the system Hamiltonian is described by $\hat{H}_S = \frac{\hbar\omega}{2}\hat{S}_Z$, where $\hat{S}_Z$ is a spin operator. The measurement of whether the atomic electron is in an upper state or lower state disturbs the quantum coherence (dipole) between the two states. This is the back action noise of the measurement. The QND measurement of atomic energy (level) is achieved by utilizing the dipole coupling between the atom and the probe optical pulse.[16] An interesting aspect of this QND measurement is that the transition between level 1 and level 2 is inhibited by a measurement-induced collapse of the wavefunction. That is, the dipole moment between level 1 and level 2 is destroyed each time by the measurement, so the transition probability from one state to the other never build up to an appreciable value. This is the quantum Zeno effect (paradox) that is the inhibition of transitions between quantum states by frequent measurements.[17] The back action noise of the measurement freezes the atomic state. Recently, such an effect has been experimentally demonstrated using $^9B_e^+$ ions trapped in a Penning trap.[18] In general, a continuous measurement collapses the wavefunction of a measured system which is dramatically different from the free evolution without the measurement. Such a quantum mechanical watch-dog effect is expected to solve the above mentioned dissipation problem for "standing photons" in a cavity.

In this paper we will study a new quantum measurement of standing photons in an open system (leaky cavity), specifically a semiconductor laser. We will demonstrate that a constant-current-driven semiconductor laser allows a continuous quantum measurement of standing photons. The pumping current compensates for the cavity loss and establishes the steady state field inside the cavity. The total number of photons inside the cavity can be inferred via the junction voltage, which is uniquely determined by the total number of carriers (population inversion). The quantum correlation is established between the photon number and the carrier number due to radiative recombination and absorption processes. The electrical circuit has the two functions of pumping and measurement. The voltage of the junction biased by a constant-current-source is free to fluctuate, which makes this measurement strategy possible. The back action noise of the measurement is imposed on the phase, i.e. the quantum coherence between different photon number states is disturbed by incoherent spontaneous emission photons. The measurement of population inversion inevitably destroys the dipole moment, so the phase of spontaneous photons is randomized. This measurement process accompanies an internal subsequent feedback stabilization function for photon number. This is accomplished by a relaxation oscillation between the cavity internal photon number and the population inversion. Such a measurement and subsequent feedback stabilization mechanism reduces the photon number noise to even below the Poisson limit.[19] However, there remains a finite photon number noise, which is determined by a subtle balance between the fluctuating forces from reservoirs and the stabilizing force of relaxation oscillation. The information about the remaining photon number noise can be extracted via the junction voltage. Therefore, if the voltage fluctuation is measured, further reduction of the photon number noise is possible. The uncertainty principle is conserved between the Schawlow-Townes phase diffusion noise and the measurement-induced sub-Poissonian photon number noise. In a sense, the cavity internal field is continuously monitored by the electron system which is connected to the pump electrical circuit even though an experimentalist does not actually read out the junction voltage. This is an example of a quantum mechanical watchdog effect and measurement-induced state reduction to a photon number state or a number-phase squeezed state. The main difference between the present scheme and the quantum Zeno effect using trapped ions is that the quantum state (the cavity internal field) in the present scheme dissipates into reservoirs and is continuously fluctuated by noise forces. Therefore, the external pumping is needed to compensate for the loss. The main thrust of this paper is the discovery of such a quantum mechanical watch-dog effect and continuous measurement-induced state reduction in a semiconductor laser.

2. QUANTUM CORRELATION BETWEEN PHOTON NUMBER AND JUNCTION VOLTAGE IN A SEMICONDUCTOR LASER

If a cavity has a finite decay rate, which is always the case, the cavity internal photon number is no more a conserved observable in a strict sense. We can, of course, pump the cavity to compensate for the loss and establish a steady state field. A laser is such an example. The internal field in such an open cavity is not a pure photon number state. Nevertheless, it is expected that there exists a negative correlation between the photon number and the excited atom number. This is because whenever one photon is generated, one excited atom disappears and vice versa. A question that we will address in this section is whether this negative correlation extends to a quantum regime in spite of various fluctuation forces from reservoirs. We will study a semiconductor laser. The first reason for choosing a semiconductor laser is that the pump noise can be suppressed by a high impedance constant-current-source.[19,20] The second reason is that the spontaneous emission noise can also be suppressed by a microcavity structure.[21] The third reason is that the excited atom (electron-hole pair) number can be directly measured via the junction voltage. This is because the junction voltage corresponds to a quasi-Fermi level separation, which is determined by the number of electron-hole pairs.

Since we are dealing with both the electrical (current and voltage) noise of a circuit and the optical (internal photon number and external photon flux) noise, the Heisenberg equations of motion for these operators are more convenient than the Schrödinger equations for the wavefunctions. We will start the analysis with the quantum mechanical Langevin equations.

2.1. *Quantum mechanical Langevin equations*

The photon number operator $\hat{n}$ of the cavity internal field obeys[20]

$$\frac{d}{dt}\hat{n} = -\gamma\hat{n} + \left(\hat{E}_{cv} - \hat{E}_{vc}\right)\hat{n} + \hat{E}_{cv} + \hat{\Gamma}_0 + \hat{\Gamma}_a \quad , \tag{1}$$

where γ is the (average) photon decay rate, $\hat{E}_{cv}$ and $\hat{E}_{vc}$ are the stimulated emission and absorption operators, $\hat{\Gamma}_0$ is the fluctuation operator representing random output coupling and $\hat{\Gamma}_a$ is the fluctuation operator representing random stimulated emission and absorption events. We assume those noise sources are Marcoffian processes. The correlation functions for $\hat{\Gamma}_0$ and $\hat{\Gamma}_a$ are determined by the quantum mechanical fluctuation-dissipation theorem,

$$\langle\hat{\Gamma}_0(t)\hat{\Gamma}_0(t+\tau)\rangle = \delta(\tau)\gamma n \quad , \tag{2}$$

and

$$\langle\hat{\Gamma}_a(t)\hat{\Gamma}_a(t+\tau)\rangle = \delta(\tau)\left[(E_{cv} + E_{vc})\,n + E_{cv}\right] \quad . \tag{3}$$

Here the parameters without carets are the average numbers of the corresponding operators, for instance, $n \equiv \langle\hat{n}\rangle$.

The (minority) carrier number operator $\hat{N}$ obeys[20]

$$\frac{d}{dt}\hat{N} = P - \left(\frac{1}{\tau_{CR}} + \frac{1}{\tau_{sp}}\right)\hat{N} - \left(\hat{E}_{cv} - \hat{E}_{vc}\right)\hat{n} + \hat{\Gamma}_p + \hat{\Gamma}_b \quad , \tag{4}$$

where P is the (average) pump rate, τ_{CR} is the CR time constant of a driving circuit, τ_{sp} is the spontaneous emission lifetime, $\hat{\Gamma}_p$ is the fluctuation operator representing a random pump process and $\hat{\Gamma}_b$ is the fluctuation operator representing random spontaneous emission, stimulated emission and absorption. The (total) spontaneous emission is split into the spontaneous emission into the lasing mode and that into all the other modes,

$$\frac{\hat{N}}{\tau_{sp}} = \beta\frac{\hat{N}}{\tau_{sp}} + (1-\beta)\frac{\hat{N}}{\tau_{sp}} = \hat{E}_{cv} + \frac{1-\beta}{\beta}\hat{E}_{cv} \quad , \tag{5}$$

where β is the efficiency of spontaneous emission coupled into the lasing mode.[21] The CR time constant of a driving circuit is given by[19]

$$\tau_{CR} = CR_s = \left(\frac{qN}{2V_T}\right) R_s \quad , \tag{6}$$

where C is the junction capacitance, $V_T = k_B T/q$ is the thermal voltage, q is the electron charge, and R_s is the source resistance of the driving circuit. We also assume the noise sources $\hat{\Gamma}_p$ and $\hat{\Gamma}_b$ are Marcoffian processes. The correlation functions for $\hat{\Gamma}_p$ and $\hat{\Gamma}_b$ are given by

$$\langle \hat{\Gamma}_p(t)\hat{\Gamma}_p(t+\tau)\rangle = \delta(\tau)\frac{N}{\tau_{CR}} = \delta(\tau)\frac{2k_B T}{q^2 R_s} \quad , \tag{7}$$

$$\langle \hat{\Gamma}_b(t)\hat{\Gamma}_b(t+\tau)\rangle = \delta(\tau)\left[(E_{cv}+E_{vc})n + E_{cv} + \frac{1-\beta}{\beta}E_{cv}\right] \quad . \tag{8}$$

The pump noise is the thermal noise generated in the source resistance R_s and eq. (7) just satisfies the fluctuation-dissipation theorem between the system (minority carrier $\hat{N}$) and the reservoir (pump circuit). $\hat{\Gamma}_a$ and $\hat{\Gamma}_b$ are negatively correlated as

$$\langle \hat{\Gamma}_a(t)\hat{\Gamma}_b(t+\tau)\rangle = -\delta(\tau)\left[(E_{cv}+E_{vc})n + E_{cv}\right] \quad . \tag{9}$$

It is worth mentioning here that the fluctuation of the minority carrier number is damped by the two relaxation time constants τ_{CR} and τ_{sp}. If τ_{CR} is much shorter than τ_{sp}, the minority carrier number is kept constant by the external pump circuit. However, the pump noise introduced by such a circuit is large. This is the constant voltage operation of a semiconductor laser. On the other hand, if τ_{CR} is much longer than τ_{sp}, the minority carrier number is modulated freely by the spontaneous emission, stimulated emission, and absorption processes. In such a case, the pump noise introduced by the external circuit is small. This is the constant current operation of a semiconductor laser. This mode of operation is suitable for obtaining a strong quantum correlation between the photon number fluctuation and the junction voltage (minority carrier number) fluctuation.

Even though we can directly probe the minority carrier number via the junction voltage, we cannot directly measure the internal photon number. However, we can extract partial information about the internal photon number via the external photon flux. The external photon flux $\hat{n}_e$ coupled out via a mirror is given by[19]

$$\hat{n}_e = \gamma\hat{n} - \hat{\Gamma}_0 \tag{10}$$

The external photon flux is not identical with the transmitted internal photon number, $\gamma\hat{n}$, but has the additional term due to a reflected vacuum field fluctuation. In the experiment, we can measure the quantum correlation between the junction voltage fluctuation and the external photon flux fluctuation.

2.2. *Quasi-linearization and noise equivalent circuit*

When the laser is pumped at well above the oscillation threshold, the photon number operator and the minority carrier number operator can be expanded as $\hat{n} = n + \Delta\hat{n}$ and $\hat{N} = N + \Delta\hat{N}$. The linearized Langevin equations are written as

$$\frac{d}{dt}\Delta\hat{n} = -\left[\gamma - (E_{cv}-E_{vc})\right]\Delta\hat{n} + \frac{\Delta\hat{N}}{\tau_{st}} + \hat{\Gamma}_0 + \hat{\Gamma}_a \quad , \tag{11}$$

and

$$\frac{d}{dt}\Delta\hat{N} = -\left[\frac{1}{\tau_{CR}} + \frac{1}{\tau_{sp}} + \frac{1}{\tau_{st}}\right]\Delta\hat{N} - (E_{cv} - E_{vc})\,\Delta\hat{n} + \hat{\Gamma}_p + \hat{\Gamma}_b \quad . \tag{12}$$

Here the stimulated emission lifetime τ_{st} is defined by

$$\frac{1}{\tau_{st}} = \frac{d}{dN}(E_{cv} - E_{vc})\,n \quad . \tag{13}$$

Steady state analysis gives $n = \frac{n_{sp}}{\beta x}$ and $\tau_{st} = \tau_{sp} \cdot x / n_{sp}$, where $n_{sp} = E_{cv}/(E_{cv} - E_{vc})$ is a population inversion parameter and $x = 1/(P/P_{th} - 1)$ is a pump parameter.[19] The boundary condition eq. (10) is also linearized as

$$\Delta\hat{n}_e = \gamma\Delta\hat{n} - \hat{\Gamma}_0 \quad . \tag{14}$$

Now we introduce an "imaginary" internal photon current fluctuation $\hat{i}_L$ and external photon current fluctuation $\hat{i}_{out}$,

$$\hat{i}_L \equiv q\gamma\Delta\hat{n} \quad , \tag{15}$$

and

$$\hat{i}_{out} \equiv q\Delta\hat{n}_e \quad . \tag{16}$$

The junction voltage fluctuation $\hat{v}_n$ is related to $\Delta\hat{N}$ by[19]

$$\hat{v}_n \equiv \frac{q}{C}\Delta\hat{N} \quad . \tag{17}$$

The Fourier transform of eqs. (10)–(12) results in the following circuit equations:

$$\tilde{v}_n = (j\Omega L + R_0 - R_a)\tilde{i}_L + R_0\tilde{i}_0 - R_a\tilde{i}_a \quad , \tag{18}$$

$$\tilde{i}_L = -\left(j\Omega C + \frac{1}{R_s} + \frac{1}{R}\right)\tilde{v}_n + \tilde{i}_p + \tilde{i}_b \quad , \tag{19}$$

and

$$\tilde{i}_{out} = \tilde{i}_L + \tilde{i}_0 \quad . \tag{20}$$

Here L is an inductance defined by

$$L = \frac{\tau_{st}}{C\gamma} \quad . \tag{21}$$

The positive and negative resistances, R_0 and $-R_a$, are defined by

$$\frac{R_0}{L} = \gamma \quad , \tag{22}$$

and

$$-\frac{R_a}{L} = E_{cv} - E_{vc} \quad . \tag{23}$$

The differential resistance R of the junction is given by[19]

$$R = \frac{1}{C}\left(\frac{1}{\tau_{sp}} + \frac{1}{\tau_{st}}\right)^{-1} \tag{24}$$

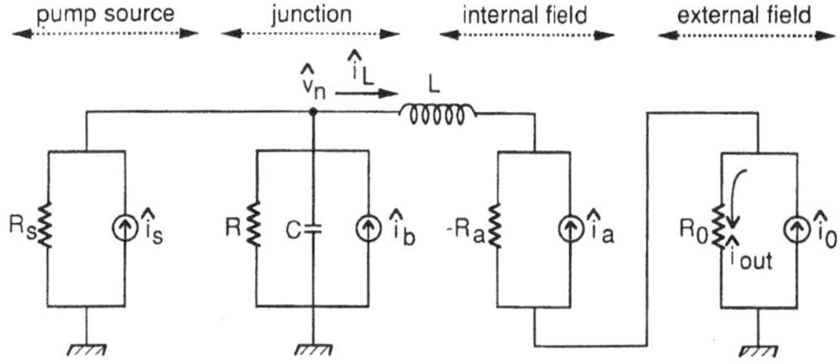

Fig. 1. A noise equivalent circuit for a semiconductor laser. $\hat{v}_n$ is the junction voltage fluctuation, $\hat{i}_L$ is the internal photon (number) current fluctuation, and $\hat{i}_{out}$ is the external photon (flux) current fluctuation.

The noise current sources are defined by $\tilde{i}_p \equiv q\tilde{\Gamma}_p$, $\tilde{i}_b \equiv q\tilde{\Gamma}_b, \tilde{i}_a \equiv q\tilde{\Gamma}_a$ and $\tilde{i}_0 \equiv -q\tilde{\Gamma}_0$.

Equations (18)–(20) suggest the noise equivalent circuit shown in Fig. 1. The whole circuit is divided into four parts, which represent a pump circuit, semiconductor laser junction, cavity internal field, and external output field. The first two parts represent a real electrical circuit but the last two parts represent an imaginary (photon current) circuit. The linearized Langevin equations (18) and (19) and boundary condition eq. (20) are just the Kirchhoff circuit equations in this noise equivalent circuit.

2.3. *Quantum correlation between the internal photon number and the junction voltage*

For eqs. (18), (19) and (20), we obtain

$$\tilde{v}_n = \frac{j\Omega L\left(\tilde{i}_p + \tilde{i}_b\right) + R_0\left(\tilde{i}_0 - \tilde{i}_a\right)}{(1 - \Omega^2 LC) + j\Omega L\left(\frac{1}{R_s} + \frac{1}{R}\right)} \quad , \tag{25}$$

$$\tilde{i}_L = \frac{\left(\tilde{i}_p + \tilde{i}_b\right) - R_0\left(\frac{1}{R_s} + \frac{1}{R} + j\Omega C\right)\left(\tilde{i}_0 - \tilde{i}_a\right)}{(1 - \Omega^2 LC) + j\Omega L\left(\frac{1}{R_s} + \frac{1}{R}\right)} \quad , \tag{26}$$

and

$$\begin{aligned} \tilde{i}_{out} = &\left\{\left(\tilde{i}_p + \tilde{i}_b\right) + R_0\left(\frac{1}{R_s} + \frac{1}{R} + j\Omega C\right)\tilde{i}_a \right. \\ &\left. + \left(\left[1 - \Omega^2 LC - R_0\left(\frac{1}{R_s} + \frac{1}{R}\right)\right] + j\Omega\left[L\left(\frac{1}{R_s} + \frac{1}{R}\right) - CR_0\right]\right)\tilde{i}_0\right\} \\ &/\left\{(1 - \Omega^2 LC) + j\Omega L\left(\frac{1}{R_s} + \frac{1}{R}\right)\right\} \end{aligned} \tag{27}$$

The tilde represents the Fourier transformed operator. A measure of the correlation between two fluctuating operators $\hat{x}$ and $\hat{y}$ is given by the normalized correlation spectrum defined by

$$C_{xy}(\Omega) \equiv \frac{\int_{-\infty}^{\infty}\langle \hat{x}^{\dagger}(t+\tau)\hat{y}(t)\rangle e^{-j\Omega\tau}d\tau}{\left[\int_{-\infty}^{\infty}\langle \hat{x}^{\dagger}(t+\tau)\hat{x}(t)\rangle e^{-j\Omega\tau}d\tau \int_{-\infty}^{\infty}\langle \hat{y}^{\dagger}(t+\tau)\hat{y}(t)\rangle e^{-j\Omega\tau}d\tau\right]^{1/2}}$$
$$= \frac{\langle \tilde{x}^{\dagger}(\Omega)y(\Omega)\rangle}{\left[\langle \tilde{x}^{\dagger}(\Omega)\tilde{x}(\Omega)\rangle\langle \tilde{y}^{\dagger}(\Omega)\tilde{y}(\Omega)\rangle\right]^{1/2}} \quad . \tag{28}$$

When there is a perfect correlation between $\hat{x}$ and $\hat{y}$, $C_{xy}(\Omega)$ is equal to 1 (positive correlation) or -1 (negative correlation). If there is no correlation, $C_{xy}(\Omega) = 0$. The normalized correlation spectrum between the junction voltage fluctuation $\hat{v}_n$ and the internal photon current fluctuation $\imath_L$ is given by

$$C_{v_n i_L}(\Omega) = \left[-R_0^2\left(\frac{1}{R_s}+\frac{1}{R}\right)(S_{i_0}+S_{i_a}) - (R_0 - R_0\Omega^2 LC)S_{i_a i_b}\right] /$$
$$\left[\Omega^2 L^2 (S_{i_p}+S_{i_b}) + R_0^2(S_{i_0}+S_{i_a})\right]^{1/2} /$$
$$\left[S_{i_p}+S_{i_b}+R_0^2\left(\Omega^2 C^2 + \left(\frac{1}{R_s}+\frac{1}{R}\right)^2\right)(S_{i_0}+S_{i_a}) + 2R_0\left(\frac{1}{R_s}+\frac{1}{R}\right)S_{i_a i_b}\right]^{1/2} \tag{29}$$

Here the spectra of the noise current forces are

$$S_{i_0} = 2q^2\gamma n \quad , \tag{30}$$

$$S_{i_a} = 2q^2\left[(E_{cv}+E_{vc})\,n + E_{cv}\right] \quad , \tag{31}$$

$$S_{i_p} = 4k_B T/R_s \quad , \tag{32}$$

$$S_{i_b} = 2q^2\left[(E_{cv}+E_{vc})\,n + E_{cv} + \frac{1-\beta}{\beta}E_{cv}\right] \quad , \tag{33}$$

$$S_{i_a i_b} = -2q^2\left[(E_{cv}+E_{vc})\,n + E_{cv}\right] \quad . \tag{34}$$

When the source resistance R_s is much larger than the differential resistance R, the pump noise eq. (32) can be neglected. If a population inversion is completely produced, the absorption rate E_{vc} is negligible. The normalized correlation spectrum in a frequency region much lower than the relaxation oscillation frequency $\Omega_r = 1/\sqrt{LC}$ is then given by

$$C_{v_n i_L}(0) = \left[-\frac{R_0}{R}(\gamma n + E_{cv}n) + E_{cv}n\right] / (\gamma n + E_{cv}n)^{1/2}$$
$$/\left\{E_{cv}n + \frac{1-\beta}{\beta}E_{cv} + \left(\frac{R_0}{R}\right)^2(\gamma n + E_{cv}n) - 2\left(\frac{R_0}{R}\right)E_{cv}n\right\}^{1/2}$$
$$= -\frac{(1+2x)^{1/2}}{\sqrt{2}(1+x)^{1/2}} \quad . \tag{35}$$

Here the following relations are used,

$$\frac{R_0}{R} \equiv \frac{\gamma L}{\frac{1}{C}\left(\frac{1}{\tau_{sp}}+\frac{1}{\tau_{st}}\right)^{-1}} = \left(1+\frac{\tau_{st}}{\tau_{sp}}\right) = 1+x \quad , \tag{36}$$

and

$$\frac{1-\beta}{\beta n} \simeq \frac{1}{\beta n} = x \quad . \tag{37}$$

At the pump rate close to the oscillation threshold ($x \gg 1$), $C_{v_n i_L}(0) = -1$ which means that the junction voltage fluctuation and the internal photon current fluctuation have a complete negative correlation. At a pump rate far above the oscillation threshold ($x \approx 0$), $C_{v_n i_L}(0) = -\frac{1}{\sqrt{2}}$, which means the junction voltage fluctuation and the internal photon number fluctuation have a partial negative correlation.

The low-frequency spectra of the internal photon current fluctuation and the junction voltage fluctuation are obtained from eqs. (25) and (26),

$$S_{i_L}(0) = 2q^2\gamma n(1+x)(1+2x) \quad , \tag{38}$$

and

$$S_{v_n}(0) = 4q^2\gamma n R^2(1+x)^2 \quad . \tag{39}$$

The low-frequency spectrum of the internal photon current fluctuation at a pump rate far above the threshold ($x \to 0$) is equal to the shot noise level $2q^2\gamma n$. The noise spectrum is a Lorentzian with a cut-off frequency $\Omega_c = \gamma$ and the integrated noise power, which is equal to the variance in photon number, is equal to $\langle \Delta\hat{n}^2 \rangle = n/2$.

This is a sub-Poissonian field with a Fano factor of 1/2. Since there is the quantum correlation between the internal photon current fluctuation and the junction voltage fluctuation, the internal photon current fluctuation can be suppressed by referring to the junction voltage fluctuation as a conditioning parameter. The low-frequency spectrum of the internal photon current fluctuation plus the junction voltage fluctuation is

$$\begin{aligned} S_{i_L+yv_n/R}(0) &= 2q^2\gamma n\left[1+x+2(1+x)^2(1-y)^2-2(1-y)\right] \\ &= q^2\gamma n(1+2x) \quad , \end{aligned} \tag{40}$$

where $y = 1-\frac{1}{2(1+x)}$ is assumed in the second equality. This is an optimum value for the parameter y to minimize the combined noise. The internal photon current fluctuation is suppressed by a factor of $(2+2x)$. The variance in photon number at the pump rate well above the threshold conditioned by the measurement result of the junction voltage fluctuation is $\langle \Delta\hat{n}^2 \rangle = n/4$. This is an example of partial reduction of the wave-packet by an unsharp measurement.

The reason the noise cancellation is not perfect is two-fold. First, the spontaneous emission into nonlasing modes, $(1-\beta)\frac{N}{\tau_{sp}}$, eliminates an electron-hole pair but does not generate a photon in a lasing mode. This is a dominant cause for the imperfect correlation at a pump rate near the threshold. Second, the vacuum field fluctuation $\tilde{i}_0$ directly perturbs the internal field but not the carrier system. This is a dominant cause for the imperfect correlation at a pump rate well above the threshold. These noise contributions can be suppressed separately but not simultaneously. We will return to this problem later.

2.4. *Quantum correlation between the external photon flux and the junction voltage*

The normalized correlation spectrum between the junction voltage fluctuation $\hat{v}_n$ and the external photon current fluctuation $\hat{i}_{out}$ is given by

$$C_{v_n i_{out}}(\Omega) =$$

$$\left\{R_0\left[1-\Omega^2 LC - R_0\left(\frac{1}{R_s}+\frac{1}{R}\right)\right]S_{i_0} - R_0^2\left(\frac{1}{R_s}+\frac{1}{R}\right)S_{i_a} - \left(R_0 - R_0\Omega^2 LC\right)S_{i_a i_b}\right\}$$

$$/\left[\Omega^2 L^2\left(S_{i_p}+S_{i_b}\right)+R_0^2\left(S_{i_0}+S_{i_a}\right)\right]^{1/2} / \left\{S_{i_p}+S_{i_b}+R_0^2\left[\left(\frac{1}{R_s}+\frac{1}{R}\right)^2+\Omega^2 C^2\right]\right.$$

$$\times S_{i_a} + \left(\left[1-\Omega^2 LC - R_0\left(\frac{1}{R_s}+\frac{1}{R}\right)\right]^2 + \Omega^2\left[L\left(\frac{1}{R_s}+\frac{1}{R}\right)-CR_0\right]^2\right)S_{i_0}$$

$$\left. +2R_0\left(\frac{1}{R_s}+\frac{1}{R}\right)S_{i_a i_b}\right\}^{1/2} \quad . \tag{41}$$

The low-frequency normalized correlation spectrum for $R_s \to \infty$ and $E_{vc} = 0$ is

$$C_{v_n i_{out}}(0) = -\sqrt{\frac{2x}{1+2x}} \quad . \tag{42}$$

At a pump rate close to the oscillation threshold ($x \gg 1$), $C_{v_n i_{out}}(0) = -1$ which means that the junction voltage fluctuation and the external photon current fluctuation have a complete negative correlation. At a pump rate far above the oscillation threshold ($x \simeq 0$), $C_{v_n i_{out}}(0) = 0$ which means that the correlation vanishes.

The low-frequency spectrum of the external photon current fluctuation is obtained from eq. (27),

$$S_{i_{out}}(0) = 2q^2\gamma n \; x(1+2x) \quad . \tag{43}$$

The noise spectrum is reduced to below the shot noise level (amplitude squeezing) at a pump rate higher than $P/P_{th} > 3$.

The low-frequency spectrum of the external photon current fluctuation plus the junction voltage fluctuation is

$$S_{i_{out}+zv_n/R}(0) = 2q^2\gamma n\left\{x + 2\left[(1+x)z - x\right]^2\right\}$$
$$= 2q^2\gamma n x \quad , \tag{44}$$

where $z = \frac{x}{1+x}$ is assumed for the second equality. This is the optimum value for the parameter z to minimize the combined noise. The external photon current fluctuation is suppressed by a factor of $(1+2x)$.

2.5. *Quantum correlation in a microcavity semiconductor laser with $\beta = 1$*

A microcavity semiconductor laser can enhance spontaneous emission rate into a single lasing mode and suppress spontaneous emission rate into all the other (non-lasing) modes.[21] The spontaneous emission coefficient for such a laser is $\beta = 1$, so the two noise driving forces $\hat{\Gamma}_a$ and $\hat{\Gamma}_b$ have a complete negative correlation, i.e.

$$S_{i_a} = S_{i_b} = -S_{i_a i_b} = 2q^2 \left[(E_{cv} + E_{vc}) n + E_{cv} \right] \quad . \tag{45}$$

One of the causes for the imperfect quantum correlation, i.e. the spontaneous emission into nonlasing modes is thus eliminated.

The low-frequency spectrum of the internal photon current fluctuation is then given by

$$S_{i_L}(0) = 2q^2 \gamma n (1 + 2x + 2x^2) \quad , \tag{46}$$

which is smaller than eq. (38) for a macrocavity laser with $\beta \ll 1$ but is reduced to the same value, $S_{i_L}(0) = 2q^2\gamma n$, at a pump rate far above the threshold ($x \to 0$). $S_{v_n}(0)$ is not altered. The normalized correlation spectrum near dc is given by

$$C_{v_n i_L}(0) = -\frac{1 + 2x}{\sqrt{2}(1 + 2x + 2x^2)^{1/2}} \quad . \tag{47}$$

We obtain $C_{v_n i_L}(0) = -1$ (complete negative correlation) at $x \to \infty$ and $C_{v_n i_L}(0) = -\frac{1}{\sqrt{2}}$ (partial correlation). The reason for the imperfect correlation at a higher pump rate is the vacuum fluctuation $\tilde{i}_0$. The low-frequency spectrum of the internal photon current fluctuation plus the junction voltage fluctuation is

$$\begin{aligned} S_{i_L + y v_n / R}(0) &= q^2 \gamma n \left\{ 1 + 4 \left[(1+x)(1-y) - \frac{1}{2} \right]^2 \right\} \\ &= q^2 \gamma n \quad , \end{aligned} \tag{48}$$

where $y = 1 - \frac{1}{2(1+x)}$ is assumed in the second equality. The variance in photon number conditioned by the measurement result of the junction voltage fluctuation is $\langle \Delta \hat{n}^2 \rangle = \frac{n}{4}$ at any pump rate. In a macrocavity laser with $\beta \ll 1$, the spontaneous emission noise into nonlasing modes can be suppressed only by the stimulated emission, which requires a higher pump rate. This is why the sub-Poissonian photon statistics $\left(\langle \Delta \hat{n}^2 \rangle = \frac{n}{4} \right)$ are achieved only at a higher pump rate. However, in a microcavity laser with $\beta = 1$, the spontaneous emission noise into nonlasing modes can be suppressed by the cavity itself.

The low-frequency spectrum of the external photon current fluctuation is given by

$$S_{i_{out}}(0) = 4q^2 \gamma n \cdot x^2 \quad , \tag{49}$$

which is smaller than eq. (44) for a macrocavity laser with $\beta \ll 1$. The normalized correlation spectrum near dc is

$$C_{v_n i_{out}}(0) = -1 \quad . \tag{50}$$

That is, there always exists a complete negative correlation between the junction voltage fluctuation and the external photon current fluctuation. This is because the vacuum fluctuation $\tilde{i}_0$ is canceled out by a destructive quantum interference at the output coupling mirror (see eq. (20)). Therefore, the low-frequency spectrum of the external photon current fluctuation plus the junction voltage fluctuation is reduced to zero, i.e.

$$\begin{aligned} S_{i_{out} + z v_n / R}(0) &= 4q^2 \gamma n \left[(1+x) z - x \right]^2 \\ &= 0 \quad , \end{aligned} \tag{51}$$

where $z = \frac{x}{1+x}$ is also assumed in the second equality. The external photon current fluctuation is completely suppressed at any pump rate.

2.6. *Quantum correlation with a squeezed incident field*

If the incident field upon an output coupling mirror is not a coherent vacuum state but is a squeezed vacuum state, the noise current force $\tilde{i}_0$ disappears. Therefore, the low-frequency spectra of the internal photon current fluctuation and the junction voltage fluctuation are given by

$$S_{i_L}(0) = 2q^2\gamma n \cdot x^2 \quad , \tag{52}$$

and

$$S_{v_n}(0) = 2q^2\gamma n R^2(1+x)^2 \quad . \tag{53}$$

The normalized correlation spectrum near *dc* is given by

$$C_{v_n i_L}(0) = -1 \quad , \tag{54}$$

which means that a complete negative correlation exists between the internal photon current noise and the junction voltage noise. Therefore, the low-frequency spectrum of the internal photon current noise plus the junction voltage noise is

$$\begin{aligned} S_{i_L+yv_n/R}(0) &= 2q^2\gamma n\left[(1+x)(1-y)-1\right]^2 \\ &= 0 \quad , \end{aligned} \tag{55}$$

where $y = \frac{x}{1+x}$ is assumed in the second equality. Because of the random spontaneous and stimulated emission processes, there remains a finite photon number noise in the internal field. However, there exists a complete negative correlation between the photon number fluctuation and the junction voltage fluctuation, so the photon number noise can be suppressed completely by referring to the junction voltage measurement result. This is an example of complete reduction of the wave-packet by an ideal measurement.

3. INTERPRETATION IN TERMS OF QUANTUM MECHANICAL WATCH-DOG EFFECT

In an ideal case with $\beta = 1$ and $\tilde{i}_0 = 0$ treated in Sec. 2-6, the internal photon number fluctuates according to eq. (52) but the internal photon number fluctuation conditioned by the junction voltage measurement result is reduced to zero as demonstrated in eq. (55). The internal wave-packet is reduced to a photon number state each time. Therefore, the quantum coherence between different photon number states must be destroyed each time. This means that the phase noise must be increased. In fact, if one quadrature amplitude noise of an incident field is squeezed to decrease the photon number noise, the other quadrature amplitude noise is inevitably enhanced. This enhanced noise randomizes the phase of the internal field. According to the analysis of an injection-locked semiconductor laser with a squeezed input field,[22] the external amplitude and normalized phase noise spectral densities near dc are obtained as

$$P_{\Delta\hat{r}}(\Omega \to 0) = \frac{1}{2\kappa}\left(\frac{\xi}{1-\xi}\right)^2 \quad , \tag{56}$$

and

$$r_0^2 P_{\Delta\hat{\psi}}(\Omega \to 0) = \frac{\kappa(1-\xi)^2 + 1 - 2\xi}{2\xi^2} \quad . \tag{57}$$

Here κ is the squeezing factor of the injection signal. One quadrature amplitude noise is reduced to below the SQL by the factor κ and the other is enhanced to above the SQL by the same factor. ξ is the parameter which depends on the internal amplitude $A_0 = \sqrt{n}$ and the injection signal F_0 according to

$$\xi = \frac{F_0}{\sqrt{\gamma} A_0} = \frac{\Delta\omega_L}{\Delta\omega_C} \quad , \tag{58}$$

where $\Delta\omega_L$ and $\Delta\omega_C$ are the locking bandwidth and the cold cavity bandwidth, respectively. In the case of strong squeezing ($\kappa \gg 1$), the increased phase noise and decreased amplitude noise satisfy the minimum uncertainty product,

$$P_{\Delta\hat{r}}(\Omega \to 0) \times r_0^2 P_{\Delta\hat{\psi}}(\Omega \to 0) = \frac{1}{4} \quad . \tag{59}$$

In the limit of $\kappa \to \infty$ ($\tilde{i}_0 \to 0$), the internal phase noise goes to infinity according to eq. (56), so the uncertainty relationship is not violated even if the internal photon number noise is reduced to zero by such an ideal measurement.

In a case with $\beta = 1$ but $\tilde{i}_0 \neq 0$ treated in Sec. 2-6, the external photon number fluctuates according to eq. (49) but the external photon number fluctuation conditioned by the junction voltage measurement result is reduced to zero as demonstrated in eq. (51). The external wave-packet is reduced to a photon number state each time. Therefore, the phase noise must go to infinity without asking for the squeezed input field in this case. The phase is kicked by the random-phase spontaneous emission and goes through a random walk. Consequently, the phase noise spectral density in the limit of low frequency ($\Omega \to 0$) goes to infinity. This Schawlow-Townes phase diffusion noise conserves the uncertainty principle. The measurement of the population inversion via the junction voltage actually destroys the quantum coherence of the upper and lower electron levels (dipole moment). This leads to the random-phase spontaneous emission. In this way, the Schawlow-Townes phase diffusion noise is considered as the back action noise of the measurement of the internal photon number.

The stabilizing force for the internal photon number fluctuation is the relaxation oscillation between the population inversion and the photon number. The destruction of the dipole coherence plays an important role for this process. If the dipole moment does not dephase, the population inversion and the photon number periodically oscillate according to the Rabi flopping. The measurement of the internal photon number via the measurement of the population inversion inhibits such a coherent oscillation and stabilizes both the photon number and the population inversion to the steady state value. The sacrifice for this case is the loss of a phase coherence mentioned above.

4. EXPERIMENTAL RESULTS

The experimental arrangement is shown in Fig. 2.[24]

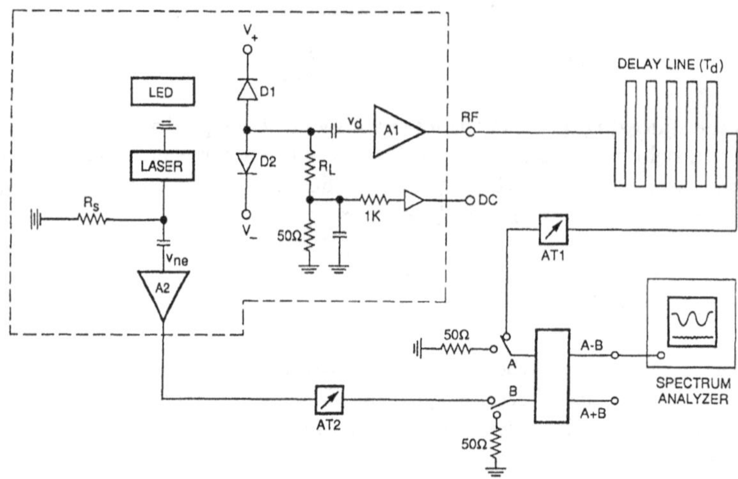

Fig. 2. The experimental arrangement for measuring the correlation $C_{v_n i_{out}}$. The apparatus inside the dotted lines was enclosed in a closed cycle refrigerator and the temperature was maintained at 66 K. The junction voltage noise v_{ne}, and the delayed laser noise voltage v_d are combined using a wideband 180° hybrid. Attenuators AT1 and AT2 were used to equalize the individual noise spectra observed on the spectrum analyzer. At the output labeled DC, the magnitude of the detector current generated by either the LED or laser light was obtained.

Two detectors D1 and D2 with quantum efficiency $\eta_d = 0.89$ and detector diameter 0.88 mm were placed less than 2 mm from the front facets of the laser and LED in a balanced configuration. Close proximity of the laser and detector reduces the deleterious effects of weak optical feedback[23] on the amplitude noise squeezing. The laser was a Mitsubishi TJS laser with AR coating ($R_1 \leq 0.03$) on the front facet and HR coating ($R_2 \geq 0.9$) on the rear facet. The measured external differential quantum efficiency at 66 K of the Mitsubishi TJS laser was 0.57, the slope efficiency was 0.88 mW/mA, the wavelength was $\lambda = 0.786$ μm and the threshold current was 0.45 mA. An LED (NEC NDL4105A) was used to calibrate the shot noise level (standard quantum limit). At low frequencies (typically below 200 MHz), the photocurrent noise power produced by light from the LED was linear for a detector current I_d up to a maximum of 7 mA. The photocurrent noise power starts to saturate at a higher detector current. Amplifiers A1 and A2 were homemade devices with bandwidths of 1.5 GHz and estimated noise figures of 3 dB. A semirigid coax-cable provided a delay T_d for RF frequency Ω. The junction voltage measured by the a.c. coupled amplifier A2 was subtracted from the photocurrent noise power at a wideband magic TEE (180° hybrid). Occasionally, a differential amplifier was used instead of the magic TEE.

For accurate comparison of the laser and LED noise levels, it is crucial that the two detector channels formed by D1, R_L, and D2, R_L, have identical responses. This was confirmed in several ways. After replacing the laser with another LED, the photocurrent noise spectra due to an LED in either channel were measured. They were identical to within 0.2 dB. The common mode suppression ratio was measured

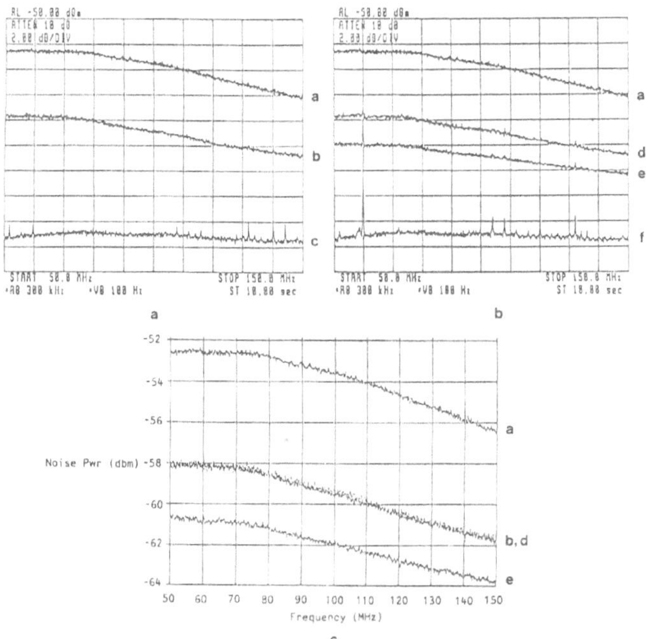

Fig. 3. (a) The noise spectra of the laser pumped at $\frac{1}{x} = 41$ and LED with the dark noise spectrum, when the laser and LED are mounted at opposite channels. (b) The noise spectra of the laser pumped at $\frac{1}{x} = 41$ or at $\frac{1}{x} = 52$ and LED with the dark noise spectrum, when the laser and LED are mounted at the same channel. (c) The laser and LED noise spectra with the dark noise subtracted.

to be higher than 18 dB by illuminating both detectors, after removing the LED and the laser, with beams of equal powers obtained from a modulated Hitachi HLP1400 laser and beam splitter combination.

4.1. *Squeezing of external photon number noise*

In Fig. 3(a), the noise spectra of the laser pumped at $\frac{1}{x} \equiv P/P_{th} - 1 = 41$ and LED are separately recorded as traces b and a, respectively, for an identical detector current of 8.6 mA. The laser and LED are coupled to the different detectors. Trace c is the dark level, i.e. the thermal noise level of the amplifier circuit.

Figure 3(b) shows the laser and LED noise spectra for the same detector current, when the laser and LED are coupled alternately to the same detector. The thermal noise is subtracted from the laser noise and LED noise, respectively, and the results are shown in Fig. 3(c). The two different measurements, corresponding to Fig. 3(a) and (b), produce identical results. The laser noise is smaller (amplitude squeezing) by 6 dB than the shot noise level produced by the LED.

The trace e in Fig. 3(b) is the laser noise at a higher pump rate $\frac{1}{x} = 52$. The LED noise level was set equal to the trace a. The laser noise is smaller than the shot noise level by 8.3 dB as shown in Fig. 3(c).

The external photon number noise measured in this way vs. relative pump rate $\frac{1}{x}$ is shown in Fig. 4. The solid line is the theoretical estimate obtained using

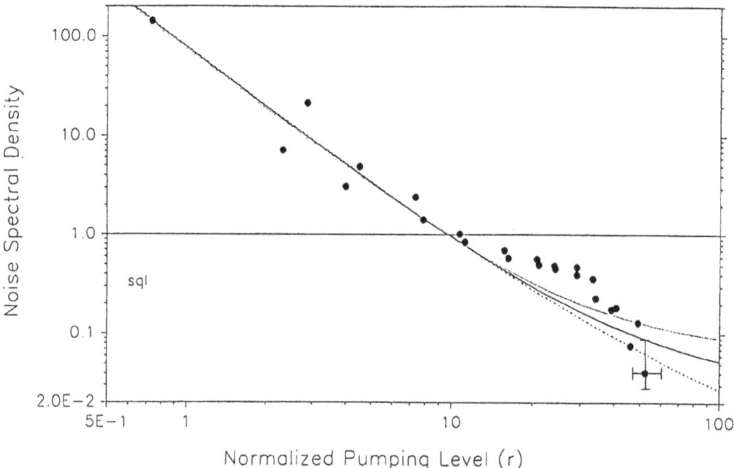

Fig. 4. The external photon flux noise spectral density normalized by the shot noise level vs. the pump rate $\frac{1}{x}$. The solid and dotted lines are the theoretical curves for $R_1 = 0.03$ and $R_2 = 0.9$ and for an internal absorption loss $\alpha_i = 0$ and $\alpha_i = 3\ \text{cm}^{-1}$, respectively. The dashed line is that for $R_1 = 0.01$, $R_2 = 0.99$ and $\alpha_i = 0$.

$R_1 = 3\%$, $R_2 = 90\%$ and no internal absorption loss, while the dotted line assumes the internal absorption loss of 3 cm^{-1}. The dashed line assumes $R_1 = 1\%$, $R_2 = 99\%$ and no internal absorption loss. The observed maximum photon number noise reduction is 85% below the shot noise limit. This corresponds to 96% squeezing at the front facet of the laser when corrected for the detector quantum efficiency. The error bar in estimating the relative pump rate $1/x$ is mainly due to the nonlinearity in the laser input current vs. output power characteristics. The error bar in estimating the normalized noise power is mainly due to the nonlinearity in the LED induced detector current vs. shot noise power characteristics.

4.2. *Quantum correlation between external photon number and junction voltage noise*

Figure 5 shows the measured junction voltage noise (trace a), laser photon number noise (trace b), and LED photon number noise (trace c). Trace d is the spectral density of the combined signal of the junction voltage noise and the delayed laser photon number noise, $(v_n(\Omega) - g_r i_{out}(\Omega) e^{i\Omega\tau_d})$. The laser bias level was $\frac{1}{x} =$ 9.7 and the detected current was 1.6 mA. The measurements were performed as follows. After the bias level of the laser was set and the drive current to the LED was turned off, the photocurrent spectrum due to the laser was recorded. During this measurement the other input to the hybrid TEE, labeled B in Fig. 2 was terminated on 50 Ω. Following measurement of the laser noise, the junction voltage noise was recorded with the input A to the hybrid TEE terminated on 50 Ohms. The combined signal was then recorded. The LED noise level, for the same detector current as that due to the laser, was then recorded with the laser off and the input B terminated on 50 Ohms. A second measurement was performed and occasionally a third measurement was too. Verifying the reproducibility of the circuit response to different terminations and confirming the SQL required multiple measurements of the LED noise level.

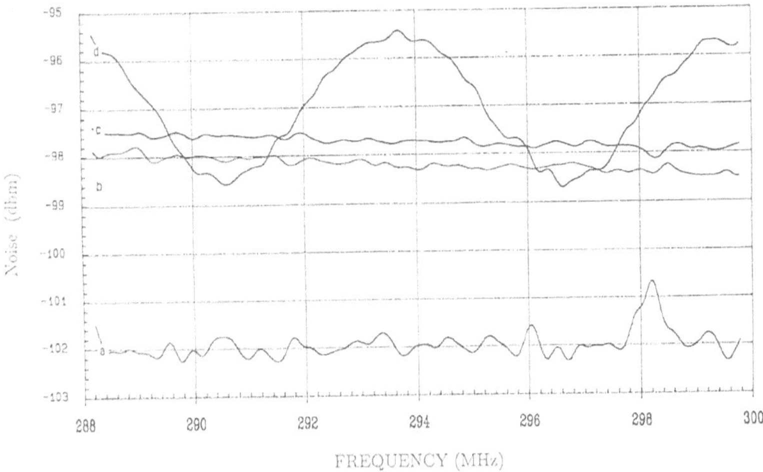

Fig. 5. Demonstration of the quantum correlation between the photon flux and the junction voltage. The traces are (a) voltage fluctuation, (b) laser noise, (c) the SQL produced by LED, and (d) combined signal $(g_2 v_{ne}(\Omega) - g_1 v_d(\Omega) e^{i\Omega\tau_d})$, where g_1 and g_2 are the voltage gains of amplifiers A1 and A2. In all the traces the respective dark noise levels were subtracted and the spectrum repeatedly filtered with a Gaussian of full width 81 kHz. Resolution bandwidth of the spectrum analyzer was 100 kHz and the video bandwidth was 30 Hz.

The sinusoidal variation of trace (d) in Fig. 5 indicates a correlation between the external photon number fluctuation and the junction voltage fluctuation. In the absence of a correlation, trace d would be flat and the noise power numerically equal to the sum of the noise powers given by traces a and b. Separate measurement of the spectral densities of the sum and difference of the combined signal $S_{v_n \pm g_r i_{out}} = \langle (v_n \pm g_r i_{out})^2 \rangle$ taken from the respective ports of the hybrid TEE – with the delay line absent – confirm that the correlation between v_n and i_{out} is *negative.* That is, the spectral density of the difference signal $S_{v_n - g_r i_{out}}$ is larger than that of the sum. The fact that the minimum of the signal in trace d, $S_{v_n + g_r i_{out}}$ is below $(\langle v_n^2 \rangle + g_r^2 \langle i_{out} \rangle)$ is indicative of a quantum correlation, because the laser amplitude noise is already below the SQL. Furthermore note that $S_{v_n + g_r i_{out}}$ is less than $g_r^2 S_{i_{out}}$. This indicates that the correlation in the quantum regime has crossed the threshold where the spectral density of some combination of the two operators is below that for one of the observable.

The correlation spectrum $C_{v_n i_{out}} = -0.40 \pm .02$, obtained from Fig. 5, compares favorably with the theoretical estimate of -0.38. A theoretical plot of $C_{v_n i_{out}}$ vs. $1/x$ for frequencies far below the relaxation oscillation frequency is shown in Fig. 6. In the frequency range shown in Fig. 6, the estimate for $C_{v_n i_{out}}$ depends sensitively on the source resistance R_s. Current leakage to a nonlasing junction may limit R_s to the value of the contact resistance of 7 Ω. The values for the independent parameters used in Fig. 6 were $\tau_{sp} = 5$ ns, $n_{sp} = 1.1$, $I_{th} = 0.45$ mA, $R_1 = 0.03$, $R_2 = 0.9$, and the spontaneous emission factor $\beta = 5.7 \times 10^{-16} n_{sp}/\tau_p$. The correlation is perfect ($C_{v_n i_{out}} = 1$) when the laser is operated at close to threshold. At the

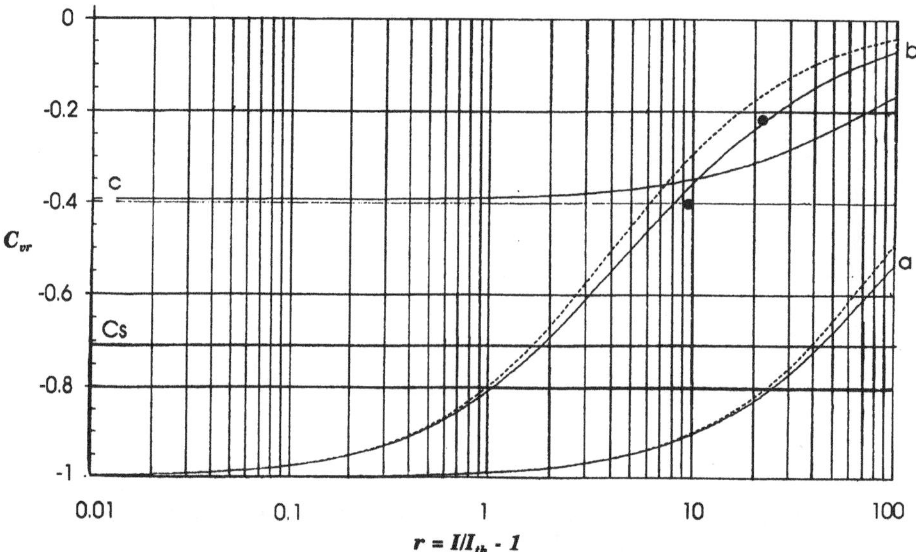

Fig. 6. Theoretical plots of the normalized cross correlation $C_{v_n i_{out}}$ versus the normalized pumping parameter $1/x$. Curves (a) and (b) are for frequencies far below the relaxation resonance frequency, while curve (c) is for $\Omega/(2\pi) = 250$ MHz. The source resistance is (a) $R_s = 1\Omega$, (b) $R_s = 750\ \Omega$, and (c) $R_s = 10\ \Omega$. The solid lines are for no internal absorption $\alpha_i = 0$ and the dotted lines are for $\alpha_i = 5\ \text{cm}^{-1}$. The line C_s is the correlation value when the output photon flux noise is at the SQL. Two experimental results are shown by filled circles.

pump rate (x_{sql}) required to produce a beam at the standard quantum limit (SQL), $x_{sql} = [2(1 + \tau_{sp}/\tau_{CR})]^{-1}$], $C_{v_n i_{out}}(0)$ is $-1/(2 + 2\tau_{pe}/\tau_{po})^{1/2}$. Here τ_{pe} and τ_{po} are the photon lifetimes due to output coupling loss and internal loss. Classically, the correlation vanishes when the output field statistics become Poissonian, while quantum mechanically $C_{v_n i_{out}}(0)$ is $-1/\sqrt{2}$, in the absence of internal absorption ($\tau_{po}^{-1} \to 0$).

5. DISCUSSION AND CONCLUSIONS

The quantum measurement of standing photons in a lossy (open) cavity has an inherent problem that the internal field decays with a finite time constant and the quantum state is not conserved. In a sense, the cavity loss is considered as a photon counting process and the internal field in such a lossy cavity is under a destructive photon number measurement (Pauli's second-kind measurement). The QND measurement of standing photons has previously been discussed only for a high-Q cavity such as a microwave superconductor cavity[14] and a whispering-Gallery-mode micro-spherical cavity[15], in which the dissipation is negligible for the repeated measurement processes. The present paper studied the possibility of compensating for cavity loss by a stimulated emission gain and of realizing the quantum measurement via counting the number of excited atoms. The electrical pump circuit has the two functions of supporting the steady state internal field and simultaneously probing the photon number. The quantum correlation between the internal photon number and the excited atom number can be established by simultaneous occurrence of one photon generation and one excited atom destruction or vice versa.

Each time, the internal photon number is "measured" via the junction voltage (excited atom number), even though an experimentalist does not actually read out the result. The conjugate observable, i.e. the phase is randomized by this measurement. This is a quantum mechanical "watch-dog effect". The random walk of the phase due to incoherent spontaneous emission is considered as the back action noise of such a measurement and preserves the uncertainty principle. For example in a microcavity semiconductor laser with $\beta = 1$, the external photon flux noise can be decreased to zero by referring to the junction voltage noise measurement, i.e. $S_{i_{out}+zv_n/R}(0) \to 0$ at an arbitrary pump rate. On the other hand, the phase noise goes to infinity at such a low frequency limit, $\Omega \to 0$, due to a random-walk phase diffusion process. In a microcavity semiconductor laser with a squeezed incident field, the internal photon number noise can be decreased to zero by referring to the junction voltage noise measurement result, i.e. $S_{i_L+yv_n/R}(0) \to 0$ at an arbitrary pump rate. This means that the variance in internal photon number, which is equal to the integrated photon number noise spectrum, goes to zero. On the other hand, the phase is disturbed by the enhanced quadrature noise component of an incident squeezed field and goes to infinity.

Since no quantum system is free of dissipation in a real world, it is important to realize a continuous quantum measurement of Pauli's first kind for an open system. Suppose there is a quantum mechanical harmonic oscillator, in which the quantum state is established by the balance between the system's ordering force and the reservoirs' fluctuating forces. Also suppose we want to measure the quantum number $\hat{n}$ of the harmonic oscillator without disturbing the self-organized quantum state, because some relevant information, such as the arrival of a gravitational wave or the coupling of a modulated signal wave, is related to the change in $\hat{n}$. The present junction voltage measurement scheme allows the internal field of a semiconductor laser (quantum mechanical harmonic oscillator) to evolve freely and still allows us to extract information about the internal photon number with an accuracy better than the standard quantum limit.

REFERENCES

1. W. Pauli, *Handbuch der Physik, vol. V* Springer (1958).
2. J. von Neumann, *Mathematical Foundations of Quantum Mechanics*, Princeton University Press (1954).
3. L. Landau and R. Peierls, *Z. Phys.* 69:56 (1931).
4. V. B. Braginsky and Y. I. Vorontsov, *Sov. Phys. Usp.* 17:644 (1974).
5. C. M. Caves et al., *Rev. Mod. Phys.* 52:341 (1980).
6. V. B. Braginsky, Y. I. Vorontsov and K. S. Thorne, *Science* 209:47 (1980).
7. V. B. Braginsky and S. P. Vyatchanin, *Sov. Phys. Dokl.* 26:686 (1981).
8. G. J. Milburn and D. F. Walls, *Phys. Rev. A* 28:2065 (1983).
9. N. Imoto, H. A. Haus and Y. Yamamoto, *Phys. Rev. A* 32:2285 (1985).
10. M. D. Levenson, R. M. Shelby, M. Reid and D. F. Walls, *Phys. Rev. Lett.* 57:2473 (1986).
11. K. Watanabe and Y. Yamamoto, *Phys. Rev. A* 42:1699 (1990).
12. J. Krause, M. O. Scully and H. Walther, *Phys. Rev. A* 36:4547 (1987).

13. P. Meystre and E. M. Wright, *Phys. Rev. A* 37:2524 (1988).
14. M. Brune, S. Haroche, V. Lefevre, J. M. Raimond and N. Zagury, *Phys. Rev. Lett.* 65:976 (1990).
15. V. B. Braginsky, *in Proc. 3rd Int. Symp. Foundations of Quantum Mechanics, Tokyo 1989,* ed. S. Kobayashi, H. Ezawa, Y. Murayama and S. Nomura (The Physical Society of Japan) p.135.
16. R. J. Cook, *Phys. Scr. T* 21:49 (1988).
17. B. Misra and E. C. G. Sudarshan, *J. Math. Phys.* 18:756 (1977).
18. W. M. Itano, D. J. Heinzen, J. J. Bollinger and D. J. Wineland, *Phys. Rev. A* 41:2295 (1990).
19. Y. Yamamoto, S. Machida and O. Nilsson, *Phys. Rev. A* 34:4025 (1986);Y. Yamamoto and S. Machida,Phys. Rev. A35:5114 (1987).
20. S. Machida, Y. Yamamoto and Y. Itaya, *Phys. Rev. Lett.* 58:1000 (1987);S. Machida and Y. Yamamoto,Phys. Rev. Lett.60:792 (1988);S. Machida and Y. Yamamoto,Opt. Lett.14:1045 (1989).
21. Y. Yamamoto, S. Machida, K. Igeta and Y. Horikoshi, *in Coherence and Quantum Optics VI,* ed. L. Mandel, E. Wolf and J. H. Eberly Plenum, New York, (1990).
22. L. Gillner, G. Björk and Y. Yamamoto, *to appear in Phys. Rev. A* (1990).
23. W. H. Richardson and R. M. Shelby, *Phys. Rev. Lett.* 64:400 (1990).
24. W. H. Richardson and Y. Yamamoto, *Phys. Rev. Lett.* (1991).

QUANTUM NON-DEMOLITION MEASUREMENT OF AN OPTICAL INTENSITY IN A THREE-LEVEL ATOMIC NON-LINEAR SYSTEM

Philippe Grangier, Jean-François Roch and Gérard Roger

Institut d'Optique - CNRS URA 14
B.P. 147 - F91403 ORSAY Cedex

1. INTRODUCTION

Two-photon non-linear systems are well-known theoretical candidates for quantum noise reduction effects [1-6]. Presently, a lot of investigations are devoted to the realization of optical "quantum non-demolition" (QND) measurements. When such a measurement is performed, it is possible to leave the observed quadrature component of an optical field unchanged, and to add the quantum back-action noise to the other (complementary) quadrature component [7-9]. Such measurements were implemented in the optical domain, where the harmonic oscillator is a mode of the electromagnetic field, and where the coupling responsible for the measurement is created using optical non-linearities in solids [10-11].

We recently proposed to use nearly-resonant two-photon non-linearities, in order to perform QND measurements, using atomic three-level systems excited by two different, initially independant, laser beams [12-13]. The basic physics of this non-linear coupling can be understood very simply, as a crossed-phase-modulation effect : in the lowest-order approximation, the refractive index for one beam will depend on the intensity of the other beam, while the usual self-phase-modulation (Kerr effect) can be made much smaller. It can be shown [12-14] that this coupling yields directly QND measurements of the intensity fluctuations of a "signal" beam, which are copied into phase fluctuations of a "meter" beam by the measurement process. The measured quantity (intensity fluctuations) is left unchanged, while the back-action noise appears on the phase fluctuations of the signal beam.

2. EXPERIMENT

We will describe here the implemention of a two-photon QND measurement scheme, using two-photon non-linearities in an atomic three-level system [15]. The overall experimental set-up is shown on Figure 1.

A sodium atomic beam provides a density of 5 10^{11} atoms/cm^3, with a Doppler width of about 200 MHz FWHM. The interaction region with the laser beams is 1 cm long. The atomic beam is surrounded by a single-ended optical cavity (8 cm long, 5 cm mirror radius),

Quantum Measurements in Optics, Edited by P. Tombesi and
D.F. Walls, Plenum Press, New York, 1992

which resonates for the IR laser light exciting the upper transition. The transmittivity of the input/output mirror is $t^2 = 0.006$, while total losses are $p^2 = 0.011$, including both mirrors transmittivities. The light beams inside the cavity are circularly polarized to optimize the non-linear coupling, and the output IR beam is separated from the input beam using a prism polarizer. Both beams are detected using PIN silicon photodiodes, followed by low-noise preamplifiers. The photocurrents are added or substracted using 0°/180° power combiners, and sent to a spectrum analyzer. When taking data, the spectrum analyzer is set at zero frequency span, and the noise of the photocurrents is analysed while the optical cavity is swept onto resonance of a noise sideband of the IR beam. This resonance is obtained when the cavity is detuned from the laser frequency by an amount equal to the noise frequency set on the spectrum analyzer. Resonance of the laser beam itself is avoided during the sweep. The optical cavity has two roles: first, it enhances the non-linear coupling (see below), and second, it turns the initial read-out phase modulation of the meter beam into an intensity modulation, which can be directly detected [14-17].

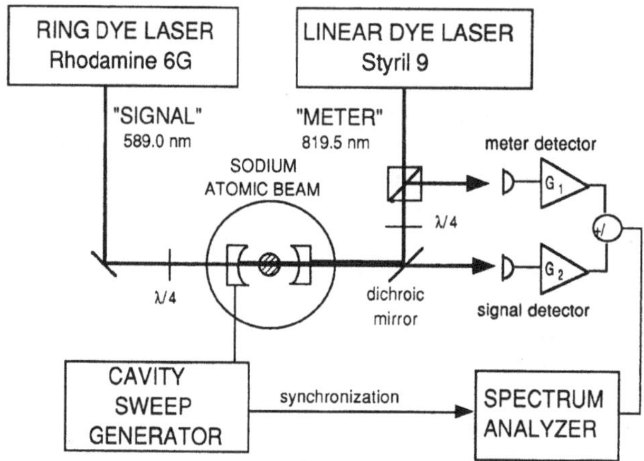

Figure 1 . Experimental set-up. Two lasers at 589.0 nm ("signal" beam) and 819.5 nm ("meter"beam) are tuned close to the resonances of the 3s1/2-3p3/2-3d5/2 cascade in a sodium atomic beam. The single-ended optical cavity resonates for a noise sideband of the meter beam. The quantum intensity noises of the output light beams are correlated, while the quantum intensity noise of the signal beam is not changed, providing evidence for a QND measurement.

Typical results are shown on Figure 2. Curves (a) and (b) show the separate noise levels of the probe and meter beams. When the cavity resonance of the sideband is achieved, the noise of the meter beam is increased, while the noise of the signal beam is left unchanged. Outside the resonance, both noise levels have been checked to be at the shot-noise-level. They are adjusted to the same electrical level using slight adjustment of the preamplifiers gains. Curves (c) and (d) show the recombined photocurrents. Outside the resonance, the noises are

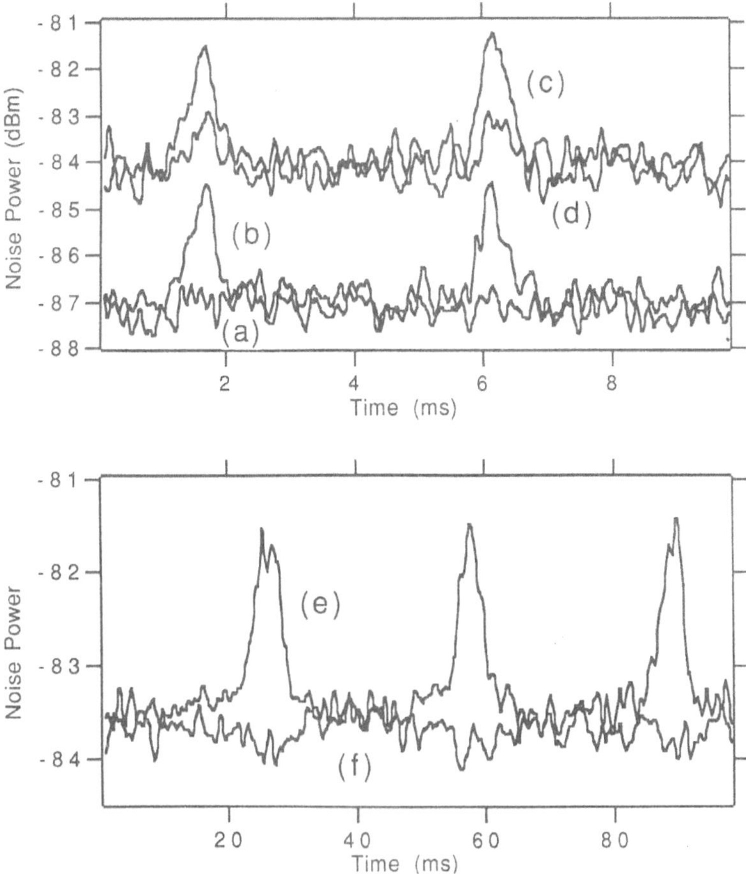

Figure 2 . Noise levels recorded while the cavity is swept onto a noise sideband of the meter beam. The spectrum analyser is set at zero frequency span, 14 MHz center frequency, 100 kHz radio-frequency bandwidth, 300 Hz video bandwidth. Curves (a) and (b) are respectively the signal and meter noises, recorded separately. Other curves are the minimum and maximum noise levels of the combined photocurrents, obtained by adding or substracting the signal and meter photocurrents. Curves (c) and (d) correspond to a one-photon detuning Δ = -1.75 GHz and a two-photon detuning δ = -1 GHz, while curves (e) and (f) are obtained for Δ = +2.25 GHz and δ = -1 GHz, with a cavity sweep ten times slower (video bandwidth 30 Hz).

obviously uncorrelated, and the noise level increases by 3 dB, independently of the relative recombination phase. On the other hand, the noise of the meter beam at resonance is correlated with the (unchanged) noise of the signal beam : adding or substracting the photocurrents results in a 1.5 dB change of the noise levels.

For this experiment, both lasers intensities on the atoms were close to 10 W/cm^2 (1 mW power). The one- and two-photon detunings were respectively $\Delta = -1.75$ GHz and $\delta = -1$ GHz, relative to the F=2 ground state hyperfine sublevel . Other experiments were done for an opposite sign of the one-photon detuning, i.e., $\Delta = +2.25$ GHz and $\delta = -1$ GHz, while the other parameters had the same values. In that case, the correlations were clearly increased, to about 2.5 dB (see curves (e) and (f)). Moreover, the differenced noise was then below the shot-noise level of the recombined photocurrent, by about 0.4 dB.

These results could be further improved by optimizing the electronic gain of the meter channel [11,15-17]. The combined photocurrents of the signal and attenuated meter are shown on Figure 3. The differenced noise is now 0.8 dB below the shot-noise level of the recombined photocurrent, and reaches the shot-noise level of the signal alone. One important advantage of this scheme is that only direct detection is needed, but we also used an homodyne detection set-up [11], in order to check qualitatively that the signal phase noise is increased when the measurement is performed.

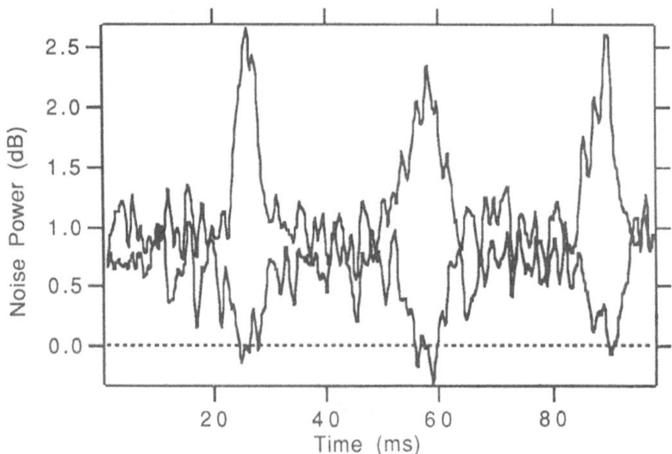

Figure 3 . Combined noises levels with 6 dB attenuation of the meter photocurrent. This curves are corrected for the amplifiers noises, and normalized so that 0 dB correspond to the shot-noise-level of the signal alone. The differenced photocurrent goes 0.8 dB below the combined shot-noise level.

According to these results, the following conclusions can be drawn:

(i): the non-linear medium establishes a correlation between the signal and meter beam, i.e., detecting the meter beam yields an information (i.e., an unperfect measurement) about the fluctuations of the signal beam[18].

(ii): the output intensity fluctuations of the signal beam remain at the shot-noise level, within our experimental precision. Therefore, the intensity noise added to the signal beam, if any, is more than 10 dB below the shot-noise-level. This is a good indication that the QND condition is satisfied.

(iii): in general, the meter extra noise is only partly correlated to the signal noise. In our experimental configuration, the correlated to non-correlated noises ratio is better for a positive value of the one-photon detuning.

3. THEORETICAL MODEL

In order to quantify these results for comparison with theory, we will introduce now a very simple phenomenological model. The measured quantities are the noise levels of the signal, meter, and recombined photocurrents, which will be denoted respectively V_s^{out}, V_m^{out}, and V_ε^{out} ($\varepsilon = \pm 1$ depending on the recombination phase). Assuming that the shot-noise levels of both beams have the conventional value one, and taking off the amplifiers noises, one obtains [13-15] :

$$V_s^{out} = 1 \quad (1)$$
$$V_m^{out} = 1 + \eta_m N \quad (2)$$
$$V_\varepsilon^{out} = 2 + \eta_m N + 2 \varepsilon C (\eta_m \eta_s)^{1/2} \quad (3)$$

where N and C are respectively the meter excess noise and the correlation introduced by the QND device. The parameter η_s and η_m (with $\eta_s \cong \eta_m \cong 0.5$) are the overall detection efficiencies of the signal and meter channels, including the linear cavity losses. According to our experimental results, we will neglect any excess noise added to the signal beam. For both negative (n) and positive (p) values of the detuning Δ, the experimental values corresponding to the results of Figure 2 are :

$$N_p^{mes} = 1.5 \qquad C_p^{mes} = 0.9 \quad (4)$$
$$N_n^{mes} = 3 \qquad C_n^{mes} = 0.7 \quad (5)$$

The two-photon crossed-phase-modulation effect can be calculated at the lowest $\chi^{(3)}$ approximation by solving the Optical Bloch Equations for a three-level system [19]. We will consider only purely radiative decay, and denote γ_1, γ_2 the intermediate and upper level transverse linewidth (half of the Einstein coefficients). The atom-laser detunings and Rabi frequencies will be denoted respectively δ_1, Ω_1 (lower transition) and δ_2, Ω_2 (upper transition). $\Delta = \delta_1$ is the one-photon detuning, and $\delta = \delta_1 + \delta_2$ is the two-photon detuning. All detunings are much larger than γ_1, γ_2 and the Doppler width (dispersive limit). Finally, the dipole matrix elements are d_1, d_2. Assuming that $\Delta >> \delta$, the lowest-order effect of the laser fields is to induce a coherence between the ground and upper levels, whose expression is simply $(\Omega_1 \Omega_2)/(\Delta \delta)$. The resulting non-linear polarizations for the lower and upper transitions are then respectively $(n_{at} d_1 \Omega_1 \Omega_2^2)/(\delta \Delta^2)$ and $(n_{at} d_2 \Omega_2 \Omega_1^2)/(\delta \Delta^2)$, where n_{at} is the atomic number density. These polarizations translate into non-linear phase shifts for the laser fields, with respective values $(n_{at} k_1 L d_1^2 \Omega_2^2)/(2 \varepsilon_0 \hbar \delta \Delta^2)$ and $(n_{at} k_2 L d_2^2 \Omega_1^2)/(2 \varepsilon_0 \hbar \delta \Delta^2)$. Here, $k_1 = 2\pi / \lambda_1$ and $k_2 = 2\pi / \lambda_2$ are wave-vectors of the laser fields, and L is the length of the non-linear medium. Using the definitions of γ_1, γ_2, one obtains convenient final forms of the non-linear phase shifts:

$$\kappa_1 = n_{at} L (3 \lambda_1^2 / 4 \pi)(\gamma_1 \Omega_2^2 / \delta \Delta^2) \quad (6)$$
$$\kappa_2 = n_{at} L (3 \lambda_2^2 / 4 \pi)(\gamma_2 \Omega_1^2 / \delta \Delta^2) \quad (7)$$

We note that, though they look non-symetric, κ_1 and κ_2 are equal if the field intensities are the same in photons units.

The value of the non-linear refractive index can then be injected into a calculation of the fluctuations using the semi-classical linear input-output technique [20]. When the noise sideband

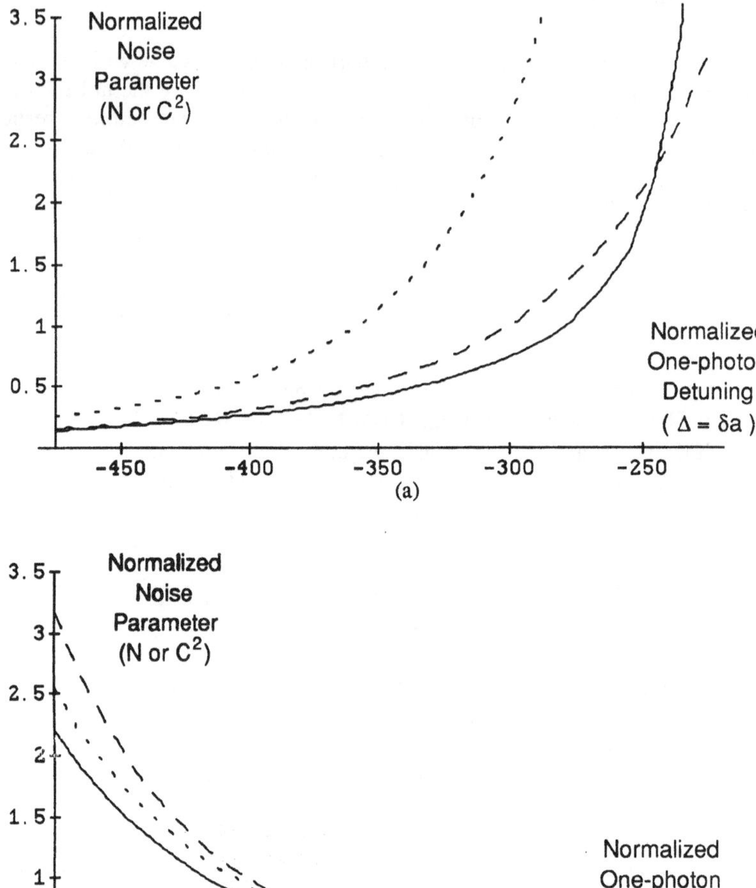

Fig.4 . Theoretical values of noise parameter N (dotted line) and correlation parameter C^2 (full line), together with their common perturbative value, obtained from a $\chi^{(3)}$ calculation of the crossed-phase modulation effect (dashed line). The value of the two-photon detuning is fixed at δ = -1 GHz. The agreement between the three curves is much better for positive one-photon detuning (4 (b)) than for negative one-photon detuning (4 (a)).

of interest is resonating in the cavity, the correlation coefficient C is simply $(2/p)\,(\kappa_1 \kappa_2)^{1/2}$, while N is $(2/p)^2 \kappa_1 \kappa_2$. The signature of an "ideal" non-linear effect ("parametric" limit) lies in the relation $N = C^2$. Deviations from this relation (i.e., $N > C^2$) will be an indication of excess noise in the non-linear process.

Typical experimental values (in γ_1 units) are $\gamma_2 = 0.8$, $\Omega_1 = \Omega_2 = 30$, $\Delta = 50$, $\delta = 200$, $n_{at} L (3 \lambda_1 \lambda_2 / 4 \pi) = 1200$, so that $C = 0.8$, which fairly agrees with the experimental values. The complete calculation, including the sodium hyperfine levels and polarization effects, does not change significantly the above value of C. However, this simple theory fails to predict the experimentally observed deviations from the relation $N=C^2$, which clearly depend on the sign of the one-photon detuning Δ, while the $\chi^{(3)}$ model depends only on Δ^2.

In order to explain this discrepancy, we have used a theory recently developed [21] for describing the case of three-level atoms inside an optical cavity. The general idea is as follow. The required frequency-dependant field correlation functions are first related to atomic correlation functions. These can be evaluated using the quantum regression theorem. The eight corresponding equations are solved using the usual Laplace-transform method, which yields directly solutions in the frequency space. The cavity is then taken into account using a filtering technique, under the condition that only the noise sideband of interest is actually resonating. We emphasize that this calculation is non-perturbative in the laser intensities and detunings, but obviously assumes that the quantum noise terms are small compared to the mean laser fields. Therefore, we check that the mean fields solutions are stable for the useful values of the parameters. Figure 4 shows the calculated values of C^2 and N , together with their perturbative value ($4 \kappa_1 \kappa_2$), as a function of the one-photon detuning Δ. Other parameters are set to experimental values corresponding to Figure 2. The two-photon detuning is fixed at $\delta = -1$ GHz, i.e, $\delta = -200$ in normalized units. All three curves should be superimposed in the parametric limit. Clearly the positive-detuning case is more favourable than the other, which is in agreement with the observed behaviour.

Summarizing, the observed value of the correlation coefficient C is correctly described using the simplest $\chi^{(3)}$ approximation of the two-photon dispersion effect. On the other hand, the value of the "excess noise" on the meter beam, (part of the added noise which is not correlated to the signal beam) can only be described using a more elaborate theory, which takes into account atomic fluctuations in three-level systems. The results of this theory, presented in Ref. 21, are in satisfactory agreement with the experimental results.

4. CONCLUSION

As a conclusion, we have observed a quantum non-demolition measurement effect using a two-photon non-linear system. Presently, the best reproducibly observed result is a reduction of 0.8 dB below the combined shot-noise levels of the two output beams, which originate from two different lasers. This result is about the same as in previously published experiments in other systems, but is obtained with very small laser powers (typically 1 mW C.W.). Optimization of the cavity and of the detection system should result in a significant improvement of these values. Moreover, very intriguing predictions can be made in the doubly resonant case, associated with bistable behaviour. Therefore many developments can be expected, both on the theoretical and experimental sides, and three-level atomic systems seem quite well suited for quantum noise studies.

We acknowledge fruitful discussions with Serge Reynaud and Jean-Michel Courty. This work is supported in part by Direction des Recherches, Etudes et Techniques, and by Centre National d'Etude des Télécommunications.

References

1. H P Yuen, Phys.Rev. **A13**, 2226, 1976
2. L A Lugiato and G. Strini, Optics Comm. **41**, 374, 1982
3. C M Savage and D.F. Walls, Phys. Rev.**A33**, 3282, 1986
4. B A Capron , D.A. Holm and M. Sargent III, Phys.Rev.**A35**, 3388, 1987
5. P Galatola , L.A. Lugiato, M.Vadacchino and N.B. Abraham, Optics Comm. **69**, 414 and 419, 1989
6. C Blockley and D.F. Walls, Optics Comm. **79**, 241, 1990
7. C M Caves , K.S. Thorne, R.W.P. Drever, V.D.Sandberg and M.Zimmermann Rev.Mod.Phys. **52**, 341,1980
8. B Yurke,J.Opt.Soc.Am **B2**, 732, 1985
9. Y Yamamoto , N.Imoto and S.Machida, Phys.Rev. **A33**, 3243, 1986
10. M D Levenson , R.M. Shelby, M.D. Reid and D.F. Walls,Phys.Rev.Lett. **57**, 2473,1986
11. A LaPorta , R.E.Slusher and B.Yurke, Phys.Rev.Lett. **62**, 28, 1989
12. P Grangier, J.F. Roch and S. Reynaud, Optics Comm.**72**, 387, 1989
13. P Grangier and J.F. Roch, Quantum Optics **1**, 17 , 1989
14. P Grangier and J.F. Roch, to appear in Optics Comm. (1991)
15. P Grangier, J.F. Roch and G. Roger, Phys. Rev. Lett.**66**,1418, 1990
16. M D Levenson and R.M. Shelby, J.Mod.Optics **34**, 775, 1987
17. J Mertz , A.Heidmann, C.Fabre, E.Giacobino and S.Reynaud, Phys.Rev.Lett. **64**, 2897, 1990
18. M.J. Holland, M.J. Collett, D.F. Walls and M.D. Levenson, Phys. Rev. **A42**, 2995,1990
19. R M Whitley and C.R. Stroud, Phys. Rev.**A14**,1498, 1976
20. S Reynaud and A.Heidmann, Optics Comm. **71**, 209,1989
21. J.M. Courty, P. Grangier, L. Hilico and S. Reynaud, to appear in Optics Comm. (1991)

THE QUANTUM MEASUREMENT PROCESS AND THE ONE-ATOM MASER

R.J.Brecha and H.Walther

Sektion Physik der Universität München and
Max-Planck-Institut für Quantenoptik
8046 Garching, Fed. Rep. of Germany

The rich variety of phenomena associated with the Jaynes-Cummings[1] model (and extensions thereof) describing the interaction of a two-level atom with a single electromagnetic field mode has resulted in a tremendous amount of theoretical work as well as concerted experimental efforts over the past 25 years. As the experiments have progressed to approximate to an ever increasing degree the ideal conditions of the Jaynes-Cummings, (i.e. no effects due to dissipation or thermal photons) theoretical predictions of new quantum behavior to be expected from this simple system have kept pace. In this paper a brief review of the previous work with the one-atom maser, the closest realization of the simple Jaynes-Cummings model, will be given and two new experiments currently in the building stage will be discussed in some more detail. Parts I and II of the paper are a review of the theory and previous experiments. In part II those previous experiments will be discussed, including the collapse and revival of Rabi oscillations and the detection of sub-Poissonian photon statistics. Parts III and IV consist of a description of the two experiments currently in progress.

I. Model

The Hamiltonian describing the interaction of a single two-level atom with a mode of the electromagnetic field is given by

$$\hat{H} = \hbar\omega\hat{a}^{\dagger}\hat{a} + \frac{1}{2}\hbar\omega\hat{\sigma}_3 + \hbar\Omega\left(\hat{a}\hat{\sigma}_+ + \hat{a}^{\dagger}\hat{\sigma}_-\right) \tag{1}$$

where ω is the common cavity mode and atomic transition frequency, $\hat{a}^{\dagger}$ ($\hat{a}$) is the field creation (annihilation) operator, and $\hat{\sigma}_{\pm}, \hat{\sigma}_3$ are the Pauli spin matrices for the two-level atomic system. Ω is the coupling constant or single photon Rabi frequency. This establishes the notation which will be used throughout the paper. The Jaynes-Cummings model as given here is the starting point for the theoretical treatment of the problem at hand.[2] In the experiment to be described, the situation is roughly as follows: Time is broken up into two different segments.[3,4] In the first, an atom from the atomic beam interacts with the cavity field as it flies through the resonator. It is during this interaction time that the Jaynes-Cummings formalism is applicable. By assuming throughout that the atomic lifetime is very long compared to the interaction time, this dissipation process can be ignored. The cavity decay time is also long compared to the interaction time but not necessarily

Quantum Measurements in Optics, Edited by P. Tombesi and
D.F. Walls, Plenum Press, New York, 1992

compared to the time interval between incoming atoms. It is this time scale which is important because it determines the development of the field in the times when it is not being "kicked" by the energy provided by the atom which enters in the excited state. During the time periods when no atom is in the cavity, the field decays via losses to the walls; the dissipation of the cavity field can be described by the standard master equation for a damped harmonic oscillator[2] where $\gamma/2$ is the cavity field decay rate and n_{th} is the mean thermal photon number, which depends of course on the temperature. The state of the field can be calculated at any given time using standard techniques.

If the atoms are injected without coherence, for example as in the experiments here in which all atoms are in the excited state, and if the initial density matrix for the field is diagonal in the number state representation, then it remains diagonal throughout the interaction and decay.

Most of the theoretical calculations begin with these conditions and prove to be in excellent agreement with the experimental results. Only in the final section will we consider a more general situation, that of coherently excited atoms entering the resonator, in which case the diagonal density matrix elements will not be sufficient to describe the phase diffusion of the maser field. More detailed expositions of the micromaser as revealed by calculations based on the above-outlined formalism can be found in references 2-4.

II. Experiments with the One-Atom Maser

In this section the experiments which have been carried out with the one-atom maser will be presented. Only a rather brief outline of each of three experiments will be given as the work has been previously published. A general sketch of the maser apparatus is shown in Fig.1. The oven is filled with rubidium and resistively heated to produce the diffusive atomic beam. A frequency-doubled ring dye laser is the source for the 297nm radiation used to excite the atoms to a high-lying Rydberg level ($63p_{3/2}$) of the ^{85}Rb isotope. The excitation occurs shortly before the atoms enter the cavity so that they are initially in the upper state of the two-level system at the beginning of the interaction. In other words, the atoms are the pump source for the maser itself since they carry the energy into the resonator.

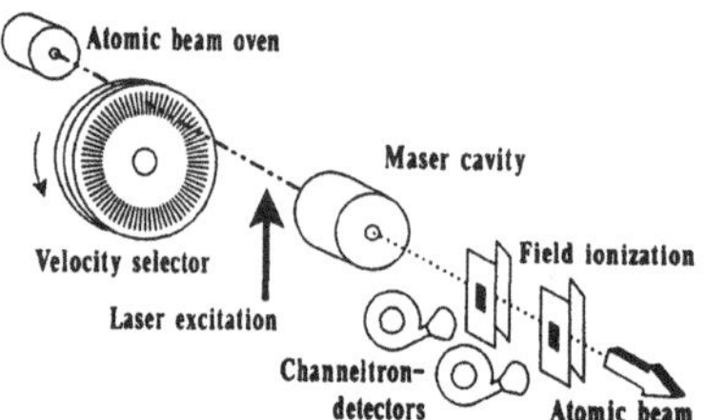

Fig. 1. Experimental setup schematic. The entire apparatus is housed in a vacuum chamber with the cavity itself contained in a ^{3}He cryostat.

Each different atomic transition requires the use of a different resonator since the frequency tuning range is limited and the two transitions of interest ($63p_{3/2} \rightarrow 61d_{5/2}$ and $63p_{3/2} \rightarrow 61d_{5/2}$)are separated by 50 MHz. The resonators are hollow niobium cylinders, 24.7 mm in diameter and 24mm long, with small apertures in both ends to allow the atoms to pass through. The field mode in the resonator used for the experiments is TE_{121}, which is nearly constant over the diameter of the atomic beam. A measurement of the cavity Q is made by monitoring the decay time for the field in the resonator. The cavity is superconducting (critical temperature 9.2K) and the end temperature determines the Q as well as the average number of black-body photons. Values for the Q lie between 1.7×10 for the first experiments and 3×10^{10} for the more recent ones.

After the atoms leave the cavity they are used as the "detector" for the properties of the maser field. For these highly excited Rydberg atoms it is easily possible to selectively field ionize the atoms such that a determination can be made of the state, i.e. whether the atom leaves the cavity in the upper or the lower state (in the earlier experiments only a detector for the upper state was used). The detection region is illustrated in Fig.1 as well. For the more recent experiments and for those now planned and under way, it is also necessary to have a well-defined velocity for the atoms, which thereby determines the interaction time, or looked at another way, the maser pump parameter, which is given by

$$\Theta = \sqrt{N_{ex}}\, \Omega \tau_{int}, \quad N_{ex} = R/\gamma \tag{2}$$

where R is the rate of atoms injected into the cavity and therefore N_{ex} is the number of atoms interacting with the field per cavity lifetime.

The next three subsections will show the results of three different experiments with the one-atom maser. Initially the maser action itself was the goal of the experiments, later the collapse and quantum revival of field Rabi oscillations was observed, and most recently the quantum statistical nature of the intracavity field was investigated with the observation of sub-Poissonian fluctuations in the number of lower state atoms emerging from the cavity.

A. Single-Atom Maser Operation[5,6]

Maser action is determined by tuning the cavity with the aid of a piezoelectric transducer (PZT) across the common cavity resonance/atomic transition frequency. The field ionization signal for atoms leaving the cavity in the upper state is recorded as a function of the detuning. If an atom deposits a photon in the resonator (through enhanced spontaneous or induced emission in the cavity; the free space lifetime is large), the result will be a decrease in the count rate for upper state atoms. Figure 2 shows the results of such measurements for varying flux of atoms through the cavity. In each case, over more than an order of magnitude difference in flux, the reduction in the rate of upper-state atoms is clearly visible and the depth of the signal is independent of the flux. The latter point is the indication that the transition is saturated, even when on the average less than 0.1 atom is in the cavity. Put another way, there is only $\sim 1\%$ probability that two atoms are present in the cavity together, so that one can say that the low pump rate of one atom every 60ms is enough to maintain the field in the cavity, i.e. to maintain maser action.

As the pump rate is increased the signal depth does not change, but its linewidth becomes broader. This power broadening is due to multiple exchanges of photons from the cavity field with the atom as it transits the interaction region. The number of photons in the cavity is related to the pump rate and to the cavity decay time in such a way that at the first maser threshold the relation $n = N_{ex}$ holds. In general, the photon number in the cavity is a complicated function as shown in Figure 3. The average photon number is almost zero for small Θ, but a finite $\langle n \rangle$ emerges when $\Theta=1$ which corresponds to the threshold of the maser. Then $\langle n \rangle$ increases rapidly and decreases later to reach a minimum at about $\Theta \approx 2\pi$, where the field abruptly jumps to a higher intensity. This general behavior occurs roughly at higher multiples of 2π, but becomes less pronounced for increasing Θ.

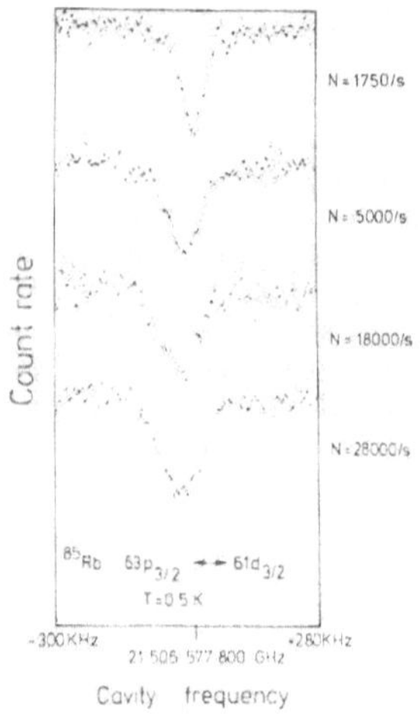

Fig. 2. Maser action. Displayed is the channeltron count rate for the atoms in the upper state as a function of cavity frequency for four different atomic flux (pump parameter) values. The traces show power broadening and evidence of AC Stark shifts caused by nearly non-resonant levels.

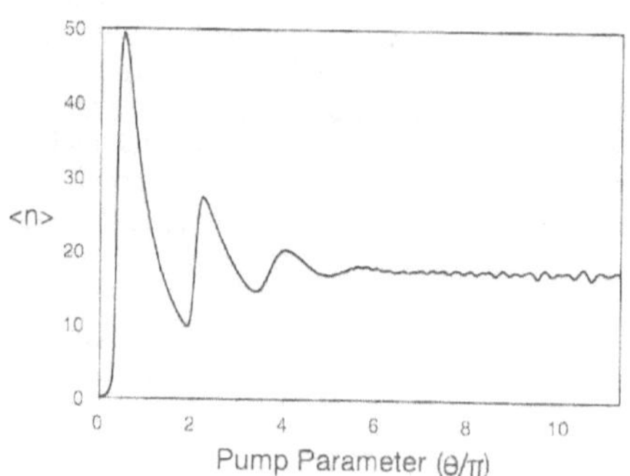

Fig. 3. Calculated average photon number in the maser as a function of the pump parameter shows the existence of several thresholds. The traces of Fig. 2 correspond to the first threshold.

It should be noted as well that at the temperatures used for the original experiments the number of black-body photons was not at all negligible as will be seen more clearly in the next subsection dealing with the collapse and revival of Rabi oscillations. Even at 0.5K ($n_{th} \sim 0.16$) the effect of thermal radiation is significant, a point which will be further discussed in Section III.

Finally, a shift in the center frequency of the resonance can be seen with increasing pump rate. This can be attributed to an AC Stark shift caused by the presence of the other $61d_{5/2}$ level 50 MHz detuned from the transition of interest, as discussed in Ref. 5. Recently Knight and coworkers[7] have investigated theoretically the effect of a third detuned level on the maser lineshape and position, as well as that of thermal photons as discussed in Ref. 5.

B. Jaynes-Cummings Collapse and Revival[8]

Following the work of Eberly et al.[9] there was a great deal of interest in trying to observe the manifestly quantum nature of the electromagnetic field by detecting the collapse and revival of coherence as a function of time. The coherence collapse itself was first discussed by Jaynes and Cummings and can be seen as simply a dephasing of the coherent atom-field interaction due to different incommensurate Rabi frequencies corresponding to different photon numbers. A classical thermal field also will exhibit such a collapse which therefore cannot be taken as a clear sign of the quantized field.[19] Rephasing, or revival of the Rabi oscillations after the collapse is however a feature which can only occur for a quantized field.

For T=0K with a Fock state field the probability of detecting an atom in the upper state as it exits the cavity shows regular cosinusoidal oscillations as a function of time. In the case of a coherent state resonator field, shown in Figure 4, the distribution of photon numbers is Poissonian, leading to the collapse and revivals.

The dephasing on collapse is seen to occur in general with an envelope which is given by exp $(-\frac{1}{2}\Omega^2 t^2)$, and the revival or rephasing of Rabi oscillations occurs after a time $T_R \sim 2\pi\sqrt{n_{coh}}/\Omega$.[9] If a thermal field is in the cavity then the pro-

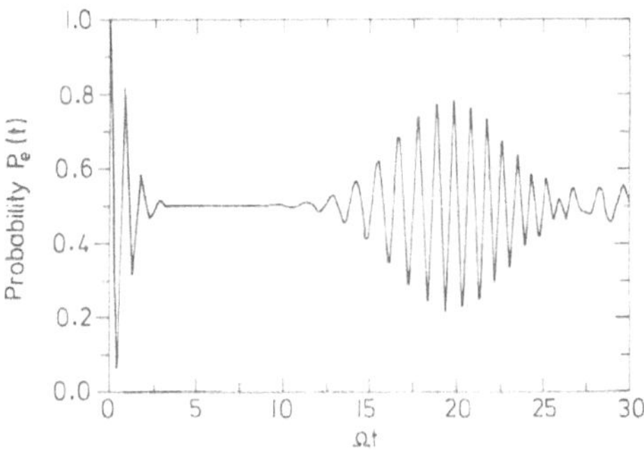

Fig. 4. For a coherent state field (here with $< n >= 9$) a plot of the probability for an atom to exit the cavity in the excited state is shown as a function of time. The collapse (dephasing) and revival (rephasing) of the Rabi oscillations is clear.

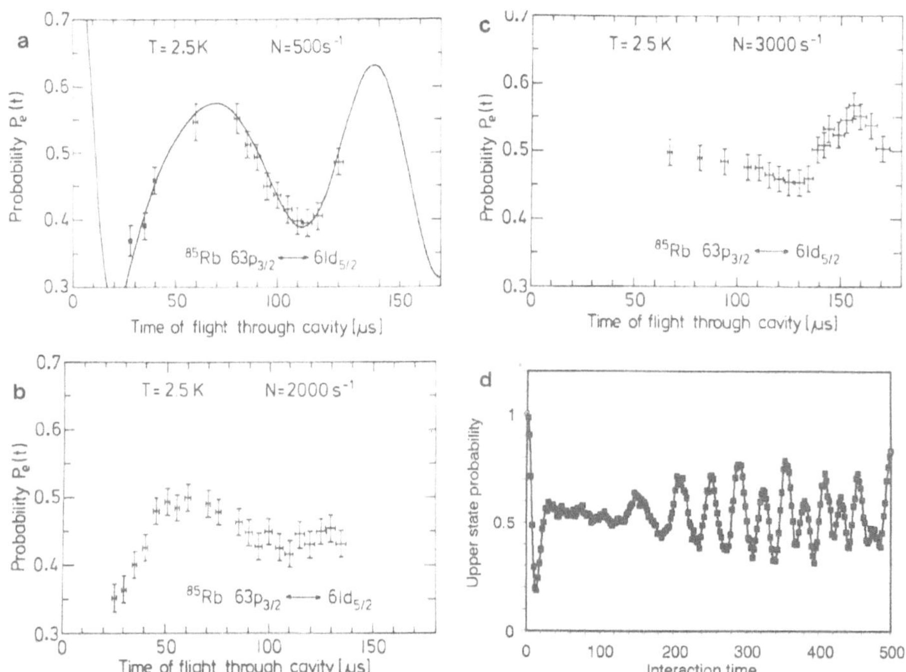

Fig. 5. The upper state atom probability as a function for the atom-cavity interaction time is shown for the experiments (a-c) and for a Monte-Carlo simulation (d). The temperature was 2.5K and therefore $n_{th} = 2$. In a-c the flux is, respectively 500 s^{-1}, 2000 s^{-1} and 3000 $s-1$. For the simulation, the parameters correspond to those of the experimental results in c, but it was possible to simulate much longer interaction times than were available for the experiment.

bability distribution (Bose-Einstein) for various photon occupation numbers must be averaged over as well.

In the experiments the situation is a bit more complicated because the field is a mixture of coherent and thermal photon distributions. In addition to the first one-atom maser work, it was necessary to select the velocity of the atoms of the atomic beam. The Fizeau velocity selector, a set of slotted discs rotating about a common axis (shown roughly in Figure 2) defines a subgroup of the modified Maxwellian distribution in the atomic beam with a spread of 4 %. Since directly following of the time dependence of the Rabi oscillations could not be done, the equivalent parameter, namely the interaction time, was changed in the experiments. Figure 5 shows the results of these measurements. The probability of atoms being detected in the excited state is shown as a function of the interaction time. For each of the sets of data the increasing flux (for constant interaction time) represents an increase in maser pump parameter. Figure 5a shows as well a theoretical curve calculated from the Jaynes-Cummings model.

It is also possible to carry out a Monte Carlo simulation of the collapse and revivals using the formalism which was described earlier. This is done in the spirit of the experiments by calculating the effect of each atom as it passes through the cavity and by varying the interaction time as was done in the actual èxperiment. Fig.5d shows one result for parameters as in Fig.5c in the real data. The simulations have the advantage that one can go to much longer interactions times than were possible for the experiments. The simulated results are seen to agree well with the experiments.

C. Sub-Poissonian Atomic and Photon Statistics[11,12]

Nonclassical states of the electromagnetic field have been investigated in several systems over more than a decade of experimental work. One of the properties of a quantum field which was first observed in the resonance fluorescence of a single atom is that of sub-Poissonian photon statistics.[20] This property is characterized by a photon distribution p(n) which is narrower than that produced by a coherent source such as a laser. Filipowicz et al.[2] have calculated the variance in photon number for the micromaser; one example is shown in Figure 6. The variance is given by

$$\sigma^2 = \frac{<n^2> - <n>^2}{<n>}. \tag{3}$$

For a coherent state we have $\sigma^2 = 1$, which marks the boundary between classically describable field states and those which cannot be described classically. One sees from the figure that near the first and subsequent maser thresholds the statistics of the field show a strong super-Poissonian nature. The thresholds can be compared to phase transitions in thermodynamic systems, where one expects to find large fluctuations at the point where the transition occurs. Between transitions the variance displays regions of sub-Poissonian photon statistics. These regions can be thought of as a sort of quantum non-demolition measurement situation in which further atoms transiting the cavity interact with the field there, but leave the field in its initial state.

As with the previously mentioned single-atom maser experiments the state of the exiting atom is probed, as it must be for microwave frequency range transitions. It is important to be able to relate the statistics of atomic lower- or upper-state fluctuations to the photon statistics of the field in the maser cavity.

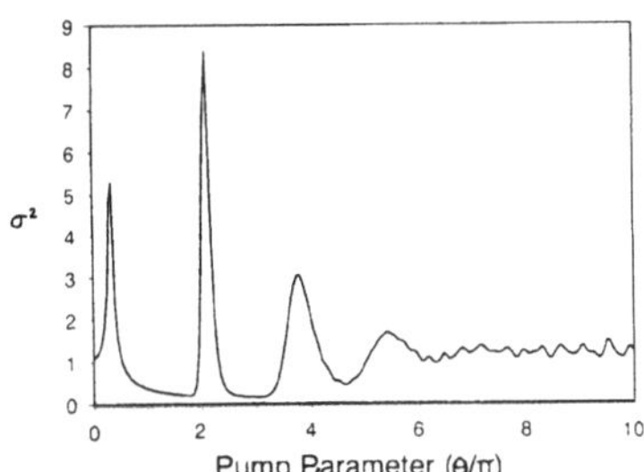

Fig. 6. Calculated variance in photon number as a function of maser pump parameter shows the different maser thresholds as well as the nonclassical character of the fluctuations away from the threshold ($\sigma <$ 1 is sub-Poissonian)

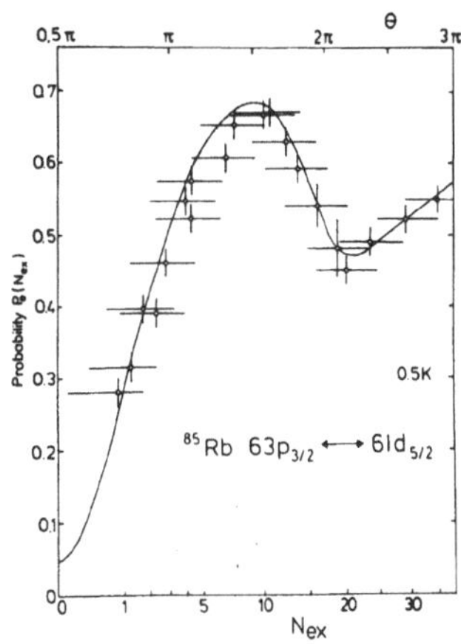

Fig. 7. Calculated as well as measured probability from atom to emerge from the cavity in the upper maser level, as a function of the flux N_{ex}.

This can be accomplished with the help of a few assumptions,[12,13] namely that the measurement time interval to be used is longer than or equal to the cavity decay time (this can be chosen easily for the experiment), and that the atomic beam flux fluctuations obey Poissonian statistics, which is found to be in fact the case. An atomic "Mandel Q-parameter "[14] can be defined as

$$Q_a = \frac{< N^2 > - < N >^2}{< N >} - 1 \tag{4}$$

which takes the value Q_a=0 for Poisson statistics. The relation between atomic and photon statistics is then given simply by[12]

$$Q_a = \epsilon P_l Q_F (2 + Q_F) \tag{5}$$

where ϵ is the detection efficiency, P_l is the probability of an atom emerging from the resonator in the lower state and Q_F is the Mandel Q-parameter for the cavity field. In Figure 7 is shown a plot of the *excited* state probability as a function of N_{ex}, or effectively of the flux (solid line), together with a set of experimentally measured values. Measurements of statistics were done by looking at *lower*-state atoms, but from the plots one sees that if $dP_e/dN > 0$ ($dP_l/dN<0$)then an increase in flux, for example from the natural atomic beam density fluctuations, leads to a decreased probability that an atom leaves the cavity in the lower state and vice-versa. The result of this is a self-driven feedback mechanism which smooths out fluctuations in the number of lower-state atoms exiting the cavity, and thereby leads to sub-Poissonian statistics for these atoms.

The measurements were made in an apparatus which could be cooled to 0.5K, which reduces the average thermal photon number to 0.16.[6] The experimentally accessible range of pump parameters (for a given interaction time) for the weaker of the two transitions, $63p_{3/2} \leftrightarrow 61d_{3/2}$, corresponds to the region around the first maser threshold. As expected the fluctuations here are super-Poissonian, as shown in Figure 8a, with a peak value of Q_a=0.11. The stronger transition, $63p_{3/2} \leftrightarrow 61d_{5/2}$ allows the investigation of the area around the second maser threshold, and therefore regions which exhibit nonclassical fluctuations. Again as expected, at the second threshold itself the fluctuations are large, but below and above that are regions of suppressed fluctuations as indicated by values of $Q_a <$

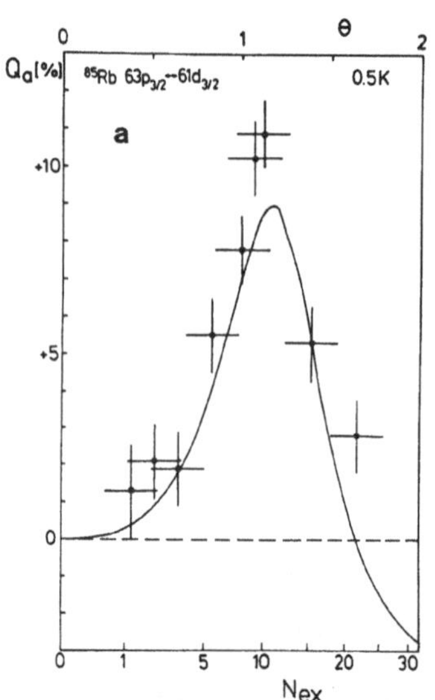

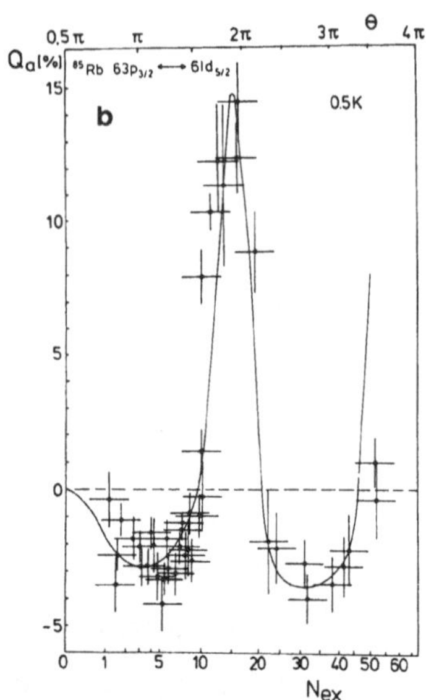

Fig. 8. Variance Q_a of the atoms in the lower maser level as a function of the flux N_{ex}. In a) is shown the region near the first maser threshold for the transition $63p_{3/2} \leftrightarrow 61d_{3/2}$, and in b) the region near the second threshold for the transition $63d_{3/2} \leftrightarrow 61p_{5/2}$ transition.

0. The measurements were made by opening a window for $T = T_{cav} = 200\,ms$ and counting the number of lower-state atoms exiting the cavity in that interval. Measurements were taken for about $2\text{x}10^4$ time intervals in order to keep the error small ($\sim$ 1%), and this was repeated for many different values of the atomic flux (variation of N_{ex}).

The largest nonclassical effect, that is the largest negative Q_a value occurs near N_{ex}=30 and agrees well with the theoretical curves, also shown in Figure 8b. The theoretical values are obtained from the calculated field Q_F parameters along with Eq. (5). It is important to note that the plotted values for the atomic Q_a parameter show sub-Poissonian fluctuations without any corrections being necessary. It is also possible to calculate back, knowing the detection efficiency to be $\sim$ 10%, and determine the statistics of the fields in the cavity which gave rise to the atomic statistics. For N_{ex}=30, P_e=0.45 (from Figure 7) we find Q_F=0.3, which says that the field fluctuations are 70% less than those of a coherent source. This finishes the review of the most important experiments carried out in the last few years with the micromaser. The next two sections will consist of a look to the future, at the experiments which are now in the process of being performed.

III. Trapping States

Due to the dependence of the excited-state atomic probability on the *sin* and *cos* of a pseudo-angle $\Theta = \Omega\sqrt{n_q + 1}\ \tau_{int}$ it is clear that for fixed coupling constant and interaction time, a situation can arise in which, for a certain photon number n_q in the field, the transiting atoms see a pulse such that $\Theta = 2q\pi$.[16] If this occurs, the initially excited atom leaves the cavity in the excited state again and nothing happens to the field. For a well-defined interaction time, the next atom experiences exactly the same as the previous one, and again the cavity field is unchanged. Thus one can say that the maser field is "trapped" in the state with n_q photons. There are three parameters which must be considered when extending this idea beyond the limit of the lossless micromaser: cavity losses, black-body photons, and a spread in interaction times (velocities). Each of these will be considered in turn.

An example including resonater dissipation is shown in Figure 9 for a situation in which the black-body photon number is negligible ($n_{th} = 10^{-5}$) and where one fixed interaction time was used. All of the dips seen in the curve of photon number as a function of maser pump parameter can be associated with two numbers, i.e. the number of photons in the field (n_q) and the number (q) of 2π pulses the atom sees as it flies through the interaction region. The trapping states are seen from this figure to have varying depths and widths. This will be important when considering an experiment since, as might be expected, any disturbing effects will most easily wash out shallow or very narrow resonances. It is important for the experiments that for larger values of the flux the resonances or trapping states become extremely narrow and tend to bunch in the region very near to a phase transition.

With a velocity spread N=4%, even at a temperature of 100μK and with a low flux of N_{ex}=10, almost no signature remains of the clear dips in photon number that were present for a monokinetic beam (shown in Fig. 9). A velocity spread of 1% is needed to be able to recover unambiguously the trapping states. This is explained by the fact that as the resonances get narrower, any slight mismatch in velocity compared to the ideal $2q\pi$ pulse velocity will allow the photon number distribution to "spring over" the barrier represented by the trapping state. A distribution of $\eta = \frac{\Delta v}{v} = 1\%$ represents an achievable improvement over the η=4% in the previous experiments.

The next problem to be addressed is that of thermal photons as a disrupting effect. For the present ^{3}He cryostat at a temperature of 500mK even an improvement in the velocity spread will not allow detection of the trapping states. Lower temperatures recover the resonance features and will make possible an unambiguous identification of these features. The temperature used above (100mK) is not randomly chosen, but rather that which is achievable with a dilution refrigerator that will replace the present ^{3}He cryostat in the very near future.

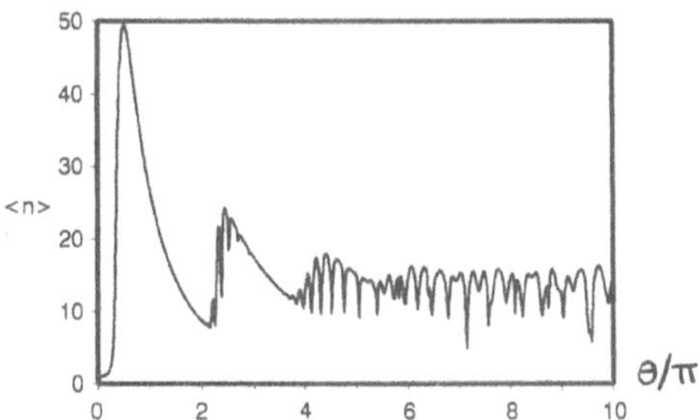

Fig. 9. Average photon number in the maser cavity as a function of the pump parameter Θ/π. The dips seen in the figure are the evidence of the trapping states. Parameters are as follows: $\Omega = 44kHz$, $N_{ex} = 50$ and $n_{th} = 0.001$.

The set of examples given here show not only the purely theoretical possibility of the existence of trapping states of the micromaser field, but also show that many experimental difficulties can be overcome with modifications to the existing apparatus. These changes should make possible the experimental verification of the trapping states.

IV. Phase Diffusion in the One-Atom Maser

The one-atom maser provides an ideal situation for investigating one of the fundamentally interesting problems in laser theory, that of the phase diffusion of the field. In all of what has been discussed up until now, the question of phase or coherence of the maser field has not been addressed. Choosing to look at the phase of the maser field and its time development will involve a more complicated calculational procedure as well as a slightly different experimental approach.

The calculations will be briefly outlined and some of the results shown, and these results then connected to the appearance of trapping states as well as to the

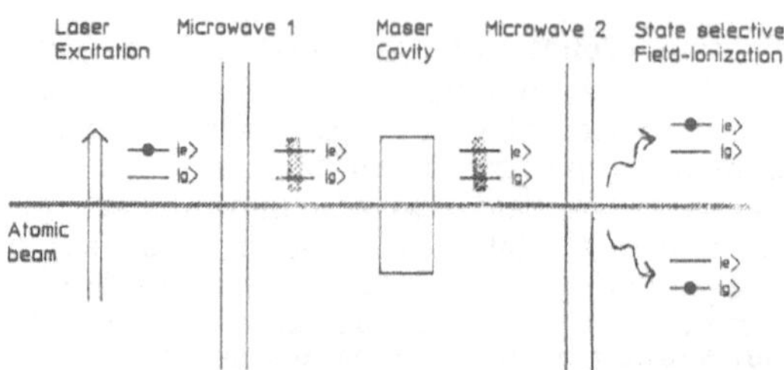

Fig. 10. Schematic of proposed experiment to observe maser field phase diffusion. Shown are the Rydberg atom excitation zone, the coherent micromaser field for preparing the atomic superposotion, the cavity itself, and the analyzing microwave field.

usual Schawlow-Townes linewidth for the laser. As was shown in Ref. 10 the coherent preparation of the atoms in a beam can lead to the transfer of that coherence to the field in the maser cavity. A rough schematic of such an experimental situation is shown in Figure 10. As in the other micromaser experiments a laser is used to prepare the Rydberg atoms in the upper of two levels. Between the excitation zone and the cavity a microwave source is used to create a superposition of upper and lower states; this is the important addition to the apparatus. The phase of this classical microwave field will be transferred to the cavity[18] and after enough atoms interact with the field mode a steady-state field will be built up. For the computer simulations an additional difficulty is that the off-diagonal elements of the density matrix are also needed because of the coherence. Such simulations of the one-atom maser can however be performed and for clarity of presentation, the Q-representation is used to plot the real and imaginary parts of the field. Figure 11a shows one example for a steady-state field distribution reached after 500 atoms have interacted with the field.

To watch the phase diffusion of the field it is simply necessary to turn off the coherent microwave field and continue pumping the laser (continue injecting atoms). Without the "orientation" provided by the coherent source, the phase of the field begins to wander around the complex-α plane, while the amplitude remains constant. Fig.11b-d illustrates one example in which the phase uncertainty diffuses slowly away from its initial well-defined value. After 40 cavity lifetimes there is still a noticeable phase associated with the Q-distribution. One might expect such behavior qualitatively since for a laser near threshold the linewidth (diffusion rate) is of the order of the cavity linewidth, above threshold, the linewidth in the phase diffusion limit is much narrower. Therefore in this above-threshold micromaser case we see a slow diffusion as well.

It is also possible to find parameters for which after five cavity lifetimes the phase is completely undefined. This is however understandable in light of the fact that the chosen pump parameter leads to a photon number distribution which is one of the many trapping states discussed earlier. Thus, the presence of such a well-defined photon number must lead to a very large variance in phase due to the field-phase uncertainty relation. When the system is released from its coherent driving field, the phase very quickly diffuses, reflecting the large uncertainty. It is also possible to calculate the diffusion coefficient as a function of pump parameter and see explicitely that the results mirror those for the photon number distribution in that the presence of a trapping state in the photon number distribution corresponds to an increase in the diffusion rate at the identical pump parameter.[17]

A brief word to the experiment being planned to observe the maser phase diffusion is necessary. It was also shown in Ref. (18) that a second microwave field, phase-locked to the initial one, but placed after the cavity, can be used to project the exiting atoms into one of the two eigenstates and then with usual field ionization techniques the population of upper- and lower-state atoms can be studied as a function of the time from the beginning of the field diffusion. The experiment described here uses a modified Ramsey fringe setup[19] as has been discussed for slightly different configurations elsewhere.[18] It will be necessary to do a time-resolved version of the experiments which have been done until now.

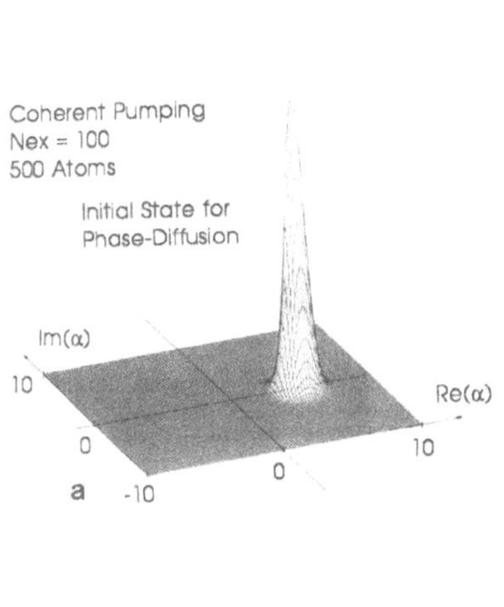

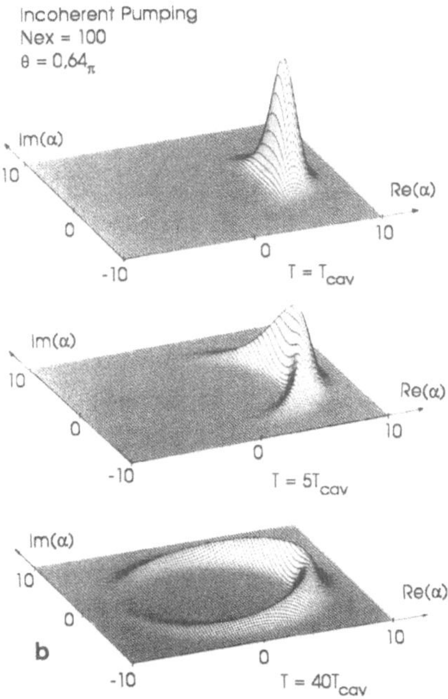

Fig. 11. Maser field Q-distribution. a) Initial maser field after coherent pumping with 500 atoms. b) After turning off the first coherent microwave source, the field diffuses on a time-scale long compared to the cavity decay time, as could be expected for a normal laser field.

The first microwave field can be turned on for a time long enough to allow ~1000 atoms to interact with the resonator field and to build up the phased field. Next, the first microwave field can be turned off and data collection begun synchronously with the microwave field switch-off-time. After one set of data is collected, the phase-determining field is turned back on, and the process repeated.

We wish to thank Christian Wagner and Achim Peters for the permission to use the results of their computer simulations before official publication.

References

1) E.T. Jaynes and F.W. Cummings, Proc. IEEE **51**, 89 (1963)
2) P. Filipowicz, J. Javanainen, and P. Meystre, Phys. Rev. **A34**, 3077 (1986)
3) P. Meystre, Opt. Lett. **12**, 669 (1987)
4) P. Meystre and E.M. Wright, Phys. Rev. **A37**, 2524 (1988)
5) D. Meschede, H. Walther, and G. Müller, Phys. Rev. Lett. **54**, 551 (1985)
6) G. Babst, Dissertation, Univ. Munich (1989)
7) P. Knight, Private Communication
8) G. Rempe, H. Walther, and N. Klein, Phys. Rev. Lett. **58**, 353 (1987)
9) J.H. Eberly, N.B. Narozhny, and J.J. Sanchez-Mondragon, Phys. Rev. Lett. **44**, 1323 (1980)
10) F.W. Cummings, Phys. Rev. **140**, A1051 (1965)
11) G. Rempe, F. Schmidt-Kaler, and H. Walther, Phys. Rev. Lett. **64**, 2783 (1990)

12) G. Rempe and H. Walther, Phys. Rev. **A42**, 1650 (1990)
13) G. Rempe, Habilitationsschrift, Univ. Munich (1989)
14) L. Mandel, Opt. Lett. **4**, 205 (1979)
15) P. Meystre, G. Rempe, and H. Walther, Opt. Lett. **13**, 1078 (1988)
16) P. Filipowicz, J. Javanainen, and P. Meystre, JOSA **B3**, 906 (1986)
17) M.O. Scully, H. Walther, G.S. Agarwal, T. Quang, and W. Schleich, in preparation
18) J. Krause, M.O. Scully, and H. Walther, Phys. Rev. **A34**, 2032 (1986)
19) P.L. Knight and P.M. Radmore, Phys. Lett. **90A**, 342 (1982)
20) R. Short and L. Mandel, Phys. Rev. Lett. **51**, 384 (1983)
21) C. Wagner, Diplomarbeit, Univ. Munich (1991)
22) A. Peters, Diplomarbeit, Univ. Munich (1991)

PHASE-SENSITIVE BACK-ACTION ON A MICROSCOPIC SYSTEM

V.Bagini[1], P.Lerner[2] and P.Tombesi[1,3]

1. Dipartimento di Fisica, Università di Roma "La Sapienza", P.le A. Moro 2, I-00185 Roma (Italy)
2. Department of Physics, University of New Mexico, Albuquerque (USA)
3. Dipartimento di Fisica, Università di Camerino, Via Madonna delle Carceri, I-62032 Camerino (Italy)

1. INTRODUCTION

In the last decade considerable interest has been devoted to a single electron in a Penning trap[1]. The Penning trap is a particular combination of constant magnetic field B and quadrupolar electrostatic potential. The motion of the electron in this trap can be considered as the superposition of three independent oscillations. The quadrupolar potential stabilizes, with respect to small perturbations, the electron's motion along the constant magnetic field B which becomes an oscillation at frequency $\omega_z/2\pi$ assuming the direction of the magnetic field as the z axis. Such a potential, however, introduces a very slow drift of the cyclotron orbit (the magnetron motion) at a frequency $\omega_m/2\pi << \omega_c/2\pi$ where $\omega_c = |e|B/mc$ is the cyclotron angular frequency with e and m respectively the charge and the rest mass of the electron while c is the speed of the light.

All the information concerning the cyclotron motion are obtained by measuring the oscillatory axial motion. The coupling between the axial and the cyclotron motions is obtained by means of a magnetic bottle. However, Kaplan showed[2] that such coupling is also obtained by considering the relativistic corrections. In particular, he showed that, due to relativistic corrections of the mass, the cyclotron motion of the electron in a magnetic field has a bistable behavior when its motion is driven by an external field. Thus, even the simplest of microscopic objects interacting with an electromagnetic field is intrinsically bistable. Gabrielse, Dehmelt and Kells[3] measured the cyclotron energy by observing the shift induced on the axial frequency ω_z. By varying the frequency of the driving field, they also observed the bistable hysteresis in the cyclotron motion of the electron. This hysteresis results entirely from a relativistic correction to the mass and is measured by monitoring the axial motion of the electron. Thus, we can say that the axial motion is a "probe" analysing the properties of the cyclotron motion which is the system subject to the measurement.

Quantum Measurements in Optics, Edited by P. Tombesi and D.F. Walls, Plenum Press, New York, 1992

Recently, a new radiation source with nonclassical properties of the emission, i.e. the squeezed light, has been generated[4]. One of the most interesting nonclassical properties of the squeezed light is the dependence on the phase of the quantum noise in contrast with the thermal and coherent light which have phase independent noises. Furthermore, it has also been shown that in an optical bistable medium, imbedded in a squeezed vacuum, the bistable curve can be influenced by varying the phase of a squeezing parameter[5].

Using parametric down-conversion techniques, La Porta, Slusher and Yurke[6] were able to perform a phase-sensitive back-action evading measurement, and the development of such techniques might be usefull for a wide range of measurements in quantum systems. In this paper we demonstrate that the implementation of the phase-sensitive back-action techniques might result in the possibility of control over the bistability of the cyclotron motion of the electron by using a phase-sensitive detection scheme. This effect is of particular interest because the situation considered provides an example of the influence of the quantum properties of the measurement apparatus on the microscopic system.

The quantum behaviour of the measurement device is modelled in the present example by the squeezed noise in the axial degree of freedom of the monitored electron. It means that the procedure of measurement is characterized by some intrinsic phase (phase of the squeezing parameter) which influence, in principle, can be measured. Then, the anomalous quantum fluctuations, induced by the phase-sensitive device on the axial oscillator, may influence the cyclotron motion.

2. THE MODEL FOR QUANTUM MEASUREMENT

One considers the motion of an electron in a uniform magnetic field B along the positive z axis driven by a circularly polarized electric field. As it is well known[1], the motions of the electron in the Penning trap are well-separated in energy scale and only the cyclotron and axial degrees of freedom should be taken into account. We neglect all the spin related terms in the Hamiltonian which can be approximated by:

$$H = \frac{(\underset{\sim}{p} - \frac{e}{c}\underset{\sim}{A})^2}{2m} - \frac{(\underset{\sim}{p} - \frac{e}{c}\underset{\sim}{A})^4}{8m^3c^2} + e\,V_0 \frac{x^2 + y^2 - 2z^2}{4d^2} \tag{1}$$

with $\underset{\sim}{A} \equiv (A_x - \frac{yB}{2}, A_y + \frac{xB}{2}, 0)$; $A_x = ic\,(\tilde{E}^* e^{i\omega_p t} - \tilde{E} e^{-i\omega_p t})/\omega_p$, $A_y = c\,(\tilde{E}^* e^{i\omega_p t} + \tilde{E} e^{-i\omega_p t})/\omega_p$

and $\tilde{E}$ is the amplitude of the driving field at frequency ω_p; d characterize the dimensions of the trap and V_0 is the potential applied to the electrodes of the trap. We have neglected all contributions of the order of c^{-3}. It is convenient to introduce the lowering and rising operators for the cyclotron motion

$$a = \frac{1}{2}\,[\,\beta(x - iy) + \frac{i}{\beta\hbar}(\hat{p}_x - i\hat{p}_y)]$$

$$a^+ = \frac{1}{2}\,[\,\beta(x + iy) - \frac{i}{\beta\hbar}(\hat{p}_x + i\hat{p}_y)] \tag{2}$$

with

$$[a, a^+] = 1 \tag{3}$$

and $\beta = \left(\frac{m\omega_c}{2\hbar}\right)^{1/2}$. For the axial motion we define

$$a_z = \left(\frac{m\omega_z}{2\hbar}\right)^{1/2} z + i\left(\frac{1}{2m\hbar\omega_z}\right)^{1/2} \hat{p}_z$$

$$a_z^+ = \left(\frac{m\omega_z}{2\hbar}\right)^{1/2} z - i\left(\frac{1}{2m\hbar\omega_z}\right)^{1/2} \hat{p}_z \tag{4}$$

with $\omega_z^2 = \frac{|e|V_0}{md^2}$.

In the dipole approximation of the electromagnetic interaction and rotating-wave approximation of the electron motion the Hamiltonian becomes

$$H = \hbar\hat{\omega}_M a^+a + \hbar\mu(a^+a)^2 + i\hbar k\,(\tilde{E}a^+e^{-i\omega_p t} - \tilde{E}^*a\, e^{i\omega_p t}) + \hbar\omega_z\,(a_z^+a_z + \frac{1}{2}) \tag{5}$$

and we have defined

$$\hat{\omega}_M = \omega_c\left(1 - \frac{\hat{p}_z^2}{2m^2c^2} - \frac{\hbar\,\omega_c}{2mc^2}\right)\ ;\ \ \mu = -\frac{\hbar\,\omega_c^2}{2mc^2}\ ,\ \ k = \frac{e}{\omega_p}\sqrt{\frac{\omega_c}{2m\,\hbar}}\ . \tag{6}$$

All terms not containing the rising or lowering operators for the cyclotron motion have been omitted. These terms should be regarded as a small correction to the electron's rest mass due to an external field. We have also neglected the anharmonicity of the axial motion $(\hat{p}_z^2/2m)^2/mc^2$ which is smaller by a factor $(\omega_z/\omega_c)^2$ with respect to the cyclotron anharmonicity.

To go further, we need more detailed inquiry into the basics of the measurement procedure. Namely, the monitoring of the cyclotron motion of the electron in the Penning trap is provided by the measurement performed on the axial degree of freedom (i.e., the oscillatory motion along the direction of the magnetic field). The shift of the axial resonance is related to the properties of the cyclotron motion. For the cyclotron motion the coupling term between the two oscillators, given by

$$-\frac{\hbar\omega_c}{mc^2}\frac{\hat{p}_z^2}{2m}\,a^+a\ , \tag{7}$$

represents the back-action of the measurement device. Indeed, it transforms the cyclotron frequency in a new

$$\omega'_c = \omega_c\left(1 - \frac{\langle\hat{p}_z^2/2m\rangle}{mc^2}\right) \tag{8}$$

The back-action of a measurement device is hard to control over. Nevertheless, the intuitive guess should be provided that in conventional types of measurement the quantum statistics of a back-action is more or less similar to that of the thermal noise. Recently, some experimental schemes were developed in which the back-action of the measurement device can be controlled on the quantum level[6]. It means that the pumping of the probe system by the measuring apparatus could be considered at best to be equivalent to the penetration of the squeezed vacuum from the output port. Then, we may consider the measuring apparatus as playing the role of the environment which measures the axial oscillation via a particular coupling[7]. Finally, by following Kennedy and Walls[8], we consider the phase-sensitive measurement device as a squeezed bath interacting with a harmonic oscillator representing the axial motion of the electron which is the probe analysing the properties of the cyclotron motion. The above mentioned wide separation of the characteristic frequencies in the system ensure that degrees of freedom other than axial are not directly effected by the back-action of the measurement.

So, we assume that the momentum of the oscillatory axial motion contains the fluctuating part as a result of the back-action of the measurement device. As it follows from the above considerations this back-action can provide fluctuations with different quantum statistical properties. To describe the influence of the squeezed back-action we consider the axial oscillation as imbedded in a squeezed bath[8] thus the master equation for the axial degree of freedom in the interaction picture becomes[9]

$$\frac{\partial\rho}{\partial t} = \frac{\gamma}{2}\{(N+1)(2a_z\rho a_z^+ - a_z^+a_z\rho - \rho a_z^+a_z + N(2a_z^+a_z - a_za_z^+\rho - \rho a_za_z^+))\}$$

$$+\ \frac{\gamma}{2}\{M(2a_z^+\rho a_z^+ - a_z^{+2}\rho - \rho a_z^{+2} + M^*(2a_z\rho a_z - a_z^2\rho - \rho a_z^2))\} \qquad (9)$$

Where N represents the number of photons associated with the bath which is assumed to have a white noise spectrum. $M = |M|\, e^{i\varphi}$ is the signature that the bath is squeezed[9]. In order to respect the Heisenberg uncertainety principle the relation $|M| \leq \sqrt{N(1+N)}$ holds. In the steady state the statistical properties of the axial operators are the following[9]

$$<a_z> = <a_z^+> = 0$$

$$<a_z^+a_z> = N,\ <a_z^2> = |M|\, e^{i\phi} \quad \text{or} \quad \frac{\langle\hat{p}_z^2\rangle}{2m} = \frac{\hbar\omega_z}{2}\,(N+1/2+|M|\cos\varphi) \qquad (10)$$

If the back-action (non-dynamical) degrees of freedom are introduced in such a way, Eq.(10) is the precise manifestation that the measuring device has nonclassical properties. This could, probably, be obtained by control over the fluctuations in the current loop of the Penning-trap.

We further assume that the radiative decay rate for the axial oscillations, due to the interaction with such a quantized back-action, is sufficiently high for the electron suddenly reaches its steady state.

Since the damping rate of the cyclotron motion, due to the spontaneous emission, is very weak and it has been shown that could be significantly reduced by using a microwave cavity[10], we can first average over the back-action degrees of freedom and the Hamiltonian becomes

$$H= \hbar\,\omega_M\,(1-\tilde{\eta}\ \ \kappa(\phi))\,a^+a + \hbar\,\mu\,(a^+a)^2 + \hbar k\, i(\tilde{E}a^+\, e^{-i\omega_p t} - \tilde{E}^*\, a\, e^{i\omega_p t}) + \Gamma a^+ + \Gamma^+ a \qquad (11)$$

with

$$\omega_M = \omega_c \left(1 - \frac{\hbar\omega_z}{4mc^2} - \frac{\hbar\omega_c}{2mc^2}\right) \tag{12}$$

$$\tilde{\eta} = \hbar\omega_z \left(1 - \hbar\omega_z/4mc^2 - \hbar\omega_c/2mc^2\right)^{-1} /2mc^2 \simeq \hbar\omega_z /2mc^2 \tag{13}$$

$$\kappa(\phi) = N + |M|\cos\phi \tag{14}$$

and we have also included the interaction with a bath, described by the bath's non-dinamical variables Γ and Γ^+, to take into account the dissipation of the cyclotron motion. The coefficient $\tilde{\eta}$ is reflecting the efficiency of the back-action of the measurement procedure which consists in control over the cyclotron motion by monitoring the axial degree of freedom.

The Hamiltonian in Eq.(11) is similar to the one discussed by Drummond and Walls for the quantum theory of the dispersive optical bistability[11].

3. PHASE-SENSITIVE BISTABILITY OF THE ELECTRON IN A TRAP

In a frame rotating at ω_p, the master equation for the reduced density matrix $\rho(t)$ can be derived by standard procedure[12] to get

$$\frac{\partial}{\partial t}\rho = \sum_{i=1}^{4} L_i(\rho) \tag{15}$$

$$L_1(\rho) = -i\,(\Delta\omega - \eta\ \kappa(\phi))\,[a^+a\,,\rho]$$

$$L_2(\rho) = -i\mu\ [(a^+a)^2\,,\rho]$$

$$L_3(\rho) = k\ [(\tilde{E}\,a^+ - \tilde{E}^*\,a)\ ,\ \rho]$$

$$L_4(\rho) = \frac{\gamma_c}{2}\,[2a\rho a^+ - a^+ a\rho - \rho a^+ a]$$

with γ_c representing the energy relaxation rate of the cyclotron motion and the temperature of the bath is considered to be zero. $\Delta\omega = \omega_M - \omega_p$ is the detuning and we have redefined $\eta{=}\omega_M\tilde{\eta}$. By standard techniques[12], the equation of motion for the density matrix operator ρ is converted into partial differential equation for the function $P(\alpha,\alpha^+)$. This is a c-number function that can be interpreted as a non necessarily definite positive measure, whose statistical moments with respect to variables α,α^+ give the normally ordered expectations of the quantum operators[12]. We follow Drummond and Walls[11] by using the generalized P-function[13] because, due to a non positive definite diffusion matrix, the usual Glauber-Sudarshan P-function turns out to become more singular than a Dirac δ-function.

The correspondence principle between operators and c-numbers reads: $\alpha \leftrightarrow a$ $\alpha^+ \leftrightarrow a^+$. However, the variables α, α^+ are complex conjugate only in the mean[11]. The Fokker-Planck equation corresponding to the master equation (15) reads as :

$$\frac{\partial}{\partial t} P(\alpha, \alpha^+) = \{\frac{\partial}{\partial \alpha} [(1 + i (\Delta\omega - \eta\ \kappa(\phi))\alpha + 2i\mu\alpha^2\alpha^+ - E] - i\mu \frac{\partial^2}{\partial \alpha^2} \alpha^2$$

$$+ \frac{\partial}{\partial \alpha^+} [(1 - i (\Delta\omega - \eta\ \kappa(\phi))\alpha^+ - 2i\mu\alpha^{+2}\alpha - E^*] + i\mu \frac{\partial^2}{\partial \alpha^{+2}} \alpha^{+2}$$

$$+ 2 \frac{\partial^2}{\partial \alpha\, \partial \alpha^+} \} P(\alpha, \alpha^+) \tag{16}$$

Where all the parameters $\Delta\omega, \eta$, μ and the driving field amplitude $E = k\tilde{E}$ are divided by $\gamma_c/2$ and we used dimensionless time $t = \gamma_c t/2$.

We now consider the equation for the mean field amplitude $\langle a \rangle = \alpha$.The energy of the system is assumed to be sufficiently high so the fluctuations are negligible. Thus, the correlation functions are factorized and the semiclassical approximation gives

$$\frac{d\alpha}{dt} = E - [1 + i (\Delta\omega - \eta\ \kappa(\phi) + 2\mu n)]\ \alpha \tag{17}$$

with $n = \langle a^+ a \rangle = \alpha^* \alpha$ representing the average quantum number of excitations in the system. The steady state equation is obtained in the form

$$|E|^2 = n\ [1 + (\Delta\omega + 2\mu n - \eta\ \kappa(\phi))^2]\ . \tag{18}$$

The equation (18) is a cubic equation whose turning points $n_\pm$ are obtained by

$$\frac{\partial |E|^2}{\partial n} = 0 \tag{19}$$

and we get

$$n_\pm = \frac{2(\Delta\omega - \eta\kappa(\phi)) \pm \sqrt{(\Delta\omega - \eta\kappa(\phi))^2 - 3}}{6|\mu|} \tag{20}$$

The bistable curve emerges for the detuning exceeding the critical value

$$(\Delta\omega - \eta\kappa(\phi))^2 \geq 3 \tag{21}$$

The turning points of the curve thus depend on the phase of the squeezing as it is shown in Fig.1.

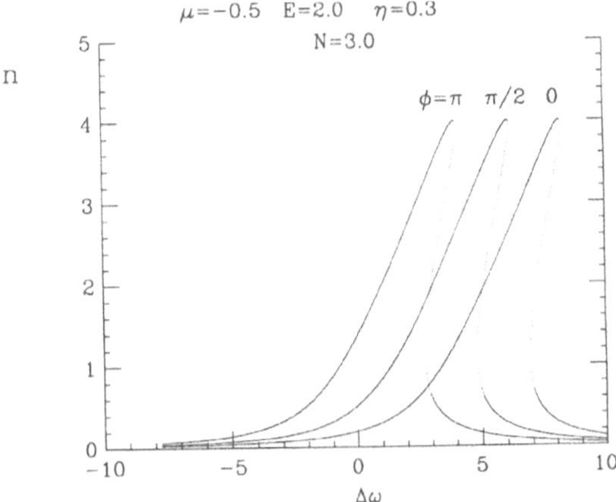

Fig.1 The excitation number n of the cyclotron motion is plotted versus the dimensionless detuning $\Delta\omega$. The curves shown refer to various values of the squezing phase angle ϕ. The values of the various dimensionless parameters are shown in the figure.

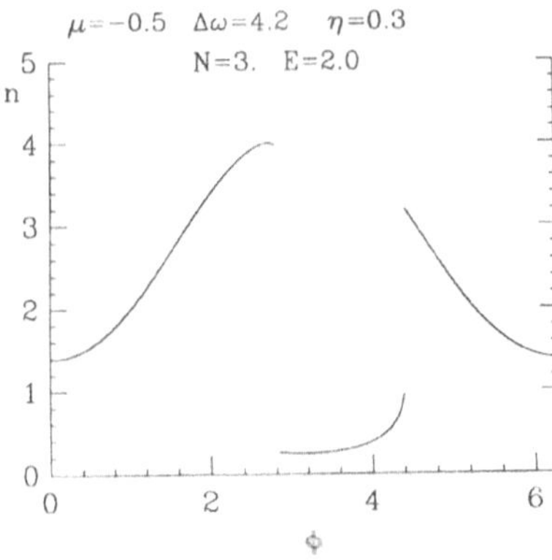

Fig. 2 The switching from the two branches of the bistable curve is obtained by only varying the phase ϕ of the squeezing parameter.

4. CONCLUSION

We have shown that the bistable curve of the cyclotron resonance can be modified by just varying the phase angle ϕ of the squeezing parameter associated with the quantum properties of the measurement apparatus. In the previous section it has been shown that the shift of the turning points depends on the strength of

$$\eta \simeq \frac{\omega_c}{\gamma_c} \frac{\hbar\omega_z}{mc^2} \tag{22}$$

and on the value of the squeezing parameter N by choosing $|M| = \sqrt{N(N+1)}$. With the values of the experiment in Ref.(3) we have $\eta \simeq 0.2$. This value could be enhanced by choosing a smaller γ_c. This can be achieved by modifying the cavity of the trap, as was shown by Gabrielse and Dehmelt[10]. The Fig. 2 shows that for a fixed value of the driving amplitude E, which can be chosen as real, and for a given detuning, the switching from one branch of the bistable curve to another can be realized by only varying the phase of the squeezing parameter.

5. ACKNOWLEDGEMENTS

This work has been supported in part by the Consorzio Interuniversitario Nazionale di Fisica della Materia.

REFERENCES

1) L.S.Brown and G.Gabrielse. Rev. Mod. Phys. **58**, 233 (1986) and references therein.
2) A.E.Kaplan; Phys. Rev.Lett. **48**, 138 (1982).
3) G.Gabrielse, H.Dehmelt and W.Kells; Phys. Rev.Lett. **54**, 537 (1985).
4) D.F. Walls, Nature (London) **306**, 141 (1983); R.Loudon and P.L.Knight, J.Mod. Opt. **34**, 709 (1987).
5) P.Galatola, L.A.Lugiato, M.Porreca and P.Tombesi, Opt.Comm. **81**, 175 (1991).
6) A.La Porta, R.E.Slusher and B.Yurke; Phys. Rev.Lett. **62**, 28 (1989).
7) D.F.Walls and G.J.Milburn, Phys.Rev. A **31**, 2403 (1985).
8) T.A.B.Kennedy and D.F.Walls; Phys.Rev. A**37**, 152 (1988).
9) C.W.Gardiner and M.J.Collet; Phys. Rev. A **31**, 3761 (1985).
10) G.Gabrielse and H.Dehmelt, Phys.Rev.Lett.55, 67 (1985).
11) P.D.Drummond and D.F.Walls; J.Phys. A: Math. Gen. **13**, 725 (1980).
12) W.H.Louisell: *Quantum Statistical Theory of Radiation,* Wiley (New York 1973).
13) P.D.Drummond and C.W.Gardiner; J.Phys. A: Math. Gen. **13**, 2353 (1980).

NOVEL EFFECTS IN QUANTUM OPTICS

Klaus M. Gheri and Daniel F. Walls

Department of Physics, University of Auckland, Private Bag
Auckland, New Zealand

1. INTRODUCTION

We shall describe two novel quantum effects which can be realized in optics. The first describes a proposal by Tan, Walls and Collett[1] of an experiment which demonstrates the nonlocality of a single photon. It differs from all other experiments and proposals on nonlocality in that only a single photon is involved and not a *correlated* photon *pair*. In previous experiments two photons maintained their mutual quantum mechanical correlations even when spatially separated. In the proposed experiment a single photon influences two spatially separated detectors. The photon is incident on a beam splitter after which the photon may follow either of two routes to detectors which include a homodyne measurement of the photons quadrature phase. The light field after the beam splitter may be described as an entangled state of one photon with the vacuum. Nonclassical features in the coincidence rate between the two detectors are predicted. Under certain conditions Bell's inequalities may also be shown to be violated.

The second topic we discuss is a proposal by Dance, Collett and Walls[2] for a quantum nondemolition measurement by means of second harmonic generation. The aim of this proposal is to measure the amplitude quadrature of a signal beam by measuring the amplitude quadrature of its second harmonic generated inside an optical cavity. Provided the probability for second harmonic generation inside the cavity is high only a small amount of the signal is requested to enter the cavity and consequently the degradation of the fundamental may be small. The usefulness of this scheme as a QND detector is evaluated by using a set of criteria introduced by Holland *et al* [3] and generalized to spectral correlation functions. It is found that for frequencies away from the central carrier frequency of the signal the QND criteria are well satisfied. This suggests the application of this system for the transmission of information by encoding the signal on a carrier wave as a modulation at the optimum frequency for QND detection.

2. NONLOCALITY OF A SINGLE PHOTON

At the beginning of this section we want to give a brief recall of the essence of Bell's inequalities. There are a number of states of light which are in conflict

Quantum Measurements in Optics, Edited by P. Tombesi and
D.F. Walls, Plenum Press, New York, 1992

with the classical ideas of locality and causality. Bell's inequalities are a criterion to test this violation. In any theory that has a thoroughly quantum mechanical axiomatic formulation a measurement carried out on an observable inevitably affects the entire system being measured. This is a consequence of the strong correlations introduced by the noncommutativity of some observables. The Einstein-Podolsky-Rosen (EPR) paradox[4] and Bell's inqualities[5] rely on creating a spatially extended "system" so that measurements at one detector depend not only on local parameters, but may depend on parameters at another detector. Bell's inequalities now limit the amount of correlation possible between two spatially separated events for a whole class of hidden variable theories. As a consequence any violation of them must be seen as a reinforcement for the assumption that the quantum mechanical description of physical events is the apropriate one.

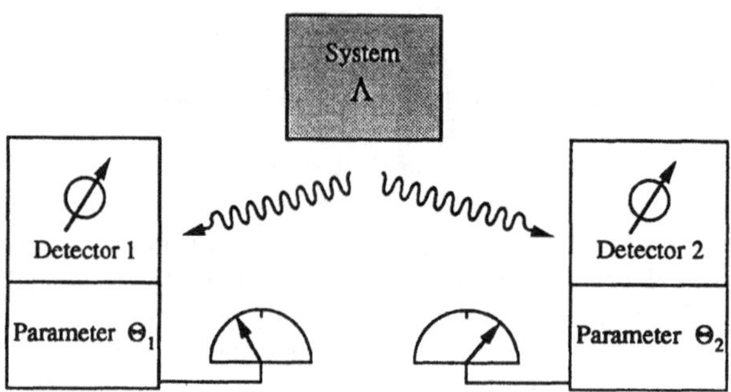

Fig. 1. Scheme for a test of Bell's inequalities using two detectors with externally adjustable parameters Θ_j.

2.1. *Local causality*

Suppose a pair of correlated systems, which have a joint state which can be described by a hidden variable Λ, separate and are measured by two unconnected detectors. The values recorded for the two systems are denoted as X_1 and X_2, dependent on externally adjustable parameters Θ_1 and Θ_2 of their respective detectors. (Cf. Fig. 1) The joint probability distribution for measuring these values can be expanded over the hidden variable Λ

$$p(X_1, X_2 \mid \Theta_1, \Theta_2) = \int \mathrm{d}\Lambda\, p(X_1, X_2 \mid \Theta_1, \Theta_2, \Lambda) p(\Lambda), \tag{1}$$

where we have assumed the independence of the specific state Λ of the local parameters Θ_1 and Θ_2

$$p(\Lambda \mid \Theta_1, \Theta_2) = p(\Lambda). \tag{1a}$$

By making use of the definition of conditional probabilities we find

$$p(X_1, X_2, \mid \Theta_1, \Theta_2) = \int \mathrm{d}\Lambda\, p(X_1 \mid X_2, \Theta_1, \Theta_2, \Lambda) p(X_2 \mid \Theta_1, \Theta_2, \Lambda) p(\Lambda). \tag{2}$$

If we assume classical local causality to be valid we will have to require that the outcome of a measurement at the one detector doesn't depend on the adjustment of the local parameter and the outcome of a simultaneous measurement at the other detector, or more precisely that

$$p(X_1 \mid X_2, \Theta_1, \Theta_2, \Lambda) = p(X_1 \mid \Theta_1, \Lambda), \tag{3}$$

and

$$p(X_2 \mid \Theta_1, \Theta_2, \Lambda) = p(X_2 \mid \Theta_2, \Lambda). \tag{3a}$$

Insertion of these two identities into the general expression for the coincidence probability $p(X_1, X_2 \mid \Theta_1, \Theta_2)$ yields

$$p(X_1, X_2 \mid \Theta_1, \Theta_2) = \int \mathrm{d}\Lambda\, p(X_1 \mid \Theta_1, \Lambda) p(X_2 \mid \Theta_2, \Lambda) p(\Lambda). \tag{4}$$

This restriction on the possible forms of the coincidence probability finally leads to Bell's inequalities.

2.2. *Non-Locality of a single photon*

In the many configurations which have been proposed to demonstrate a violation of Bell's inequalities, a common feature is that a pair of particles is involved, generated by some interaction which imposes a high degree of correlation between them. In the experiments in optics which have demonstrated a violation of Bell's inequalities *correlated pairs* of photons have been generated either by a two-photon atomic cascade[6,7] or by non-degenerate parametric amplification[8,9,10,32].

It is possible , however, to find a configuration which does *not* rely on having two correlated particles. Even the field of *single* photon can have a pronounced non-local effect on two homodyne detectors giving rise to EPR-correlations and a violation of Bell's inequalities. The non-local property of the photon field is evident in the Youngs's interference experiment performed by Grangier *et al* [11] where only a single photon is incident on the two slits. The interference fringes observed are explained quantum mechanically by the interference of the two paths the photon may take. However the observed interference fringes may be explained by a classical field theory which is non-local. In the experiment now proposed by S. M. Tan *et al* [1], the effects predicted may *not* be duplicated by any classical theory.

We consider a pair of homodyne detectors (indexed by subscript k), each of which consists of a 50-50 beam-splitter, a coherent local oscillator with amplitude $\alpha_k = \alpha e^{i\theta_k}$ and two photodetectors in the output ports, as shown in Fig. 2. Other schemes involving homodyne detection have been proposed to investigate the EPR paradox[12] and violations of Bell's inequalities[13].

Referring to Fig. 2, we see that homodyne detector k may be regarded as making a measurement of mode b_k, with a local parameter Θ_k. This local parameter is analogous to the analyser used in the conventional two-particle EPR experiment.

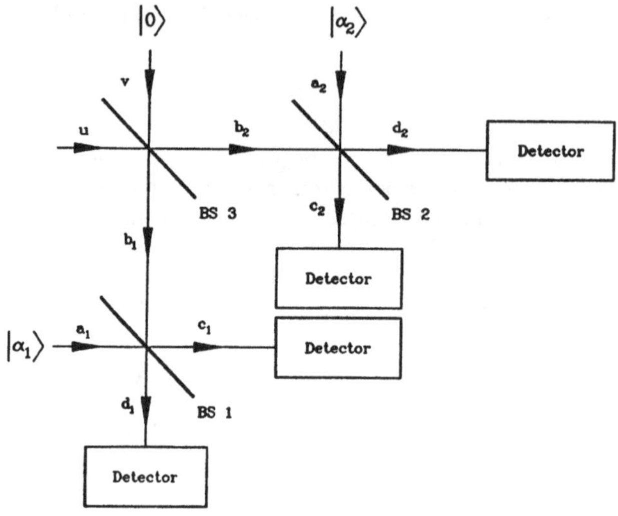

Fig. 2. Proposed experimental configuration.

We wish to determine the probabilities with which the individual detectors respond, and the coincidence probabilities for pairs of photodetectors, one in each homodyne detector.

Remark: The reader who is not entirely satisfied (and indeed he is right) with what will subsequently be called "coincidence probability" is referred to the *Appendix A* at the end of this article. Therein we will give expressions for the actual probabilities for counting a certain number of photons in one or two detectors and relate them to the coincidence probabilities in the case of weak local oscillators.

The transformation between the mode operators shown in Fig. 2 is given by

$$\begin{bmatrix} \hat{c}_1 \\ \hat{d}_1 \\ \hat{c}_2 \\ \hat{d}_2 \end{bmatrix} = \begin{bmatrix} \frac{1}{\sqrt{2}} & \frac{i}{2} & 0 & -\frac{1}{2} \\ \frac{i}{\sqrt{2}} & \frac{1}{2} & 0 & \frac{i}{2} \\ 0 & -\frac{1}{2} & \frac{1}{\sqrt{2}} & \frac{i}{2} \\ 0 & \frac{i}{2} & \frac{i}{\sqrt{2}} & \frac{1}{2} \end{bmatrix} \begin{bmatrix} \hat{a}_1 \\ \hat{v} \\ \hat{a}_2 \\ \hat{u} \end{bmatrix} . \tag{5}$$

This enables us to calculate the coincidence probabilities between the detectors directly in terms of the input fields. We begin first by considering vacuum inputs to the modes $\hat{u}$ and $\hat{v}$. The local oscillators are assumed to be in the coherent states $|\alpha e^{i\theta_1}\rangle$, $|\alpha e^{i\theta_2}\rangle$. As it is to be expected the intensities at all detectors are found to be equal

$$\langle I \rangle = I_0 = \frac{1}{2}\,\alpha^2. \tag{6}$$

The two-photon coincidence rates due to *rare* chance coincidences between local oscillators are also equal between the pairs of detectors

$$\langle I_{c_1} I_{c_2} \rangle = \langle I_{d_1} I_{d_2} \rangle = \langle I_{c_1} I_{d_2} \rangle = \langle I_{d_1} I_{c_2} \rangle = \frac{1}{4}\,\alpha^4. \tag{7}$$

We now consider the input of a single photon in mode $\hat{u}$ while the mode $\hat{v}$ is the vacuum. The state of the two-mode field $\hat{b}_1$ and $\hat{b}_2$ after the first beam-splitter is then an entangled state[14,15] of a one photon state and the vacuum

$$|\psi\rangle = \frac{1}{\sqrt{2}}\left(|0\rangle_{\hat{b}_1}\,|1\rangle_{\hat{b}_2} + \mathrm{i}\,|1\rangle_{\hat{b}_1}\,|0\rangle_{\hat{b}_2}\right) \tag{8}$$

which is precisely the same state as one gets (except for a phase factor) for a one photon state incident on the two slits of a Young's interference experiment.

The photon count probabilities at the individual detectors are now increased by $\frac{1}{4}$ being the probability that the one photon input is detected by any given detector. The coincidence count probability, however, between the pairs of detectors differ, now depending on *which* pair is considered. One finds

$$\langle I_{c_1} I_{c_2}\rangle = \langle I_{d_1} I_{d_2}\rangle = \frac{1}{4}\left[\alpha^4 + \alpha^2(1+\sin(\theta_1-\theta_2))\right] \tag{9}$$

and

$$\langle I_{c_1} I_{d_2}\rangle = \langle I_{d_1} I_{c_2}\rangle = \frac{1}{4}\left[\alpha^4 + \alpha^2(1-\sin(\theta_1-\theta_2))\right]. \tag{9a}$$

The coincidence probabilities depend on the phase difference between the local oscillators $\theta_1 - \theta_2$, if this is set to $-\pi/2$, we get the minimum possible coincidence probability of $I_0^2 = \frac{1}{4}\alpha^4$ between detector pairs $(\hat{c}_1, \hat{c}_2)$ and $(\hat{d}_1, \hat{d}_2)$ and the maximum coincidence probability of $I_0(1+I_0)$ between pairs $(\hat{c}_1, \hat{d}_2)$ and $(\hat{d}_1, \hat{d}_2)$.

We will now contrast these purely quantum mechanical results with those expected from a naïve particle theory and a classical wave theory. From the naïve particle viewpoint the enhancement of the singles count probability over that with vacuum inputs I_0 is easily understood. On the other hand, a coincidence between two detectors is expected to be a rare event since there is only one incident photon, and a coincidence can only occur if an additional photon is generated by the (weak) oscillator of the homodyne detector which the photon does *not* reach (cf. the *Appendix A*). Since these two photons are detected at two spatially separated detectors and have apparently arisen from different sources, we would not expect any correlation between the paths of these photons within each homodyne detector. In fact this correlation is so great that for the choice of phases given above, no additional coincidences (above the vacuum level) occur for particular detector pairs, whereas there is a relatively large coincidence probability (proportional to the local oscillator intensity) for the other pairs.

A classical analogue to the single photon input is a wave of low amplitude and unspecified phase. We may formally obtain the results for the classical wave theory from the quantum mechanical calculation by substituting the wave amplitude $\beta e^{\pm i\phi}$ for $\hat{b}$ and $\hat{b}^\dagger$ respectively, and averaging over the random phase ϕ. It is easy to check that the predicted average intensities and intensity correlations are given by

$$\langle I_{c_j,d_k}\rangle = I_0 + \frac{1}{4}\beta^2 \stackrel{\text{def}}{=} I_{0,class}, \tag{10}$$

$$\langle I_{c_1} I_{c_2}\rangle = \langle I_{d_1} I_{d_2}\rangle = I_{0,class}^2 + \frac{1}{4}\sin(\theta_1-\theta_2)\alpha^2\beta^2, \tag{10a}$$

$$\langle I_{c_1} I_{d_2}\rangle = \langle I_{d_1} I_{c_2}\rangle = I_{0,class}^2 - \frac{1}{4}\sin(\theta_1-\theta_2)\alpha^2\beta^2. \tag{10b}$$

If we consider the coincidence probability as a function of $\theta_1 - \theta_2$, we see that they vary between $I_{0,class}$ to $I_{0,class} + \frac{1}{2}\alpha^2\beta^2$. This corresponds to a "visibility" of

$$\mathcal{V}_{class} = \frac{\rho}{(\rho+\frac{1}{2})^2}, \tag{11}$$

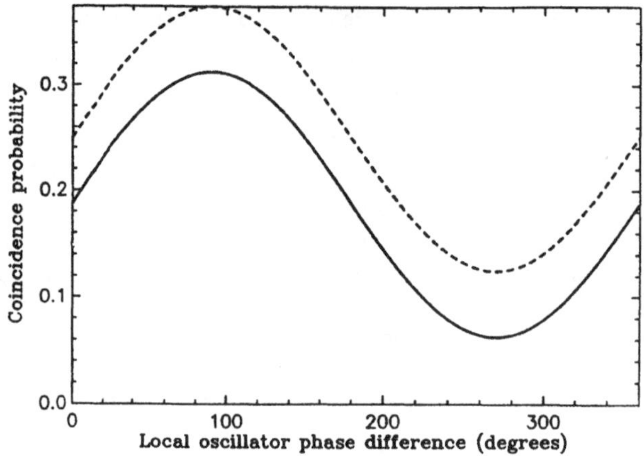

Fig. 3. Coincidence probabilities for quantum-mechanical model (solid line) and classical wave model (dashed line) for $\beta = 1$.

where $\rho = (\alpha/\beta)^2$. The visibility attains a maximum value of $\frac{1}{2}$ when $\rho = \frac{1}{2}$. By contrast the visibility as calculated from the quantum mechanical result is

$$\mathcal{V}_{qm} = \frac{1}{1+\alpha^2}. \tag{12}$$

This can be made arbitrarily close to unity by choosing a sufficiently small value of α.

Fig. 3 shows the coincidence probabilities $\langle I_{c_1} I_{c_2} \rangle = \langle I_{d_1} I_{d_2} \rangle$ as a function of the local oscillator phase difference for the quantum mechanical and classical results with $\beta = 1$ and $\alpha = 1/\sqrt{2}$. This gives the same singles count probability of $\frac{1}{2}$ in each detector, and the local oscillator amplitudes are optimized for maximum visibility in the classical result.

However the quantum mechanical visibility is considerably larger than that expected classically. This is clearly seen in Fig. 4 where the visibility $\mathcal{V}$ is plotted as a function of the coherent local oscillator amplitude α for the quantum mechanical single photon state and the classical wave model with $\alpha = 1$.

We thus see that by measuring the coincidence probability in a pair of detectors, it is possible to distinguish between the classical and quantum mechanical models. In a less idealized calculation detector efficiencies less than unity must be taken into account. This means that certain coincidences will be missed, but the *ratio* of minimum to maximum coincidence rates as the relative phase of the local oscillator is varied remains unaffected, provided that spurious coincidences due to the dark count rates of the detectors are removed. Even without compensating for the dark count rates, it is possible to test the quantum mechanical prediction that for the correct choice of local oscillator phases, there is no increase in the coincidence count in one of the detector pairs when the one photon input is applied.

So far we have not shown that Bell's inequalities may be violated. This is, however, necessary to rigorously rule out classical explanations for the quantum mechanical result. The use of phase-sensitive detectors for showing violations of Bell's

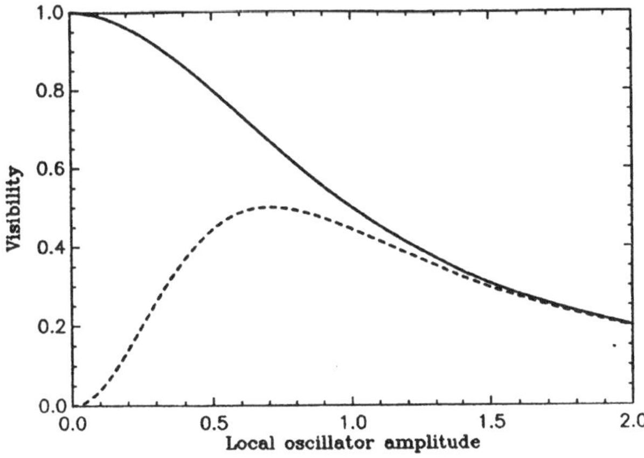

Fig. 4. Variation of visibility with local oscillator amplitude for quantum-mechanical model (solid line) and classical wave model (dashed line) for $\beta = 1$.

inequalities has previously been discussed by P.Grangier *et al* [13] and more recently by S.M.Tan *et al* [16]. An intensity correlation coefficient is used which involves all four photodetectors

$$E(\theta_1, \theta_2) = \frac{\langle (I_{d_1} - I_{c_1})(I_{d_2} - I_{c_2}) \rangle}{\langle (I_{d_1} + I_{c_1})(I_{d_2} + I_{c_2}) \rangle}. \tag{13}$$

Evaluating this in terms of the statistics of the input mode $\hat{u}$, where $\hat{v}$ is the vacuum yields

$$E(\theta_1, \theta_2) = -\frac{\alpha^2 \left[\langle \hat{u}^\dagger \hat{u} \rangle \sin(\theta_2 - \theta_1) + |\langle \hat{u}^2 \rangle| \sin(\theta_2 + \theta_1 - \zeta) \right]}{\alpha^4 + \langle \hat{u}^\dagger \hat{u} \rangle \alpha^2 + \frac{1}{4} \langle \hat{u}^{\dagger 2} \hat{u}^2 \rangle}, \tag{14}$$

where $\langle \hat{u}^2 \rangle = R\, e^{i\zeta}$. When a single photon input is used for $\hat{u}$, this reduces to

$$E(\theta_1, \theta_2) = \mathcal{V}_{qm}(\alpha) \sin(\theta_1 - \theta_2), \tag{14a}$$

where $\mathcal{V}_{qm}(\alpha)$ is the visibility (as defined in Eq.(12)). If $\mathcal{V}$ is greater than $1/\sqrt{2}$, it is well-known that this functional form for the correlation allows a violation of Bell's inequalities. This is clearly possible if α is made sufficiently small. More over it has been shown[16] that such a violation of Bell's inequalities is *not* possible if $\hat{u}$ is in a *coherent* state, no matter how small the input amplitude may be.

3. QND MEASUREMENTS VIA SECOND HARMONIC GENERATION

3.1. *The general QND measurement*

A measurement carried out on an observable of a quantum mechanical system influences the state of the system usually in such a way that any subsequent measurement on the same observable will yield a different result. In other words the measurement alters the previous state of the system therefore nearly no information (except for that it is not in the same state as before) is obtained about the state after the measurement. One may also say that noise is introduced by the measurement process. Obviously this is a rather inconvenient fact, especially if one thinks of tapping information (measuring the amplitude or intensity of an electromagnetic field) from a network and thereby destroying the information. The fact of noise introduction through measurements is inevitable, but one may think of more clever ways of measuring. If the measurement is carried out not on the signal observable itself but on one of the quadratures of an auxiliary observable which is highly correlated with the signal observable most of the noise can be fed into this auxiliary observable while the signal remains nearly unperturbed. The class of measurement schemes[17] for which the "back action" noise arising from the measurement may be evaded is called "*quantum nondemolition* (QND)" schemes. The key feature of such a nondemolition measurement is that it must be possible to assemble a sequence of precise measurements of an observable of the system such that that the result of each measurement is completely predictable from the result of the preceding measurement.

A number of possible QND schemes in optics employing a probe beam to measure the properties of a signal have been suggested[18,19,20,21]. Milburn, Lane and Walls[22] have proposed QND schemes using parametric amplifiers and frequency converters. An exact QND scheme balancing a parametric mixing interaction with a mode conversion interaction has been analysed by Hillery and Scully[23], and by Yurke[24]. These schemes involved a measurement of the quadrature phase of the optical signal field. A very recent proposal of a QND sceme using a ring laser cavity containing a near resonant $\chi^{(3)}$ Raman active medium comes from M. J. Holland[3]. Experimental demonstrations have been given by Levenson *et al* using four-wave mixing in optical fibres[25], and by La Porta *et al* [26] employing non-linear mixing in a $\chi^{(2)}$ medium.

3.2. *Theoretical criteria for QND measurements*

In the QND schemes we are interested in one of the quadratures of the output probe is strongly correlated with one quadrature of the signal input, output or both of them. In a *good* QND scheme it must be possible to obtain information on the signal by channeling all the noise introduced by the measurement into the variable conjugate to the signal quadrature of interest. Since there are good and not so good QND schemes one has to find criteria to assess the properties of the individual schemes. Suppose a QND device as depicted in Fig. 5.

A set of three criteria were found sufficient to characterize any such device[3]:

1. *How good is the scheme as a measurement device?* A measure of this is given by the correlation function

$$C^2_{X_s^{in} X_p^{out}} = \frac{\left|\langle X_s^{in} X_p^{out}\rangle - \langle X_s^{in}\rangle\langle X_p^{out}\rangle\right|^2}{\mathrm{Var}(X_s^{in})\mathrm{Var}(X_p^{out})}. \tag{15}$$

   In the case of a perfect measurement device both amplitude quadratures are proportional to each other by the QND gain and the correlation coefficient is one.
2. *How much does the scheme degrade the signal field?* Or, how strong are signal

input and output correlated. This may be measured by the correlation function

$$C^2_{X_s^{in} X_s^{out}} = \frac{\left|\langle X_s^{in} X_s^{out} \rangle - \langle X_s^{in} \rangle \langle X_s^{out} \rangle \right|^2}{\mathrm{Var}(X_s^{in}) \mathrm{Var}(X_s^{out})}. \tag{16}$$

This correlation coefficient

(i) allows for attenuation of the signal field

$$X_s^{in} = \epsilon X_s^{out}, \qquad \epsilon < 1, \tag{16a}$$

which still gives perfect correlation,

and

(ii) assesses the ability of the scheme to isolate quantum noise introduced by the measurement process from the observable.

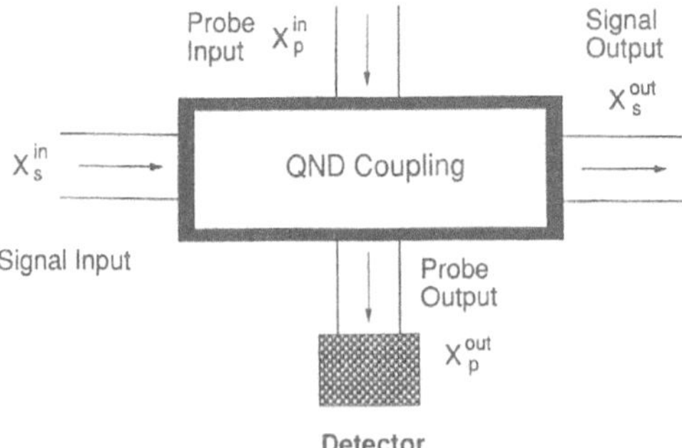

Fig. 5. A schematic QND device for a correlation between the amplitude quadratures X_s, X_p of signal and probe beam[27].

3. *How good is the scheme as a state preparation device?* This may be characterized by the variance of X_s^{out} given a measured value of X_p^{out}, i.e. the conditional variance of the signal output

$$\mathrm{Var}(X_s^{out} \mid X_p^{out}) = \mathrm{Var}(X_s^{out})(1 - C^2_{X_s^{out} X_p^{out}}). \tag{17}$$

This special form for the conditional variance is valid only under the assumption of Gaussian statistics for the out fields[3]. This requirement, however, is always satified by a linearization of the non-linear operator equations, which describe the time evolution of the signal and probe fields.

3.3. *QND measurements via second harmonic generation (SHG)*

3.3.1. *Introductory remarks.* This new scheme proposed by M. Dance *et al* [2] involves determing the amplitude quadrature of a signal by making a measurement on

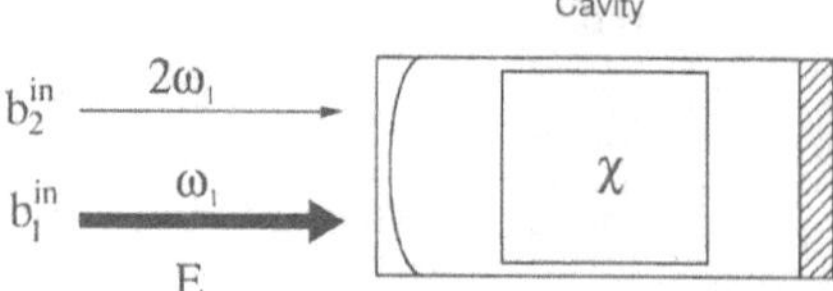

Fig. 6. Intracavity second harmonic generation using a single ended cavity .

its second harmonic. It can be shown that while the signal is slightly degraded by the measurement a high correlation may be obtained between the measured probe amplitude and the signal amplitude. The scheme shows much promise for experimental realisation since high efficiency intracavity second harmonic generation has been demonstrated[28]. Squeezing has been predicted[29] and observed in intracavity SHG[30,31].

Consider a system (as depicted in Fig. 6) in which two cavity modes interact via a second-harmonic-generating crystal. By the subscript 1 the fundamental mode with frequency ω_1 is labelled, the subscript 2 corresponds to the exact second harmonic at frequency $\omega_2 = 2\omega_1$.

3.3.2. *Basic equations.* The Hamiltonian for a single ended cavity consists of the following three parts

$$\mathcal{H} = \mathcal{H}_{Sys} + \mathcal{H}_{Int} + \mathcal{H}_{Res}\,, \tag{18}$$

$$\mathcal{H}_{Sys} = \hbar\{\omega_1 a_1^\dagger a_1 + \omega_2 a_2^\dagger a_2 + \mathrm{i}\frac{\chi}{2}(a_2^\dagger a_1^2 - a_1^{\dagger 2} a_2)\}\,, \tag{18a}$$

$$\mathcal{H}_{Res} = \hbar \sum_{j=1}^{2} \int_{\omega_j - \Delta_j}^{\omega_j + \Delta_j} \mathrm{d}\nu\, \nu\, b_1^\dagger(\nu) b_1(\nu)\,, \tag{18b}$$

$$\mathcal{H}_{Int} = \mathrm{i}\hbar \sum_{j=1}^{2} \sqrt{\frac{\gamma_j}{\pi}} \int_{\omega_j - \Delta_j}^{\omega_j + \Delta_j} \mathrm{d}\nu\, \nu \{b_1^\dagger(\nu) a_j - h.c.\}. \tag{18c}$$

All operators are assumed to obey boson commutation relations. The spectral ranges Δ_j are large compared with the cavity linewidths γ_k, but small on an optical frequency scale, i.e. $\Delta_j/\omega_1 \ll 1$. Using the input-output formalism developed by Collett and Gardiner we end up with the non-linear quantum Langevin equations for the slowly varying operators $\bar{a}_j(t) = \mathrm{e}^{i\omega_j t} a_j(t)$. We obtain

$$\frac{\mathrm{d}\bar{a}_1}{\mathrm{d}t} = -\gamma_1 \bar{a}_1 - \chi \bar{a}_1^\dagger \bar{a}_2 + \sqrt{2\gamma_1}\bar{b}_1^{in}, \tag{19}$$

$$\frac{\mathrm{d}\bar{a}_2}{\mathrm{d}t} = -\gamma_2 \bar{a}_2 + \frac{1}{2}\chi \bar{a}_1^2 + \sqrt{2\gamma_2}\bar{b}_2^{in}. \tag{20}$$

The state of the input signal is assumed to be a coherent state with real amplitude E for frequency $\nu = \omega_1$ and a vacuum state for the other modes within the spectral range of b_1. The state of the input probe is the vacuum state for all modes. The Eqs.(19),(20) may be rewritten in terms of phase quadrature operators $X_j = \bar{a}_j + \bar{a}_j^\dagger$ and $Y_j = -i(\bar{a}_j - \bar{a}_j^\dagger)$ and similar expressions for the associated sl in and *out*-fields. A solution of the Eqs. (19),(20) may be obtained by linearization around the steady-state semiclassical mean values of the quadratures, denoted by $X_{j,s}, Y_{j,s}$. They are found to be $Y_{j,s} = 0$, $X_{2,s} = \frac{\chi}{4\gamma_2} X_{1,s}^2$, where $X_{1,s}$ is a real solution of the polynomial

$$\underbrace{[\frac{(\chi X_{1,s})^2}{8\gamma_1\gamma_2}}_{\stackrel{\text{def}}{=}\beta} + 1] X_{1,s} - 2\sqrt{\frac{2}{\gamma_1}} E = 0. \tag{21}$$

The scaled quantity β provides a measure for the strength of the non-linear interaction and is therefore called non-linearity parameter. A phase instability occurs for values of $\frac{\chi}{2\gamma_1} X_{2,s}$ larger than the threshold value, given by $\beta_{th} = 1 + \frac{\gamma_2}{\gamma_1}$. At the threshold we encounter an oscillatory instability, with the initial frequency of oscillation being the critical frequency ω_c, where $\omega_c^2 = \gamma_2(2\gamma_1 + \gamma_2)$. The first order correction terms to the semiclassical solutions are the fluctuation operators $X_{j,f} = X_j - X_{j,s}$. Neglecting terms that are quadratic in the fluctuation operators, we find that amplitude and phase fluctuations are *decoupled*. We finally arrive at

$$\frac{\mathrm{d}}{\mathrm{d}t} \begin{bmatrix} X_{1,f} \\ X_{2,f} \\ Y_{1,f} \\ Y_{2,f} \end{bmatrix} = \begin{bmatrix} -(\gamma_1 + \frac{\chi}{2} X_{2,s}) & -\frac{\chi}{2} X_{1,s} & 0 & 0 \\ \frac{\chi}{2} X_{1,s} & -\gamma_2 & 0 & 0 \\ 0 & 0 & (\frac{\chi}{2} X_{2,s} - \gamma_1) & -\frac{\chi}{2} X_{1,s} \\ 0 & 0 & \frac{\chi}{2} X_{1,s} & -\gamma_2 \end{bmatrix} \begin{bmatrix} \sqrt{2\gamma_1}\, X_{1,f}^{in} \\ \sqrt{2\gamma_2}\, X_{2,f}^{in} \\ \sqrt{2\gamma_1}\, Y_{1,f}^{in} \\ \sqrt{2\gamma_2}\, Y_{2,f}^{in} \end{bmatrix}. \tag{22}$$

As a next step one has to Fourier-transform the above equations for the amplitude quadratures $X_{j,f}$ and express the *out*-fields $X_{j,f}^{out}(\omega)$ as functions of the *in*-fields $X_{j,f}^{in}(\omega)$ and the frequency ω by making use of the boundary conditions (cf. *Appendix B*). This will finally enable us to calculate the various correlation coefficients listed in Sec. 3.2 in order to assess the quality of the new scheme.

3.3.3. *The correlation coefficients.* In the following we will use that $\langle X_{k,f}^{in}(\omega) X_{j,f}^{in}(\nu)\rangle = \delta_{k,j}\delta(\omega + \nu)$ to calculate the correlations for the individual frequency components. (Cf. *Appendix B* for a derivation of the coefficients below.) The coefficient determing the quality of the scheme as a measurement device (cf. Eq. (15)) is found to be

$$C^2_{X_1^{in} X_2^{out}}(\omega) = \frac{8\beta}{8\beta + (\beta - 1 - \frac{\omega^2}{\gamma_1\gamma_2})^2 + (\frac{\omega}{\gamma_1\gamma_2})^2(\gamma_2 - \gamma_1 - \beta\gamma_1)^2}. \tag{23}$$

This correlation function is plotted as a function of ω for three different values of the non-linearity parameter β in Fig. 7. The measurement correlation always peaks in the region of the critical frequency of the system. It is found to be extremely good very close to ω_c if the signal is dampened much stronger than the second harmonic, $\frac{\gamma_1}{\gamma_2} \ll 1$. But the behaviour is sensitive to the proximity of β to its threshold value and a linearized analysis must fail very close to critical operation. However if $\frac{\gamma_1}{\gamma_2}$ is slightly increased (to about 0.01), so that more of the fundamental gets into the cavity, we obtain the best correlation when operating at β values some way below the threhhold value. In fact, perfect correlation can be achieved for $\beta = \beta_{th} - 2$ (if $\beta_{th} > 2$) at $\omega^2 = \gamma_2(\gamma_2 - 2\gamma_1)$; while, as can be seen from Fig. 7, good results can be achieved over a wide range of β and ω. This obviously has practical advantages for stability of operation. For larger values of $\frac{\gamma_1}{\gamma_2}$ the peak moves towards the origin.

The correlation coefficint describing signal degradation is given by

$$C^2_{X_1^{in}X_1^{out}} = 1 - \frac{8\beta}{8\beta + (3\beta - 1 - \frac{\omega^2}{\gamma_1\gamma_2})^2 + (\frac{\omega}{\gamma_1\gamma_2})^2(\gamma_1 - \gamma_2 - \beta\gamma_1)^2}. \tag{24}$$

The analytical expression for this correlation coefficient reveals that if $\frac{\gamma_1}{\gamma_2}$ is reasonably small (e.g. 0.01) and hence β_{th} is much larger than one, we may obtain almost perfect correlation for all values of ω and a sufficiently large range of values for $\beta < \beta_{th}$.

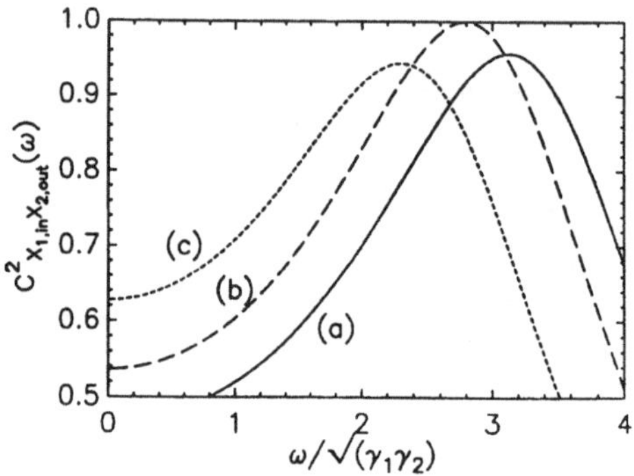

Fig. 7. Spectra of the measurement correlation $C^2_{X_1^{in}X_2^{out}}(\omega)$ for $\gamma_1/\gamma_2 = 0.1$. The three curves a (solid),b (dashed) and c (dotted) correspond to $\beta = \beta_{th}$, $\beta = 0.8\beta_{th}$ and $\beta = 0.6\beta_{th}$.

For $\gamma_1/\gamma_2 = 0.1$, this spectrum is shown in Fig. 8 for three different values of β (cf. Fig. 7). It is seen that the correlation between X_1^{in} and X_1^{out} is reduced when more of the input signal is allowed to pass into the cavity, as one would expect physically. Also as expected, the correlation goes to one as $|\omega|$ tends to infinity: far from the cavity resonance most of the signal input is reflected off the cavity front mirror, and has no chance of being depleted by the non-linear interaction.

To assess the state preparation ability of the system, we use the residual variance given by Eq. (17). (As noted, by using a linearised analysis we automatically satisfy the assumption of Gaussian statistics.) We find

$$\mathrm{Var}(X_1^{out}|X_2^{out},\omega) = 1 - \frac{4\beta(1 + \frac{\omega^2}{\gamma_2{}^2})}{(\beta + 1 - \frac{\omega^2}{\gamma_1\gamma_2})^2 + (\frac{\omega}{\gamma_1\gamma_2})^2(\gamma_1 + \gamma_2 - \beta\gamma_1)^2 + 4\beta(1 + \frac{\omega^2}{\gamma_2{}^2})}. \tag{25}$$

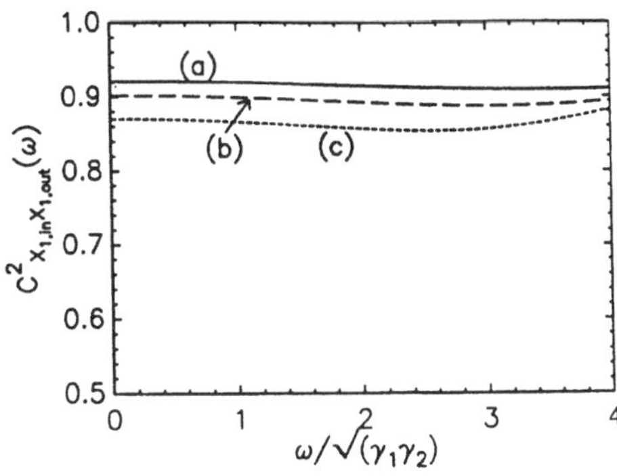

Fig. 8. Spectra of the correlation coefficient for signal degradation in the cavity $C^2_{X_1^{in}X_1^{out}}$ with $\gamma_1//\gamma_2 = 0.1$, plotted for the same values of β as in Fig. 7.

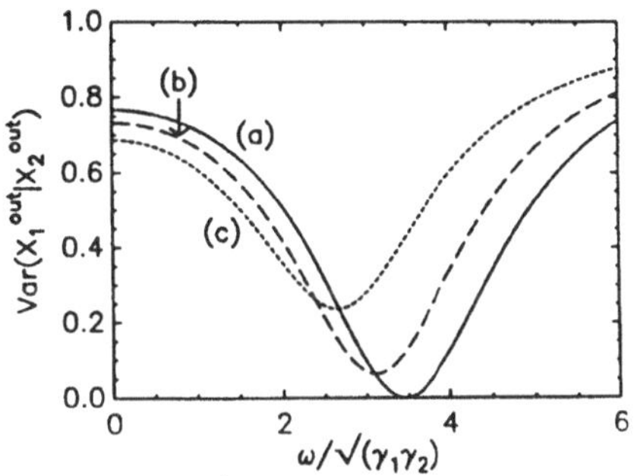

Fig. 9. Spectra of conditional variance $\text{Var}(X_1^{out}(\omega) \mid X_2^{out}(\omega))$. Curves correspond to the same parameters as in Figs. 7,8.

The behaviour of the conditional variance is shown in Fig. 9, for $\gamma_1/\gamma_2 = 0.1$ and the same values of β as in Figs. 7,8.

In this case the optimum behaviour is achieved as one approaches the critical point. However, the variance can be made close to zero over a range of sideband frequencies depending on the values of the input amplitude and nonlinearity parameter χ. There is a great deal of flexibility in the choice of parameters in order to obtain good state preparation. It should be pointed out that good QND measurement correlation and good state preparation occur over similar ranges of parameters and frequencies (compare Figs. 7 and 9). This system appears to offer several advantages for an experimental investigation of QND measurements. In particular, the low frequency excess noise which hinders measurements made at finite frequency is *avoided* in this case, since the maximum correlation occurs at a finite frequency. This would have direct application for the transmission of information by encoding the signal as a modulation, at this finite frequency, of a carrier wave.

ACKNOWLEDGMENTS

This work was supported by the University of Auckland Research Committee. One of us K.M.G. wants to express his special thanks to Prof. Tombesi and Prof. Walls, whose support made the attendance of this meeting possible.

APPENDIX A

As has already been indicated in Sec. 2, the socalled coincidence probabilities are no probabilities in the true sense but expectation values. In the general case of local oscillators it is not possible to give an analytical expression for the probabilities of measuring exactly N_j photons in detector j, according to some numbering scheme. Fortunately we are only interested in the case of weak local oscillators, therefore we can make the approximation

$$|\alpha_k\rangle \approx e^{-|\alpha_k|^2/2}\left(|0\rangle + \alpha_k\,|1\rangle + \frac{\alpha_k^2}{\sqrt{2}}\,|2\rangle\right). \tag{A1}$$

According to the standard rules of quantum mechanics, the probability of finding an observable in a special state is defined as expectation value of the projection operator on that state.

If we want to calculate the probability of finding detector mode $\hat{c}_1$ in a N-photon state we have to calculate the infinitely-fold degenerate eigenstates $|\,N\rangle_\mu$, where μ is the degeneracy parameter, of $\hat{c}_1^\dagger\hat{c}_1$ and subsequently evaluate the trace over a three mode Fock-space $\mathcal{F}_3$, that is

$$\text{prob}(N \text{ photons in mode } \hat{c}_1) = \text{Tr}_{\mathcal{F}_3}\{\sum_{\mu=0}^{\infty} |N\rangle_\mu \otimes {}_\mu\langle N|\ \rho^{in}\}, \tag{A2}$$

where ρ^{in} denotes the density operator on the Fock-space defined by $\{\hat{a}_1, \hat{b}_1, \hat{b}_2\}$ for a coherent and an entangled input state (cf. Eq. (8)). We find up to $o(\alpha^6)$

$$\text{prob}(0) = \frac{3}{4} - \frac{\alpha^2}{4} + \frac{\alpha^4}{32}, \tag{A3}$$

$$\text{prob}(1) = \frac{1}{4} + \frac{\alpha^4}{32}, \tag{A4}$$

$$\text{prob}(2) = \frac{\alpha^2}{4} - \frac{5}{32}\alpha^4, \tag{A6}$$

$$\text{prob}(3) = \frac{3}{32}\alpha^4. \tag{A7}$$

In a similar way one has to proceed in order to calculate the probability of having N_1 photons in detector mode $\hat{c}_1$ and N_2 photons in detector mode $\hat{c}_2$. The only difference is that now all calculations have to be carried out in a 4-mode Fock-space $\mathcal{F}_4$. We obtain the following results for weak local oscillators with phases θ_1, θ_2 and amplitudes of comparable size α_1 and α_2, respectively

$$\text{prob}(0,0) = \frac{1}{2}\{1 - \frac{\alpha_1^2+\alpha_2^2}{4} + \frac{1}{4}\alpha_1\alpha_2 \sin\Theta\,(2-a_1^2-\alpha_2^2)\}, \tag{A8}$$

$$\text{prob}(1,0) = \frac{1}{4}\{1 - \frac{1}{2}(\alpha_1^2+\alpha_2^2) + \frac{1}{8}(3\alpha_1^4+\alpha_2^4) + \frac{1}{2}\alpha_1^2\alpha_2^2 - \alpha_1\alpha_2 \sin\Theta\,(1-\alpha_1^2-\frac{1}{2}\alpha_2^2)\}, \tag{A9}$$

$$\text{prob}(1,1) = \frac{1}{4}\{\frac{1}{2}(\alpha_1^2+\alpha_2^2) - \frac{1}{4}(\alpha_1^4+\alpha_2^4) - \alpha_1^2\alpha_2^2 + \alpha_1\alpha_2 \sin\Theta\,(1-\alpha_1^2-\alpha_2^2)\}, \tag{A10}$$

$$\text{prob}(2,0) = \frac{\alpha_1^2}{4}\{1 - \frac{\alpha_2^2}{2} - \frac{3}{4}\alpha_1^2 - \frac{1}{2}\alpha_1\alpha_2 \sin\Theta\}, \tag{A11}$$

$$\text{prob}(2,1) = \frac{\alpha_1^2}{8}\{\alpha_2^2 + \alpha_1\alpha_2 \sin\Theta + \frac{\alpha_1^2}{4}\}, \tag{A12}$$

$$\text{prob}(3,0) = \frac{3}{32}\alpha_1^4 \quad . \tag{A13}$$

Θ denotes the phase difference $\theta_1 - \theta_2$. One immediately realizes that the "real" probabilities for measuring photons in two detectors also show the characteristic sinusoidal interference terms. The terms given in Sec. 2 are related to them in the following way

$$\langle I_{\hat{c}_1} I_{\hat{c}_2} \rangle = \sum_{N_1=1}^{\infty} \sum_{N_2=1}^{\infty} N_1 N_2 \, \text{prob}(N_1, N_2). \tag{A14}$$

In the case of a weak oscillator this is almost equal to the probability of detecting exactly two photons, one in each detector.

APPENDIX B

After having Fourier-transformed the time-evolution equations Eq. (22) we have to use the boundary conditions at the front mirror

$$X_{j,f}^{out}(\omega) - X_{j,f}^{in}(\omega) = \sqrt{2\gamma_j}\, X_{j,f}(\omega), \tag{B1}$$

which enable us to express $X_{j,f}$ in terms of the *in* and *out*-fields. The subsequent solution for the out-fields yields

$$X_{1,f}^{out}(\omega) = -\frac{\sqrt{8\beta}\, X_{2,f}^{in}(\omega) + [3\beta - 1 - \frac{\omega^2}{\gamma_1\gamma_2} + \frac{i\omega}{\gamma_1\gamma_2}(\gamma_1-\gamma_2-\beta\gamma_1)]X_{1,f}^{in}(\omega)}{1 + 3\beta - \frac{\omega^2}{\gamma_1\gamma_2} - \frac{i\omega}{\gamma_1\gamma_2}(\gamma_1+\gamma_2+\beta\gamma_1)}, \tag{B2}$$

$$X_{2,f}^{out}(\omega) = -\frac{-\sqrt{8\beta}\, X_{1,f}^{in}(\omega) + [\beta - 1 - \frac{\omega^2}{\gamma_1\gamma_2} + \frac{i\omega}{\gamma_1\gamma_2}(\gamma_2-\gamma_1-\beta\gamma_1)]X_{2,f}^{in}(\omega)}{1 + 3\beta - \frac{\omega^2}{\gamma_1\gamma_2} - \frac{i\omega}{\gamma_1\gamma_2}(\gamma_1+\gamma_2+\beta\gamma_1)} \tag{B3}$$

Note that $\omega = 0$ corresponds to the absolute frequencies ω_1 for the fundamental and $2\omega_1$ for the second harmonic. Since we are not considering any detunings, the correlation coefficients are symmetric about $\omega = 0$.

The correlation coefficients $C_{X,Y}^2$ are obtained by calculating the Fourier-transforms of the correlations $\langle XY \rangle$ as well as of the variances $\mathrm{Var}(X)$ and $\mathrm{Var}(Y)$ and inserting them into the defining Eqs. (15),(16).

REFERENCES

1. S. M. Tan, D. F. Walls and M. J. Collett, *Phys. Rev. Lett.* **66(3)**, 252 (1991).
2. M. Dance, M. J. Collett and D. F.Walls, *Phys. Rev. Lett.* **66**, 1115 (1991).
3. M. J. Holland, *M.Sc. Thesis*, University of Auckland 1990.
4. A. Einstein, B. Podolsky and N. Rosen, *Phys. Rev.* **47**, 777 (1935).
5. J. S. Bell, in: *Foundations of Quantum Mechanics*, edited by B. d'Espagnat (Academic, New York, 1971), p.171.
6. A. Aspect, P. Grangier and G. Roger, *Phys. Rev. Lett.* **47**, 460 (1981), and **49**, 91 (1982).
7. A. Aspect, J. Dalibard and G. Roger, *Phys. Rev. Lett.* **49**, 1804 (1982).
8. Z. Y. Ou and L. Mandel, *Phys. Rev. Lett.* **61**, 50 (1988).
9. Y. H. Shih and C. Alley, *Phys. Rev. Lett.* **61**, 2921 (1988).
10. J. G. Rarity and P. R. Tapster, *Phys. Rev. Lett.* **64**, 2495 (1990).
11. P. Grangier,G. Roger and A. Aspect, *Europhys. Lett.* **1**, 173 (1986).
12. M. D. Reid and P. D. Drummond, *Phys. Rev. Lett.* **60**, 2447 (1988).
13. P. Grangier, M. J. Potasek and B. Yurke, *Phys. Rev. A* **38(6)**, 3132 (1988).
14. M. A. Horne, A. Shimony and A. Zeilinger, *Phys. Rev. Lett.* **62**, 2209 (1989).
15. M. D. Reid and D. F. Walls, *Phys. Rev. A* **34**, 1260 (1986).
16. S. M. Tan, M. J. Holland and D. F. Walls, *Optics Communications* **77(4)**, 285 (1990).
17. C. M. Caves, K. S. Thorne, R. W. P. Drever, V. D. Sandberg and M. Zimmerman, *Rev. Mod. Phys.* **52**, 341 (1980).
18. G. J. Milburn and D. F. Walls, *Phys. Rev. A* **28**, 2065 (1983).
19. N. Imoto, H. A. Haus and Y. Yamamoto, *Phys. Rev. A* **32**, 2287 (1985).
20. R. M. Shelby and M. D. Levenson, *Optics Communications* **64**, 553 (1987).
21. P. Alsing, G. J. Milburn and D. F.Walls, *Phys. Rev. A* **37**, 2970 (1988).
22. G. J. Milburn, A. S. Lane and D. F. Walls, *Phys. Rev. A* **27**, 2804 (1983).
23. M. Hillery and M. O. Scully, *Phys. Rev. D* **25**, 3137 (1982).
24. B. Yurke, *J. Opt. Soc. Am. B* **2**, 732 (1985).
25. M. D. Levenson, R. M. Shelby, M. Reid and D. F. Walls, *Phys. Rev. Lett.* **57**, 2473 (1986).
26. A. La Porta, R. E. Slusher and B. Yurke, *Phys. Rev. Lett.* **62**, 28 (1989).
27. Obviously any other combination of signal and probe quadratures is also possible.
28. W. J. K. Kozlovsky, C. D. Nabor and R. L. Byer, *IEEE J. Q. E.* **24**, 913 (1988).
29. M. J. Collett and D. F. Walls, *Phys. Rev. A* **32**, 2887 (1985).
30. S. F. Pereira, Min Xiao, H. J. Kimble and J. L. Hall, *Phys. Rev. A* **38**, 4931 (1988).
31. A. Sizmann, R. J. Horowicz, E. Wagner and G. Leuchs, contribution to: *International Quantum Electronics Conference*, Anaheim, Calif., May 1990, unpublished.
32. P. G. Kwiat, W.A. Vareka, C. K. Hong, H. Nathel and R. Y. Chiao, *Phys. Rev. A* **41**, 2910 (1990).

CAVITY Q.E.D.: FUNDAMENTAL THEORY OF THE MICROMASER AND MEASUREMENTS OF ITS CAVITY FIELD

R.K. Bullough, N.M. Bogoliubov†, N. Nayak and B.V. Thompson

Department of Mathematics, UMIST, P.O. Box 88
Manchester M60 1QD, UK

1. INTRODUCTION

This paper reports a new theoretical investigation, numerically based, of the action of the one-atom one-mode micromaser. In the recent experiments[1,2] on the Garching micromaser[1-4] there is at most one ^{85}Rb atom in the maser cavity during an atomic transit time $t_{int} \sim 35\mu$ sec. But atoms enter with a repetition rate T_p (say) $\sim 7000\mu$ sec, so that the cavity is empty of atoms for more than 99% of the time. The main result of this paper is that if the empty-of-atoms time $T_p - t_{int}$ is reduced towards the values of t_{int} itself, the system acts more like a maser with infinite cavity Q_o *even at the now realised* Q *and temperature* T[1,2] $Q = 3 \times 10^{10}$, $T = 0.5^oK$. Thus in distinction to previous discussions, e.g.[5], we suggest that it may be possible to create Fock states at this Q and T. However, as T_p is reduced, the theoretical problem becomes complicated by the possibility of more than one atom being in the cavity at the same time.

In a second analysis the paper takes up the recent idea[6] that the micromaser may be used to distinguish between quantum mechanical systems quantised by normal canonical quantisation as opposed to the quantisation imposed by the recently discovered 'quantum groups'[7-10]. The Jaynes-Cummings model[11] of one 2-level atom and one quantised cavity mode is fundamental to the theory of the micromaser[12-14]. We show how the model can be given a quantum group quantisation by using the q-deformed bosons ('q-bosons'[10]) recently described[15-17]. We show that q-dependent features derived from the q-bosons would appear in the quantum revivals (as already suggested[6]) and in the second and higher Rabi split energies. The vacuum field Rabi splitting is not changed. New measurements are needed to determine whether there is any *physics* in this form of quantum group quantisation.

The material is presented in the following way. We make a few remarks on the *measurement* of the cavity field in high-Q microwave cavities and on aspects of cavity q.e.d. in high and low-Q cavities in the §§2 and 3 following. In §4 we give the theory of the q-deformed Jaynes-Cummings model. In §§5-7 we show how the theory of the one-atom micromaser based on conventional quantum mechanics in cavities of finite

Quantum Measurements in Optics, Edited by P. Tombesi and D.F. Walls, Plenum Press, New York, 1992

$Q < \infty$ moves towards the *qualitatively* different $Q = \infty$ theory as T_p reduces in magnitude towards t_{int}.

2. CAVITY FIELD MEASUREMENT IN MICROWAVE CAVITIES

Evidently this topic fits in with that of the paper by R. Brecha[18] and is particularly appropriate for this ARW on 'Quantum measurements in optics'. With this in mind our paper was originally offered under the title 'Atomic statistics via photon statistics : measurement of the cavity field in the ^{85}Rb atom maser'. Unfortunately, at the time of writing, we are still in the process of including a *stochastic* input of the atoms to the maser cavity in our numerical calculations. In our programme[19-21] of simulating the micromaser[1-4] at Garching atoms have so far entered successively at regular intervals T_p. A stochastic input is essential if the state of the outcoming atoms is to be used as in References 1, 2, 18 to determine the state of the cavity field with which they have interacted. Fluctuations in the atomic input induce responses in the cavity field which react back on the atomic statistics changing that statistics in a significant and measurable way[1,2].

At the time of writing we are able to calculate the evolution of the total density matrix ρ for atom and field continuously coupled to a heat bath at temperature $T > 0$ as atoms successively enter. We find the field density matrix ρ_f as $\rho_f = Tr_a \rho$ and the atomic density matrix as $\rho_a = Tr_f \rho$; ρ_a is for the outcoming atoms and can be observed. We are thus able to correlate ρ_a, as calculated, against the state of the cavity field, which determines it, described by ρ_f. Atoms enter successively, and in the Fig. 1 shown, $t_{int} = T_p$ i.e. the empty-of-atoms cavity time

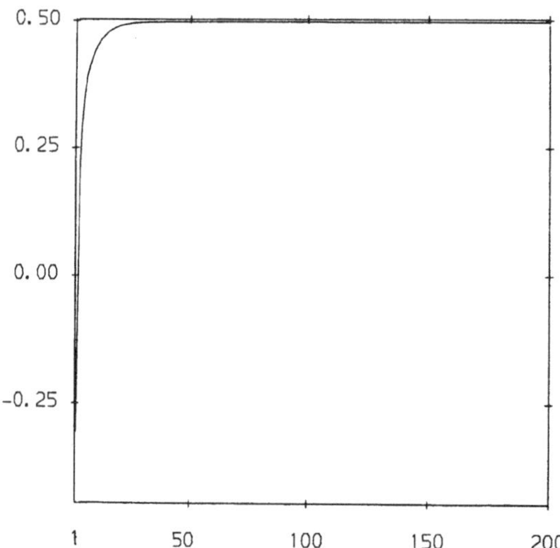

Fig. 1. Plot (for the trapping state $gt_{int} = \frac{1}{2}\pi$ - see text, §6) of the probability (minus $\frac{1}{2}$) of finding the outcoming atom in its upper state as a function of the number of atoms which have traversed the cavity.

is actually zero. The curve is certainly featureless, but it does show that atoms drop their photons into the cavity initially and that by the time some 40 atoms have entered they cease to do this. They then leave in the same upper state in which they entered; but they have made what is predominantly a 2π-rotation of the Bloch vector. This evolution of ρ_a corresponds to an evolution of ρ_f towards the state with Fock state $|3\rangle$, the 3-photon state: after the 40 atoms have entered this state has probability $P_3 = 0.988$ as explained (in §6) below. This figure is *not* a steady state value. We shall show that a steady-state situation is never reached which, because $T_p \approx t_{int}$, is very different from that envisaged in the previous theoretical work[5,12,13]. These theories assume $T_p >> t_{int}$ and beautifully predict the recent observations[1,2,18]. These are for $T_p \sim 200\ t_{int}$ and steady states are apparently achieved.

3. CAVITY QUANTUM ELECTRODYNAMICS

In this article we are primarily concerned to report a theory of the ^{85}Rb atom micromaser which both extends and significantly changes the earlier theories[12,13], and this theory is presented in §§ 5-7. We are primarily concerned there to discover optimum conditions for creating and observing Fock states $|n\rangle$ of the cavity field. This means we focus on the so-called 'trapping states'[5,11,14] and the means of reaching them in the micromaser. The papers[11,14] are concerned with cavities with $Q = \infty$: there are no finite temperature effects. But a comparison of the $Q = \infty$ theory[14] with the $Q < \infty$, $T > 0$ theories[12,13] shows immediately that the recent experiments[1,2] (where $Q = 3 \times 10^{10}$, $T = 0.5^oK$) involve theoretical descriptions *qualitatively* different from the $Q = \infty$ theory, as mentioned. This is despite the fact that at $T = 0.5^oK$, the Planck function $\bar{n} = (e^{\hbar\beta\omega} - 1)^{-1}$, $\beta^{-1} = k_BT$, for the black-body field in the cavity at the mode frequency $\omega = 2\pi \times 21.456$ GHz for the 63 $P_{3/2} \longrightarrow 61\ D_{5/2}$ transitions of ^{85}Rb used[1,2] is only $\bar{n} = 0.15$. This result indicates the importance of small amounts of black-body radiation in theoretical and experimental microwave cavity q.e.d. In this §3 we quickly summarise some of our results[19-28] in the microwave cavity q.e.d. of high *and* low-Q cavities with this feature in mind.

There is a wholly *qualitative* change in behaviour for one or more Rydberg atoms in microwave cavities at finite temperatures as the cavity Q changes from low-Q ($Q \sim 10^3$) through intermediate values[25-27] ($Q \sim 10^6$) to very high Q ($Q > 10^8$)[28]. At low-Q there are remarkable[22-24] collective phenomena for $N_A >> 1$ atoms in black-body radiation which have been observed[29]: the experiments[29] showed that although single 2-level atoms act like fermions, $N_A > 1$ such atoms form one giant collective quantum oscillator (one giant collective boson) as $N_A \longrightarrow \infty$. At $Q = 5 \times 10^3$ and $T = 300^oK$ ($\bar{n} = 47$) and $T = 900^oK$ ($\bar{n} = 137$) the mean number of 2-level atoms N_+ leaving the cavity in their upper states can be measured[29] and was found to become the *bose* occupation number $N_+ = \bar{n}$ as $N_A \longrightarrow \infty$. It is the *fermi* value $N_+ = \bar{n}\ [1 + 2\bar{n}]^{-1}$ for $N_A = 1$[22-24]. We have shown elsewhere[9] how this is an aspect of the 'fermi-bose equivalence' of the 'quantum integrable systems'. We say something about quantum integrable systems in §4 next. We show there how a giant q-deformed collective boson would be formed if the system were quantised by the quantum groups rather than canonically.

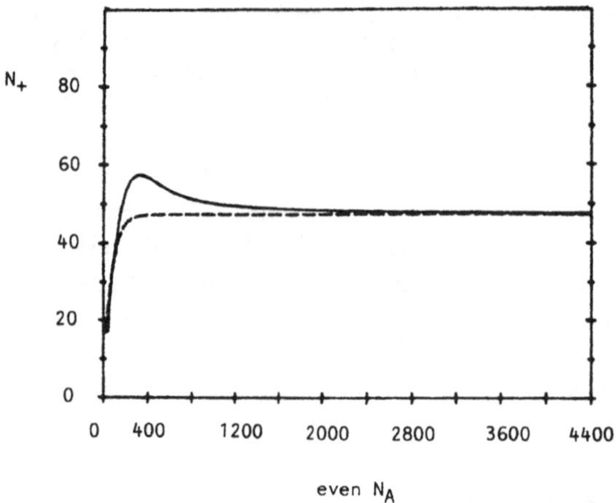

Fig. 2. Plot of N_+, as a function of N_A for a low-Q cavity with $N = \langle a^\dagger a\rangle = 47$: dashed line thermal case, hard line in a squeezed vacuum.

There is a similar prediction under normal quantum mechanics[30,31] for the situation in which a squeezed vacuum replaces the black-body field: the Fig. 2 (dashed line) shows N_+ in equilibrium as a function of the number of atoms in the cavity N_A; the bold line in Fig. 2 is for the squeezed vacuum [30,31] with mean photon number $N = \langle a^+a\rangle = 47$ and minimum uncertainty $|M|^2 = N(N+1)$ ($M = \langle a^+a^+\rangle$: the squeezed vacuum is broad band and M,N are evaluated at the resonant mode [30,31]). Again $N_+ \longrightarrow N$ as $N_A \longrightarrow \infty$, but the curve is no longer monotonically rising, and an experiment to observe this feature, as well as the $N_A \longrightarrow \infty$ limit $N_+ = N$, is still to be carried out. We recommend it to experimentalists with a good enough squeezed vacuum as part of an obvious programme to do for Rydberg atoms in squeezed vacua what has so successfully been done for the cavity q.e.d. of black-body radiation and the ordinary vacuum.

For single atoms we have computed *spectra* for $10^3 \le Q \le 10^8$. These include both resonance fluorescence and cavity field spectra[25-27] and spectral shapes depend critically on $\bar{n}$ and Q. At the larger Q value it is not possible to measure the fluorescence spectrum directly. However, S. Haroche and colleagues[32] have now measured the vacuum field Rabi splitting of the cavity field for the $39\ S_{1/2} \longrightarrow 39\ P_{3/2}$ transitions of ^{85}Rb at 68.5 GHz : $Q = 10^8$ and $\bar{n} = 0.17$. The Fig. 3 shows our predictions for the spectrum. Widths and shapes are entirely due to the black-body radiation. Note the small additional sidebands corresponding to transitions between the first two pairs of the Rabi split excited states of the J-C model (§4).

Haroche's method is to probe the atom-field system with a weak, tuneable microwave field of frequency ν and measure the mean number of atoms leaving the cavity in their upper states as a function of ν. Linear response theory indicates that the absorption spectrum of the probe (which is also the mean number of atoms emerging in their upper states) is proportional to the Fourier transform $W(\nu) = \int_0^\infty W(\tau)\, e^{-i\nu\tau} d\tau$ of $W(\tau) \equiv - \lim_{t\to\infty} \langle [S^+(t+\tau), S^-(t)]\rangle$: $S^\pm(t)$ are spin-$\frac{1}{2}$ operators (eqn.

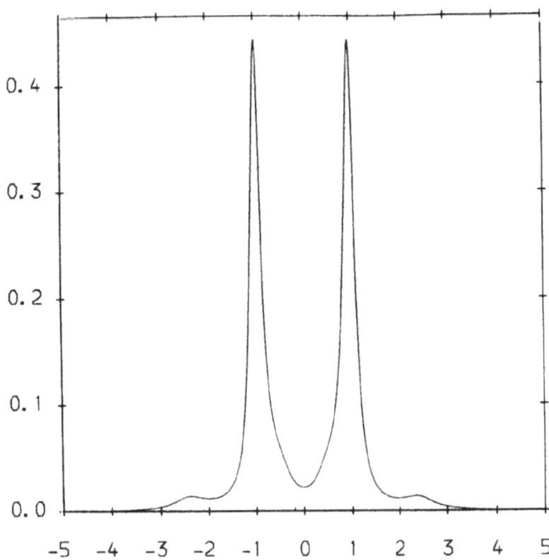

Fig. 3. Calculated fluorescence spectrum for vacuum field Rabi splitting on the 39 $S_{1/2} \rightarrow 39P_{3/2}$ transitions of ^{85}Rb at $Q = 10^8$ and $\bar{n} = 0.17$.

(2) below)[25]. The transform $W(\nu)$ can be related[25] to $S(\nu)$, the fluorescence spectrum, the transform of $S(\tau) = \lim_{t\to\infty} \langle S^+(t+\tau)S^-(t)\rangle$, as $W(\nu) = S(\nu)/\bar{n}$ (and[25] $S(\nu) \rightarrow 0$ as $\bar{n} \rightarrow 0$ and $W(\nu) \neq 0$ as $\bar{n} \rightarrow 0$: $S(\nu) \sim \bar{n}\,[1 + 2\bar{n}]^{-1} f(\nu)$ and $W(\nu) \sim [1 + 2\bar{n}]^{-1} f(\nu)$). Thus Haroche's experiment[32] is actually measuring the *fluorescence* spectrum $S(\nu)$, and it is exactly this quantity we calculate in Fig. 3 (ν in units of g). Note that this method is using the atoms as their own detectors. It circumvents the fundamental problem that the *continuum* of modes needed to measure $S(\nu)$ is not available in a microwave cavity (it also detects the very few soft microwave photons energies $\sim 10^{-23}$ J). It is therefore disappointing that splitting could be observed for $N_A = 6$ (there is a $\sqrt{N_A}$ factor on the vacuum field Rabi splitting for $N_A > 1$) but not for smaller N_A because of 'uncontrollable fields'[33] in the cavity.

From these few remarks on measurement and cavity q.e.d. (§§2,3) we turn to $Q = 3 \times 10^{10}$ and the theory of the micromaser.

4. THE MICROMASER AS A FUNDAMENTAL QUANTUM DEVICE

The J-C model[11] of one 2-level atom coupled to a single cavity mode and cavity $Q = \infty$ is an example of a "completely integrable quantum system"[9]. The terminology follow that for classical Hamiltonian systems which are completely integrable in the sense of Liouville[34] if there are, for N degrees of freedom, N independent first integrals I_k (say) ($k = 1, \ldots, N$) which commute under the Poisson bracket : $\{I_K, I_\ell\} = 0$. A quantum integrable system with N quantum degrees of freedom has N independent operator constants $\hat{I}_K$ which commute: $[\hat{I}_K, \hat{I}_\ell] = 0$.

The J-C model has

$$\hat{H} = \omega_o S^z + \omega a^+a + g\,(a^\dagger S^- + S^+a), \qquad \hat{N} = S^z + a^\dagger a + \tfrac{1}{2} \quad , \tag{1}$$

and $[\hat{H}, \hat{N}] = 0$: the single cavity mode (frequency ω) with bose operators $a, a^{\dagger}$ and $[a, a^{\dagger}] = 1$ provides one quantum degree of freedom: the single atom (frequency ω_0) with *three* spin-$\frac{1}{2}$ operators $S^{\pm}$, S^z satisfying the su(2) Lie algebra

$$[S^{\pm}, S^z] = \mp S^{\pm}, \ [S^+, S^-] = 2S^z, \tag{2}$$

but with $S = \frac{1}{2}$, provides a second quantum degree of freedom. The system can indeed be solved[11] for its energy eigenvalues and eigenstates : the problem of diagonalising an infinite dimensional matrix for $\hat{H}$ is reduced to diagonalising a sequence of 2×2 matrices characterised by the eigenvalues n of $\hat{N}$: $\hat{N}|n, \pm\rangle = n\,|n, \pm\rangle$, $n = 0, 1, 2, \ldots$; $\hat{H}|n, \pm\rangle = (n-\frac{1}{2})\omega \pm\frac{1}{2}[(\omega-\omega_0)^2 + 4g^2 n]^{\frac{1}{2}}$.

The Figs. 1 in Refs.[8,9] show many connections (many of them canonical transformations) between different aspects of both quantum, and classical, integrability. They are referred to here because they introduce the idea of 'quantum groups' and quantum group quantisation (§ 1). The integrable dynamical systems are associated with the Lie algebras $\tilde{g}$[8,9]. In the case of classical integrable dynamical systems in one space and one time dimensions these are extended to the 'loop' algebras $\tilde{g} \otimes [\lambda, \lambda^{-1}]$, $\lambda \in \mathbb{C}$[8,9] (an element of the algebra is an element of $\tilde{g}$ multiplied by a Laurent polynomial in λ). The classical system can be *quantised* by 'deforming' $\tilde{g}$: the route is[8,9] that $\tilde{g}$ is deformed by a parameter $q = e^t$ ($t \sim \hbar$) to $\tilde{g}_q$: $\tilde{g}_q$ is a Hopf algebra[8,9] : dual to $\tilde{g}_q$ is the 'quantum group' $\hat{T}(\lambda)$[8,9] : the quantum group $\hat{T}(\lambda)$ is also a Hopf algebra. The elements $T(\lambda)$ of the quantum group are matrices, and satisfy the matrix equation

$$R(\lambda,\mu)\hat{T}(\lambda) \otimes \hat{T}(\mu) = \hat{T}(\mu) \otimes \hat{T}(\lambda)\, R(\lambda,\mu): \tag{3}$$

$\otimes$ is direct i.e. Kronecker product, and if $\hat{T}$ is 2×2, $R(\lambda,\mu)$ is a 4×4 matrix characteristic of the dynamical system. A quantum integrable system in one space and one time has an infinity of degrees of freedom. The matrix equation (3) is evidently a set of commutation relations for the elements of T. These are the commutation relations induced by the quantum groups. The system *is* quantum integrable because the matrix trace of (3) yields $[\hat{\Delta}(\lambda), \hat{\Delta}(\mu)] = 0$, a large number of commuting operator constants : $\hat{\Delta}(\lambda) \equiv \mathrm{Tr}\ \hat{T}(\lambda)$, and $\ell n\ \Delta(\lambda)$ is generator of quantum integrable Hamiltonians H describing integrable models, typically nonlinear integrable models, in one space and one time dimension[8,9].

A deformed Lie algebra $\tilde{g}_q$ is exemplified[8,9] by the deformation $su_q(2)$ of the Lie algebra su(2) : $su_q(2)$ is

$$[S^{\pm}, S^z] = \mp S^{\pm}, \ [S^+, S^-] = [2S^z]. \tag{4}$$

The notation in the second expression is $[x] \equiv (q^x - q^{-x})/(q - q^{-1})$, so, with $q = e^t$

$$[2S^z] = (e^{2S^z t} - e^{-2S^z t})/(e^t - e^{-t}) = (\sinh 2S^z t)/(\sinh t)\ . \tag{5}$$

When $q = 1$ ($t=0$) $su_q(2)$ is su(2). The deformed loop algebra $\tilde{g}_q \otimes [\lambda, \lambda^{-1}]$ has the quantum group with elements $\hat{T}(\lambda)$ its dual. The $\hat{T}(\lambda)$ satisfy the commutation relations (3) and quantise an integrable quantum nonlinear (in general) field theory. However,[8] this quantum group quantisation proves to be identical with the familiar canonical quantisation. This seems to be true for quantum field theories in one space and one time dimension, and we can show it is true for the quantised Maxwell equations of normal q.e.d.

For discrete systems (integrable lattices[10]) and dynamical systems with a finite number of degrees of freedom this is not necessarily so. The J-C model has two quantum degrees of freedom only. In [6] it is shown that a form of J-C model quantised in terms of q-bosons and depending on q will display quantum revivals which depend on the value ($\neq 1$) of the q-parameter. The suggestion is implicitly made that observation of the revivals in the atomic inversion in the micromaser will indicate whether this quantum system is quantised by q-deformation or it is quantised canonically. The analysis is marred by an assumption about the model that the coupling constant g (eqn. (1)) is field intensity dependent. Obviously it *is* important to see if there is any physics in the q-deformation theory. Therefore in this section we work out the q-boson theory of the deformed J-C model.

The q-quantised (q deformed) J-C model consists of one q-deformed 2-level atom and one q-deformed mode of the e.m. field, a 'q-boson'[10,15-17]. Although the q-bosons[15-17] are new, the ideas were already available some 10 years ago[35-37]. The q-deformed 2-level atom has the $su_q(2)$ algebra (4) with the notation (5). But because it is *one* $S = \frac{1}{2}$ system $(2\ S^z)^2 = 1$ and $[2\ S^z]$, eqn (5), is $2\ S^z$. The single q-deformed 2-level atom is therefore the ordinary 2-level atom. This is no longer true of N_A 2-level atoms displaying collective action in black-body radiation as observed in Ref[29] (§3) : for $N_A = 2$, $[S^+, S^-] = 2(q+q^{-1})S^z$, and it will not be true for the micromaser if two or more atoms are present in the cavity at one time (§7).

It remains to deform the Heisenberg algebra $[a,a^\dagger] = 1$ of the cavity mode: this deformed algebra defines the q-bosons[10,15-17]. The algebra of q-bosons has the three elements a^+, a^- and n

$$[a^{\pm}, \hat{n}] = \mp\, a^{\pm} \quad , \tag{6}$$

$$a^- a^+ - q\, a^+ a^- = q^{-\hat{n}} \quad . \tag{7}$$

The vacuum state is $|0\rangle$: $a^-|0\rangle = 0$. Then $\hat{n}$ has eigenvalues n and eigenstates

$$|n\rangle,\ n = 0,\ 1, \ldots\ .\ :\ \hat{n}|n\rangle = n|n\rangle\ , \tag{8}$$

$$|n\rangle = ([n]!)^{-\frac{1}{2}}\, (a^+)^n\, |0\rangle\ , \tag{9}$$

with $[n]! \equiv [1].[2] \ldots .[n]$ in which

$$[n] \equiv (q^n - q^{-n}) \,/\, (q-q^{-1})\ . \tag{10}$$

For $q \longrightarrow q^{-1}$, (7) is

$$a^- a^+ - q^{-1}\, a^+ a^- = q^{\hat{n}}\quad , \tag{11}$$

and from (7) and (11)

$$a^+a^- = [\hat{n}] \equiv (q^{\hat{n}} - q^{-\hat{n}})/(q - q^{-1}) \quad . \tag{12}$$

One has $a^+|n\rangle = [n+1]^{\frac{1}{2}} \mid n+1\rangle$ and $a^-|n\rangle = [n]^{\frac{1}{2}}|n-1\rangle$ and $[n]$ replaces n in the usual (canonical) bose theory. There is now the Primakov-Holstein realisation of $su_q(2)$ which is

$$S^+ = a^+\sqrt{[2S-\hat{n}]}\ ,\ S^- = \sqrt{[2S-\hat{n}]}\ a^-,$$
$$S^z = \hat{n} - S \tag{13}$$

which satisfies (4) for any S. For the single 2-level atom, $S = \frac{1}{2}$, and there is the q-boson representation for the 2-level atom which is identical with the ordinary boson representation[9,24]. For $S = \frac{1}{2}$ there is also a q-fermion representation identical with the fermion representation[9,24]. When the number of deformed 2-level atoms $2S = N_A \longrightarrow \infty$, $\bar{S}^+ = [2S]^{-\frac{1}{2}}q^{\frac{1}{2}\hat{n}}S^-$ and $\bar{S}^- = [2S]^{-\frac{1}{2}}S^+q^{\frac{1}{2}\hat{n}}$ become a^- and a^+ respectively and

$$\bar{S}^+\ \bar{S}^- - q\ \bar{S}^-\ \bar{S}^+ = q^{-\hat{n}} \tag{14}$$

so the N_A atoms become one giant collective q-boson. The Hamiltonian is $\hbar\omega_0 S^z = \hbar\omega_0(\hat{n}-\frac{1}{2}N_A)$ where ω_0 is the atomic resonance frequency. The energy levels are therefore equally spaced with spacing $\hbar\omega_0$. If these are on resonance with the black-body radiation in a low-Q cavity the collective q-bose oscillator will have the mean number of atoms N_+ excited given by $\bar{n}$ for the black-body radiation as before (§3) : $\bar{n} = (e^{\beta\hbar\omega_0}-1)^{-1}$ whether the black-body radiation is made up of q-bosons or normal bosons. Thus an experiment like that of Ref 29 (§3) would not detect the q-deformation. Notice however that the system *is* a q-boson in the sense that $\langle\bar{S}^+\bar{S}^-\rangle = (e^{\beta\hbar\omega_0} - 1)/(e^{\beta\hbar\omega_0} - q)\ (e^{\beta\hbar\omega_0} - q^{-1})$ which coincides with $\bar{n}$ only when $q \to 1$. (This analysis solves a 'homework problem' left open in Ref 9!).

However, the q-deformed J-C model leads to observationally different results from those of the ordinary J-C model[11]. The q-deformed J-C model studied here has Hamiltonian

$$\hat{H} = \omega_0 S^z + \omega\hat{n} + g(S^+a^- + a^+S^-), \qquad \hat{N} = S^z + \hat{n} + \tfrac{1}{2} \quad . \tag{15}$$

Since $[S^\pm, S^z] = \mp S^\pm$ and $[a^\pm, \hat{n}] = \mp a^\pm$ are unchanged from the usual J-C model $[\hat{H}, \hat{N}] = 0$. Thus the deformed J-C model is still a completely integrable quantum system. Note that at first guess the q-deformed bosons of the electromagnetic field replace $\hat{n}$ by $\frac{1}{2}(a^+a^- + a^-a^+)$ in (15) either in $\hat{H}$ alone or in both of $\hat{H}$ and $\hat{N}$. For the latter model $[\hat{H}, \hat{N}] \neq 0$ and we shall not explore these further complications here. Evidently (15) as chosen is the *simplest* q-deformed J-C model we can consider.

From (15) for eigenvalue n of $\hat{N}$ and $n > 0$ the eigenenergies of $\hat{H}$ are now the roots of

$$(\tfrac{1}{2}\ \omega_0+\omega(n-1)-\lambda)(-\tfrac{1}{2}\omega_0 + \omega n - \lambda) - g^2\ [n] = 0 \quad . \tag{16}$$

These are the eigenenergies of the undeformed J-C model when $[n] \longrightarrow n$

for $q \longrightarrow 1$ [11]. When $n = 0$, $(-\frac{1}{2}\omega_o - \lambda) = 0$, so $\lambda = -\frac{1}{2}\omega_o$. This is the unchanged ground state energy of the J-C model. When $n = 1$, $[1] = 1$ so

$$\lambda = +\tfrac{1}{2}\omega \pm \tfrac{1}{2}[(\omega_o-\omega)^2 + 4g^2]^{\frac{1}{2}} . \tag{17}$$

The vacuum field Rabi splitting is therefore unchanged in the q deformed J-C model : for zero detuning $\Delta \equiv (\omega_o - \omega) = 0$, $\lambda = \frac{1}{2}\omega_o \pm g$ and the splitting is 2g as in the J-C model. No check on the q-deformation theory is therefore possible in an experiment like that of Haroche and colleagues[32] (§3, and Fig.3) to measure the vacuum field Rabi splitting. However, the four transition frequencies between the first two pairs of excited levels are, at resonance, $\omega_o \pm g(q+q^{-1})^{\frac{1}{2}} \pm g$ and depend on q. Fig.3 shows that two of these sidebands, at $\omega_o \pm (g(q+q^{-1})^{\frac{1}{2}}+g)$, might just about show up in a vacuum field Rabi splitting experiment with $Q = 10^8$, searching for any q-dependence. Moreover, in a 2-photon experiment, as is now being suggested [33], the vacuum field Rabi splitting between the ground state and the $n = 2$ Rabi split pair is $g(q + q^{-1})^{\frac{1}{2}}$ and this also depends on q.

The expectation value of the atomic inversion at resonance is

$$\langle S^z(t)\rangle = \tfrac{1}{2}\sum_{n=0}^{\infty} p_n(1-2\sin^2(g[n+1]^{\frac{1}{2}}t)); \tag{18}$$

p_n is the probability of finding n photons in the mode at $t = 0$. Note that, at $t = 0$, $\langle S^z(0)\rangle = +\frac{1}{2}$ and the appropriate eigenvalue of $\hat{N}$ is $n + 1$, not n, and the resonant Rabi splitting is $2g\,[n+1]^{\frac{1}{2}}$. Thus for $n = 1, 2, \ldots$ in the sum in (18) the terms depend on q. If there is any physics in quantum group quantisation it may show up in the quantum revivals of the micromaser.

In this suggestion made already[6] the authors introduce an intensity dependent coupling constant $g([\hat{n}])^{\frac{1}{2}}$ which does not correspond to the physics of the micromaser. The Fig. 1 in Ref. 6 shows a *qualitative* change in predicted revivals at $Q = \infty$ as q increases. For $q = 1$ revivals are of 2-photon type (see eg. Fig. 2 in Ref. 38) but become of typical 1-photon type as q increases. We shall report comparable results for the q-deformed J-C model (15) elsewhere.

The possibility of detecting quantum group quantisation this way depends of course on the size of q and the effects of black-body radiation (which as noted may also be q-deformed). We already know that small changes in the observable revivals are induced by finite temperature effects. The Fig. 5 of Ref. 20 shows revivals in $\langle\hat{n}\rangle = \langle a^\dagger a\rangle$ for the *undeformed* J-C model for $Q = 2 \times 10^{10}$ and an initial coherent state field $|\alpha\rangle$ with $|\alpha|^2 = 2$; $\bar{n} = 0.15$ and $\omega_o = 2\pi \times 21.456$ GHz corresponding to the $63P_{3/2} - 61D_{5/2}$ transitions for ^{85}Rb. That the few black-body photons move the positions of the revivals in a detectable fashion was already reported in Ref. 19. The Fig. 6 of Ref. 20 shows 1-photon revivals in $C(t) \equiv [\langle a^\dagger(t)a^\dagger(t)a(t)a(t)\rangle/\langle a^\dagger(t)a(t)\rangle^2 - 1$ for the same parameters as the Fig.5 ($C(t) \equiv P_2P_1^{-2} - 1$ in the notation of §5 following). The results show that $C(t) < 0$ for much of the resonant motion and is < 0 off resonance during the actual revivals. Here the field is *anti-bunched* (sub-Poissonian) : the Mandel Q-parameter $Q = (\langle a^\dagger a^\dagger aa\rangle/\langle a^\dagger a\rangle) - \langle a^\dagger a\rangle < 0$.

We show how these calculations were performed in the next section.

5. SINGLE ATOM CAVITY Q.E.D. AT FINITE TEMPERATURE

The model is the *undeformed* J-C model in which the single cavity mode is coupled to a heat bath at T > 0. The well-known master equation is[23,24]

$$\frac{d\rho}{dt} = -i\,[H,\rho] - (\bar{n}+1)\,\kappa\,[a^\dagger a\rho - 2a\rho a^\dagger + \rho a^\dagger a]$$

$$- \bar{n}\kappa\,[a\,a^\dagger\rho - 2a^\dagger\rho a + \rho a a^\dagger] \qquad (19)$$

in which $\kappa = \frac{1}{2}\,\omega Q^{-1}$ and $\bar{n} = (e^{\beta\omega}-1)^{-1}$ ($\hbar$=1). The calculational problem rests in the infinite number of photon states $|n\rangle$ spanning the Hilbert space : a complete basis is $|e\rangle|n-1\rangle$, $|g\rangle|n\rangle$ with n = 1,2,..., and $|g\rangle|0\rangle$. Our procedure has been[14] to calculate the four one-body quantities $P_n = \langle a^{\dagger n} a^n\rangle$, $Q_n = \langle a^{\dagger n} a^n S^z\rangle$ (and $n \geq 0$), $S_n = \langle S^+ a^{\dagger n-1} a^n + a^{\dagger n} a^{n-1} S^-\rangle$, $R_n = i\,\langle S^+ a^{\dagger n-1} a^n - a^{\dagger n} a^{n-1} S^-\rangle$ (and $n > 0$), which satisfy a closed coupled system of equations infinite in number : $Q_0 = \langle S^z\rangle$ and $P_1 = \langle a^\dagger a\rangle$, $P_2 = \langle a^\dagger a^\dagger a a\rangle$. The problem reduces[19] to the solution of matrix equations

$$\frac{d\psi_n}{dt} = \mathcal{A}_n\psi_n + B_n\psi_{n+1} + C_n\psi_{n-1}\;,\quad n \geq 0\;,\quad C_0 = 0\;, \qquad (20)$$

in which $\psi_n(t) = [P_n, Q_n, R_n, S_n]^T$ and $\mathcal{A}_n$, B_n and C_n are 4 × 4 matrices. The Laplace transform on t yields a 3-term matrix recurrence relation solved by the matrix continued fraction techniques first introduced by Risken[39]. We also calculate the 2-body correlation functions for spectra at thermal equilibrium in a similar way[25-27]. These correlations at t and t+τ satisfy a closed system and at t ⟶ ∞ (equilibrium) the problem reduces to solving a problem of the form of (20) with larger matrices and τ replacing t[25-27]. The fluorescence spectrum Fig. 3 is calculated this way. Moreover,[25-27] we have calculated fluorescence and cavity field spectra for $10^3 \lesssim Q \leq 10^8$ and watched their *qualitative* changes as Q increases (§3): vacuum field Rabi splitting is first discernible for Q's ~ 10^6. One can also expect to see the breakdown of the Born and Markhoff approximations with increasing Q[25].

We turn to the micromaser. An instructive result found by the methods just described is Fig 4. This shows the evolution of the Fock state $|10\rangle$ in the presence of the single atom for Q = 2 × 10^{10} and $\bar{n}$ = 0.15 : κg^{-1} = 0.0000767, and 63 $P_{3/2}$ ⟶ 61$D_{5/2}$ transitions of ^{85}Rb are assumed : detuning Δ = 0. The scale in gt is that 9 units ~ one atomic transit time crossing the cavity (gt_{int}, §1). The Fig. 4 shows that after 8 such transit times the amplitude of the inversion has fallen by some 10%. In the empty-of-atoms cavity time $g(T_p - t_{int})$ of the micromaser (§1 and Refs. 1,2) this amplitude would be ~ $(0.9)^{36}$ = 0.02. Evidently if the Fock state $|10\rangle$ *were* created it would decay substantially in each period T_p. This suggests that a smaller T_p is needed for the creation and maintenance of Fock states in the micromaser. We investigate this suggestion next.

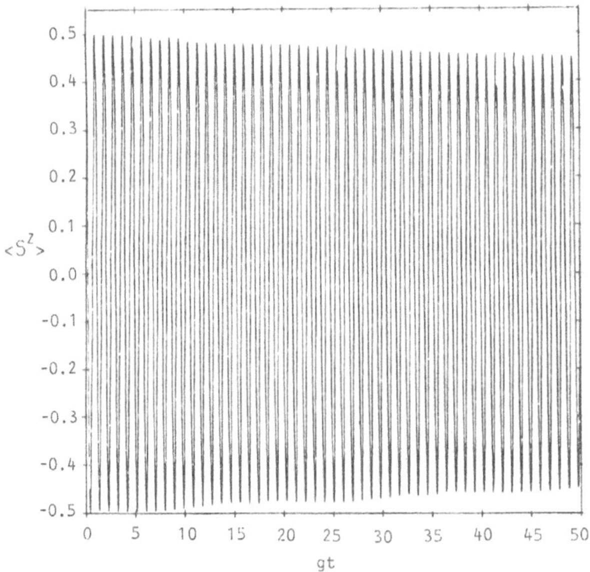

Fig.4. Decay of an initial Fock state $|10\rangle$, as described by $\langle S^z(t)\rangle$, on resonance: $Q = 2 \times 10^{10}$, $\bar{n} = 0.5$.

6. FOCK STATE CREATION IN THE MICROMASER

In this connection the significant objects are the so-called 'trapping states'[5,11,14]. For $Q = \infty$, Refs 11, 14 show that for single inverted atoms entering and leaving the cavity successively in a time t_{int} the probability $P(n,k)$ of finding n photons in the cavity after k atoms have passed necessarily satisfies

$$P(n,\ k+1) = a(n)P(n,k)+b(n-1)P(n-1,k) \tag{21}$$

where $b(n-1)$ is the *a priori* probability that an atom emits its photon into the cavity in the presence of n-1 photons already in the cavity and $a(n-1)$ is the *a priori* probability that it does not : $a(n) + b(n) = 1$. Moreover,[11,14]

$$b(n,k) = \sin^2 \{(n+1)^{\frac{1}{2}} gt_{int}\} \quad . \tag{22}$$

Thus if $b(n) = 0$, $P(n,k+1) = P(n,k)$ and the cavity field ceases to build up as further atoms pass. The crucial result however is that if $n = n_o$ is the first zero of (22), then $P(n_o,k) = P(n_o,k+1) \longrightarrow 1$ ie a Fock state $|n_o\rangle$ is created in the cavity[14].

It is convenient to consider the 'trapping states' as the zeros of (22) labelled by $n = n_o$ and r: $(n_o+1)^{\frac{1}{2}}\ gt_{int} = r\pi$ (Ref 5 uses q for our use of r here). Thus, as example, $gt_{int} = \frac{1}{2}\pi$ and $r = 1$, $n_o = 3$ defines a 'first' trapping state. Then there are the second, third, etc. trapping states with $n_o = 15, 35, 63, \ldots$. There are trapping states for *any* $gt_{int} = r\pi(n_o+1)^{-\frac{1}{2}}$ i.e. the gt_{int} for trapping states have the measure of the irrationals : the gt_{int} for trapping states are therefore 'dense' and there is a large number of such states. Without attempting to make this argument rigorous one can, in short, expect that the cavity field of the micromaser 'almost always' evolves to a Fock state. The Fock state is characterized by the variance $\sigma = 0$: following[12] $\sigma^2 = (\langle\hat{n}^2\rangle - \langle\hat{n}\rangle^2)/\langle\hat{n}\rangle$. The experiments[1,2] show that such evolution does not necessarily occur : *sub*-Poissonian fields with $\sigma \gtrsim 0.3$ were resolved for $Q = 3 \times 10^{10}$ and $T = 0.5^{o}$ K, but not $\sigma = 0$; moreover at the trapping states defined by $gt_{int} = \frac{1}{2}\pi$ super-Poissonian fields with $\sigma>1$ were detected. These observations conform to the predictions of the theoretical work[12,13].

There are *two* reasons why evolution to a single Fock state does not occur. The first of these is the fundamental one [14] that (at $Q = \infty$) if gt_{int} defines a first trapping state (n_o, r), then if (n_1, r), (n_2, r), etc. are the second, third, etc. trapping states, each of these will evolve as long as the initial value of the cavity field is $\sum_{n=0}^{N_{max}} \alpha_n|n\rangle$ $(\alpha_{N_{max}} \neq 0)$ and the n_i in the trapping states have $n_i < N_{max}$. In the case of black-body radiation $N_{max} = \infty$ so all of the sequence of trapping states (n_1, r) will evolve, *and no steady state will ever be resolved!* However, in Ref 14 no time scale is given for this evolution. From a computer calculation with $Q = \infty$ we find that after 91 inverted atoms have entered and exited successively $p_3 = 0.9997105$, $p_{15} = 0.2659983 \times 10^{-3}$ and $p_{35} = 0.197007 \times 10^{-14}$. After 500 inverted atoms this becomes $p_3 = 0.9997105$, $p_{15} = 0.2894501 \times 10^{-3}$ and $p_{35} = 0.701391 \times 10^{-14}$. After 2000 inverted atoms only p_{35} changes (to $0.7019316 \times 10^{-14}$). All the figures then remain unchanged as 100,000 atoms have passed!

The second reason why Fock states do not occur is that at $Q = 3 \times 10^{10}$ and $T = 0.5^{o}K$ (as in[1,2]) the finite value $\bar{n} = 0.15$ induced by coupling to a heat bath at that T *qualitatively* changes the action of the micromaser compared with its action at $Q = \infty$. Observations[1,2] are consistent with the predictions[12,13]. Both theories[12,13] admit black-body radiation. The experiments have $gT_p \sim 200\ gt_{int}$. Ref. 12, in particular, had assumed already that $gT_p \gg gt_{int}$ and coupled in the heat bath only during the long empty-of-atoms cavity time $T_p - t_{int}$. During the short time t_{int} atom and field are allowed to evolve together as a unitary transformation without any interaction with the heat bath. This simplification leads to an *additive* (linearly additive) gain and loss theory as in conventional laser theory [13]. In particular Ref. 12 finds, for normalisation C that

$$p_n = C\left(\frac{\bar{n}}{1+\bar{n}}\right)^n \prod_{k=1}^{n}\left(1+\frac{N_{ex}}{k\bar{n}}\ \sin^2\ \sqrt{k}\ gt_{int}\right) \qquad (23)$$

Here $N_{ex} = R\gamma^{-1}$ in the notation of Ref 12 : $\gamma = 2\kappa$ from (19) and R is the mean number of inverted atoms traversing the cavity per sec : thus $R^{-1} = T_p$ and N_{ex} is the mean number traversing in the cavity decay time

$T_c = (2\kappa)^{-1}$. There is a natural tipping angle $\Theta = (N_{ex})^{\frac{1}{2}} gt_{int}$ which plays the role of pump parameter[12]. In the steady state assumed for (23) $R = 2\kappa N_{ex}$ and gain equals loss (linearly). Formula (23) will be good for $T_p \gg t_{int}$, the "linear" gain and loss regime. It depends significantly on $\bar{n}$ which enters the theory during the long time $T_p - t_{int}$ in which the heat bath is switched in. That this changes the theory qualitatively compared with $Q = \infty$ follows from the $T \to 0$ ($\bar{n} \to 0$) limit of (23) which is $p_n = C \prod_{k=1}^{n} N_{ex} k^{-1} \sin^2 \sqrt{k} gt_{int}$: $p_n = 0$ for $n \geq n_o + 1$ where $\sin(n_o+1)^{\frac{1}{2}} gt_{int} = 0$, though the theory here is evidently incomplete since no Fock state $|n_o\rangle$ with $p_{no} = 1$ is predicted. Nor does the theory show the higher trapping states will emerge. Both of the theories[12,13] are steady state theories but there is no steady state at $Q = \infty$ as we have explained.

The Fig. 4 suggested that if we wish to create Fock states in the micromaser at finite $Q < \infty$ and $T > 0$, we should reduce T_p to $T_p \approx t_{int}$. Evidently it then becomes necessary to couple in the heat bath realistically during both $T_p - t_{int}$ and t_{int}, namely throughout the motion. This coupling introduces an intrinsically *nonlinear* gain and loss mechanism.

A simple minded approach to see this, which however is still concerned largely with the empty-of-atoms time $T_p - t_{int}$, is as follows:- Assume there *is* a steady state varying during T_p but repeating each T_p. Suppose $\langle\hat{n}\rangle$, the mean number of photons in the cavity, is n_1 as the inverted atom enters, is n_2 when it leaves after time t_{int}, and is n_3 before the next atom enters at T_p. For the steady state $n_3 = n_1$, and

$$\bar{n} + (n_2 - \bar{n})\, e^{-2k(T_p - t_{int})} = n_3 = n_1 \tag{24}$$

where the case n = 1 of the exact expression (26) below has been used. Since $n_2 \leq n_1 + 1$ (only one photon can be dumped) then, for small enough $\bar{n}$ compared with n_1, n_2, n_3 (a numerically valid assumption in the present context though this removes $\bar{n}$ from the theory)

$$n_2 \leq \left[1 - e^{-2\kappa(T_p - t_{int})}\right]^{-1} . \tag{25}$$

It follows that for $t_{int} \ll T_p$ (the assumption of[12,13]) *and* $2\kappa T_p \ll 1$, $n_2 \leq (2\kappa T_p)^{-1} = N_{ex}$. The equality is the equality of linearly additive gain and loss $R = T_p^{-1} = 2\kappa n_2$. More generally this simple minded analysis shows the gain-loss mechanism is nonlinear. But it tells us nothing about the evolution during t_{int}, only that n_2 could become large for $T_p \approx t_{int}$.

We have used the methods sketched in the previous section to investigate this nonlinear gain and loss regime: we solve (19) for the total density operator ρ during t_{int} when the atom is in the cavity: the heat bath is thus coupled in. We then solve the same equation with the atom omitted during the time $T_p - t_{int}$ and the heat bath coupled in. This is done numerically or we can make use of the exact analytic formula derived from (19) when no atom is present: we define $P_n = (n!)^{-1}\langle a^{\dagger n} a^n \rangle$ and find (by solving equation (29) below) that

$$P_n(\tau) = (\bar{n})^n + \sum_{r=1}^{n} e^{-r\tau} M_{nr}(\bar{n})^{n-r} \sum_{j=1}^{r} (-1)^{r-j} M_{rj}(\bar{n})^{r-j} \{P_j(0) - (\bar{n})^j\} \qquad (26)$$

in which $M_{ij} = i!\ (j!(i-j)!)^{-1}$, $i \geq j$; $M_{ij} = 0$, $i<j$: κ enters through $\tau = 2\kappa\ (T_p - t_{int})$ and (26) generalizes (24).

So far in our calculations both t_{int} and $T_p - t_{int}$ repeat regularly with fixed T_p (a stochastic input of the atoms is not yet included, §2). The empty of atoms cavity has initial condition $\rho_f(0) = \mathrm{Tr}_a \rho(t_{int})$ where $\rho(t_{int})$ is ρ as the atom of current interest leaves the cavity and $\rho_f(0)$ evolves to $\rho_f(T_p - t_{int})$ in the empty of atoms cavity time either by numerical solution for $\rho_f(t)$ from (19) with no atom present or through (26): $\rho_f(T_p - t_{int})$ now serves as initial condition for the cavity field as the next atom enters so $\rho = |e\rangle\ \rho_f(T_p - t_{int})\ \langle e|$ is the initial condition for ρ as the next atom enters.

The equations of motion during the interaction time t_{int} can be reduced to a 3-term recurrence relation solved once again by matrix continued fraction methods. Details will be published elsewhere[28,40]. A significant point is that although off-diagonal elements of ρ develop, $\rho_f(T_p - t_{int})$ is diagonal if $\rho_f(-t_{int}) = \mathrm{Tr}_a \rho(0)$ is initially diagonal. But this is the case since the first atom enters a cavity containing only black-body radiation.

Obviously the smallest empty-of-atoms time is zero, i.e. $T_p - t_{int} = 0$ and $T_p = t_{int}$. The Fig. 5 shows the evolution of the probabilities p_n for finding n photons in the cavity as a function of the number of inverted atoms which have passed through the cavity in successive intervals $gt_{int} = \frac{1}{2}\pi$. This is the condition for evolution to a first trapping state $n_0 = 3$, $r = 1$ at $Q = \infty$ and the next trapping state has $n_0 = 15$. However, in these calculations $Q = 2 \times 10^{10} < \infty$ and $T = 0.5°K$ much as in the experiments[1,2]. It will be seen that after 40 atoms have passed p_3 evolves to 0.988, while p_{15} now evolves and after 500 atoms have passed p_3 and p_{15} have become 0.924 and 0.056 respectively. The Fig. 2 of Ref. 21 also shows comparable results for $gt_{int} = \frac{1}{4}\pi$ and the first trapping state is at $n_0 = 15$: $p_{15} = 0.774$ after 400 atoms have passed and is still growing. Thus at this finite Q and T the system displays the action predicted for $Q = \infty$ in Ref. 14. Presumably no steady state is reached as the higher trapping states at $n_0 = 35, 63, \ldots$ begin significantly to grow.

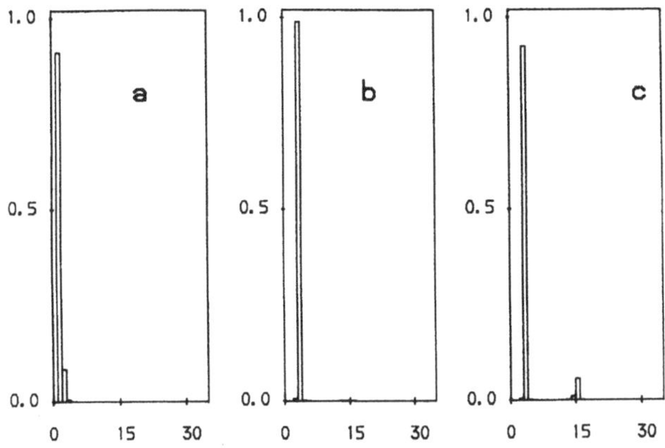

Fig.5. Histogram of $p_n = \rho_f(n)$ after 1(a), 40(b) and 500(c) atoms have passed through the cavity: $\omega = \omega_0$ and $t_{int} = \pi/2g$.

The major conclusion is that as $T_p - t_{int}$ becomes smaller the finite Q finite T system behaves more and more like a $Q = \infty$ system as described in Refs 11, 14.

We support this conclusion by the further work described in the next section.

7. ACTION OF THE MICROMASER AS THE EMPTY OF ATOMS CAVITY TIME CHANGES

The numerical calculations described in §5 are slow even on the Amdahl 5890 system available in Manchester. The matrix continued fraction methods used offer the very high accuracy and good stability necessary for the detailed investigation of black-body dynamics displayed in Fig. 5, but calculations are very long. We have developed faster, less accurate and less stable, methods for the PC. We describe some of the results found so far through these.

These particular calculations determine the moments P_n, Q_n, R_n and S_n mentioned below eqn. (19). On resonance we need P_n, Q_n and R_n only, and the equations of motion derived from (19), with the atom present in the cavity, are

$$\begin{aligned}
\dot{Q}_0 &= -R_1 \\
P_n &= n\, R_n - 2\,\kappa\, g^{-1} n\, (P_n - \bar{n}\, P_{n-1}) \\
Q_n &= -\tfrac{1}{2} n\, R_n - (n+1)\, R_{n+1} - 2n\, \kappa\, g^{-1}\, (Q_n - \bar{n}\, Q_{n-1}) \\
R_n &= 4Q_n + P_{n-1} + 2Q_{n-1} - \kappa\, g^{-1}\, \{(2n-1)R_n - 2(n-1)R_{n-1}\, \bar{n}\}\ .
\end{aligned} \tag{27}$$

Conveniently we have rescaled $P_n \longrightarrow (n!)^{-1} P_n$, etc. so with time t for gt

$$\begin{aligned}
P_n(t) &\equiv \mathrm{Tr}\,\{\rho(t)(a^\dagger)^n a^n\}\ /n! \\
Q_n(t) &\equiv \mathrm{Tr}\,\{\rho(t)(a^\dagger)^n a^n S^z\}\ /n! \\
R_n(t) &\equiv \mathrm{Tr}\,\{\rho(t)\ S^+\ (a^\dagger)^{n-1} a^n - \rho(t)(a^\dagger)^n a^{n-1} S^-\}\ /n!\ \ .
\end{aligned} \tag{28}$$

The empty of atoms cavity field evolves as

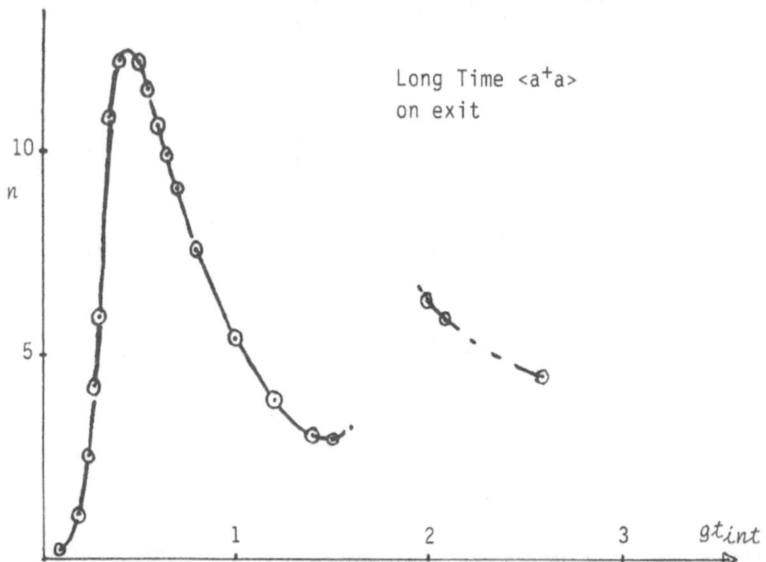

Fig.6. Plot from the PC of the 'long time' value $\langle\hat{n}\rangle = \langle a^{\dagger}a\rangle$ on atomic exit as a function of gt_{int} for $Q = 10^8$ $(\kappa g^{-1} = 0.01)$ and $\bar{n} = 0.1$:$gT_p = 4$ and $\Theta = (3.53)gt_{int}$.

$$\dot{P}_n = -2n\,\kappa\, g^{-1}\,(P_n - \bar{n}\,P_{n-1}) \tag{29}$$

and the analytical solution of this is (26). When the atom is in, we

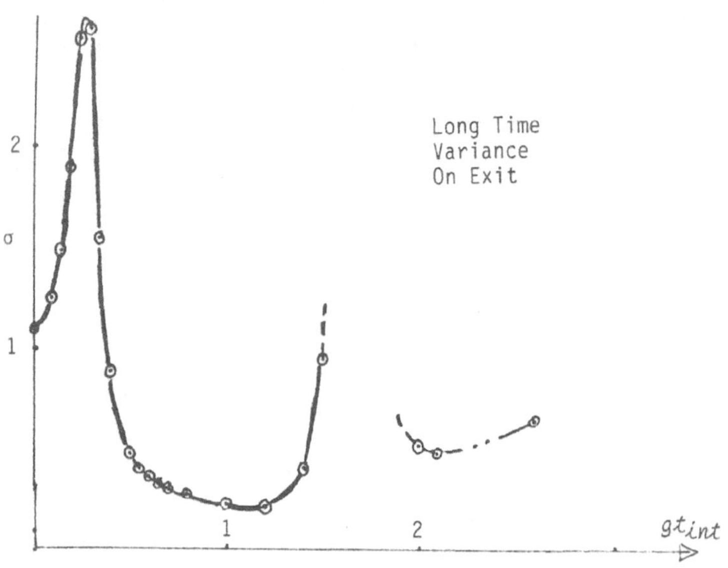

Fig.7. Plot of σ from the PC as a function of gt_{int} for the parameters of Fig.6. Note the super-Poissonian region where $gt_{int} \sim 2$ $(\Theta \sim 2\pi)$.

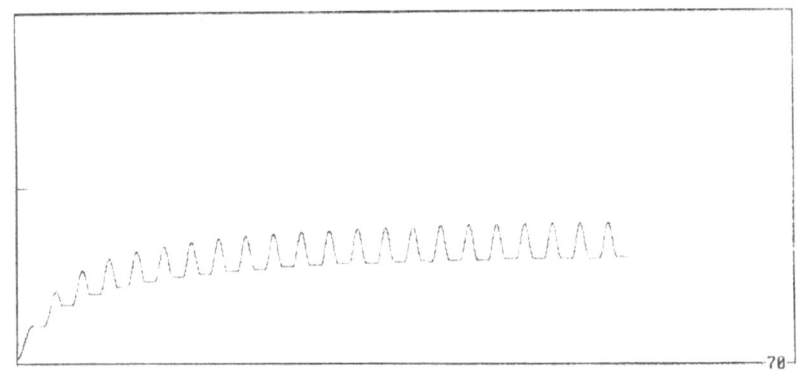

Fig.8. Evolution of cavity photon number: $\kappa g^{-1} = 7.67 \times 10^{-5}$, $gt_{int} = 1.57$, $\bar{n} = 0.15$, $gt_e = 1$. After 22 atoms have passed through, $\langle n \rangle = 2.98$, $\sigma = 0.053$, $\langle S^z \rangle = 0.4914$.

close the system by $R_{m+1} \approx (m+1)^{-1} P_1 R_m$. We need $m >> \langle a^\dagger a \rangle \equiv P_1$ and there are 3m+1 equations.

The Figs. 6 and 7 show results achieved this way for $Q = 10^8$ and $gt_{int} = \frac{1}{2}\pi$: $\bar{n} = 0.1$ and $\omega = \omega_0$. In the Fig. 7 in particular there are pronounced sub-Poissonion regions where $\sigma < 1$ and there is a general qualitative agreement in this respect with the results predicted in Refs. 12 and 13 and confirmed in the experiments[1,2]. Oualitative agreement over the plotted range $gt_{int} \lesssim 3.0$ can be reached through $\Theta = (N_{ex})^{\frac{1}{2}} gt_{int} = (2\kappa T_p)^{-\frac{1}{2}} gt_{int} = 3.53\, gt_{int}$. Thus there is also the super-Poissonian region in Fig. 7 which occurs after the threshold and where $\Theta \approx 2\pi$. The significance of these results compared with the previous work[1,2,12,13] however is that $gt_{int} \lesssim 3.0$ and $gT_p = 4$, so the empty of atoms cavity time reaches up to as little as 25%. Agreement with Refs 1, 2, 12, 13 is thus reached by the rather low value (10^8) chosen for Q. At this Q there is no evidence of the trapping state n_0 =

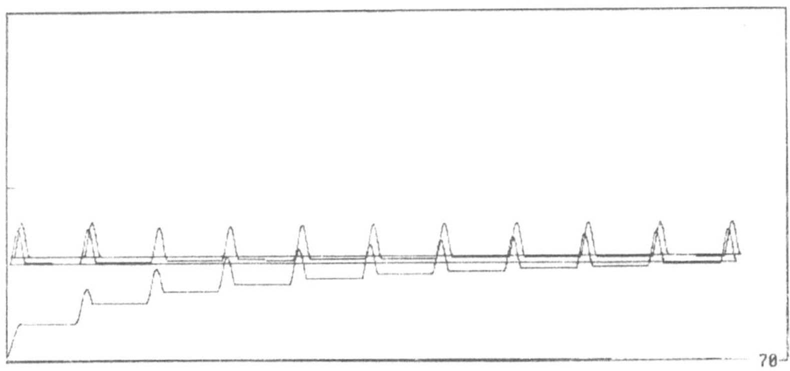

Fig.9. Evolution of cavity photon number. Same parameters as Fig.8, but now $gt_e = 5$. For convenience gt(mod 72) is plotted rather than gt. After 24 atoms, $\langle n \rangle = 3.01$, $\sigma = 0.119$, $\langle S^z \rangle = 0.4889$.

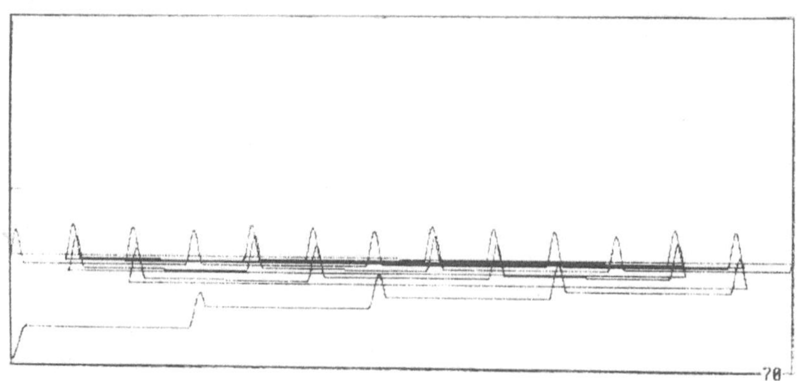

Fig.10. Evolution of cavity photon number. Parameters as for Fig.8, but now $gt_e = 15$. After 25 atoms, $\langle n \rangle = 3.05$, $\sigma = 0.277$, $\langle S^z \rangle = 0.4795$.

3 at $gt_{int} = \frac{1}{2}\pi$ in agreement with the ideas of Ref. 5. Instead the super-Poissonian region at $\Theta \approx 3.53(\frac{1}{2}\pi)$ occurs. Thus there are *three* important parameters: the-empty-of-atoms cavity time $T_p - t_{int}$, $\bar{n}$, *and* Q.

The remaining Figs. 8, 9, 10 and 11 show that, as $t_e \equiv T_p - t_{int}$ is increased from zero, the final values of σ as some number $N_A \sim 20$-30 of inverted atoms has passed through the cavity systematically increases. These figures have $Q = 2 \times 10^{10}$ and $\bar{n} = 0.15$ ($T = 0.5\mathring{K}$) much as in the experiments and plot $\langle \bar{n} \rangle = \langle a^\dagger a \rangle$ as a function of time as successive atoms pass through. The value of $gt_{int} = \frac{1}{2}\pi$ and the trend is that, as $gt_e \rightarrow 72$, σ on final exit has moved from 0.053 (almost a Fock state) for $gt_e = 1$, to $\sigma = 1.106$. This is already super-Poissonian ($\sigma > 1$) much as in the observations - though very many more atoms were used in a steady state for these.

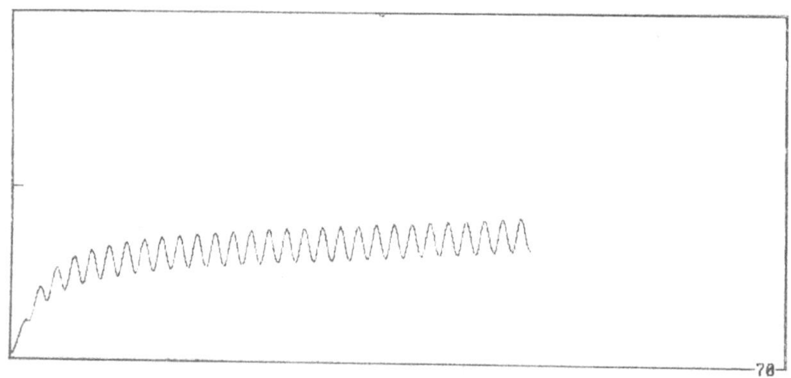

Fig.11. Evolution of cavity photon number. Parameters as for Fig.8, but now $gt_e = 72$. After 29 atoms, $\langle n \rangle = 3.31$, $\sigma = 1.106$, $\langle S^z \rangle = 0.4302$.

Our conclusion is that a relatively short empty of atoms cavity time $T_p - t_{int}$ will help in the creation of Fock states at finite Q and T. Evidently the "nonlinear" gain and loss mechanism enhances the *effective* cavity Q towards $Q = \infty$.

Of course this brings into question the validity of the model. Once the stochastic input of atoms is included the probability of finding more than one atom in the cavity becomes rather significant unless $T_p \gg t_{int}$. For example in the experiments like[4], with $Q = 5 \times 10^9$ and $T = 2.0^\circ K$, $T_p^{-1} = 2,800\ sec^{-1}$ led to 1.4 atoms on average in the cavity and 420 photons[41]. At the parameters of the experiments[1] which were $Q = 3 \times 10^{10}$ and $T = 0.5^\circ K$, $T_p^{-1} = 4500\ sec^{-1}$ and $t_{int} \sim 70\ \mu\ sec$ induces 0.3 atoms on average and 370 photons in the cavity[41]: when $T_p^{-1} = 8000\ sec^{-1}$ the probability of finding one atom in the cavity is ~ 0.6. For $T_p \approx t_{int} \lesssim 50\ \mu$ sec we need $T_p^{-1} \gtrsim 20,000$ atoms per sec so the probability of more than one atom in the cavity is large (the probability of two atoms ~ 0.63).

If there were a regular input of precisely two atoms into the cavity, the micromaser would become based on the 2 atom equivalent of the J-C model. For this model one finds that $\hat{N} = S^z + a^\dagger a$ is a constant of the motion and $S^z = \sum_{i=1}^{2} S_i^z$ for the two atoms. For the eigenvalue n of $\hat{N}$ there is the basis set $|n-1,+1\rangle$, $|n,0\rangle$, $|n+1,-1\rangle \equiv |n+1\rangle|+1\rangle$, $|n\rangle|0\rangle$, $|n+1\rangle|-1\rangle$, and at resonance the three eigenstates of $\hat{H}$ for fixed n are[42]

$$(n/2(2n+1))^{\frac{1}{2}}|n-1,+1\rangle + (2)^{-\frac{1}{2}}|n,0\rangle + ((n+1)/2(2n+1))^{\frac{1}{2}}|n+1,-1\rangle,$$
$$-((n+1)/(2n+1))^{\frac{1}{2}}|n-1,+1\rangle + (n/(2n+1))^{\frac{1}{2}}(n+1,-1\rangle,$$
$$(n/2(2n+1))^{\frac{1}{2}}|n-1,+1\rangle - (2)^{-\frac{1}{2}}|n,0\rangle + ((n+1)/2(2n+1))^{\frac{1}{2}}|n+1,-1\rangle, \quad (29)$$

with corresponding eigenvalues $n\omega + g(2(2n+1))^{\frac{1}{2}}$, $n\omega$ and $n\omega - g(2(2n+1))^{\frac{1}{2}}$. The mean inversion proves to be[42]

$$\langle S^z(t)\rangle = \sum_{n=0}^{\infty} p_n(2n+3)^{-2}\Big[(1+(n+1)(1 + \cos g\sqrt{2(2n+3)}t))^2 - (n+1)(n+2)(1 - \cos g\sqrt{2(2n+3)}t)^2\Big] \quad (30)$$

to be compared with (see eqn. (18).

$$\langle S^z(t)\rangle = \tfrac{1}{2}\sum_{n=0}^{\infty} p_n \cos(2g(n+1)^{\frac{1}{2}}t) \quad (31)$$

in the 1-photon case. The revivals are therefore different, the Rabi frequencies are more complicated and change, and the trapping state conditions are changed. We have still to determine how far this considerable complication will change the opportunity for Fock state creation for the realised Q's ~ 3×10^{10}.

8. SUMMARY AND CONCLUSION

Apart from a number of remarks on the measurement of cavity field statistics and cavity field spectra made in the introductory sections §§2 and 3, the paper carries out two investigations: in the §4 it solves the 'q-deformed' Jaynes-Cummings model which corresponds to quantum group rather than canonical quantisation, and shows how the parameter q of the deformation changes both the eigenspectrum and the quantum revivals. The vacuum field Rabi splitting is not changed but transitions involving the next and higher Rabi split pairs of levels depend on the value of q. It may be possible to investigate both features experimentally and so determine whether such quantum group quantisation has any physics in it.

In the §§5-7 we develop a theory of the micromaser involving normal canonical quantisation, ie. based on normal quantum mechanics. The basic model is the undeformed J-C model but as inverted atoms enter and leave the cavity the system is coupled to a finite temperature heat bath *throughout the motion*. It is shown that the micromaser theory for cavity $Q = \infty$ becomes fundamentally and *qualitatively* changed by the very small number $\bar{n}$ of black-body photons ($\bar{n} = 0.15$) achieved in the recent experiments. Two times are of importance: the interaction time of single atoms with the cavity field t_{int} and the repetition time T_p of atoms entering the cavity. The cavity is empty of atoms during $T_p - t_{int}$. The main result is to show that as $T_p \gg t_{int}$ becomes $T_p \approx t_{int}$ the micromaser system behaves more like a $Q = \infty$ system and that there is no steady state even at the $Q = 3 \times 10^{10}$ and $T = 0.5^{\circ}K$ achieved in the recent experiments. The qualitative change in the behaviour is induced by the advent of an intrinsically nonlinear gain-loss mechanism as T_p approaches t_{int}. The consequent problem of more than one atom in the maser cavity is addressed but not solved.

---

† On leave from Steklov Mathematical Institute, Fontanka 27, 191011 Leningrad, USSR. We are grateful to the SERC for providing a Visiting Fellowship for him.

REFERENCES

1. G. Rempe, F. Schmidt-Kaler and H. Walther, Phys. Rev. Lett. 44: 2783 (1990).
2. Gerhard Rempe and Herbert Walther, Phys. Rev. A 42: 1650 (1990).
3. D. Meschede, H. Walther and G. Müller, Phys. Rev. Lett. 54: 551 (1985).
4. G. Rempe, H. Walther and N. Klein, Phys. Rev. Lett. 58: 353 (1987).
5. P. Meystrè, G. Rempe and H. Walther, Opt. Lett. 13: 1078 (1988).
6. M. Chaichian, D. Ellinas and P. Kulish, Phys. Rev. Lett. 65: 980 (1990).
7. L.A. Takhtajan, Adv. Stud. Pure Math. 19: 435 (1989).
8. R.K. Bullough, S. Olafsson, Yu-zhong Chen and J. Timonen, in: "Physics and Geometry", NATO ARW and Proc. XVIIIth Intl. Conf. on Diff. Geom. Methods in Theor. Phys., Ling-Lie Chau and Werner Nahm eds., Plenum, New York (1990) and references.
9. R.K. Bullough and J. Timonen, "Quantum groups and quantum complete integrability: theory and experiment" in: Proc. 19th Intl. Conf. on Diff. Geom. Methods in Theor. Phys., C. Bartocci, U. Bruzzo and R. Cianci eds., Springer Lecture Notes in Physics and the references in Ref. 6 there. To appear in 1991.
10. N.M. Bogoluibov and R.K. Bullough. To be published.
11. E.T. Jaynes and F.W. Cummings, Proc. IEEE 51: 89 (1963).
12. P. Filipowicz, J. Javainen and P. Meystrè, Phys. Rev. A 34: 3077 (1986).
13. L.A. Lugiato, M.O. Scully and H. Walther, Phys. Rev. A 36: 740 (1987).

14. F.W. Cummings and A.K. Rajagopal, Phys. Rev. A 39: 3414 (1989).
15. A.J. Macfarlane, J. Phys. A: Math. Gen. 22: 4581 (1989).
16. L.C. Bidenharn, J. Phys. A: Math. Gen. 22: L873 (1989).
17. P.P. Kulish and E.V. Damaskinsky, J. Phys. A: Math. Gen. 33: L415 (1990).
18. R. Brecha. This meeting.
19. N. Nayak, R.K. Bullough, B.V. Thompson and G.S. Agarwal, IEEE J. Quant. Electronics QE-24:1331 (1988).
20. N. Nayak, R.K. Bullough and B.V. Thompson in: "Coherence and Quantum Optics", J.H. Eberly, L. Mandel and E. Wolf eds., Plenum, New York, (1990), pp.809-813.
21. N. Nayak, B.V. Thompson and R.K. Bullough in: "Proc. ECOOSA '90 meeting", Rome, 7-11 November 1990, M. Bertolotti and E.R. Pike eds. To appear in "Quantum Optics", June 1991.
22. R.K. Bullough, M.R.B. Wahiddin, R.R. Puri, S.S. Hassan and G.P. Hildred, in: as in Reference 20.
23. R.K. Bullough, Hyperfine Interactions 37: 71 (1987) and references.
24. R.K. Bullough and 10 co-authors in "Squeezed and Nonclassical Light", P. Tombesi and E.R. Pike eds., Plenum Publ. Corp., New York (1989) and references.
25. G.S. Agarwal, R.K. Bullough and G.P. Hildred, Optics Commun., 39: 23 (1986).
26. G.S. Agarwal, R.K. Bullough, G.P. Hildred and N. Nayak, "Cavity quantum electrodynamics at arbitrary Q and temperature T". To be published.
27. G.S. Agarwal, R.K. Bullough and N. Nayak, "Probing the dressed states of an atom interacting with a quantized field". Submitted to Opt. Commun., March 1991.
28. N. Nayak, B.V. Thompson and R.K. Bullough, "Cavity quantum electrodynamics: finite temperature theory of the ^{85}Rb atom maser". To be published.
29. J.M. Raimond, P. Goy, M. Gross, C. Fabre and S. Haroche, Phys. Rev. Lett. 49: 117 (1982).
30. M.R.B. Wahiddin, S.S. Hassan, J. Timonen and R.K. Bullough in "Coherence and Quantum Optics", J.H. Eberly, L. Mandel and E. Wolf eds., Plenum, New York (1990), pp.1195-1200.
31. M.R.B. Wahiddin, S.S. Hassan, R.R. Puri, J. Timonen and R.K. Bullough in: as in Reference 21.
32. S. Haroche. Private Communication at the Davos Meeting, October 1990.
33. M. Brune. Private Communication. This meeting.
34. V.I. Arnold, "Mathematical Methods of Classical Mechanics", Springer-Verlag, Heidelberg (1978).
35. A. Janussis and 5 co-authors, Hadronic J. 6: 1563 (1983) and references.
36. A. Jannusis, Quantum Group and the Lie-Admissible Q-Algebra, in "Proc. 5th Intl. Conf. on Hadronic Mechanics and Nonpotential Interactions", and references. To apear 1991.
37. A. Jannusis, G. Brodimas, S. Baskontas and D. Vavougios "Quantum Group for Linear Combination of Bose and Fermi Operators", and references. Preprint 1990. Dept. of Phys., Univ. of Patras, Greece.
38. R.R. Puri and R.K. Bullough, J. Opt. Soc. Am. B 5: 2021 (1988).
39. H. Risken "Fokker-Planck Equations", Springer-Verlag, Berlin (1984).
40. N. Nayak, B.V. Thompson and R.K. Bullough. To be published.
41. Taken from the lecture by H. Walther at the Hyderabad meeting, January 5-10, 1991.
42. Tarik M. Al-Zubaydy "The collapse and revival of Rydberg atoms in a microwave cavity". M.Sc. Thesis, UMIST, Manchester, April 1989.

THE ACTION UNCERTAINTY PRINCIPLE

FOR CONTINUOUS QUANTUM MEASUREMENTS

Michael B. Mensky

P.N.Lebedev Physical Institute
117924 Moscow, USSR

1. INTRODUCTION

The path-integral approach to quantum theory of continuous measurements has been developed in preceding works of the author. According to this approach the measurement amplitude determining probabilities of different outputs of the measurement can be evaluated in the form of a restricted path integral (a path integral "in finite limits"). Given the measurement amplitude, the maximum deviation of the measurement outputs from the classical one can be easily determined. The aim of the present paper is to express this variance in a more simple and transparent form of an inequality of the type of the uncertainty principle. Symbolically it can be written as

$$\Delta(\text{Equation})\,\Delta(\text{Path}) \lesssim \hbar.$$

This means that deviation of the observed (measured) motion from that obeying the classical equation of motion is reciprocally proportional to the uncertainty in a path (the latter uncertainty resulting from the measurement error). The precise form of the inequality includes an integral over the measurement time interval:

$$\int_{t'}^{t''} \frac{\delta S[q]}{\delta q(t)} \Delta q(t)\, \mathrm{d}t \lesssim \hbar,$$

where the paths $[q]$ and $[\Delta q]$ stand correspondingly for the measurement output and for the measurement error. This inequality is valid for sufficiently fine measurements (those performed in the quantum regime). The consequence of the action uncertainty principle is that improving the measurement precision beyond the threshold of the quantum regime leads to decreasing information resulting from the measurement.

The question of measurements was one of the central conceptual problems of quantum mechanics from the very beginning and has not been solved completely so far. In recent years this question became even more burning first of all because of general rise of interest to foundations of quantum mechanics and secondly because of practical need in high-precision measurements requiring quantum effects to be taken into account (see for example[1] and references therein). A strong stimulus for development practical aspects of quantum theory of measurements arose due to development theory of gravitational-wave detectors.[1,2] Quantum measurements can be considered from a more wide point of view as influence of an "environment" on a quantum system.[3,4]

An important example of a complicated regime of measurement is continuous (protracted in time) measurement. For simplicity this type of measurement can be understood as a limit of sequential (repeated) instantaneous measurements with the time interval between successive instantaneous measurements tending to zero. This type of measurement attracted recently much attention in connection with the so-called quantum Zeno effect (paradox) known also under name of the watch-dog effect.[5,6] This effect consists in that the state of a quantum system "freezes" (its change is prevented) if it is observed (measured) continuously. If the measurement is not strictly continuous but successive measurements are often, then probability of transition of the measured system into another state decreases. This effect has been recently experimentally confirmed.[7]

As a result of the quantum Zeno effect a continuous measurement should give the trivial output (a constant function). However this is valid only for precise measurement of an observable having a discrete spectrum. If an observable having a continuous spectrum is measured approximately (an unsharp measurement), then the continuous measurement leads to a nontrivial output (expressed by a function of time). Of course this type of measurement is spread much wider in practice. It is this type of continuous measurement that will be considered in the present paper. We shall use the term "continuous measurement" for it without making wording more precise.

The formalism of Feynman path integrals was applied to continuous quantum measurements,[8–14] and proved to be natural and efficient. An amplitude A_α for each output α of a continuous measurement can be presented in the form of a path integral in such a way that $P_\alpha = |A_\alpha|^2$ be a probability density for different outputs of the measurement. The most probable is of course the output $\alpha_{\mathbf{class}}$ predicted by classical theory. However deviation of outputs α from classical one $\alpha_{\mathbf{class}}$ is possible. The variance $\delta\alpha$ of probable outputs around $\alpha_{\mathbf{class}}$ is one of the most important conclusions of the theory. Of course the variance $\delta\alpha$ of the outputs depends on the error $\Delta\alpha$ of the measurement.

It turns out that (partly simplifying the situation) there is a quantum threshold $\Delta\alpha_{\mathbf{quant}}$ such that for rough measurements ($\Delta\alpha \gg \Delta\alpha_{\mathbf{quant}}$) a classical regime of measurement realizes while for fine measurements ($\Delta\alpha \ll \Delta\alpha_{\mathbf{quant}}$) the quantum regime arises. In the classical regime $\delta\alpha \simeq \Delta\alpha$, i.e. the variance of the measurement outputs is determined completely by the measurement errors. However in the quantum regime $\delta\alpha \simeq \Delta\alpha^2_{\mathrm{quant}}/\Delta\alpha$. In this case the variance of outputs is a consequence of quantum fluctuations (or of an inavoidable back reaction of a measuring device on the measured quantum system). This "quantum measurement noise" increases with the measurement error diminishing. This rather paradoxical phenomenon is the cause why there is no sense in decreasing the measurement error $\Delta\alpha$ beyond the quantum threshold $\Delta\alpha_{\mathbf{quant}}$. An optimal regime of measurement is on the frontier between the classical and quantum regimes, when $\Delta\alpha \simeq \Delta\alpha_{\mathbf{quant}}$. In this regime the variance of outputs is minimum and equal to $\Delta\alpha_{\mathbf{opt}} \simeq \Delta\alpha_{\mathbf{quant}}$. This is an absolute limit for the given type of measurement.

Of course the scheme of the preceding paragraph is only rough qualitative outline for much more complicated actual situation varying from one measurement to another. The two main corrections required for this rough scheme are following: 1) the quantum threshold $\Delta\alpha_{\text{quant}}$ and definition of the measurement regimes are to be defined separately for each frequency range of the process or for the given form of a force acting on the system under consideration; 2) there is a type of measurement[1,2,16,17] called quantum nondemolition (QND) measurement, which has no quantum threshold at all. QND measurements are in fact always performed in the classical regime.[18,19]

Any continuous measurement can be completely analyzed if the measurement amplitudes A_α are evaluated for all measurement outputs α. However this requires calculation of a path integral. The aim of the present paper is to give another and more simple way of such an analysis. It turns out that (at least for linear systems) the variance of the measurement outputs $\delta\alpha$ can be found from a rather simple inequality including the action functional. This inequality resembles the known uncertainty principle in many aspects. This is why it could be called "the action uncertainty principle".

2. AMPLITUDES AND ALTERNATIVES

From some point of view the main object in quantum mechanics is a probability amplitude because it expresses the principal difference between quantum and classical theory. Quantum mechanics differs in that not probabilities but probability amplitudes should be summed up for a quantum system. Dirac argued[20] that the most important distinction of quantum theory is not in operators but in amplitudes.

Suppose that some event can occur through one of two alternative channels and the probability amplitudes for realizing these channels are correspondingly A_1 and A_2. Then the complete probability amplitude for the event under consideration is

$$A = A_1 + A_2 \tag{1}$$

and its probability is

$$p = |A|^2 = |A_1 + A_2|^2.$$

The typical example is a particle passing trough one of two slits in an opaque screen with A_1 being an amplitude for the particle to pass through one of the slits and A_2 an amplitude for passing through the other slit. It is because of the law of amplitude summation (1) that passing through one of the slits cannot be considered independently from passing through the other slit. The consequence of this law is interference pattern at the scintillating screen when a series of particles passes through two slits.

The formula (1) is valid however only if there is no means to make known what one of two possible alternative channels has been actually realized. In this case *alternatives* are called, according to Feynman, *interfering*. They may be called also *quantum alternatives.*

If an additional observation (measurement) is performed giving information about what channel has been realized, then the amplitude summation rule changes into the probability summation rule:[21]

$$p = p_1 + p_2 = |A_1|^2 + |A_2|^2.$$

As a consequence, no interference pattern will arise in the two-slit experiment if for example the flow of photons falls on the opaque screen so that scattering of the photons shows what of two slits the particle has passed through. In such a situation Feynman called the alternatives *incompatible*. It seems also convenient to use the term *classical alternatives*.

Of course, one can say that in the latter case quite another physical system is dealt with. But equally right is to talk about additional observation of the same system. This is typical for the quantum theory of measurement: there is arbitrariness in what part of real world is included into the system and what is included into measuring device.

The amplitude summation rule is valid also for many alternative channels,

$$A = A_1 + A_2 + \ldots + A_n \tag{2}$$

provided that there is no possibility to find out what of the channels has been actually realized. If some observation (measurement) is performed giving information about the channel realized, then the amplitude summation rule is to be corrected. The way of correction depends on the information the measurement provides.

The information may be complete so that the realized channel is known precisely. Then probabilities of separate channels should be summed up instead of their amplitudes:

$$p = p_1 + p_2 + \ldots + p_n, \quad p_i = |A_i|^2.$$

For another type of observation information may be only partial. This means that the measurement is more rough. Let for example the measurement permits one to know whether the number of the realized channel belongs to one of the following pairs:

$$(1,2);\ (3,4);\ \ldots(n-1,n)$$

(we suppose the complete number of channels is even). Then probabilities corresponding to separate pairs are to be summed up but amplitudes should be summed up inside the pairs:

$$p = p_1 + p_2 + \ldots + p_{n/2}, \quad p_i = |A_{2i-1} + A_{2i}|^2. \tag{3}$$

In this case alternatives inside each pair are interfering (quantum), while different pairs are incompatible (classical) alternatives for the final event. In the case of yet more rough measurement all channels may be divided into triplets:

$$p = p_1 + p_2 + \ldots + p_{n/3}, \quad p_i = |A_{3i-2} + A_{3i-1} + A_{3i}|^2. \tag{4}$$

The values p_i in the last two formulas are probabilities for different results (outputs) of the measurement. For example the value

$$p_i = |A_{2i-1} + A_{2i}|^2$$

in Eq.(3) is the probability for the i-th pair to emerge as a result of the measurement. The value

$$A_{i-\text{th pair}} = A_{2i-1} + A_{2i} \tag{5}$$

is nothing else than the probability amplitude for the measurement to give the result expressed by the i-th pair. More precisely, this is the amplitude for the event under consideration to occur and the measurement of pairs of channels to show the i-th pair.

The situation described by the formulas (3), (4) can be modeled in a many-slit experiment. For this aim the photon flow should be directed at the opaque screen in such a way that pairs or triplets of slits could be distinguished by the photon scattering rather than individual slits. It might be interesting to realize an experiment of this type for direct experimental investigation of quantum and classical alternatives and the modified amplitude summation rules (3), (4).

3. PATHS AND CONTINUOUS MEASUREMENTS

This argument may be applied to *Feynman paths* considered as quantum alternatives. The amplitude $A(q'', q')$ for a particle to go over from the point q' to the point q'' (called a *propagator*) has been expressed by Feynman[22,21] in the form of sum (or rather integral) of the amplitudes $A[q]$ corresponding to all possible paths

$$[q] = \{q(t) | t' \leq t \leq t''\}$$

connecting the points q', and q'':

$$A(q'', q') = \int A[q] \, \mathrm{d}[q]. \tag{6}$$

The formula (6) is quite analogous to Eq. (2) but for paths in the role of quantum alternatives. And analogously to the above argument Eq. (6) for the propagator is valid only if there is no possibility to find out what path is realized as a channel for transition of a particle from q' to q''. This is usually the case. Let however suppose that a *continuous measurement* is performed simultaneously with this transition. Let this measurement give some information about the path of transition. Such an information can be expressed by some set of paths I_α. If the measurement gives the result α then the transition is realized through one of the paths $[q]$ belonging to the set I_α. Then, in analogy with Eq. (5), the amplitude for the transition from q' to q'' can be expressed[8–10] as an integral over paths belonging to I_α:

$$A_\alpha(q'', q') = \int_{I_\alpha} A[q] \, \mathrm{d}[q]. \tag{7}$$

If α is fixed, the amplitude $A_\alpha = A_\alpha(q'', q')$ can be considered as the propagator of a particle undergoing the continuous measurement (with the given output). If q', q'' are fixed, the same amplitude can be thought of as the probability amplitude for the continuous measurement to give the result α. Taking a square modulus $|A_\alpha|^2$ of the amplitude one can obtain the probability density for different results of the continuous measurement.

Generalization of the formula (7) on the case of an arbitrary quantum system and an arbitrary continuous measurement is straightforward. Generalization is also possible for the measurement of configuration of a quantum field. (In the latter case the measurement can be called *continual* because it is not only prolonged in time but also protracted in space).

4. CLASSICAL REGIME OF MEASUREMENT

The Heisenberg uncertainty principle $\Delta q \Delta p \gtrsim \hbar$ is convenient for expressing the features of instantaneous quantum measurements. However it is inconvenient for continuous measurements. Indeed, if the measurement is for example monitoring the coordinate q up to some error Δq, then it gives some information about momentum too, but it is difficult (though possible) to express this information quantitatively and apply the uncertainty principle.

Therefore the task arises to obtain restrictions of the type of the uncertainty principle but for continuous measurements. In all cases these restrictions can be derived from the path integrals of the type of Eq. (7). Let us now analyze some general features of such a calculation and try to formulate the uncertainty principle for continuous measurements.

The amplitude $A[q]$ of a separate path is, according to Feynman,

$$A[q] = e^{\frac{i}{\hbar} S[q]}$$

where $S[q]$ is the action functional for the system under consideration. The path integral (7) can therefore be rewritten as follows:

$$A_\alpha = \int_\alpha e^{\frac{i}{\hbar} S[q]} \, d[q] \tag{8}$$

where we omitted the arguments q', q'' and identified the result of the measurement α with the corresponding set of paths I_α.

The value A_α can be interpreted as an amplitude for the measurement to give the result α. This means that only those measurement results are probable for which A_α is large in absolute value. We shall try to estimate this value analyzing the behavior of the action functional $S[q]$.

In what follows we shall suggest that all sets α have the same "width" differing only by their forms. Then the value of A_α depends only on how rapidly $S[q]$ varies along α.

Remark 1. The idea is that the "measure"

$$d[q] = \prod_t dq(t)$$

is the same for all α so that A_α depends only on oscillations of the factor $\exp\left(\frac{i}{\hbar} S[q]\right)$ when $[q] \in \alpha$. The statement about the "measure" $d[q]$ can be given sense with the aid of the limiting process. More precise formulation is that $d[q]$ is invariant under the group G of time-localized shifts[23]

$$g = \{c(t) | t' \leq t \leq t''\}, \quad g[q] = \{q(t) + c(t) | t' \leq t \leq t''\},$$

and all sets α can be presented in the form $\alpha = g\alpha_0$ with different $g \in G$ and the same α_0.

It is known that only paths close in a sense to the classical trajectory contribute the unrestricted Feynman integral

$$A(q'', q') = \int e^{\frac{i}{\hbar} S[q]} \, d[q]. \tag{9}$$

This is because the action functional $S[q]$ is extremal on the classical trajectory $[q_{\text{class}}]$:

$$\left.\frac{\delta S[q]}{\delta q(t)}\right|_{[q]=[q_{\text{class}}]} = 0.$$

Indeed, $S[q]$ changes slowly in the vicinity of $[q_{\text{class}}]$ and all partial amplitudes $A[q] = \exp\left(\frac{i}{\hbar}S[q]\right)$ have almost equal phases in this vicinity giving "constructive interference".

The constructive interference maintains in the limits of the set

$$I_{\text{class}} = \{[q] | \, |S[q] - S[q_{\text{class}}]| \lesssim \hbar\}$$

and changes into "destructive interference" outside this set when oscillation becomes rapid. Therefore, the main contribution into the Feynman path integral is given by the set of paths I_{class}:

$$A = \int_{I_{\text{class}}} d[q] \, e^{\frac{i}{\hbar}S[q]} \, .$$

It is evident (by an analogous argument) that the measurement amplitude (8) has maximum value for $\alpha = \alpha_{\text{class}}$ where the set α_{class} contains the classical trajectory as its middle path.

Remark 2. The latter statement can be made strict with the help of the above mentioned group G of time-localized shifts. In terms of this group $\alpha_{\text{class}} = V[q_{\text{class}}]$ where the subset $V \subset G$ is symmetrical in respect with unity, i.e. contains g^{-1} together with any g.

Now the classical regime of measurement can be defined by the requirement that

$$\alpha_{\text{class}} \supset I_{\text{class}}. \tag{10}$$

In this case the integral $A_{\alpha_{\text{class}}}$ is practically equal to complete Feynman integral A. It is evident that A_α remains the same for $\alpha \supset I_{\text{class}}$. However A_α decays as an intersection $\alpha \cap I_{\text{class}}$ shrinks becoming almost null for $\alpha \cap I_{\text{class}} = \emptyset$ since the amplitude $A[q]$ oscillates rapidly in this case in all regions of the set α.

From the argument of the preceding paragraph one can see that the probable outputs of the measurement performed in the classical regime (10) may be characterized by the condition

$$[q_{\text{class}}] \in \alpha. \tag{11}$$

This condition takes into account the sets α including I_{class} and those having significant intersection with I_{class}. The only sets excluded by the condition are those having small or null intersection with I_{class}.

The condition (11) completely corresponds to conclusions of classical measurement theory, as could be expected in the case of classical regime of measurement. In fact, all predictions of classical measurement theory are valid for continuous measurements performed in the classical regime.

5. QUANTUM REGIME OF MEASUREMENT

Consider now the quantum regime of measurements when no set α includes $I_{\rm class}$. In this case different frequencies should be considered separately. A continuous measurement gives information about each frequency component of a path $[q]$. Therefore the measurement can be thought of as a number of measurements, each measuring a definite frequency component q_Ω. Typical is the situation when some frequency components q_Ω are measured in classical regime ($\alpha_\Omega \supset I_\Omega^{\rm class}$), while the other components are measured in quantum regime. Of course this analysis is applicable only to a linear system.

Let the system under consideration be linear i.e. the Lagrangian $L(q, \dot{q})$ and the action $S[q]$ quadratic. Then, expanding a path $[q]$ in frequency components with frequencies Ω and amplitudes q_Ω, we have an expansion for the action too:

$$S[q] = \sum_\Omega S_\Omega(q_\Omega).$$

The measure for integrating over paths factorizes as follows:[21]

$$\mathrm{d}[q] = \prod_\Omega \mathrm{d}q_\Omega.$$

As a result, the integral (9) factorizes in the following way:

$$A = \prod_\Omega A_\Omega, \quad A_\Omega = \int \mathrm{d}q_\Omega \mathrm{e}^{\frac{i}{\hbar} S_\Omega(q_\Omega)}.$$

Restriction of a path $[q]$ inside the path set α can be expressed in terms of frequency components too,

$$q_\Omega \in \Delta_\Omega(a_\Omega) = [a_\Omega - \Delta a_\Omega, a_\Omega + \Delta a_\Omega].$$

Here the set of intervals Δ_Ω for frequency components defines (at least for some measurements) the measurement output α.

The restricted integral (8) takes then the following factorized form:

$$A_\alpha = \prod_\Omega A_\Omega(a_\Omega), \quad A_\Omega(a_\Omega) = \int_{\Delta_\Omega(a_\Omega)} \mathrm{d}q_\Omega \, \mathrm{e}^{\frac{i}{\hbar} S_\Omega(q_\Omega)}. \tag{12}$$

In this formula the measurement is interpreted as the measurement of frequency components (that is possible not for arbitrary, but for a large class of measurements). The output of the measurement of the frequency component q_Ω is denoted by a_Ω, and the error of this measurement is denoted by Δa_Ω. An interval $\Delta_\Omega(a_\Omega)$ is equivalent to the pair $(a_\Omega, \Delta a_\Omega)$.

The most probable are classical outputs $a_\Omega^{\rm class}$ forming the expansion for the classical path $q_{\rm class}$. This means that $A_\Omega(a_\Omega)$ has maximum absolute value for $a_\Omega = a_\Omega^{\rm class}$. This is because the partial action $S_\Omega(q_\Omega)$ has extremum at $q_\Omega = a_\Omega^{\rm class}$ and oscillation of the integrand in (12) are minimum in this point.

Denote

$$I_\Omega^{\rm class} = \{q_\Omega | \, |S_\Omega(q_\Omega) - S_\Omega(a_\Omega^{\rm class})| \lesssim \hbar\}.$$

Then classical regime of Ω-measurement is defined by

$$\Delta_\Omega(a_\Omega^{\text{class}}) \supset I_\Omega^{\text{class}}$$

and quantum regime emerges otherwise. In classical regime probable are outputs a_Ω such that $a_\Omega^{\text{class}} \in \Delta_\Omega(a_\Omega)$. The latter condition may be written also as $a_\Omega^{\text{class}} \in \alpha$ since $\Delta_\Omega(a_\Omega)$ is a frequency component of α.

Consider now quantum regime of measurement of the Ω-component, when $\Delta_\Omega(a_\Omega^{\text{class}})$ is narrower that I_Ω^{class}. In this case an amplitude of the most probable output a_Ω^{class} is approximately equal to the measure of the integration range:

$$|A_\Omega(a_\Omega^{\text{class}})| \simeq \Delta a_\Omega. \tag{13}$$

Going over to another measurement output a_Ω differing from a_Ω^{class} one sees that the partial measurement amplitude A_Ω remains of the same value until oscillation of the exponent in the limits of Δ_Ω becomes small. The condition for this is that

$$\Delta S_\Omega \lesssim \hbar$$

for

$$\Delta S_\Omega = S_\Omega(q_\Omega) - S_\Omega(a_\Omega)$$

with any q_Ω in the limits of Δ_Ω.

The latter condition should be summed up over all Ω to give the following condition for the measurement output α to be probable:

$$\sum \Delta S_\Omega \lesssim \hbar.$$

Here summation is taken only over those Ω for which $a_\Omega \neq a_\Omega^{\text{class}}$.

The expression under the sum can be given the form

$$\left.\frac{\partial S_\Omega(q_\Omega)}{\partial q_\Omega}\right|_{q_\Omega = a_\Omega} \Delta q_\Omega,$$

with $\Delta q_\Omega = q_\Omega - a_\Omega$. This has the advantage that this expression is automatically equal to zero for $a_\Omega = a_\Omega^{\text{class}}$, so that the preceding sum may be taken over all Ω:

$$\sum_\Omega \left.\frac{\partial S_\Omega(q_\Omega)}{\partial q_\Omega}\right|_{q_\Omega = a_\Omega} \Delta q_\Omega \lesssim \hbar. \tag{14}$$

The deflections Δq_Ω here can be arbitrary providing that they do not drive out from the given measurement output α. If the inequality is valid for all possible choice of deflections then the output α is probable.

6. THE ACTION UNCERTAINTY PRINCIPLE

Going over back from frequency components to whole paths, one can rewrite the condition (14) in the form of an integral:

$$\int_{t'}^{t''} \frac{\delta S[q]}{\delta q(t)} \Delta q(t)\, \mathrm{d}t \lesssim \hbar. \tag{15}$$

By $[q]$ the middle path of the set of paths α is here denoted, and $[\Delta q]$ is a deviation of $[q]$ which does not drive it from α so that $[q + \Delta q] \in \alpha$ too. If the inequality (15) fulfils for a given set α and for any deflection $[\Delta q]$ not driving out from α, then the corresponding measurement output α emerges with high probability. f The inequality (15) resembles the uncertainty principle and plays the same role for continuous measurements. Since this inequality includes the action functional, it can be called the action uncertainty principle.

The entity

$$F(t) = \frac{\delta S[q]}{\delta q(t)} = \frac{\partial L(q, \dot{q})}{\partial q(t)} - \frac{\mathrm{d}}{\mathrm{d}t}\left(\frac{\partial L(q, \dot{q})}{\partial \dot{q}(t)}\right) \tag{16}$$

is nothing else than the left-hand side of the classical equation of motion:

$$\frac{\delta S[q]}{\delta q(t)} = 0.$$

This is why the inequality (15) can be symbolically rewritten as follows:

$$\Delta(\text{Equation})\, \Delta(\text{Path}) \lesssim \hbar.$$

It means that deviation from classical picture is insignificant for rough measurements (when Δ(Path) is large) and is significant for fine measurements (with small Δ(Path)). Fine measurements performed in quantum regime give rise to nonclassical behavior of the system as a picture arising from the measurements.

If the left-hand-side (16) of a classical equation is not equal to zero, then one could say that a fictitious force $F(t)$ has arisen equal to this nonzero value. The inequality (15) can therefore be rewritten as follows:

$$\int_{t'}^{t''} F(t) \Delta q(t)\, \mathrm{d}t \lesssim \hbar. \tag{17}$$

Here left-hand-side of the action uncertainty principle is presented as an integral of left-hand-sides $\Delta p(t) \Delta q(t)$ of an ordinary uncertainty principle with

$$\Delta p(t) = F(t) dt$$

defined by a fictitious force $F(t)$ arising as a result of the measurement noise.

7. AN EXAMPLE

The formula (17) is convenient also because it allows in a simple and transparent way to calculate minimum force observable with the help of the given continuous measurement. Let for example the system under investigation be a harmonic oscillation with the mass m and the frequency ω. Then $F = m(\ddot{q} + \omega^2 q)$. If the continuous measurement consists in monitoring coordinate q with the error Δa, then $|\Delta q(t)| \leq \Delta a$. The strongest inequality corresponds to the choice $\Delta q(t) = \Delta a$ for $F(t) \geq 0$ and $\Delta q(t) = -\Delta a$ for $F(t) \leq 0$. This brings the inequality (17) into the form

$$\Delta a \int |F(t)| dt \lesssim \hbar. \tag{18}$$

The fictitious force $F(t)$ satisfying to this inequality can arise as a measurement noise. To be observable, the actual force acting on the oscillator should be larger than this fictitious force. To estimate the minimum observable force one has to know (or suggest) its form. Let for example the force is harmonic with the frequency Ω. The fictitious harmonic force

$$F(t) = F_0 \cos(\Omega t + \phi)$$

can arise, according to Eq. (18), if $F_0 \lesssim \hbar/\tau\Delta a$ where $\tau = t'' - t'$ is the time of measurement. This gives an estimation for the minimum observable force in quantum regime of measurement: $\Delta F_{\text{quant}} = \hbar/\tau\Delta a$. The corresponding estimation for classical regime can be obtained in the framework of classical measurement theory starting from $F = m(\ddot{q} + \omega^2 q)$. This gives $\Delta F_{\text{class}} = m|\Omega^2 - \omega^2|\Delta a$.

Taking both classical and quantum measurement noises into account, one has

$$\Delta F_{\text{total}} = \Delta F_{\text{class}} + \Delta F_{\text{quant}} = m|\Omega^2 - \omega^2|\Delta a + \frac{\hbar}{\tau\Delta a}.$$

In the classical regime of measurement (when Δa is large enough) the second term of this sum is negligible while in the quantum regime (small Δa), on the contrary, the second term dominates. The optimal regime of measurement corresponds to $\Delta a = \Delta a_{\text{opt}}$, where

$$\Delta a_{\text{opt}}^2 = \frac{\hbar}{\tau m|\Omega^2 - \omega^2|}.$$

In the optimal regime the minimum is achieved of an observable force equal (up to the order of value) to

$$\Delta F_{\text{opt}} \simeq \sqrt{\frac{m\hbar|\Omega^2 - \omega^2|}{\tau}}.$$

Remark 3. According to this formula the minimum observable force is finite far from resonance frequency and tends to zero near the resonance frequency. The latter is connected with the suggestion accepted in the present consideration that initial and final positions of the system q', q'' are known precisely. The resonance frequency force can indeed be observed with any precision with the help of continuous monitoring position accompanied by the precise measurement of the initial and final positions[10] (this is in fact one of possible types of quantum nondemolition measurements). If these positions are not controlled so that the oscillator is in coherent states before and after the continuous measurement, then the standard quantum limit

$$\Delta F_{\text{res}} \simeq \frac{\sqrt{m\hbar\omega}}{\tau}$$

can be obtained for the observable resonance force.[10] This formula can be derived from the preceding one in suggestion that $|\Omega - \omega|$ cannot be smaller than τ^{-1}.

8. CONCLUDING REMARKS

A simple inequality (15) has been derived above serving as a condition for the continuous measurement outputs to be probable. This inequality expresses a sort of an uncertainty principle for continuous measurements. It can be called the action uncertainty principle.

Qualitative consequences of this principle arising in typical situation can be formulated as follows:

1) For a rough measurement (when the error of the measurement is large and the sets α wide) only those measurement outputs are probable which are predicted by classical theory. This is the classical regime of measurement.

2) For a fine measurement (when the error is small and the sets α narrow) probable are even those measurement outputs that are far from classical predictions. The more precise is the measurement, the wider is range of probable measurement outputs (this may be called quantum measurement noise). This is the quantum regime of measurement.

3) Optimal is the regime between classical and quantum regimes of measurement. More rough measurements are inefficient because of classical measurement errors. More fine measurements are inefficient because of the quantum measurement noise.

However this simple scheme is invalid for quantum nondemolition (QND) measurements. For QND measurement there is in fact no quantum regime of measurement.[18,19] Therefore no quantum threshold arises in this case and no absolute restriction on observable force exists. The QND measurement can be considered with the help of the action uncertainty principle too.

ACKNOWLEDGMENTS

The author is grateful to Professor P.Tombesi for kind hospitality at the Workshop in Cortina d'Ampezzo.

REFERENCES

1. C.M.Caves, in: *Quantum Optics, Experimental Gravity and Measurement Theory,* P.Meystre and M.O.Scully, eds., Plenum, New York (1983), p.567.
2. V.B.Braginsky, in: *Topics in Theoretical and Experimental Gravitation Physics,* V.DeSabbata and J.Weber,eds., Plenum, New York (1977), p.105.
3. W.H.Zurek, *Phys. Rev.* D24: 1516 (1981).
4. W.H.Zurek, *Phys. Rev.* D26: 1862 (1982).
5. C.B.Chiu, E.C.G.Sudarshan and B.Misra, *Phys. Rev.* D16: 520 (1977).
6. A.Peres, *Amer. J. Phys.* 48: 931 (1980).
7. W.M.Itano, D.J.Heinzen, J.J.Bollinger and D.J.Wineland, *Phys. Rev.* A41: 2295 (1990).

8. M.B.Mensky, *Phys. Rev.* D20: 384 (1979).
9. M.B.Mensky, *Sov. Phys.-JETP* 50: 667 (1979).
10. M.B.Mensky, *The Path Group: Measurements, Fields, Particles*, Nauka, Moscow (1983) (in Russian; Japan translation: Yosioka, Kyoto (1988)).
11. R.Ya.Khalili, *Vestnik Mosk. Universiteta* Ser. 3, Phys., Astr., 22: 37 (1981).
12. A.Barchielli, L.Lanz and G.M.Prosperi, *Nuovo Cimento* B72: 79 (1982).
13. C.M.Caves, *Phys. Rev.* D33: 1643 (1986).
14. C.M.Caves, *Phys. Rev.* D35: 1815 (1987).
15. V.B.Braginsky, Yu.I.Vorontsov and R.Ya.Khalili, *Zh. Eksp. Teor. Fiz.* 73: 1340 (1977).
16. W.G.Unruh, *Phys. Rev.* D19: 2888 (1979).
17. C.M.Caves, K.S.Thorne, R.W.P.Drever *et al* , *Rev. Mod. Phys.* 52: 341 (1980).
18. G.A.Golubtsova and M.B.Mensky, *Intern. J. Modern Phys.* A4: 2733 (1989).
19. M.B.Mensky, Continuous quantum nondemolition measurements and measurability of electromagnetic field from path integrals, in: *Proceed. Intern. Workshop on Gravitational Wave Signal Analysis and Processing, Amalfi, Italy, 1-5 July, 1988*, World Scientific, Singapore, to be published.
20. P.A.M.Dirac, *Fields and Quanta*, 3: 139 (1972).
21. R.P.Feynman and A.R.Hibbs, *Quantum Mechanics and Path Integrals*, McGraw-Hill, New York (1965).
22. R.P.Feynman, *Reviews of Modern Physics*, 20: 367 (1948).
23. M.B.Mensky, *Lett. Phys.*, A150: 331 (1990).

CAVITY QED AND CLASSICAL ANTENNA THEORY

Jonathan P. Dowling

National Research Council Associateship Program
United States Army Missile Command
Weapons Sciences Directorate, AMSMI-RD-WS
Redstone Arsenal, Alabama, 35898 USA

1. INTRODUCTION

We present a classical model for the radiation patterns of atomic dipoles confined between parallel mirrors. Taking a classical radiating dipole to model the atom, we then replace the parallel mirrors with an array of image dipoles that reproduce the correct boundary conditions on the electromagnetic field. Using elementary antenna theory we then compute the radiation pattern of the resultant antenna array. The patterns of this antenna are then plotted for various plate spacings and dipole positions. This model allows one to probe the mode structure of the cavity and to observe the directionality of the emitted radiation from the confined dipoles. In addition, the experimentally observed phenomena of spontaneous emission quenching and enhancement are predicted. These results show that the mode structure of a cavity is essentially a classical phenomenon and it effects the radiative properties of even classically radiating systems.

2. THEORY OF SPONTANEOUS EMISSION IN CAVITIES

As first pointed out by Purcell,[1] the spontaneous emission rate of an atom is not an intrinsic property of the atom, but also depends on the atom's surroundings. This can most easily be seen by considering Fermi's Golden Rule:

$$w=\frac{2\pi}{\hbar}\rho(\hbar\omega)\left|\langle f|H_{int}|i\rangle\right|^2 \qquad (1)$$

Quantum Measurements in Optics, Edited by P. Tombesi and
D.F. Walls, Plenum Press, New York, 1992

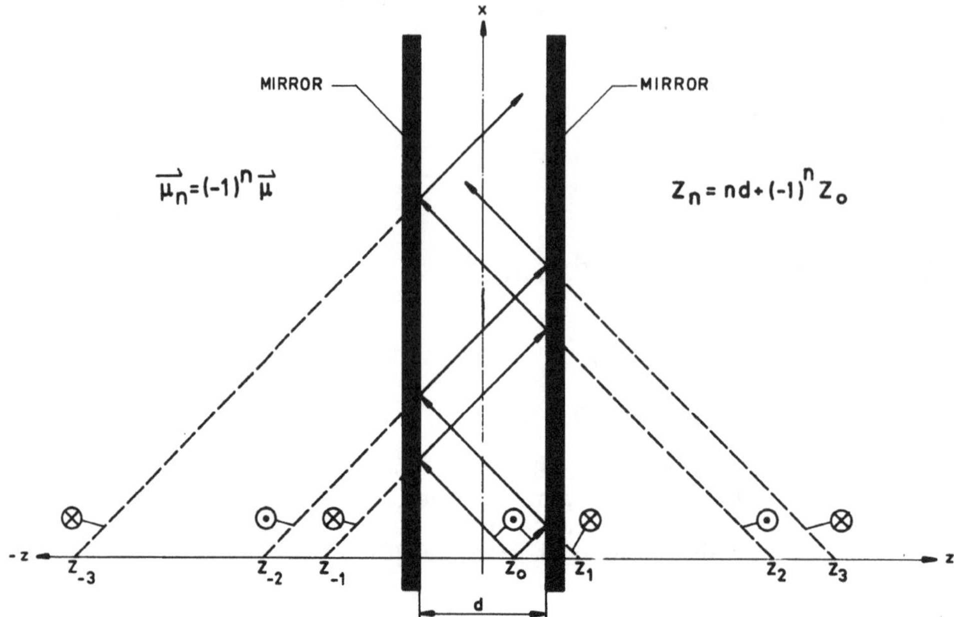

Fig. 1. An oscillating dipole is located between two mirrors on the z axis at z_0 with an instantaneous dipole moment μ pointing out of the page. Equivalent boundary conditions are obtained by removing the mirrors and replacing them with an array of image dipoles at locations z_n with moments μ_n. The radiation pattern of the resultant antenna can be computed, and then be interpreted as that of the original dipole. The total number of images $2N$ is related to the number of reflections a photon undergoes before it is lost, and hence to the Q value of the cavity, by $Nd \cong cQ/\omega$.

The transition probability w per unit time between the initial atom-field state i and the final state f depends upon the matrix element of the interaction, H_{int}, and also the mode density ρ. The mode density depends explicitly on the geometry of the cavity in which the electromagnetic field is quantized and hence the spontaneous emission rate depends on the cavity also. For parallel mirrors of separation d with an enclosed atom aligned transversely to the mirrors and radiating as a dipole at wavelength λ, as in figure (1), spontaneous emission is completely suppressed for a plate spacing of $d < \lambda/2$. For this range of plate spacings there is no vacuum mode of the correct wavelength available to stimulate the dipole transition and hence the atom never decays. A spontaneous emission enhancement by a factor of 3/2 is predicted by QED at $d=\lambda/2$; the decay rate then approaches the freespace value asymptotically as d approaches infinity. [2,3]

3. EXPERIMENTAL CONSIDERATIONS

Cavity-induced changes in the spontaneous emission rate between parallel mirrors can be seen in experiments such as those of Kleppner and co-workers,[4] Haroche and colleagues,[5] or De Martini et. al.[6] for example. In the experiment of Kleppner, atomic dipoles are aligned transversely to the mirrors and then launched in the excited state between the mirrors on trajectories parallel to the mirror surface. The lifetime of the atom in the cavity can be then determined by counting the number of atoms in the excited state as they emerge from between the mirrors. A very large suppression of spontaneous emission is seen for plate spacings of less then one half of a wavelength and the enhancement factor of 3/2 at $d=\lambda/2$ is also seen — as predicted by theory. The precise position of the atoms between the plates is not known and an averaging of the atom's location z_0 [see figure (1)] over the plate separation d is necessary.

In the experiments of De Martini et. al.[6] an active medium inside an adjustable-spacing parallel mirror cavity is pumped by an external laser, and the relaxation time of the intracavity atoms is measured. Spontaneous emission quenching is seen in this experiment also. Furthermore, one can control the location of excited atoms between the mirrors by varying the angle of the incoming pump beam. In a more recent version of this experiment[7] De Martini has seen correlations between confined atoms that are separated transversely, i.e. in the x direction in figure (1). The observation of such correlations leads naturally to the question "What is the radiation pattern of an atom confined between parallel mirrors?" The present model was developed in part to answer that question.

4. THE METHOD OF IMAGES AND ELEMENTARY ANTENNA THEORY

A simple way to reproduce the boundary conditions of parallel mirrors on the electromagnetic field is to use the method of images. This technique has been used in full QED calculations of spontaneous emission in cavities as a means of constructing the photon propagator or Green's function for the field in the cavity.[2,3] In the present model we will assume that we have a classical radiating dipole between the mirrors and we use the image method to reproduce the correct boundary conditions on the classical radiation field of the dipole. The mirrors are replaced by an array of $2N$ image dipoles whose phases alternate in sign, as shown in figure (1). If the Q of the cavity were infinite, the number of images would also be infinite. We can, however, model the finite Q of the cavity by restricting the number of dipoles to a finite quantity. One can see in figure (1) that the number N corresponds to the number of bounces a photon undergoes before it is lost from the cavity; which can be related roughly to the cavity Q by $Nd \cong cQ/\omega$ if we assume that the the photons dominating emission enhancement and suppression are those moving in the z direction. One can easily show that the image dipole locations are on the z axis at $z_n=nd+(-1)^n z_0$, and the relative instantaneous dipole moments are given by $\mu_n=(-1)^n\mu_0$ $(n=0,\pm1,\pm2,\ldots)$ where z_0 and μ_0 are, respectively, the location and moment of the actual dipole.

Now with the mirrors removed we are left with a linear array of radiating classical dipoles. This array can be treated as a classical antenna and the far field, normalized, radiation intensity pattern is then given by

$$I = \left| \sum_{n=-N}^{N} k_n \exp\left\{ -i \left[\frac{2\pi d_n}{\lambda} \cos\phi + \theta_n \right] \right\} \right|^2 \tag{2}$$

The pattern is normalized by dividing through by $(2N)^2 I_o$ where I_o is the intensity of the actual dipole and $2N$ is the number of image atoms. (Image atoms do not actually radiate power and division by this factor is necessary to maintain overall conservation of energy.) We have defined $d_n \equiv z_n - z_o$ and the alternating phase $\theta_n = \pi(1 - \cos n\pi)/2$ gives the correct flip-flopping of the image dipoles. The factor $k_n \leq 1$ can be used to model mirrors that are made of dielectric materials. We shall assume in this work the limit of perfectly conducting mirrors, or $k_n = 1$. The polar angle ϕ is measured in the counterclockwise direction from the positive z axis in figure (1).

5. A MULTI-LOBED MODE NODE PROBE

In figure (2) we plot the radiation pattern for an atom located midway between the mirrors, (2a), and also for an atom located near the right mirror, (2b), using formula (2). For this particular example we have chosen a mirror separation of $d = \lambda/2$ which corresponds to atomic resonance with the cavity if the dipole is located right in the middle. In order to touch upon the predictions of QED, we have indicated schematically by a dashed curve the vacuum mode of wavelength λ whose wave number is normal to the mirrors. In figure (2a) the atom would be located at a maximum of the vacuum mode and hence spontaneous emission would be stimulated at a rate that would be maximal. In figure (2b), however, the atom is located at a node of the vacuum mode an hence one would expect spontaneous emission from vacuum stimulation to be minimal. Both of these predictions show up in the classical radiation pattern.

In terms of our classical antenna model these results are interpreted as follows. When the dipole is located symmetrically at $z_o = 0$ the elements of the image array are all equally spaced half-a-wavelength apart, but neighboring elements are out of phase. [See figure (1).] Hence, a wavefront emitted in the z direction from an element n arrives at element $n+1$ in phase with the wavefront that is being emitted there and the two interfere constructively. Such an array configuration produces strong intensity lobes along the z axis, a situation known in the antenna business as an "endfire" pattern. Now if we move the dipole over to the surface of the right mirror the antenna array changes drastically. The images tend to bunch up into pairs of dipoles that are out of phase with one another. These antenna element pairs hence radiate very weakly as quadrapoles. The intensity of the antenna pattern in the $\pm z$ direction is strongly suppressed. Thus we see that our classical dipole and its associated multi-lobed radiation pattern acts as a sensitive probe of the nodal structure of the vacuum field mode.

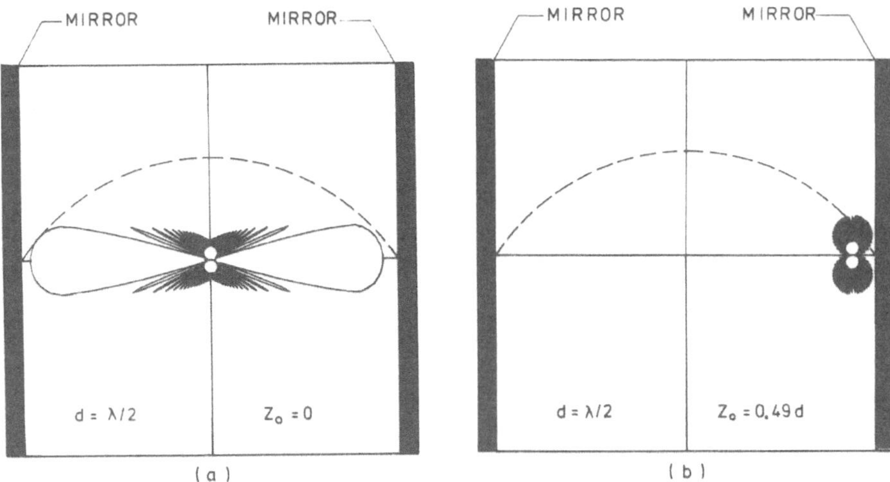

Fig. 2. This is a polar plot of the radiation pattern of a confined dipole, given by the normalized intensity formula in equation (2). The dipole is oriented out of the page as in figure (1). When the dipole is centered between the mirrors, (a), the image-antenna of figure (1) forms a uniform linear array phased for maximum endfire radiation that is resonant with the fundamental mode of the cavity. This situation corresponds to the maximum spontaneous emission rate of an actual atom — so oriented — located at the maximum of the fundamental vacuum mode of the cavity. (This mode is illustrated by the dashed curve.) If we now move the dipole over next to the right mirror, (b), the dipoles in the image-antenna join up in pairs to form an array of weakly radiating quadrapole elements and the radiation pattern is hence quenched in the z direction. (There is no mode confinement and hence little pattern change orthogonal to the z direction.) In QED this would be interpreted as the spontaneous emission suppression of an atom that is placed at the node of the vacuum mode.

6. SPONTANEOUS EMISSION QUENCHING

Now we wish to present a model of spontaneous emission quenching as seen in the experiments of Kleppner, Haroche, and De Martini.[4-6] In particular, in Kleppner's experiment, as was mentioned above, the atoms sample the entire range of z_0 values from the left mirror to the right mirror. We can include this lack of knowledge by averaging formula (2) for the radiation pattern over the range $-d/2$ to $d/2$ in the argument z_0. The averaged pattern is then plotted in figure (3) for various plate spacings d. When $d=\infty$ the dipole should radiate freely with a circular radiation pattern in the plane of the page. A relic of this circular pattern is seen even at $d=2\lambda$ in figure (3a). As one decreases the mirror separation the shrinking number of available modes is reflected in the decreasing number of radiation lobes. Finally, in figure (3d), there is only one available mode at

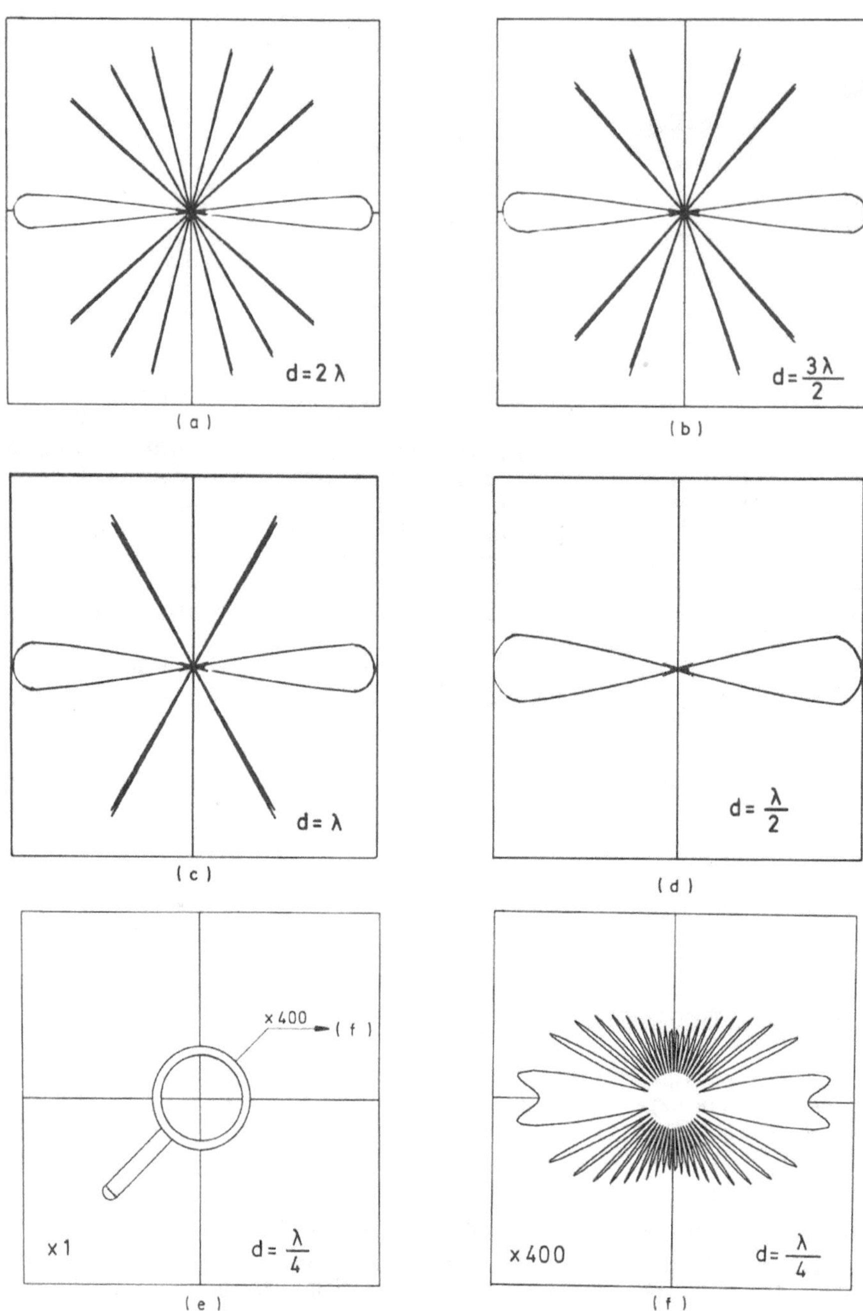

d = 2 λ
(a)
d = 3λ/2
(b)
d = λ
(c)
d = λ/2
(d)
x 400 → (f)
x 1
d = λ/4
(e)
x 400
d = λ/4
(f)

$d=\lambda/2$ — and the average dipole is in resonance with the cavity. However, if the separation is taken below $\lambda/2$ down to, say, $\lambda/4$ as in figure (3e), there are no modes left for the dipole to radiate into and the radiation pattern is quenched by a factor of about 400, as seen in figure (3f).

In the terminology of QED, there is no vacuum mode available to stimulate spontaneous emission of the atom and so the emission ceases. In our antenna model the result is interpreted as an antenna that is not phased properly for radiating efficiently.

7. SUMMARY AND CONCLUSIONS

As a complement to the usual full QED account of spontaneous emission in cavities, our classical model of a radiating dipole between parallel mirrors offers a simple picture of the salient features of cavity -induced spontaneous emission suppression and enhancement. Our approach highlights the fact that the mode structure of a cavity is really a classical property. A radiation process taking place inside a cavity — whether considered classically or quantum electrodynamically — will depend strongly on this mode structure. The radiative properties of any isolated system of charges are not only local functions of the system's own internal structure but also have a global dependence on the geometry of the surrounding space.

A simple mock-up experiment to demonstrate the classical nature of emission between mirrors is being planned at the University of Alabama at Huntsville. We will take a dipole antenna ($\lambda \cong 1$m) attached to an LC circuit and place it between two movable chalkboards covered with aluminum foil. As the chalkboards are pushed together — we hope to see an enhanced lifetime for our macroscopic "atom".

Fig. 3. In this figure we show the response of the confined dipoles to the global mode density of the cavity by averaging equation (2) over the mirror spacing. This averaged polar radiation intensity is plotted here for various mirror spacings d. For large d we would expect to recover the circular radiation pattern of an unconfined dipole — this circular pattern is apparent in (a) and (b). (We are using a fairly high Q value here, corresponding to $N=41$, in order for the cavity to match that used in the experiment of Kleppner and co-workers. An antenna with such a large number of elements is extremely directional for certain angles, explaining the "spikeyness" of the graphs.) As we take the mirror spacing down to $d=\lambda$ and $d=\lambda/2$, (c) and (d) respectively, the mode structure of the cavity becomes clearly apparent in the patterns. Below $d=\lambda/2$ there are no modes available for the dipoles to radiate into, (e), and the pattern is strongly quenched — appearing to vanish. There is, however, a small residual pattern that is reduced in power by a factor of about 400, (f).

AKNOWLEDGEMENTS

The authors would like to thank A. O. Barut, R. Chaio, R. Glauber, J. Lyman, P. Meystre, W. Schleich, A. Sizmann, and S. Zubairy for interesting discussions which contributed to this work. One of us (JPD) would like to acknowledge H. Walther and the Max-Planck-Institut für Quantenoptik for hospitality and financial support. Another of us (MOS) would like to acknowledge the Office of Naval Research for support.

REFERENCES

1. E. M. Purcell, "Spontaneous Emission Probabilities at Radio Frequencies," *Phys. Rev.* **69** : 681 (1946).

2. P. W. Milonni and P. L. Knight, "Spontaneous Emission Between Mirrors," *Opt. Comm.* **9** : 119 (1973).

3. A. O. Barut and J. P. Dowling, "Quantum electrodynamics based on self-energy: Spontaneous emission in cavities," *Phys. Rev. A* **36** : 649 (1987); J. P. Dowling, M. O. Scully, and F. De Martini, "Radiation Pattern of a Classical Dipole in a Cavity," *Opt. Comm.* (in press).

4. R. G. Hulet, E. S. Hilfer, and D. Kleppner, "Inhibited Spontaneous Emission by a Rydberg Atom," *Phys. Rev. Lett.* **55** : 2173 (1985).

5. W. Jhe, A. Anderson, E. A. Hinds, D. Meschede, L. Moi, and S. Haroche, "Suppression of Spontaneous Decay at Optical Frequencies: Test of Vacuum-Field Anisotropy in Confined Space," *Phys. Rev. Lett.* **58** : 666 (1987).

6. F. De Martini, G. Innocenti, G. R. Jacobovitz, and P. Mataloni, "Anomalous Spontaneous Emission Time in a Microscopic Optical Cavity," *Phys. Rev. Lett.* **59** : 2955 (1987).

7. F. De Martini, M. Marrocco, and D. Murra, *Phys. Rev. Lett.* **65** : 1853 (1990).

MACROSCOPIC DISTINGUISHABLE STATES OF MECHANICAL OSCILLATORS GENERATED BY QUANTUM NONDEMOLITION MEASUREMENTS

Roberto Onofrio

Dipartimento di Fisica, Università di Roma "La Sapienza"
and INFN, Sezione di Roma, P.le Aldo Moro 2, 00185 Roma, Italy

1. INTRODUCTION

The implications of quantum mechanics on the realism in the macroscopic world have been outlined from the early development of the theory.[1-3] In more recent times the feasibility of experiments capable to answer some questions originated from the Schroedinger-Einstein debate has been discussed with particular emphasis on the possibilities offered in the quantum optics framework.[4] It has been shown that a quantum non demolition monitoring can generate macroscopically states in the optical range, i.e. optical Schroedinger's cats.[5] On the other hand quantum non demolition strategies have been originated for monitoring macroscopic mechanical oscillators and thus may open the unique opportunity for monitoring a single, macroscopic system at the quantum level of sensitivity. We discuss here the possibility to generate macroscopic distinguishable states through quantum non demolition measurements on mechanical oscillators monitored by means of electrical quantities. In section 2 we recall the quantum non demolition strategies for a phonon-photon coupling with a particular care to the stroboscopic techniques.[6] In section 3 we discuss electromechanical transducers based upon electron tunneling. The high speed of such a kind of transducers allows to implement quasi-stroboscopic quantum non-demolition techniques. In section 4 we discuss the feasibility of experiments in which distinguishable states are generated in a macroscopic mechanical oscillator through a quasi-stroboscopic quantum non demolition coupling to an electromechanical vacuum tunneling transducer and some of the problems that can arise in such a detection strategy. Potential advantages compared to the optical situation are discussed, in particular regarding the mechanism to generate the coupling between the two modes and the destruction of the coherence through coupling to a thermal bath. General considerations on the relevance of these experiments for understanding quantum mechanics and its applications to the mesoscopic systems are finally given in section 5.

Quantum Measurements in Optics, Edited by P. Tombesi and
D.F. Walls, Plenum Press, New York, 1992

2. QUANTUM NONDEMOLITION STRATEGIES FOR PHONON-PHOTON COU-

The need for quantum nondemolition techniques arose in the analysis of the quantum limits of the sensitivity for detecting small displacements in macroscopic mechanical oscillators. High sensitivity monitoring of mechanical oscillators is of crucial importance mainly for detecting post-Newtonian effects predicted by general relativity, like gravitational waves.[7] We recall briefly the QND theory strictly required to understand the following considerations, reminding the interested reader the extensive, self consistent reviews already available.[8,9]

A measurement, either classical or quantum, is a particular kind of interaction between at least two systems, the 'measured' system and the 'meter' system, this last we suppose to be better understood or more easily controllable in practice than the first one. The information process, in general, is a reciprocal one. This means that due to the interaction both the measured system and the meter will change the values of the monitored quantities. The latter effect is desired, being the process of information on the measured system through the induced changes in the meter, but unfortunately the former effect is also present, i.e. the perturbation in the measured system. This effect is already present in a classical measurement and it is known as back-action. In the quantum regime this is well known within the framework of the Heisenberg uncertainty principle. However it was observed very soon that the Heisenberg principle does not forbid to repeat high precision measurements of only one observable provided that some constraints on the observables and the coupling of the meter are satisfied. In particular, QND measurements of the position of a mechanical oscillator (frequency $\omega_1/2\pi$) through monitoring of the electric charge of an electric oscillator (frequency $\omega_2/2\pi$) are possible if stroboscopic measurements are performed through the interaction Hamiltonian

$$\hat{H}_1 = E_0 \delta(t - \frac{n\pi}{\omega_1}) \hat{x}\hat{q} \tag{1}$$

where E_0 is the amplitude of the coupling electric field. It is also interesting to use the stroboscopic measurements of the complex amplitude X_1, this last defined as

$$\hat{X}_1 = \Re e[(\hat{x} + i\frac{\pi}{m\omega_1})e^{i\omega_1 t}] \tag{2}$$

obtainable through an interaction Hamiltonian

$$\hat{H}_1 = E_0 \delta(t - \frac{n\pi}{\omega_1})(-1)^n \hat{x}\hat{q} = E_s(t)\hat{x}\hat{q} \tag{3}$$

The time dependent electric field $E_s(t)$ can be written as a Fourier expansion such that the first term gives the interaction Hamiltonian [8,10]

$$\hat{H}_i = E_0 \cos\omega_1 t \cos\omega_2 t \hat{x}\hat{q} \tag{4}$$

provided a modulation at the electrical frequency is performed. This interaction Hamiltonian is quantum nondemolition and this may be easily seen with the help of the transition picture in fig.1. The transduction effect consists in the stimulated creation of photons at frequency $\omega_2/2\pi$ through the simultaneous parametric conversion of a phonon at frequency $\omega_1/2\pi$ due to the two driving, classical pumps working

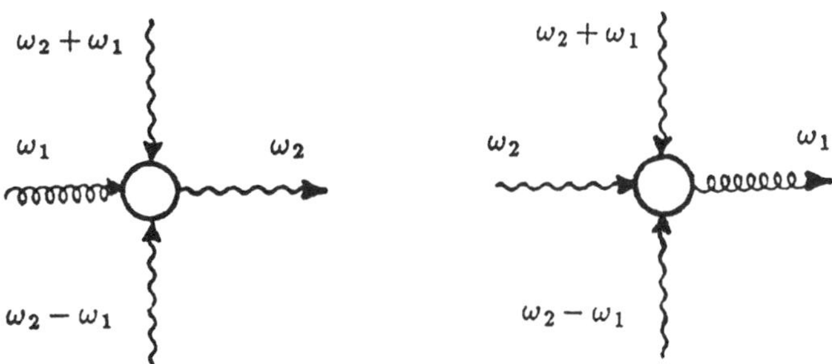

Fig. 1. Phonon-photon interaction in a back action evasion scheme: up-conversion with stimulated emission of photons at ω_2 due to phonons at ω_1 (transduction effect) and down conversion with stimulated emission of phonons at ω_1 due to photons at ω_2 (back-action effect).

at frequencies $(\omega_2 - \omega_1)/2\pi$ and $(\omega_2 + \omega_1)/2\pi$. This modulation produces an effect analogous to the one appearing in the Raman spectroscopy, i.e. lateral bands with amplitudes proportional to the modulation are produced. At the same time quantum noise present as photons at ω_2 will be parametrically down-converted into phonons at ω_1 through the same two pumps. However due to the phase inversion produced by only one of the two pumps the down converted phonon noise will be squeezed along a privileged axis. Thus the signal-to-(quantum) noise ratio is phase sensitive, i.e. time dependent, and instants of time in which this ratio is infinite are present. In the classical limit this pumping mode is able to give back action evasion of the classical noise at ω_2 and the down-converted noise is named, more properly, demodulated instead of squeezed like happens in the quantum regime. Of course also noise at $\omega_2 + 2\omega_1$ and $\omega_2 - 2\omega_1$ are down converted into phonons at ω_1, but without phase inversion (both the pumps lie on the same side with respect to those frequencies). By adding two other pumps at $\omega_2 \pm 3\omega_1$ and by properly matching the relative phases we obtain squeezing or demodulation of the noise at $\omega \pm 2\omega_1$ along the same direction. By repeating the procedure we obtain, as a process of limit, the stroboscopic Hamiltonian in (3).

In order to perform quasi-stroboscopic measurements a high speed transducer, i.e. characterized by a small time constant, is required. It would be also welcome to reach quantum noise limits with a device quite shielded by the characteristics of the electronic chain used to amplify and to process the signal, i.e. which does not need demanding requirements on the amplifier noise.

3. THE TUNNELING TRANSDUCER AS A QUANTUM LIMITED DEVICE

The requirements cited above are satisfied by a class of electromechanical transducers recently proposed and currently under experimental investigation based upon electron tunneling. The modulation of the gap between a tip and a test mass (see fig.3) gives a modulation of the tunneling current in which the natural lenght for the displacement sensitivity is the De Broglie wavelength of the tunneling electrons, of

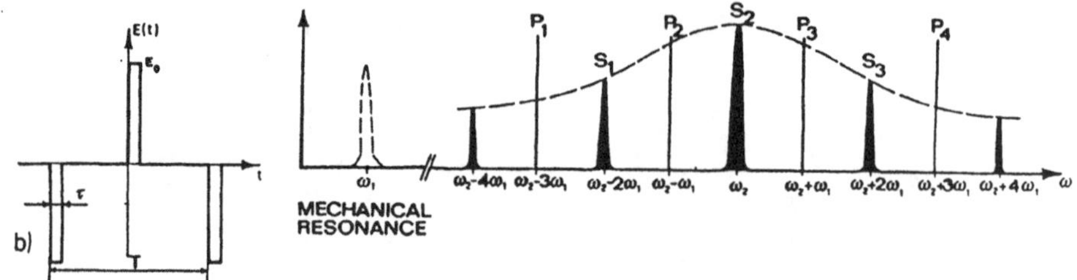

Fig. 2. Electric field versus time and related Fourier spectrum for a stroboscopic measurement of X_1 in the limit of $\tau \to 0$.

the order of $10^{-10}m$. The effective resistance of the tunneling transducer is given by the formula

$$R = R_0 e^{-2k_0 x} \tag{5}$$

in the limit of $k_0 l \gg 1$, where $\hbar k_0 = \sqrt{2m(V_0 - E)}$. The key point in this class of electromechanical transducers is the small effective output capacitance.[11] This small value is obtained maintaining at the same time a high sensitivity because this last depends upon the tunneling current and it is responsible for the two desired effects. The neglection of the back-action force noise due to the amplifier following the transducer and a time constant of the transducers small compared to the mechanical frequency, being of the order $\tau = RC \simeq 10^{-8} - 10^{-11}s$. The first effect means that there are no stringent requirements on the amplifier noise, furthermore the second effect allows to implement high-speed switching voltages to perform the quasi stroboscopic measurements described above. Of course quantum noise will arise earlier in another part of the device, namely in the electron-barrier interaction. Detailed calculations have been carried out both in a second quantization approach [12] and in a first quantization approach.[13] The final result is that the noise spectral density of the force fluctuations due to the Heisenberg principle is expressed for a rectangular barrier as

$$S_{f_Q} = 2\frac{I_0}{e}\Delta p^2 = \frac{I_0}{e}\hbar k^2 \frac{T}{2}\{[1 + (\frac{k_0}{k})^2]^2 - [1 - (\frac{k_0}{k})^2](1 - T)\} \tag{6}$$

where $\hbar k = \sqrt{2mE}$, I_0 is the tunneling current, Δp the uncertainty momentum and T the transmission coefficient. To see the conditions under which the quantum noise will be dominant, we write the other fundamental noise coming from statistical mechanics, the Brownian noise force spectral density

$$S_{f_B} = 4m\omega_1 \frac{K_B T_m}{Q_m} \tag{7}$$

where Q_m is the quality factor of the mechanical oscillator and T_m its thermodinamical temperature. To assure dominance of quantum noise over Brownian noise the

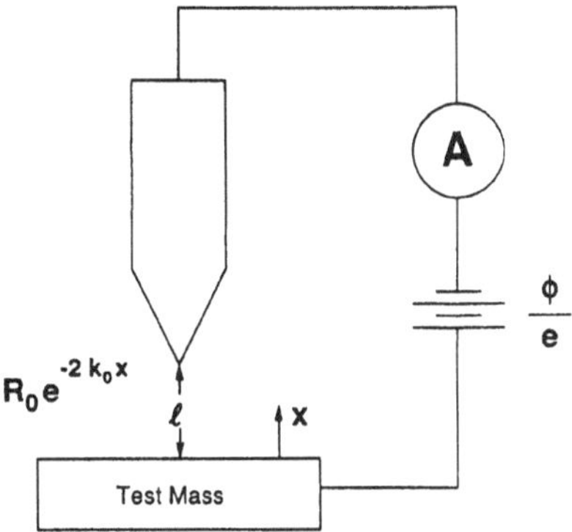

Fig. 3. Detection of displacement through a vacuum tunneling transducer. The tunneling probe is voltage biased and the current is sensed by a current amplifier.

following inequality has to be satisfied

$$\frac{10^{-6}}{I_0}\frac{m}{10^{-10}Kg}\frac{T_m}{10mK}\frac{\nu_1}{10^5 Hz}\frac{10^7}{Q_m} < 1 \tag{8}$$

where we assume $k_0 = 10^{10}m^{-1}$. It is possible to satisfy (8) by cooling at very low temperature a micromachined mass of silicon having resonance modes at high frequency, provided currents of the order of $1\mu A$ can be obtained without adding technical noise.

The other, technical, sources of noise may be decreased below the Brownian noise level with the current technology. Thus the tunneling transducer seems optimal to monitor very small masses with enough sensitivity to reach the quantum limit using quasi-stroboscopic techniques. Two problems for optimizing the sensitivity of this scheme have to be stressed. Firstly, there is an optimal value of the duration of the signals in the quasi-stroboscopic measurements.

If the duration is very small the effective coupling is very low too, but on the other hand only a very short duration of the impulses simulates a series of equal amplitude peaks. In fig.4 the Fourier spectrum for a signal like that shown in fig.2 is shown for four different values of the duration-period ratio, showing that a close simulation of the multipump scheme is obtained for very short durations. Secondly, the use of large electric fields in the small intervals of time increases the fundamental uncertainty product, producing a correlation which has to be avoided. In fig. 5 the uncertainty product versus the drop voltage is shown for a tunnel barrier biased with a constant electric field.

4. GENERATION OF SCHROEDINGER CATS THROUGH ELECTROMECHANICAL DEVICES

Generation and detection of Schroedinger cats, a topic originated from a debate between Einstein and Schroedinger on the consequences of quantum measurement theory in the macroscopic world, has been recently discussed in the light of the improvements of the technology to design feasible experiments. In particular recently it has been discussed a kind of QND scheme able to generate distinguishable

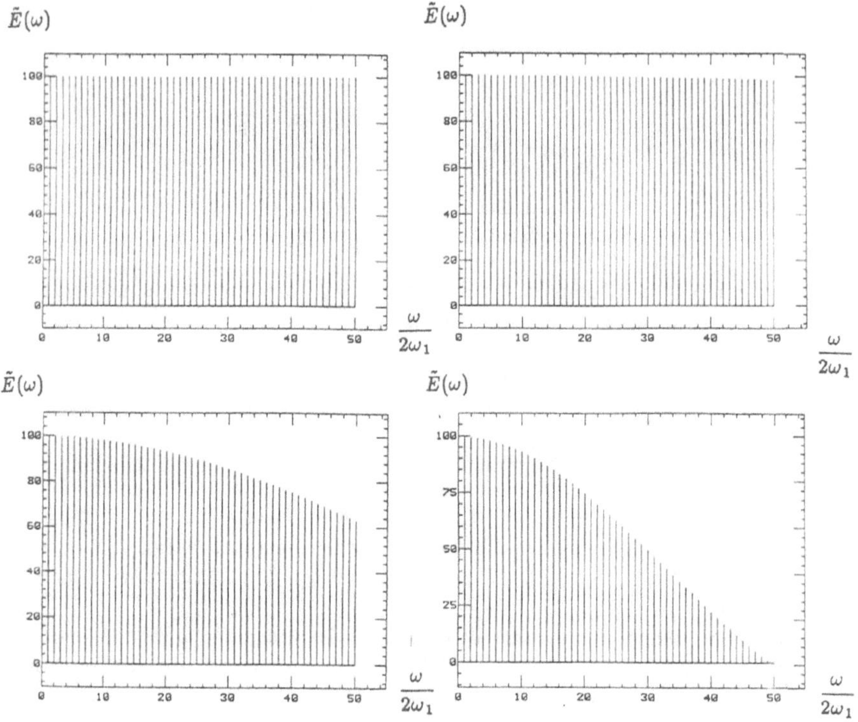

Fig. 4. Fourier spectrum for quasi-stroboscopic schemes for four different situations corresponding to $\tau/T = 10^{-4}, 10^{-3}, 5 \cdot 10^{-3}$ and 10^{-2}.

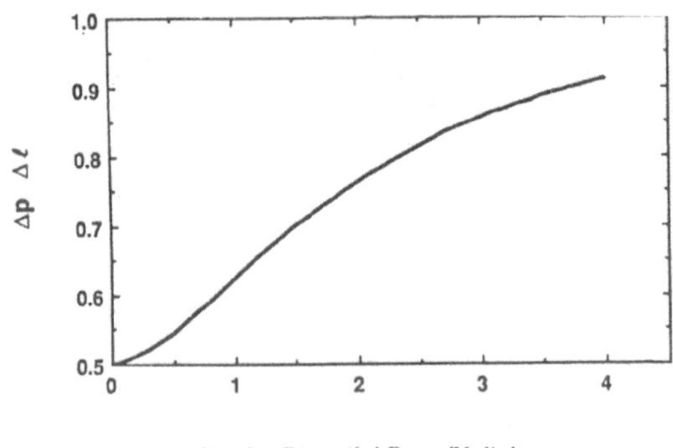

Fig. 5. Uncertainty product versus the drop voltage in a constant electric field biased barrier.

macroscopic states.[5] The idea is to correlate two harmonic oscillators in a QND way and to measure the energy in one oscillator. In such a way the oscillator is put in an energy eigenstate and the QND coupling originates a distinguishable state in the other oscillator. The situation consisting of two optical modes has been discussed pointing out also two major problems which make difficult to reach the sensitivity at which distinguishable states can be generated. The first is due to the low value of the non linearities present in the medium used to mix the beams. To overcome this problem they propose to inject squeezed light at one input port, relaxing in such a way the requirements on the strenght of the non linearities. Furthermore, to distinguish a macroscopic distinguishable state from a statistical mixture they propose to use a homodine detection in which the phase in quadrature to the one of maximum distinguishability is detected. In that phase the presence of interference fringes is the signature of the presence of a macroscopically distinguishable pure state. The presence of dissipations, eg. due to the detector efficiency, is a source of destruction for the fringes as calculated by means of an application of the Feynman-Vernon model for open quantum systems.[14,15] The result is a very high sensitivity to the presence of dissipations, even for a detector efficiency close to 1 the fringes are destroyed. The effect of the destruption of the interference fringes is taken into account by means of a detection efficiency η $(0 < \eta < 1)$. Even for η close to 1 but different from this value the fringes disappear. The two problems described here have to be re-examined in the case of generation of macroscopically distinguishable states with a quantum limited electromechanical transducer. Concerning the first aspect the non linearity required in the optical range, needed to have mixing of the electric field, is not required in the case of a phonon-photon coupling. What is important in that case is the value of the electric field, and this can be indefinitely increased for a vacuum tunneling transducer. Concerning the second point, dissipations in a mechanical oscillator are quite low compared to electrical or optical oscillators. The efficiency of the detection of fringes η is here related to the amount of thermal noise introduced during the measurement time like $\eta = 1 - \gamma\delta t$, where γ is the damping coefficient of the mechanical oscillator, and the corresponding mechanical quality factor is

$$Q_m = \frac{\pi\omega_1\delta t}{1-\eta} \tag{9}$$

By choosing $\omega_1 = 10^8 s^{-1}, \delta t = 10^{-2}s, \eta = 0.98(0.95)$ we obtain respectively $Q = 1.5 \cdot 10^8 (6 \cdot 10^7)$. Mechanical quality factors up to $Q \simeq 10^9$ have been obtained at low temperature using monocrystals of sapphire with low levels of impurities. Methods to measure position of a mechanical oscillator at high sensitivity are also available, as well as sensitive methods to measure the energy in an electrical oscillator have been proposed.[16] Thus the actual realization of a distinguishable state using electromechanical systems seems to present some advantages compared to the optical case and is worth to be investigated experimentally.

5. CONCLUSIONS

Quantum mechanics has been developed as a dynamical theory of the motion of microscopic degrees of freedom like the ones of atoms and molecules interacting through electromagnetism. Although the first efforts have been made for understanding interactions appearing on a smaller scale in terms of the same dynamical structure, it is also required to enlarge the actual applicability of quantum mechanics to the macroscopic world.[17] Perhaps the most fascinating possibility in this direction is the detection of macroscopic distinguishable states and we have discussed the use of coupled mechanical and electrical oscillators to pursue this goal. Provided this very difficult experiment can be successfully performed, a confirmation of quantum mechanical predictions can make the current interpretation of quantum measurement theory sharper, opening new experiments also in the so-called mesoscopic regime.[18] Possible departures from quantum mechanical predictions could also play a crucial role for understanding the procedure of quantization in itself and the possibility to apply it to interactions, like gravity, which phenomenologically only manifest at the macroscopic level.

ACKNOWLEDGMENTS

I am grateful to Paolo Tombesi for the kind invitation.

REFERENCES

1. E. Schrodinger, *Naturwissenschaften*, 23:807 (1935);23:823 (1935);23:844 (1935), English translation of J. P. Trimmer in Proc. Am. Philos. Soc.124:323 (1980).
2. K. Przibram (editor), *Letters from Einstein to Schrodinger in* Letters on wave mechanics, (Philosophical Library, New York, 1967), pp.35-40.
3. J. A. Wheeler and W. H. Zurek (editors), *Quantum Theory and Measurement*, (Princeton University Press, Princeton, 1983).
4. B. Yurke, D. Stoler, *Phys. Rev. Lett.* 57:13 (1986).
5. B. Yurke, W. Schleich, D.F. Walls, *Phys. Rev. A* 42:1703 (1990).
6. R. Onofrio, *Phys. Lett. A* 148:1 (1990).
7. V.B. Braginsky, Yu. Vorontsov, K.S. Thorne, *Science* 209:547 (1980).
8. C.M. Caves, K.S. Thorne, R.W.P. Drever, V. Sandberg, M. Zimmermann, *Rev. Mod. Phys.* 52:341 (1980).
9. V.B. Braginsky, *Usp. Fiz. Nauk.* 156:93 (1988), (Sov. Phys. Usp.31:836 (1989)).
10. M.F. Bocko, W.W. Johnson, *Phys. Rev. Lett.* 47:1184 (1981).
11. M.F. Bocko, K.A. Stephenson, R.H. Koch, *Phys. Rev. Lett.* 61:726 (1988).
12. B. Yurke, G.P. Kochanski, *Phys. Rev. B* 41:1703 (1990).
13. C. Presilla, R. Onofrio, M.F. Bocko, *submitted to Phys. Rev. B.*
14. R.P. Feynman, F.L. Vernon, *Ann. Phys.* 24:118 (1963).
15. A.O. Caldeira, A.J. Legett, *Physica A* 121:587 (1983).
16. V.B. Braginsky, S.P. Viatchanin, *IEEE Trans. Magnetics*, Vol. MAG-19:570 (1983).
17. D.M. Greenberger (editor), *New Techniques and Ideas in Quantum Measurement Theory*, Ann. N.Y. Acad. Sci., vol. 480 (1986), in particular the contributions of A.J. Legett, S. Chakravarty and C.D. Tesche.
18. B.L. Al'tshuler, P.A. Lee, R.A. Webb (editors), *Mesoscopic Phenomena in Solids*, (Elsevier, Amsterdam,1990).

III. QUANTUM NOISE REDUCTION AND SQUEEZING

PROGRESS AND PERSPECTIVES IN SQUEEZING

Elisabeth Giacobino, Thierry Debuisschert, Jerome Mertz,
Laurent Hilico, Antoine Heidmann, Serge Reynaud and Claude Fabre

Laboratoire de Spectroscopie Hertzienne de l'ENS, associé au CNRS
Université Pierre et Marie Curie, BP 74,
F75252 Paris Cedex 05, France

1. INTRODUCTION

A very successful way to produce squeezed light has been to use parametric generation in a crystal having a second order type non-linearity. In parametric generation, a single pump photon is down-converted into a pair of simultaneous photons. Degenerate parametric generation in a cavity has been shown to yield a high amount of quadrature squeezed light[1]. Non degenerate parametric generation gives rise to "twin" beams, presenting a high degree of quantum intensity correlation. The fluctuations in the intensity difference between two such fields is reduced below the standard quantum limit[2].
Recent investigations of the parametric emission in single pass in the low intensity cw regime have demonstrated a degree of correlation up to 67% below the standard quantum limit[3]. Higher intensities were obtained with pulsed pumping where a degree correlation up to 75% below the standard quantum limit was observed[4]. Alternative experiments were conducted where the crystal was placed in an optical cavity and cw parametric oscillation was investigated. The noise reduction in the signal and idler intensity difference spectrum is limited in this case to the bandwidth of the optical cavity[2]. Noise reduction using an optical parametric oscillator was first observed using a type II phase-matched crystal[5], obtaining a 69% noise reduction[6], and in later experiments using a type I phase-matched crystal, obtaining about 50% noise reduction[7-8]. Recently, an improved result in the type II phase matching experiment has allowed to observe a 86% (8.5 db) noise reduction, owing to particular care in reducing the influence of the loss mechanisms that deteriorate the pairwise detection of the twin beams[9]. One of the main interests of this system is that this noise reduction is robust with respect to the OPO parameters[10]. The high correlation can then be utilized to correct the quantum fluctuations in one beam by monitoring the other one, so as to obtain a sub-Poissonian beam[11].

Quantum Measurements in Optics, Edited by P. Tombesi and
D.F. Walls, Plenum Press, New York, 1992

2. TWIN BEAM GENERATION

In theory there is no fundamental limitation to the amount of noise reduction at zero frequency that may obtained in the intensity difference spectrum of twin beams in parametric generation. The noise spectrum is a Lorentzian dip, which starts from zero at zero frequency and returns to the standard quantum limit (or shot noise limit) at noise frequencies well outside the cavity bandwidth. The limitations are all technical in nature and, in particular, are related to how well the twin beam photons are generated and observed as simultaneous photon pairs. The optical losses in the apparatus occur both in the optical cavity where the twin beams are generated, and in the photodiodes where they are detected. When those are taken into account, the measured signal and idler intensity difference spectrum, normalized to its associated shot noise level, is [10]:

$$S_{1-2}(\Omega) = 1 - \frac{\xi\eta}{1+\Omega^2} \tag{1}$$

where Ω is the noise frequency normalized to the cavity bandwidth, η is the photodiode quantum efficiency, and ξ is the OPO output coupling efficiency given by:

$$\xi = \frac{T}{T+L} \tag{2}$$

where T is the transmission coefficient of the cavity coupling mirror and L is the extraneous cavity loss coefficient. The noise remaining at zero frequency is simply $1-\eta\xi$, ie the ratio of the lost photons to the total number of emitted photons. It must be noticed that the noise reduction is independent of the pump noise and of the pump power. Improvements in the noise reduction are concomitant to improvements in both ξ and η.

The intracavity loss coefficient L is fixed by unavoidable absorption in the crystal and imperfections in the optical coatings. A large output coupling efficiency ξ is obtained then by an increase in the output transmission T of the coupling mirror. A constraint is imposed by the fact that the OPO must remain above threshold for the available pumping power. An increase in T must be accompanied, therefore, by an increased recycling of the pump field in the cavity. The threshold pumping power scales approximately as

$$P_{th} \propto T_p(T+L)^2 \tag{3}$$

where T_p is the mirror transmission coefficient for the pump. As the threshold pump power P_{th} depends on the square of the transmission for the signal and idler wave, the optimum value is not easy to adjust. Moreover, this formula supposes that the pump and down-converted modes are resonant for the same cavity length, which is difficult to obtain in experiment. In particular, the oscillation condition in the OPO imposes that the detuning from the cavity resonance is necessarily the same for the signal and idler waves, but the pump wave may have a very different detuning from resonance, which is not favorable for oscillation[12]. If the cavity finesse for the pump is kept low enough, one can always find a cavity length for which all three modes are close to resonance. However this assumes that there is no phase shift on reflection for the three modes. The cavity mirrors are coated for this purpose.

The OPO in our experiment consists of a 7-mm long potassium-titanyl phosphate (KTP) crystal inside an optical cavity. The intracavity loss coefficient L is estimated at 0.6% (double pass). The cavity is triply resonant with a pump finesse of 70 (the finesse in previous experiments was of the order of 5). The output mirror transmission coefficient is 6.3% for the infrared, giving $\xi = 0.91$. The input mirror is high reflection coated for the infrared.
The detectors in this experiment are Epitaxx 300 *pin* photodiodes with quantum efficiencies η of 0.9. An improvement in these quantum efficiencies has been achieved by tilting the photodiodes in the polarization planes of the separated IR beams, to approach the Brewster angle. With a tilt angle of 45°, η was determined to be close to 0.94. Larger tilt angles gave rise to spurious edge effects.

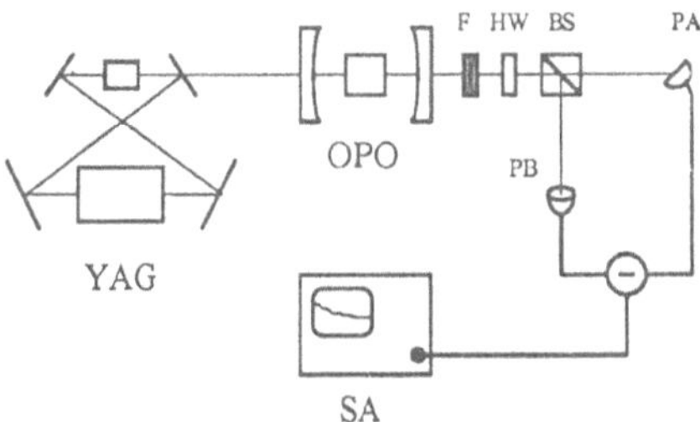

Fig. 1. Experimental layout: an Optical Parametric Oscillator OPO is pumped by an intracavity doubled YAG. F is a dichroic filter. HW is a half-wave plate. BS is a polarizing beamsplitter. PA and PB are tilted photodiodes. SA is a spectrum analyzer.

The layout of the experiment is shown in fig (1). The optical parametric oscillator is pumped above threshold by an intracavity doubled Nd:YAG laser. The KTP crystal in the oscillator is temperature stabilized and type II phase matched so that the down-converted twin infrared beams are cross polarized with nondegenerate wavelengths separated by less than one nanometer. The OPO cavity mirrors are 35 mm apart, with radii of curvature 20 mm. The pump beam is single mode cw, with wavelength 532 nm.
The pump beam passing through the OPO is blocked by dicroic filter F. The outgoing down-converted beams are separated by polarizing beamsplitter BS, and monitored by photodiodes PA and PB.
The DC portions of the photodiode signals (DC - 30 kHz) are added and used to actively stabilize the intensity of the twin beams. This intensity is dependent on OPO cavity detuning, and the beams are held at 3 mW by piezoelectric control of the cavity length. The gain balance of the DC summing network was verified and found to be better than 1% .

The AC portions of the photodiode signals (0.5 - 20 MHz) are subtracted and the resultant AC intensity difference is recorded on spectrum analyzer SA. The balance of the AC differencing network was verified by modulating the twin beam intensities and observing the reduction in the modulation peak when the intensities were subtracted. The total optical and electrical common mode rejection was found to be -35 dB up to 6 MHz and -30 dB up to 20 MHz.

A half wave plate HW is inserted before the beamsplitter BS to rotate the polarizations of the twin beams. When these are turned 45° with respect to the axes of BS, the polarizing beamsplitter becomes effectively a 50% beamsplitter, which allows a measurement of the shot noise spectrum[5].

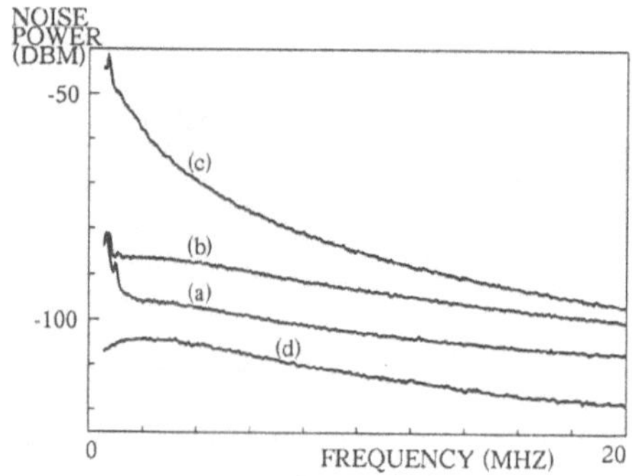

Fig. 2. Experimental results: curve (a) is the twin beam intensity difference power spectrum, curve (b) is the associated shot noise power spectrum, curve (c) is a single beam power spectrum, and curve (d) is the electrical noise floor.

The results of the experiment are shown in fig (2). Curve (a) represents the intensity difference power spectrum, curve (b) represents its associated shot noise power spectrum, curve (c) represents the single beam intensity power spectrum, and curve (d) represents the electrical noise. Large single beam excess noise is observed, due to OPO pumping close to threshold. The intensity difference spectrum, normalized to the shot noise spectrum (after both are corrected for the electrical noise) is presented on a linear scale in fig (3). The spectrum has a minimum corresponding to 86% ($\pm$ 1%) observed noise reduction below the shot noise level (curve (c)). To our knowledge, this is the largest amount of quantum noise reduction observed to date, comparable to the results reported by Yamamoto et al[13]. This result is presented along with our previous results from 1987 and 1989 ($\xi = 0.57$, $\eta = 0.9$, curve (a) and $\xi = 0.77, \eta = 0.9$, curve (b)), which illustrate the dependence of the noise reduction on the parameters ξ and η. Note that the progressive improvement of ξ is accompanied by a progressive improvement in the bandwidth of the noise reduction. The rise in the noise reduction at low frequency is due to the imbalance in the twin beam intensities.

The error quoted above for the observed noise reduction is based on errors estimated for the calibration of the shot noise level. Several auxiliary experiments were made to verify this calibration, detailed in references [5] and [6]. Here, we concentate on illustrating the quantum nature of the noise reduction by introducing losses in the twin beams and observing the change in the resultant noise spectrum. It is well known that the intensity noise spectrum normalized to the shot noise level tends linearly towards 1 with attenuation. A signature of noise below the shot noise level is that the normalized noise spectrum should *increase* with attenuation. To verify this, calibrated neutral density filters wereintroduced in the twin beams before the beamsplitter BS. Two conditions were examined: first, the intensity of the twin beams incident on the photodiodes was held constant; second, the intensity of the twin beams immediately after the OPO was held constant. Both provide complementary information.

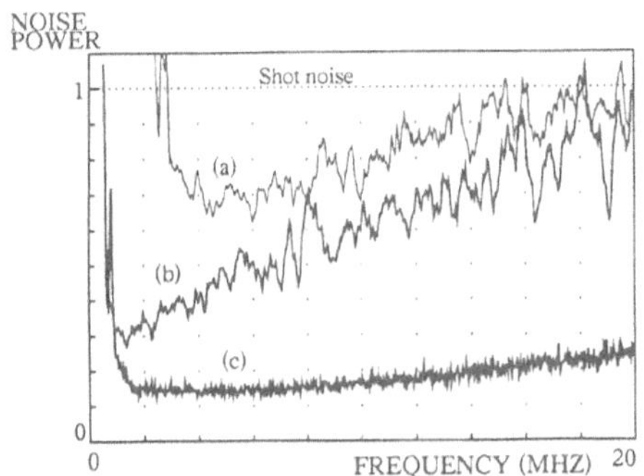

Fig. 3. Progressive increase in quantum noise reduction: curve (a) represents our results from 1987($\xi = 0.57$, $\eta = 0.90$), curve (b) our results from 1989 ($\xi = 0.77$, $\eta = 0.90$), and curve (c) our present results ($\xi = 0.91$, $\eta = 0.94$).

In the first case, the OPO cavity detuning is adjusted to compensate for the attenuation in the twin beams. This changes the OPO oscillation threshold and effectively changes the OPO pumping level. One observes that the noise reduction below the shot noise level is exactly halved on attenuation (above 5 MHz), verifying that the correlation in the twin beams is independent of OPO detuning or pumping level, as predicted by theory. Although the photodiodes were verified not to saturate at 3 mW, the possibility of any such effects is reduced here, since the intensity incident on the photodiodes is held constant.

The second case examined is where the OPO detuning (and therefore the pumping level above threshold) is held constant. Insertion of auxiliary loss at the OPO output changes only the properties of the twin beams, since the source is unchanged. Again, one observes a linear dependence of the spectrum on attenuation.

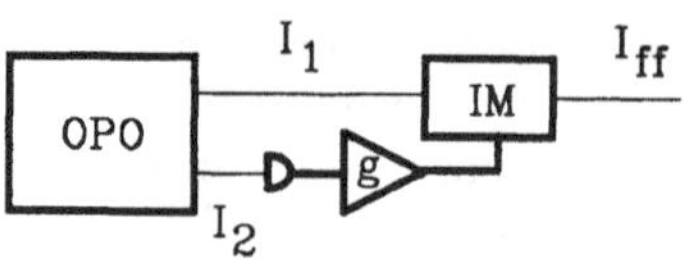

Fig. 4. Schematic diagram of the intensity noise reduction system in the feedforward configuration (IM: intensity modulator, g: amplifier).

3. SUB-POISSONIAN LIGHT GENERATION

The high correlation of the twin beams has been used for the generation of non-classical light by active control of one beam with the other[11]. This technique relies on external control of the amplitude fluctuations, using feedback or feedforward correction. Here, the control signal from one beam is fed forward directly to the second beam using an electrooptic amplitude modulator.

The feedforward correction loop consists of a variable gain amplifier (g) driven by the AC signal (1 - 20 MHz) from the detector which monitors the intensity I_2 of the control beam. The control signal is applied to an intensity modulator (IM) inserted in the first beam, of intensity I_1 (fig.4). This modulator (an electro-optic modulator followed by a polarizer) acts as an analog shutter with a transmission proportional to the applied voltage from the amplifier. Total optical losses through path 1, including losses due to the modulator transmission bias, are about 20% .

To calibrate the noise in the first beam to its shot-noise reference level, a double-balanced detector is used in beam 1, made of a 50% beamsplitting mechanism and two photodetectors. The AC signals (1 - 20 MHz) from each photodetector can be either added or subtracted using an amplifier which is commuted between a summing or differencing configuration, and the resultant signal is observed on a spectrum analyzer (SA). The total noise in beam 1 and its associated shot-noise level are measured almost simultaneously by switching from one configuration to another.

To evaluate the quantum noise reduction, we divide the noise power spectrum by the shot-noise spectrum, having corrected both for electronic noise. The resulting normalized noise power spectrum $S_{ff}(\Omega)$ is shown in fig.5 (curve (a)). $S_{ff}(\Omega)$ drops below 1 over a 700 kHz frequency range around 5 MHz. The reduction in noise power is up to 24% ($\pm 2\%$) below the shot-noise level (this result has been obtained with 69% correlation between the twin beams).

These experimental results can be compared with a theoretical model using a semi-classical input-output formalism[2,14]. Writing the semiclassical field amplitude fluctuations of the two infrared beams at the OPO output as $\delta\alpha_1$ and $\delta\alpha_2$, the significant effect of the feedforward loop is to correct the fluctuations $\delta\alpha_1$ of beam 1 by a term proportional to the fluctuations $\delta\alpha_2$ of beam 2. Neglecting losses in arm 1, the field amplitude fluctuations $\delta\alpha_{ff}$ after correction become:

$$\delta\alpha_{ff} = \delta\alpha_1 - g\delta\alpha_2 \tag{4}$$

where g is a parameter related to the gain of the feedforward loop. The intensity noise spectrum $S_{ff}(\Omega)$, normalized to its shot-noise reference, is given directly by the variance of the field fluctuations $\delta\alpha_{ff}(\Omega)$ at frequency Ω.

The optimum gain and optimum noise reduction can easily be computed from

$S_{1-2}(\Omega)$, the noise spectrum in the difference between the intensities of the twin beams, and $S(\Omega)$, the noise spectrum in the intensity of one beam [15]:

$$g^{opt}(\Omega) = 1 - \frac{S_{1-2}(\Omega)}{S(\Omega)} \tag{5}$$

$$\tilde{S}_{ff}^{opt}(\Omega) = 2S_{1-2}(\Omega)\left(1 - \frac{S_{1-2}(\Omega)}{2S(\Omega)}\right) \tag{6}$$

When the excess noise in each beam is large ($S(\Omega) >> 1$), the optimum gain is equal to one and the optimally corrected intensity noise becomes simply $\tilde{S}_{ff}^{opt}(\Omega) = 2S_{1-2}(\Omega)$. An initial quantum correlation of at least 50% ($S_{1-2}(\Omega) < 0.5$) is necessary then to obtain noise in the corrected beam below the shot noise level. If the beams are not initially correlated at the quantum level ($S_{1-2}(\Omega) \geq 1$), then the control mechanism cannot correct below the shot noise and reacts only on the classical excess noise in beam 1. When the excess noise in each beam is small ($S(\Omega) \sim 1$), noise levels less than $2S_{1-2}(\Omega)$ can be reached for optimum gains smaller than 1. An initial quantum correlation of at least 50% is no longer necessary for sub-shot noise correction. These results can be understood alternatively by separating the beam fluctuations into their symmetric ($\delta p_+ = (\delta p_1 + \delta p_2)/\sqrt{2}$) and anti-symmetric ($\delta p_-$) components. One finds then:

$$\tilde{S}_1(\Omega) = [\mid 1 - g \mid^2 S_{1+2}(\Omega) + \mid 1 + g \mid^2 S_{1-2}(\Omega)]/2 \tag{7}$$

where $S_{1+2}(\Omega)$ is the noise spectrum of the symmetric fluctuations δp_+, defined in the same way. The fluctuations in beam 1 can be reduced to zero by subtracting the incident fields from one another ($g = 1$) when they are perfectly correlated ($\delta p_- = 0$), or by adding them ($g = -1$) when they are anti-correlated ($\delta p_+ = 0$). For intermediate cases when the beams are neither perfectly correlated nor perfectly anti-correlated, such a subtraction or addition of the two fields leaves a residual noise in beam 1 equal respectively to twice the symmetric or anti-symmetric noises. In general, this is not the optimal noise reduction. The noise can optimally be reduced further by making a compromise ($\mid g \mid < 1$), leaving a residual noise in beam 1 containing both symmetric and anti-symmetric noise components.

The theoretical optimum intensity noise $S_{ff}(\Omega)$ is traced in fig.5 (curve (c)), as a function of frequency Ω, where factors such as intensity modulator loss, non-ideal detection efficiency, beam imbalance at the OPO output, and electrical noise are taken into account[11]. The experimental result (curve (a)), shown superposed, effectively attains this optimum. It is interesting at this point to consider a separate case where beams 1 and 2 are not correlated($S_{1-2} = 1$). As shown in fig.5 (curve (d)) the optimum feedforward gain for this case, while still correcting for much of the excess noise in beam 1, does not reduce the noise even to the shot-noise level. This situation is readily attainable experimentally by inserting a half-wave plate at the output of the OPO to rotate the polarization of each twin beam by 45°. Instead of separating the two cross-polarized infrared beams, the beamsplitter then randomly distributes them into the two arms. The resulting experimental noise reduction, with the feedforward gain optimized for a minimum noise around 5 MHz, is shown in fig.5 (curve(b)). As in the case of correlated beams, the experimental results are in good agreement with theory.

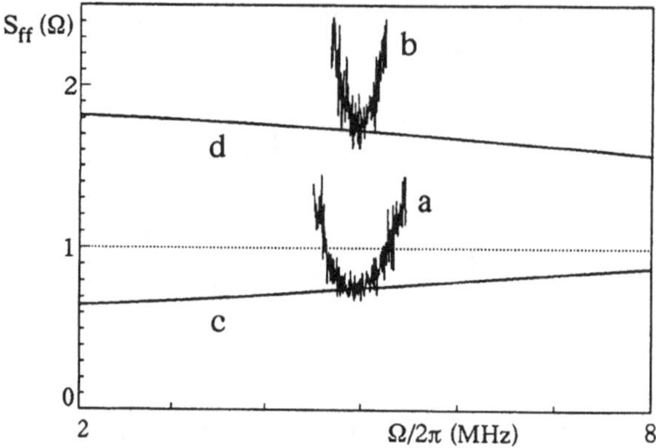

Fig. 5. Curves (a) and (b): experimental noise spectra of beam 1 corrected by the feedforward loop and normalized to the shot noise level; curve (a), using the two correlated twin beams at the output of the OPO (the largest noise reduction is 24% below the shot noise); curve (b), when the two beams are decorrelated by a beam mixer (no quantum noise reduction). The theoretical curves (c) and (d) represent the minimum noises obtainable for these two cases when the gain of the loop is optimized at all frequencies.

In conclusion, a quantum noise reduction of 86% in the intensity difference of the two beams emitted by a parametric oscillator was observed, corresponding to a noise reduction of 91% (11 db) at the output of the parametric oscillator. The fact that the generated output beams are laser-like and intense make OPOs particularly convenient for experimental applications, such as spectroscopy[3,16–17] or the generation of non-classical light by active control of one beam with the other. The generation of amplitude squeezed light with 24% noise reduction was obtained by opto-electronic feedforward correction, using the quantum correlation properties of twin beams generated by the optical parametric oscillator. Such a technique seems attractive for potential applications, particularly with the advent of monolithic OPO technology[7].

REFERENCES

1. L.A. Wu, H.J. Kimble, J.L. Hall, H. Wu, *Phys. Rev. Lett.* 57, 2520 (1986).
2. S. Reynaud, C. Fabre, and E. Giacobino, *J. Optical Society Am.* B4, 1520 (1987).
3. J.G. Rarity and P.R. Tapster, in International Conference on Quantum Electronics Technical Digest Series Vol. 8, p 8 (Optical Society of America, Washington D.C., 1990).
4. O. Aytur and P. Kumar, *Phys. Rev. Lett.* 65, 1551 (1990).
5. A. Heidmann, R.J. Horowicz, S. Reynaud, E. Giacobino, C. Fabre, and G. Camy, *Phys. Rev. Lett.* 59, 2555 (1987).
6. T. Debuisschert, S. Reynaud, A. Heidmann, E. Giacobino, and C. Fabre, *Quantum Opt.* 1, 3 (1989).
7. C.D. Nabors and R.M. Shelby, *Phys. Rev.* A40, 1428 (1989).
8. N.C. Wong, K.W. Leong, and J.H. Shapiro, *Opt. Lett.* 15, 891 (1990).
9. J. Mertz, T. Debuisschert, A. Heidmann, C. Fabre and E. Giacobino, *submitted to Opt. Lett.*
10. C. Fabre, E. Giacobino, A. Heidmann, and S. Reynaud, *J. de Physique* 50, 1209 (1989).
11. J. Mertz, A. Heidmann, C. Fabre, E. Giacobino, and S. Reynaud, *Phys. Rev. Lett.* 64, 2897 (1990).

12. L.A. Lugiato, C. Oldano, C. Fabre, E. Giacobino, R.J. Horowicz, *Nuovo Cimento,* D10, 959 (1988).
13. W.H. Richardson, S. Machida, and Y. Yamamoto, in International Conferenceon Quantum Electronics Technical Digest Series *Vol. 8, p 395* (Optical Society of America, Washington D.C., 1990).
14. S. Reynaud and A. Heidmann, *Opt. Commun.* 71, 209 (1989).
15. J. Mertz, A. Heidmann, C. Fabre, submitted to Phys. Rev. A.
16. E. Giacobino, C. Fabre, S. Reynaud, A. Heidmann, R.J. Horowicz, in "Photons and Quantum Fluctuations" ed. E.R. Pike anh H. Walther, Adam Hilger, Bristol UK (1988).
17. J.J. Snyder, E. Giacobino, C. Fabre, A. Heidmann, and M. Ducloy, *J. Opt. Soc. Am* 7, 2132 (1990).

QUANTUM NOISE REDUCTION AND INSTABILITIES IN NONDEGENERATE FOUR-WAVE-MIXING

M. Brambilla, F. Castelli, L. A. Lugiato, F. Prati and G. Strini

Dipartimento di Fisica dell' Università di Milano
via Celoria 16, 20133 Milano, Italy

1. INTRODUCTION

Four-wave-mixing (FWM) is one of the basic mechanisms in nonlinear optics. We consider the interaction of a pump field of frequency ω_1 and two signal fields (or more precisely, a signal and an idler field) of frequencies ω_2 and ω_3 such that $\omega_2 + \omega_3 = 2\omega_1$; hence the FWM process corresponds to the conversion of the two photons of frequency ω_1 into two photons with frequencies ω_2 and ω_3, and vice versa. Especially interesting is the case of a nonlinear material placed in a resonant optical cavity ; the two signal fields, propagating along the longitudinal axis of the cavity , can be spontaneously generated by the pump field. The mechanism is basically as follows: the pump field saturates the medium and, under appropriate conditions, any probe field with a frequency lying in a suitable interval experiences gain[1,2]. In a cavity , the modes work as probe fields, and one can obtain that a mode has a frequency that lies in the gain region; in addition, the pump gives rise to a FWM coupling of the mode with another mode whose frequency lies on the other side with respect to the pump frequency [3,4]. This process is an example of laser without inversion and has been experimentally realized few years ago[5,6].

The theoretical description of the system, for values of the pump intensity above the threshold for signal generation, is complicated because one must take into account the pump depletion, and therefore one must describe the attenuation of the pump field during its propagation in the nonlinear medium. The problem becomes simpler if the pump field is injected into an optical cavity , because in this case the dynamics of the system is described in terms of cavity modes. The simplest configuration is that of pump and signal fields collinear and copropagating in a unidirectional ring cavity (Fig. 1), and quasi resonant with three longitudinal cavity modes. Precisely, the pump field is near resonance with a mode of frequency ω_c , the signal fields are close to resonance with the two sidemodes of frequencies $\omega_c \pm \alpha$, where α is the free spectral range.

This configuration of the system coincides with that typical of optical bistability (OB), and the phenomenon of sidemode generation corresponds to the so-called multimode instability of OB. This phenomenon was predicted several years ago[7],

Quantum Measurements in Optics, Edited by P. Tombesi and
D.F. Walls, Plenum Press, New York, 1992

was then extensively analyzed theoretically [8-13] and was observed experimentally much more recently[14,15]. The multimode model was formulated in Ref. 16, both in semiclassical terms and as a fully quantum statistical theory in terms of a multimode master equation; the Fokker-Planck approximation of the master equation was worked out in Ref. 17.

The phenomenon of FWM is also one of the most important mechanisms to produce nonclassical states of the electromagnetic field[18-28] and, as a matter of fact, the first experimental observation of squeezing was achieved in a FWM configuration[29], see also Refs. 30,31. A recent paper[32] reports on the observation of quantum noise reduction in the intensity difference between the signal and idler fields, similar to the phenomenon first studied and experimentally observed in nondegenerate parametric oscillator[33,34].

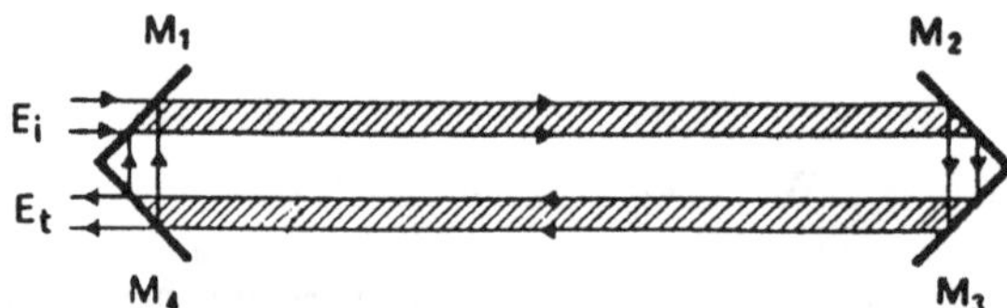

Fig. 1. Ring cavity with plane mirrors. Mirrors M_2 and M_3 are assumed to be ideal reflectors, while mirrors M_1 and M_4 have power transmission coefficient T. E_i is the incident field and it has a stationary intensity and phase; the output field E_t is composed, in general, by a field E_1 with the same frequency ω_1 of the input field, and the two signal fields with frequency ω_2 and ω_3 such that $\omega_2 + \omega_3 = 2\omega_1$.

Nondegenerate FWM in a cavity can be described by various models with different levels of completeness. Attractive for its simplicity is the model formulated by Savage and Walls[23], which describes the phenomenon as a parametric process with an interaction Hamiltonian of the form

$$H_{FWM} = \hbar g \, (a_1^2 a_2^\dagger a_3^\dagger + a_1^{\dagger 2} a_2 a_3) \quad , \tag{1.1}$$

where a_1, a_2, a_3 $(a_1^\dagger, a_2^\dagger, a_3^\dagger)$ are the annihilation (creation) operators of photons of the three cavity modes associated with the pump field (a_1) and with the two signal fields (a_2, a_3), respectively. This description is similar to that which is commonly used for optical parametric oscillators. However, in the case of FWM the nonlinear material is not a crystal but an atomic medium. In order to describe the system adequately one must include the atomic variables in the picture. Models which fulfil this requirement have been given in Refs. 21 and 22; they neglect, however, pump depletion and therefore hold only below the threshold for signal field generation. These models coincide with the approximation of the multimode OB model[16], obtained by treating the pump field as a given constant and the signal fields to lowest order. The model[21] has been generalized[24] to include also inhomogeneous broadening and the case of pump noncollinear with the signals. A model which is valid also above threshold has been formulated recently[26], and includes also the possibility of inhomogeneous broadening and noncollinear pump. In the case of homogeneous broadening and collinear pump, the model of Ref. 26 coincides with the approximation of the multimode OB model of Ref. 16, obtained by neglecting all terms which are of order higher than second in the signal modes. Because of this approximation in the above-threshold case the theory[26] can be used only in the neighbourhood of the threshold.

The analysis of this paper proceeds along two parallel lines. On the one hand we want to use the model of Ref. 16 to describe the quantum fluctuations and the instabilities above threshold, in principle over the whole parameter space of the system. On the other hand we determine under which conditions the complex model of Ref. 16 reduces exactly to a model which describes the phenomenon in terms of parametric processes involving the three modes, providing a picture with a simplicity comparable to that of the model of Ref. 23. The model we derive, however, is not assumed 'ab initio' but is derived exactly from the theory[16] by introducing appropriate limits and, furthermore, it includes all the physical processes that are relevant in this limit.

Our analysis concerns both semiclassical and quantum statistical features. In the semiclassical theory we study the steady states , the instabilities and the dynamical aspects. In the quantum statistical treatment we calculate the spectrum of the intensity difference between the signal fields, which is the quantity observed in the experiment of Ref. 32.

2. THE MULTIMODE MODEL

We consider (Fig. 1) a coherent, linearly polarized electric field E_i of frequency ω_1, which is injected into a ring cavity of total length $\mathcal{L}$, containing a homogeneously broadened system of two-level atoms. We call ω_a the atomic transition frequency , $\gamma_\perp$ the relaxation rate of the atomic polarization and $\gamma_\parallel$ the relaxation rate of the atomic population difference. The configuration of the system is described by the electric field $\mathcal{E}$, the atomic polarization $\mathcal{P}$ and the difference D between the population of the lower and the upper level, normalized to the number of atoms. In the plane wave approximation, one introduces the envelopes F and P defined by the relations ($k_1 = \omega_1/c$)

$$\mathcal{E}\,(z,t) = \frac{\hbar\sqrt{\gamma_\perp\gamma_\parallel}}{\mu}\frac{1}{2}\left\{F(z,t)e^{i(k_1 z-\omega_1 t)} + \text{c.c.}\right\} \qquad , \tag{2.1a}$$

$$\mathcal{P}\,(z,t) = \sqrt{\frac{\gamma_\perp}{\gamma_\parallel}}\frac{\mu}{\rho}\left\{P(z,t)e^{i(k_1 z-\omega_1 t)} + \text{c.c.}\right\} \qquad , \tag{2.1b}$$

where μ denotes the modulus of the dipole moment of the atomic transition and ρ is the atomic density. The coefficients appearing in Eqs. (2.1) are introduced in order to minimize the number of parameters which appear in the final equations.

The dynamics of the system can be formulated in terms of the modal amplitudes by introducing the expansions

$$F(z,t) = \sum_{n=-\infty}^{+\infty} \mathrm{e}^{\mathrm{i}n\alpha(t-\frac{z}{c})} f_n(t) \quad , \tag{2.2a}$$

$$P(z,t) = \sum_{n=-\infty}^{+\infty} \mathrm{e}^{\mathrm{i}n\alpha(t-\frac{z}{c})} p_n(t) \quad , \tag{2.2b}$$

$$D(z,t) = \sum_{n=-\infty}^{+\infty} \mathrm{e}^{\mathrm{i}n\alpha(t-\frac{z}{c})} d_n(t) \quad , \tag{2.2c}$$

The frequency of the mode labeled by n ($n = 0, \pm 1, \pm 2, ...$) is given by $\omega_c + n\alpha$, where

$$\alpha = \frac{2\pi nc}{\mathcal{L}} \quad , \tag{2.3}$$

is the free spectral range. Because the population difference D is real one has that $d_{-n}(t) = d_n^*(t)$.

The time evolution of the modal amplitudes is governed by a master equation formulated in Ref. 16 (see also Ref. 12). When the number of atoms is large, the master equation can be approximated by a Fokker-Planck equation for the quasi-probability distribution of the system in the generalized P-representation[35,36]. The Fokker-Planck equation, derived in Ref. 17, is equivalent to a set of Langevin equations for the modal amplitudes, which read[12,16,17]

$$\frac{\mathrm{d}f_n}{\mathrm{d}t} = -\kappa\left[(1+i\theta)f_n - Y\delta_{n,0} + 2Cp_n\right] \quad , \tag{2.4a}$$

$$\frac{\mathrm{d}f_n^*}{\mathrm{d}t} = -\kappa\left[(1-i\theta)f_n^* - Y\delta_{n,0} + 2Cp_n^*\right] \quad , \tag{2.4b}$$

$$\frac{\mathrm{d}p_n}{\mathrm{d}t} = \gamma_\perp\left\{\sum_{n'} f_{n'}d_{n-n'} - \left[1+\mathrm{i}\left(\Delta - \frac{\alpha_n}{\gamma_\perp}\right)p_n\right]\right\} + \Gamma_{p_n}(t) \quad , \tag{2.4c}$$

$$\frac{\mathrm{d}p_n^*}{\mathrm{d}t} = \gamma_\perp\left\{\sum_{n'} f_{n'}^*d_{n'-n} - \left[1-\mathrm{i}\left(\Delta - \frac{\alpha_n}{\gamma_\perp}\right)p_n^*\right]\right\} + \Gamma_{p_n^*}(t) \quad , \tag{2.4d}$$

$$\frac{\mathrm{d}d_n}{\mathrm{d}t} = -\gamma_\parallel\left[\frac{1}{2}\sum_{n'}(f_{-n'}^*p_{n-n'} + f_{n'}p_{n'-n}^*) + d_n\left(1 - \mathrm{i}\frac{\alpha_n}{\gamma_\parallel}\right) - \delta_{n,0}\right] + \Gamma_{d_n}(t) \quad , \tag{2.4e}$$

where :

T is the power transmissivity coefficient of the mirrors (see Fig. 1);

the parameter Y denotes the normalized input field, defined by

$$E_i(t) = \frac{\hbar\sqrt{\gamma_\perp\gamma_\parallel T}}{\mu}\frac{1}{2}Y\left[\mathrm{e}^{-\mathrm{i}\omega_1 t} + \mathrm{c.c.}\right] \quad , \tag{2.5}$$

where, for the sake of definiteness, we assumed that Y is real and positive; C is the bistability parameter

$$C \equiv \frac{\alpha_{ab} L}{2T} \quad , \tag{2.6}$$

and α_{ab} denotes the linear absorption coefficient for the electric field ; θ is the cavity detuning parameter

$$\theta \equiv \frac{\omega_c - \omega_1}{\kappa} \quad , \tag{2.7}$$

with ω_c being the frequency of the cavity mode lying closest to the input field frequency ω_1 and κ being the cavity damping constant

$$\kappa \equiv \frac{cT}{\mathcal{L}} \quad , \tag{2.8}$$

Δ is the atomic detuning parameter

$$\Delta \equiv \frac{\omega_a - \omega_1}{\gamma_\perp} \quad , \tag{2.9}$$

in the following we will always select for definiteness $\Delta \geq 0$.

The quantities Γ represent quantum noise terms and describe stationary , Gaussian stochastic processes with zero average and having time correlation functions whose explicit expression are given in Ref. 37. The semiclassical equations are obtained by dropping the fluctuation terms Γ.

For the description of the system above the threshold for generation of the signal modes, we will limit our considerations to the three field modes f_0, f_{+1}, f_{-1} and accordingly in Eqs. (2.4a,b) we will consider only the case $n = 0, \pm 1$ and in Eqs. (2.4c,d,e) the sums will be restricted to $n' = 0, \pm 1$. This model, where we have only three field modal variables f_0, f_{+1}, f_{-1} and an infinite set of atomic modal variables p_n, p_n^*, d_n , $(n = 0, \pm 1, \pm 2...)$ will be called in the following model A. We will consider also another model (called model B), in which one includes only the modal variables with $n = 0, -1, +1$ also for the atomic quantities. In agreement with the nomenclature most commonly used in the FWM literature, in the following we will label the modal variables by the indices 1,2,3 instead of 0,-1,+1, respectively.

3. THE PARAMETRIC LIMIT AND THE PARAMETRIC MODEL

In order to obtain a parametric model, it is necessary to perform the so-called cubic approximation, so that the final dynamical equations for the modal amplitudes, derived by performing the adiabatic elimination of the atomic variables under the so called good-cavity condition

$$\kappa \ll \gamma_\perp \, , \gamma_\parallel \quad , \tag{3.1}$$

display only cubic nonlinearities. According to the singlemode theory of OB, the conditions for the cubic approximation are [38,39]

1) The atomic detuning parameter Δ must be large, in the sense that

$$\Delta \gg 1 \quad , \quad \frac{|f_i|^2}{\Delta^2} \ll 1 \quad , \tag{3.2}$$

2) In order to ensure a meaningful nonlinearity in spite of the large detuning, the bistability parameter C must be large, in such a way that

$$\frac{2C}{\Delta} \gg 1 \qquad \text{with} \qquad \left(\frac{2C}{\Delta}\right)\frac{|f_i|^2}{\Delta^2} \qquad \text{arbitrary} \qquad , \tag{3.3}$$

3) The cavity detuning parameter θ must be large, so that

$$\theta \gg 1 \qquad \text{with} \qquad \theta_1 \equiv \frac{2C}{\Delta} - \theta \qquad \text{arbitrary} \qquad . \tag{3.4}$$

Furthermore in order to make the absorptive effects negligible, we require that
4)

$$\frac{2C}{\Delta^2} \ll 1 \tag{3.5}$$

We observe that the situation of large values for the parameters C and Δ, satisfying conditions (3.2) to (3.5) is met, for example, in the experiment of Ref. 32.

In the multimode case, we must ask that the same conditions hold also when Δ is replaced by the detuning parameters for the sidemodes $\Delta \mp \tilde{\alpha}$. This requires the further condition

$$\frac{\tilde{\alpha}}{\Delta} \ll 1 \qquad , \tag{3.6}$$

where

$$\tilde{\alpha} = \frac{\alpha}{\gamma_\perp} \tag{3.7}$$

which ensures also that the nonlinear terms of the final dynamical equations do not depend on α. All the previous conditions can be summarized and made precise by introducing a smallness parameter ε (a posteriori, one can verify that the final parametric model remains valid also when ε is not very small, for example $\varepsilon = 1/4$). We assume that

$$\frac{\tilde{\alpha}}{\Delta} \sim \varepsilon \quad , \tag{3.8a}$$

$$\frac{2C}{\Delta} \sim \varepsilon^{-2} \quad , \tag{3.8b}$$

$$\theta_1 \equiv \frac{2C}{\Delta} - \theta \sim \varepsilon^0 \quad , \tag{3.8c}$$

$$\Delta \sim \varepsilon^{-3} \quad , \tag{3.8d}$$

and look for solutions such that $|f_i|^2/\Delta^2 \sim \varepsilon^2$, which implies that $(2C/\Delta)(|f_i|^2/\Delta^2)$-$\sim \varepsilon^0$; these conditions make the cubic approximation rigorous. By combining Eqs. (3.8a) and (3.8b), one obtains that $(2C/\Delta)(\tilde{\alpha}^2/\Delta^2) \sim \varepsilon^0$; this condition implies that the parameter α appears in the final equations up to second order. The presence of the second order contribution in $\tilde{\alpha}$ is necessary to ensure the possibility of a multimode instability.

Let us consider first the model B. Under conditions (3.1) we perform the adiabatic elimination of the atomic variables by dropping the derivatives with respect to time in Eqs. (2.4). Then we calculate the polarization p_i ($i = 1, 2, 3$) to order zero, first and third in the field amplitudes f_i and we insert these expressions into the field equations. In the following, we will consider only the purely radiative case $\gamma_\parallel = 2\gamma_\perp$ which is the most interesting from the viewpoint of quantum noise reduction; in the case $\gamma_{\parallel} \ll \gamma_{\perp}$ there is no multimode instability in the limit (3.8). If we introduce the change of variables

$$\begin{aligned} \overline{Y} &= \left(\frac{2C}{\Delta^3}\right)^{1/2} Y \quad , \\ x_1 &= \left(\frac{2C}{\Delta^3}\right)^{1/2} f_1 \quad , \\ x_2 &= \left(\frac{2C}{\Delta^3}\right)^{1/2} \exp\left\{+\mathrm{i}\frac{2C\kappa\tilde{\alpha}}{\Delta^2} t\right\} f_2 \quad , \\ x_3 &= \left(\frac{2C}{\Delta^3}\right)^{1/2} \exp\left\{-\mathrm{i}\frac{2C\kappa\tilde{\alpha}}{\Delta^2} t\right\} f_3 \quad , \end{aligned} \tag{3.9}$$

the final equations read

$$\frac{\mathrm{d}x_1}{\mathrm{d}t} = \kappa\left\{\overline{Y} - x_1\left[1 - \mathrm{i}\theta_1 + \mathrm{i}(|x_1|^2 + 2|x_2|^2 + 2|x_3|^2)\right] - 2\mathrm{i}x_1^* x_2 x_3\right\} + \Gamma_1(t) \quad , \tag{3.10a}$$

$$\frac{\mathrm{d}x_2}{\mathrm{d}t} = \kappa\left\{-x_2\left[1 - \mathrm{i}\theta_2 + \mathrm{i}(2|x_1|^2 + |x_2|^2 + \eta|x_3|^2)\right] - \mathrm{i}x_3^* x_1^2\right\} + \Gamma_2(t) \quad , \tag{3.10b}$$

$$\frac{\mathrm{d}x_3}{\mathrm{d}t} = \kappa\left\{-x_3\left[1 - \mathrm{i}\theta_2 + \mathrm{i}(2|x_1|^2 + \eta|x_2|^2 + |x_3|^2)\right] - \mathrm{i}x_2^* x_1^2\right\} + \Gamma_3(t) \quad , \tag{3.10c}$$

where we defined

$$\theta_2 = \theta_1 + \frac{2C}{\Delta}\frac{\tilde{\alpha}^2}{\Delta^2} \quad , \tag{3.11}$$

and η is equal to 1 (the reason for introducing the parameter η will be clear in the following of this section). The quantum noise terms Γ_i obey the time correlation functions

$$\langle\, \Gamma_1(t_1)\, \Gamma_1(t_2)\, \rangle = \langle\, \Gamma_2(t_1)\, \Gamma_3(t_2)\, \rangle = -\mathrm{i}g(x_1^2 + 2x_2 x_3)\, \delta(t_1 - t_2) \quad , \tag{3.12a}$$

$$\langle \Gamma_i(t_1)\Gamma_i(t_2)\rangle = -\mathrm{i}g x_i^2\, \delta(t_1 - t_2) \qquad (i = 2,3) \quad , \tag{3.12b}$$

$$\langle \Gamma_1(t_1)\Gamma_i(t_2)\rangle = -2\mathrm{i}g x_1 x_i\, \delta(t_1 - t_2) \qquad (i = 2,3) \quad , \tag{3.12c}$$

$$\langle \Gamma_i^*(t_1)\Gamma_j(t_2)\rangle = 0\,, \qquad (i,j = 1,2,3) \quad , \tag{3.12d}$$

where we set

$$g = \frac{2C\kappa}{\Delta^3 N_S} \quad , \tag{3.13}$$

and N_S is the saturation photon number. The time evolution equations for the variables x_i^* $(i = 1,2,3)$ are obtained from Eqs. (3.11) by complex conjugation; in the same way one obtains the remaining time correlation functions from Eqs. (3.12 a-c).

The parametric equations (3.10) with (3.12) are an exact consequence of the initial model B in the limit defined by Eqs. (3.8).

If one starts from model A instead of model B and follows the same procedure, one arrives exactly at the same parametric model (3.10), (3.12) with the only difference that in this case $\eta = 2$. Let us now consider the density operator ρ of the three-mode system and the master equation

$$\frac{d\rho}{dt} = \sum_{i=1}^{3} \Lambda_i \rho - \frac{i}{\hbar}[H, \rho] \quad , \tag{3.14}$$

where the Liouvillians Λ_i are given by

$$\Lambda_i = \kappa \left\{ [a_i, \rho a_i^\dagger] + [a_i \rho, a_i^\dagger] \right\} \tag{3.15}$$

and the Hamiltonian H is defined as

$$H = H_{ext} + H_0 + H_{int} \quad , \tag{3.16}$$

with

$$H_{ext} = i\hbar\kappa\alpha_I(a_1^\dagger - a_1) \quad , \tag{3.17}$$

$$H_0 = -\kappa\{\theta_1 a_1^\dagger a_1 + \theta_2(a_2^\dagger a_2 + a_3^\dagger a_3)\} \quad , \tag{3.18}$$

$$H_{int} = H_{FWM} + H_{PM} + H_{CPM} \quad . \tag{3.19}$$

The Hamiltonian H_{FWM} is defined by Eq. (1.1), while H_{PM} and H_{CPM} are given by

$$H_{PM} = \hbar\frac{g}{2}\sum_{i=1}^{3} a_i^{\dagger 2} a_i^2 \quad , \tag{3.20}$$

$$H_{CPM} = 2\hbar g(a_1^\dagger a_1 a_2^\dagger a_2 + a_1^\dagger a_1 a_3^\dagger a_3 + a_2^\dagger a_2 a_3^\dagger a_3) \quad , \tag{3.21}$$

with the coupling constant g defined by Eq. (3.13).

The master equation can be converted into a classical-looking Fokker-Planck equation by using the P-representation; the equivalent set of Langevin equations coincides with Eqs. (3.10) with $\eta = 2$ (i.e. it coincides with the parametric model derived from model A); the correlation functions coincide with those given in Eqs. (3.12). This fact can be easily verified by taking into account the relations

$$x_i = \frac{\alpha_i}{\sqrt{N_S}}\left(\frac{2C}{\Delta^3}\right)^{1/2} \quad , \quad Y = \frac{\alpha_I}{\sqrt{N_S}}\left(\frac{2C}{\Delta^3}\right)^{1/2} \quad , \tag{3.22}$$

where α_i $(i = 1, 2, 3)$ denote the c-number variables associated with the operators a_i.

The Hamiltonians H_{PM} and H_{CPM} describe self-phase modulation and cross phase modulation, respectively, as one normally has in $\chi^{(3)}$ media. The cross phase modulation terms have the same form of those which arise in the analysis of three-level media[40].

The simplest parametric model of Ref. 23 amounts to Eq. (3.19) without the phase modulation contributions H_{PM} and H_{CPM}. In our theory Eq. (3.13) provides the microscopic expression of the coupling constant; the same is true for the parameter θ_2 defined in Eq. (3.11).

4. STATIONARY SOLUTIONS OF THE PARAMETRIC MODEL, LINEAR STABILITY ANALYSIS AND DYNAMICAL SOLUTIONS

As one sees from Eqs. (3.10), our parametric model depends on three effective parameters: the normalized input field $\overline{Y}$ and the detuning parameters θ_1 and θ_2.

In the remaining part of this section we consider Eqs. (3.10) in the semiclassical approximation, i.e. we drop the quantum noise terms Γ_i $(i = 1, 2, 3)$. The semiclassical equations admit one singlemode stationary solution given by the well-known cubic steady-state equation[41]

$$\overline{Y} = x_1^S \left\{1 - \mathrm{i}(\theta_1 - |x_1^S|^2)\right\} \quad , \tag{4.1}$$

which predicts bistability for $\theta_1 > \sqrt{3}$. The signal fields x_2 and x_3 vanish in this solution , which corresponds to the case below threshold for signal generation. The linear stability analysis of the singlemode stationary solution is discussed in Ref. 37.

Let us now calculate the multimode stationary solutions , by setting $dx_1/dt = dx_2/dt = dx_3/dt = 0$ in Eqs. (3.10) with $\Gamma_i = 0$. By using Eqs. (3.9), (2.2a), (2.4), (3.7) and (2.1a) we see that above threshold the output field displays three components with frequencies

$$\begin{aligned} &\omega_1 \quad , \\ &\omega_2 = \omega_1 - \alpha\left(1 + \frac{2C\kappa}{\gamma_\perp \Delta^2}\right) \quad , \\ &\omega_3 = \omega_1 + \alpha\left(1 + \frac{2C\kappa}{\gamma_\perp \Delta^2}\right) \quad ; \end{aligned} \tag{4.2}$$

note the frequency shifts due to the atomic dispersion.

By some algebraic manipulation one obtains that the stationary values of the moduli of the signal fields are equal:

$$|x_3| = |x_2| \quad ; \tag{4.3}$$

this result arises from the fact that in the limit (3.7) the absorptive effects are negligible. Furthermore one obtains that in the multimode stationary state the value of $|x_1|$ is linked to that of $|x_2|$ as follows:

$$|x_1|^2 = H(|x_2|^2) \quad ,$$

$$H(z) = \frac{1}{3}\left\{2\theta_2 - 2(1+\eta)z \pm \left[\theta_2^2 - 3 + (1+\eta)z^2 - 2\theta_2(1+\eta)z\right]^{1/2}\right\} \quad , \tag{4.4}$$

and finally the modal intensity $|x_2|^2$ is linked to the input intensity by the steady-state equation[42]

$$\begin{aligned} \overline{Y}^2 &= 4|x_2|^2\left[1 - \theta_1\theta_2 + 4\theta_2|x_2|^2 + \theta_1(1+\eta)|x_2|^2 - 4(1+\eta)|x_2|^4\right] + \\ &+ H(|x_2|^2)\left[1 + \theta_1^2 - 2\theta_1 H(|x_2|^2) + H^2(|x_2|^2) + 4\theta_2|x_2|^2 - 4(4+\eta)|x_2|^2\right] \quad . \end{aligned} \tag{4.5}$$

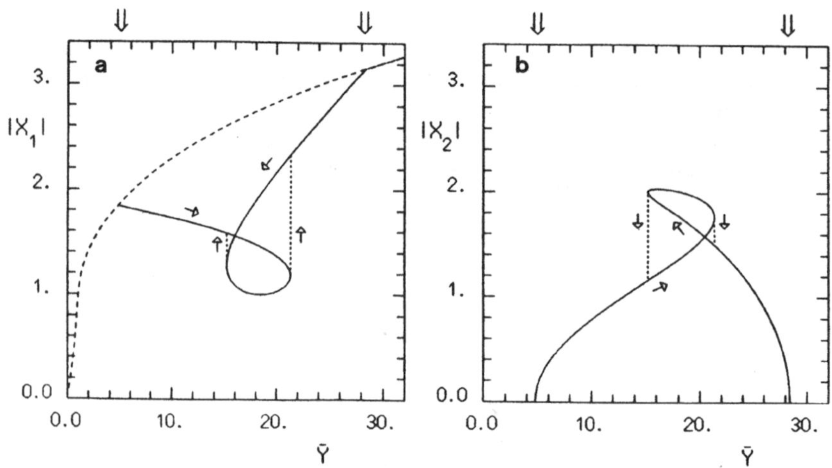

Fig. 2. Steady-state curve from the parametric model (3.10) for a) the field $|x_1|$ and b) the field $|x_2|$. The parameters are $\theta_1 =' 1$, $\theta_2 = 10$, $\eta = 1$. The broken line in a) corresponds to the singlemode stationary solution. The arrows indicate the jumps when the input field is increased or decreased. The big arrows on top show the boundaries of the instability domain of the singlemode stationary solution; at the boundaries of this domain the multimode stationary solution connects itself in a continuous way (but with a discontinuity of the derivative) with the singlemode stationary solution.

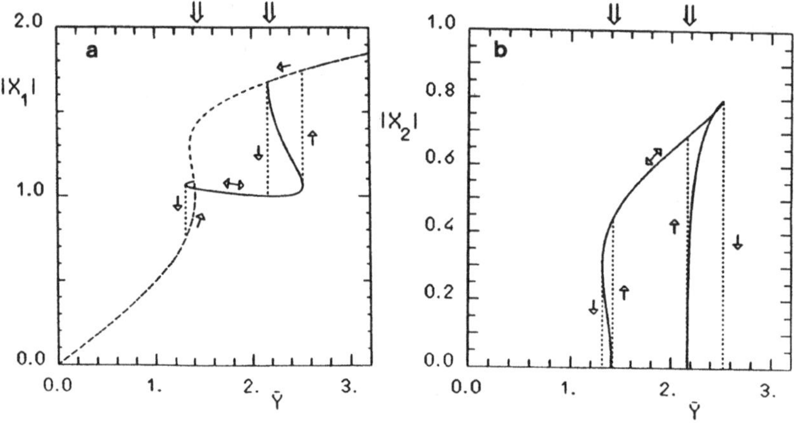

Fig. 3. Same as Figs. 2a) and 2b), but for $\theta_1 = 2$, $\theta_2 = 3$, $\eta = 1$.

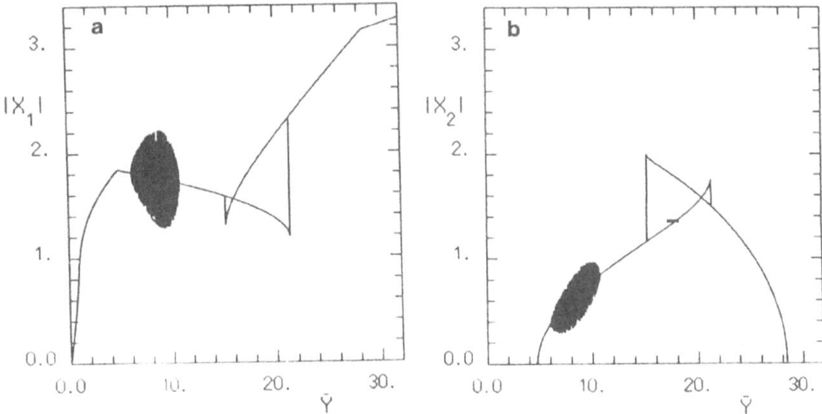

Fig. 4. The equations (3.10) of the parametric model are solved numerically with a slow forward and backward sweep of the input field $\overline{Y}$. Figs. a) and b) show the behaviour of $|x_1|$ and $|x_2|$, respectively. Parameters as in Fig. 2.

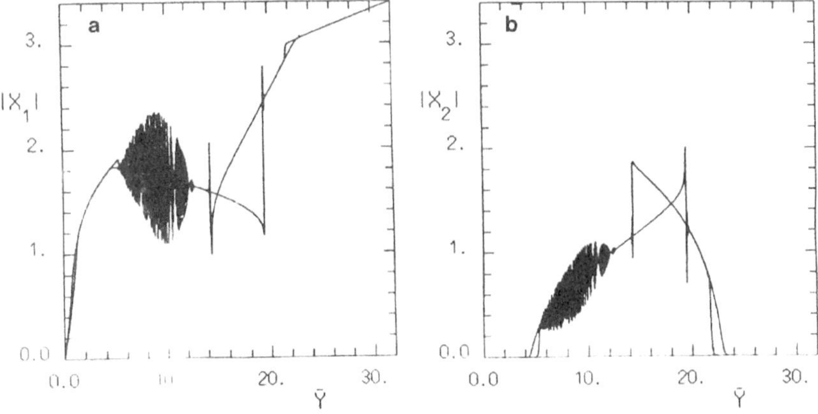

Fig. 5. The equations of model B are solved numerically with a slow forward and backward sweep of the input field Y. Figs. a) and b) show the behaviour of $|x_1|$ and $|x_2|$, respectively. The parameters are $C = 5 \times 10^4$, $\Delta = 10^3$, $\theta_1 = 1$, $\gamma_{\parallel} = 2\gamma_{\perp}$, $\kappa = 0.01\gamma_{\perp}$, $\tilde{\alpha} = 300$ (hence $\theta_2 = 10$). The broken parts correspond to the backward sweep.

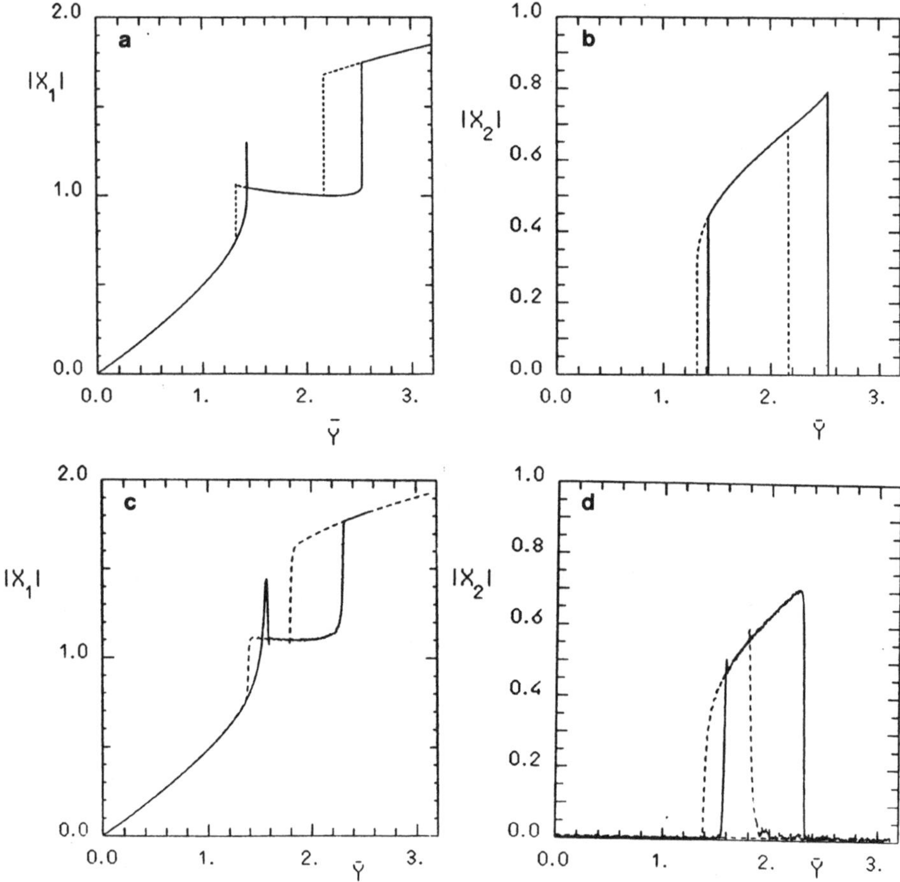

Fig. 6. a) + b) same as Figs. 4 a) + b), but for the parameters of Fig. 3; c) + d) same as Figs. 5 a) + b) but with $C = 5 \times 10^3$, $\Delta = 400$, $\theta_1 = 2$, $\gamma_\parallel = 2\gamma_\perp$, $\kappa = 0.01\gamma_\perp$, $\tilde{\alpha} = 80$ (hence $\theta_2 = 3$). The broken parts correspond to the backward sweep.

The difference between the phases of x_2 and x_3 is completely arbitrary in the multimode stationary state; this phenomenon is well-known and is identical to the behaviour of nondegenerate parametric oscillators[43].

The multimode stationary solution governed by Eq. (4.5) is substantially more complicated than that of Ref. 23, especially for the presence of the double sign in the definition of the function H, given by Eq. (4.4). This circumstance gives rise to a complex picture which includes several possibilities. A complete classification of the various behaviours will be given in a separate publication[44]. Here we discuss only two examples, illustrated by Figs. 2 and 3 which show the steady state curve of the fields $|x_1|$ and $|x_2|$ as a function of $\overline{Y}$ obtained from Eq. (4.5), for $\eta = 1$. These figures indicate also the singlemode stationary solution (broken line) and

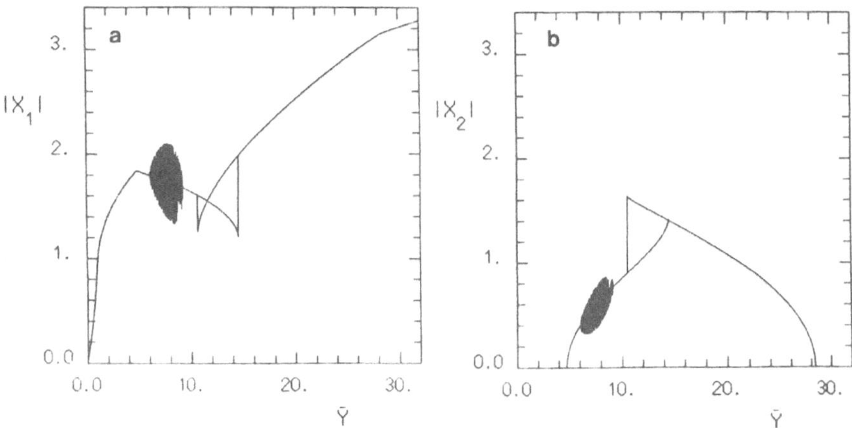

Fig. 7. Same as Figs. 4a) and 4b) but for $\eta = 2$, while the other parameters are unchanged.

the jumps among the various branches of the steady state curve, which occur when the input field is increased or decreased. In the case of Fig. 2 (3) the behaviour of $|x_2|$ at threshold exhibits a second (first) order phase transition with monostability (bistability) between the multimode stationary solution and the trivial solution $x_2 = 0$, in qualitative accord with the picture predicted by the model of Ref. 23. On the other hand the butterfly hysteresis cycle, appearing in Fig. 2 for larger values of the input field, arises from the double sign in the definition of H and it is not possible in the model of Ref. 23. The same holds for the clockwise hysteresis cycle in Fig. 3. For the parameters of Fig. 2, there is also an instability that produces undamped oscillations of the moduli. This is shown in Fig. 4, which was obtained by numerically solving Eqs. (3.10) with a slow sweep (forward and backward) of the input field $\overline{Y}$; the shaded regions display the oscillations. The presence of the oscillations is also in qualitative agreement with the simple model of Ref. 23; however, in the case of our parametric model when the input intensity is increased enough the oscillations disappear, the butterfly hysteresis cycle is met and finally the system is back to the singlemode stationary solution .

Figure 5 exhibits the numerical results of a slow sweep of the input field Y, obtained from the equations of model B, for values of the parameters corresponding to those of Fig. 2. Clearly there is a good accord with the result of the parametric model shown in Fig. 2; the boundaries of oscillatory domain and of the butterfly cycle display a shift with respect to Fig. 2. Similarly Figs. 6a, 6b (6c, 6d) show a slow sweep of the input filed obtained from the parametric model (model B) for values of the parameters corresponding to those of Fig. 3; note that the parameter values of Figs. 6c and 6d are compatible with those of the experiment reported in Ref. 32.

Figure 7 is obtained by performing a slow sweep of $\overline{Y}$ with the parametric model (3.10), for the same values of the parameters of Fig. 2 but for $\eta = 2$. The qualitative behaviour is the same of Fig. 2, but there are significant quantitative differences.

5. SPECTRUM OF THE FLUCTUATIONS OF THE INTENSITY DIFFERENCE BETWEEN THE SIGNAL FIELDS ABOVE THRESHOLD

This spectrum is defined by[45]

$$S(\omega) = 1 + \frac{2\kappa N_S}{|x_2|^2 + |x_3|^2} \left(\frac{2C}{\Delta^3}\right) \int_{-\infty}^{+\infty} dt' \, e^{-i\omega t'} \, \langle \delta I_-(t') \delta I_-(0) \rangle \quad , \tag{5.1}$$

where $|x_2|$ and $|x_3|$ denote the stationary values in a multimode stationary solution, I_- is defined as the intensity difference

$$I_- = x_2^* x_2 - x_3^* x_3 \tag{5.2}$$

and δI_- indicates the fluctuation of I_- around the multimode stationary state. It is understood that the time correlation function is calculated using the P-representation. With this definition, the shot-noise level corresponds to $S = 1$ and complete suppression of quantum noise corresponds to $S = 0$.

If we use the parametric model (3.10), the calculation of the spectrum can be performed analytically. One reformulates the Langevin equations (3.10) in terms of modulus and phase of the variables x_i ($i = 1, 2, 3$), and then one linearizes the equations around a multimode stationary solution. The linearization is performed only with respect to the moduli, to the phase of x_1 and to the sum of the phases of x_2 and x_3, following the same procedure described in Ref. 27. Details will be shown elsewhere[44]; the result is very simple and reads[42]

$$S(\omega) = 1 - \frac{4\kappa^2}{\omega^2 + 4\kappa^2} \tag{5.3}$$

One notes immediately that Eq. (5.3) coincides with the expression of the spectrum first derived for the nondegenerate parametric oscillator above threshold[33,34] which predicts complete suppression of quantum noise at zero frequency. This formula holds also for the simple parametric model of Ref. 23, as shown in Ref. 27 and, as we prove here, it remains valid even after including the terms which describe phase modulation and cross phase modulation. Hence the expression (5.3) seems to possess a sort of general validity. Note that Eq. (5.3) follows from the Langevin equations both for $\eta = 1$ and $\eta = 2$.

With the help of the equations of model B we can calculate the spectrum (5.1) also out of the limit of validity (3.8) of the parametric model (3.10). To this aim, we reformulated the equations in terms of moduli and phases of the variables f_i, p_i, d_i and we linearized the Langevin equations around a multimode stationary solution. This linearization is performed only in the variables which have a nonarbitrary stationary value. Again, the details of the calculations are described in a separate publication[44]. The spectrum $S(\omega)$ is obtained by combining elements of the spectral matrix defined as usually[46,47].

In the resonant case $\Delta = 0$, $\theta_1 = 0$ one meets at most a small reduction of quantum noise[48]. This fact is presumably due to linear and nonlinear absorption. On the other hand, Fig. 8 exhibits the spectrum $S(\omega)$ calculated from model B, for values of the parameters that satisfy conditions (3.8), and are compatible with the values of the experiment[32], in which the spectrum was observed. Clearly the curve is everywhere very close to that prescribed by the analytic formula (5.3).

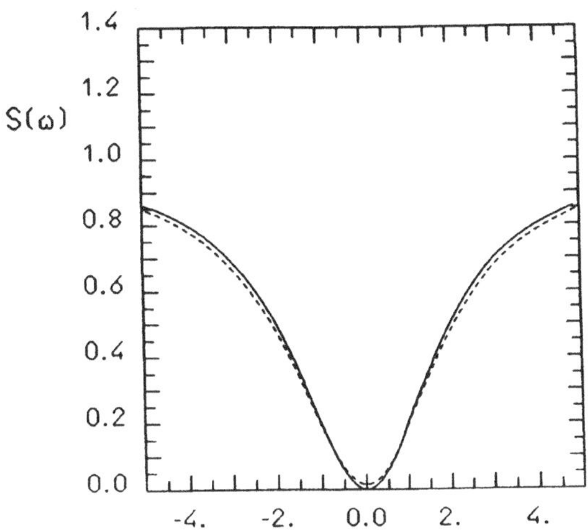

Fig. 8. The spectrum of the intensity difference between the signal fields, obtained from model B, is shown for the parametric values of Figs. 3 and 6, and $\overline{Y} = 1.69$. The full line describes the analytic curve (5.3).

More details concerning the analytical calculations and the results can be found in Ref. 37.

ACKNOWLEDGMENTS

We are grateful to Prof. Walls for sending us a copy of Refs. 24 and 26 prior to publication. This research was worked out in the framework of the ESPRIT Basic Research Action on " Quantum Noise Reduction in Optical Systems" (NOROS).

REFERENCES

1. B. R. Mollow, *Phys. Rev. A* 5:2217 (1972).
2. F. Y. Wu, S. Ezekiel, M. Ducloy and B. R. Mollow, *Phys. Rev. Lett.* 38:1077 (1977).
3. T. Fu and M. Sargent III, *Opt. Lett.* 4:366 (1979).
4. R. W. Boyd, M. G. Raymer, P. Narum and D. I. Marten, *Phys. Rev. A* 24:411 (1981).
5. D. Grandclement, G. Grynberg and M. Pinard, *Phys. Rev. Lett.* 59:40,44 (1987).
6. G. Khitrova, J.F. Valley and H. M. Gibbs, *Phys. Rev. Lett.* 60:1126 (1988).
7. R. Bonifacio and L. A. Lugiato, *Lett. Nuovo Cimento* 21:510 (1978).
8. L. A. Lugiato, *Opt. Comm.* 33:108 (1980).
9. R. Bonifacio, M. Gronchi and L. A. Lugiato, *Opt. Comm.* 30:129 (1979).
10. M. Gronchi, V. Benza, L. A. Lugiato, P. Meystre and M. Sargent III, *Phys. Rev. A* 24:1419 (1981).
11. L. A. Lugiato, V. Benza, L. M. Narducci and J. D. Farina, *Z. Phys. B* 49:351 (1983).

12. L. A. Lugiato, in: *Progress in Optics*, vol XXI, ed. by E.Wolf, North-Holland, Amsterdam , p. 69 ff.
13. M. L. Asquini, L. A. Lugiato, H. J. Carmichael and L. M. Narducci, *Phys. Rev. A* 33:360 (1986).
14. B. Segard and B. Macke, *Phys. Rev. Lett.* 60:412 (1988).
15. B. Segard, B. Macke, L.A. Lugiato, F. Prati and M. Brambilla, *Phys. Rev. A* 39:706 (1989).
16. L. A. Lugiato, *Z. Phys. B* 41:85 (1981).
17. L. A. Lugiato and F. Castelli, *Z. Phys. B* 64:375 (1986).
18. H. P. Yuen and J. H. Shapiro, *Opt. Lett.* 4:334 (1979).
19. P. Kumar and J. H. Shapiro, *Phys. Rev. A* 30:1568 (1984).
20. M. D. Reid and D. F. Walls, *Phys. Rev. A* 31:1622 (1985); R. E. Slusher, B. Yurke, P. Grangier, A. La Porta, D. F. Walls and M. D. Reid,Journ. Opt. Soc. Am. B4:1453 (1987).
21. M. D. Reid and D. F. Walls, *Phys. Rev. A* 34:4929 (1986).
22. D. A. Holm and M. Sargent III, *Phys. Rev. A* 35:2150 (1987).
23. C. M. Savage and D. F. Walls, *J. Opt. Soc. Am. B* 4:154 (1987)
24. W. Zhang and D. F. Walls, *Phys. Rev. A* 41:6385 (1990).
25. L. A. Lugiato, P. Galatola and L. M. Narducci, *Opt. Comm.* 76:276 (1990).
26. W. Zhang and D. F. Walls, *Opt. Comm.* in press.
27. B. C. Sanders and M. D. Reid, *Phys. Rev. A* in press.
28. Wang Kaige and L. A. Lugiato, *Annalen der Physik*, 200th anniversary issue, in press.
29. R. E. Slusher, L. W. Hollberg, B. Yurke, J. C. Mertz and J. F. Valley, *Phys. Rev. Lett.* 55:2409 (1985).
30. R. M. Shelby, M. D. Levenson, S. H. Perlmutter, R. G. De Voe and D. F. Walls, *Phys. Rev. Lett.* 57:691 (1986).
31. M. W. Maeda, P. Kumar and J. H. Shapiro, *Opt. Lett.* 12:161 (1986).
32. M. Vallet, M. Pinard and G. Grynberg, *Europhys. Lett.* 11:739 (1990).
33. A. Heidmann, R. J. Horowicz, S. Reynaud, E. Giacobino, C. Fabre and G. Camy, *Phys. Rev. Lett.* 59:2555 (1987).
34. S. Reynaud, C. Fabre and E. Giacobino, *J. Opt. Soc. Am. B* 4:1520 (1987).
35. R. J. Glauber, *Phys. Rev.* 130:2529 (1963); R. J. Glauber,Phys. Rev.131:2766 (1963).
36. H. Haken, in: *Laser Theory*, Encyclopaedia der Physik, Vol. XXV/2c, ed. by L. Genzel, Springer-Verlag, Berlin 1970.
37. M. Brambilla, F. Castelli, L. A. Lugiato, F. Prati and G. Strini, submitted to Opt. Comm.
38. R. Bonifacio, M. Gronchi and L. A. Lugiato, *Nuovo Cimento B* 53:311 (1979).
39. L. A. Lugiato, L. M. Narducci and R. Lefever, in: *Lasers and Synergetics*, ed. by R. Graham and A. Wunderlin, Springer-Verlag, Berlin 1987, pp 59,60.
40. P. Grangier, J. F. Roch and G. Roger, in: *Proceedings of the ECOOSA Conference*, Rome, November 7-9 1990, ed. by M. Bertolotti, Hilger, Bristol, in press.
41. H. M. Gibbs, S. L. McCall and N. C. Venkatesan, *Phys. Rev. Lett.* 36:113 (1976).
42. M. Brambilla, F. Castelli, L. A. Lugiato, F. Prati and G. Strini, in: *Proceedings of the ECOOSA Conference*, Rome, November 7-9 1990, ed. by M. Bertolotti, Hilger, Bristol, in press.
43. R. Graham, *Z. Phys.* 210:319 (1968); R. Graham,Z. Phys.211:469 (1968).
44. M. Brambilla, F. Castelli, L. A. Lugiato, F. Prati and G. Strini, in preparation.
45. A. S. Lane, M. D. Reid and D. F. Walls, *Phys. Rev. A* 38:788 (1988).
46. B. Yurke, *Phys. Rev. A* 29:408 (1984).
47. M. J. Collett and C. W. Gardiner, *Phys. Rev. A* 30:1386 (1984); M. J. Collett and D. F. Walls,Phys. Rev. A32:2887 (1985).

48. M. Brambilla, G. Strini, F. Castelli, L. A. Lugiato and F. Prati, in: *Proceedings of the Conference Nonlinear Dynamics of Optical Systems*, Afton, June 4-8 1990, ed. by N. Abraham, E. Garmire and P. Mandel, Optical Society of America, in press.

SEMI-CLASSICAL INPUT-OUTPUT LINEARIZATION TECHNIQUES FOR QUANTUM FLUCTUATIONS AND BEYOND

Claude Fabre[1], Jean-Michel Courty[1], Elisabeth Giacobino[1], Antoine Heidmann[1], Laurent Hilico[1], Paul Mandel[2], Nicolas Pettiaux[2] and Serge Reynaud[1]

1. Laboratoire de Spectroscopie Hertzienne de l'ENS, associé au CNRS, Université P. et M. Curie, BP 74, F-75252 Paris 05 (France).
2. Université Libre de Bruxelles, Campus Plaine CP 231, B-1050 Bruxelles (Belgium).

1. INTRODUCTION

In the recent years, different methods have been proposed to determine the spectrum of quantum fluctuations at the output of a given optical sytem. Among them, the semi-classical input-output method[1,2,3] has been successfully applied to steady-state problems involving intracavity three-wave[4,5,6] or four-wave mixing [7] in pure parametric media, taking into account the possible existence of linear losses in the system. After a brief overview of this method (part2), we will show how it can be extended to treat more complicated situations : firstly (part3), the case of a system without a steady-state solution (more precisely, in the self-pulsing regime), and secondly (part4), the case of a non-linear medium having its own dynamics and excess noise sources (more precisely, for atomic systems inserted in an optical cavity).

2. THE SEMI-CLASSICAL INPUT-OUTPUT METHOD

The method consists in considering the quantum optical fields as classical stochastic variables. The classical equations of electrodynamics are used to determine how a given optical system reacts to input field fluctuations, i.e. to compute the fluctuations of the outgoing fields as a function of the input fluctuations. The statistical moments of the input fields are chosen in such a way that they mimic the quantum fluctuations, either for input fields having a non-zero mean value (pump fields), or a null mean value (vacuum fluctuations). More precisely the fields $\alpha_i(t)$ in the optical system are written as :

$$\alpha_i(t) = \bar{\alpha}_i + \delta\alpha_i(t) \quad (1)$$

$\bar{\alpha}_i$ being the steady state solution of the dynamical equations. Assuming that the

Quantum Measurements in Optics, Edited by P. Tombesi and D.F. Walls, Plenum Press, New York, 1992

fluctuating part $\delta\alpha_i(t)$ is small, the output field fluctuations are obtained by linearizing the equations around the steady-state solutions $\bar{\alpha}_i$.

This method, which is related to other input-output[8,9] or semi-classical[10] approaches, has been shown to give exactly the same analytical results as the standard method based on the positive P representation[1,6] in the case of the Optical Parametric Oscillator. More generally, and as expected when dealing with connexions between classical and quantum evolution, it is closely related to writing the quantum evolution of the system using the *Wigner representation*[11]. It can be shown that such an approach is valid when dealing with ideal parametric interaction[12,13,14] and with multi-wave mixing of any order provided that the computed fluctuations are small compared to the mean fields[3]. The method allows also for linear losses to be included in the system.

In this paper, we will show how the method can be extended to treat optical systems out of the previously quoted set, and more precisely to treat the case of unstable situations, and of non parametric media.

3. BEYOND THE HOPF BIFURCATION

Because they both originate from the same optical nonlinearities, the squeezing properties of an optical system and its dynamical behaviour are closely related phenomena.

More precisely, in a lot of cases, the approach of a bifurcation between two different dynamical regimes leads to important amounts of squeezing in one of the variables of the considered problem. This fact has a simple explanation : a bifurcation point is a kind of phase transition, and therefore a divergence occurs for the fluctuations of some variable in the system. If the parameters of the problem are chosen so that the area in phase space is conserved for the corresponding field, then the fluctuations of the conjugate variable will be reduced. This simple argument can be precisely elaborated on in the case of intracavity $\chi^{(2)}$ processes[15].

The degenerate Optical Parametric Oscillator (OPO) provides a classical example of such a feature. Below a threshold value I_t for the pump power I_2, the subharmonic output field I_1 has a zero mean intensity, and can be proved to be actually a squeezed vacuum[16,17]. When approaching the bifurcation point from below, the squeezing tends to be perfect at zero frequency in one of the field quadratures (this has been experimentally confirmed[18]). For pump powers larger than I_t, the OPO starts emitting a field with a non-zero mean intensity I_1, which turns out to have reduced phase fluctuations[17,6]. Here also the phase squeezing tends to infinity when the bifurcation point is approached from above (this prediction has never been experimentally checked up to now). When the threshold is crossed, there is actually a continuous variation of the noise spectrum, and one gets good squeezing on both sides.

Another interesting example is intracavity second-harmonic generation (SHG). The mean output second harmonic field intensity I_2 is in this case non-zero for any value of the pump field intensity I_1. But there exists a threshold I_{sp} for the pump intensity[19] above which the c.w. solution for the second harmonic field is no longer stable. A phase instability develops, and a self-pulsing regime begins to take place, with a stable solution oscillating at a frequency Ω_0. Below this Hopf bifurcation, because of the phase instability, the second harmonic output field has increased phase fluctuations at noise frequencies ω close to the self-pulsing frequency Ω_0, and therefore reduced intensity fluctuations in the same region. When the cavity finesse for the pump field is very large, the amplitude field component at frequency Ω_0 is perfectly squeezed when one approaches the Hopf bifurcation from below[17] (this squeezing has recently been observed, either in the fundamental[20] or in the second harmonic[21] mode). The question which then naturally arises and that we will try to answer is: does this noise reduction property still remains in one way or another above the bifurcation, like in the OPO case ?

The equations for the intracavity fundamental (A_1) and second harmonic (A_2) fields, in reduced units and in the resonant case, are :

$$\tau dA_1/dt = -\gamma_1 A_1 + A_1^* A_2 + t_1 E_1 \tag{2a}$$

$$\tau dA_2/dt = -\gamma_2 A_1 - A_1 A_1 + t_2 \delta E_2 \tag{2b}$$

where τ is the cavity round trip time. $r_1 = 1 - \gamma_1$, $r_2 = 1 - \gamma_2$ and t_1, t_2 are respectively the amplitude reflexion and transmission coefficients of the coupling mirror for the two fields in the cavity. E_1 is the pump field at the fundamental frequency, and δE_2 is the input vacuum field in the second harmonic mode. We will only consider here the case $\gamma_1 \ll \gamma_2 \ll 1$, where a perfect amplitude squeezing is achieved at the Hopf bifurcation. In this case, above the self-pulsing threshold $(E_1)_{sp}$, but close to it, the self-pulsing stable solution has a simple series expansion in terms of the parameter ϵ[22]:

$$\epsilon = [E_1/(E_1)_{sp} - 1]^{1/2} \tag{3}$$

which is, for the second harmonic field:

$$A_2(t) = -1 + i\epsilon[(i-1)e^{i\Omega_0 t} + c.c.] - \epsilon^2[(12i+3)e^{2i\Omega_0 t} + c.c. + 1]/17 + 0(\epsilon^3) \tag{4}$$

Note that the first order term in ϵ is imaginary : the phase instability at threshold gives rise to an oscillating phase above threshold.

The noise properties of this system are now obtained by linearizing the equations of motion around the stable solution, which is now oscillating in time:

$$A_i(t) = \bar{A}_i(t) + \delta A_i(t) \quad (i = 1, 2) \tag{5}$$

Because the linearization is performed around a time dependent point, the output field fluctuations $\delta A_i(t)$ will be linear combinations of the input field fluctuations $\delta E_1(t)$ and $\delta E_2(t)$, and of their complex conjugates, with time dependent coefficients. As a result, the Fourier spectrum of these fluctuations will be given by a convolution product, and therefore will depend on the input noise at the same frequency ω, but also at frequencies $\omega \pm \Omega_0$ and $\omega \pm 2\Omega_0$. The amplitude noise spectrum of the second harmonic field, up to second order in ϵ, is given by:

$$<| Re\delta A_2^{out}(\omega) |^2>= S_0 + \epsilon^2 S_{a2} \tag{6}$$

In this expression, S_0 is the amplitude noise spectrum at threshold:

$$S_0 = (\omega^2 - \Omega_0^2)^2/(\omega^4 - 2\omega^2\Omega_0^2 + \Omega_0^4) \tag{7}$$

which can be also obtained by a below threshold calculation[19] and shows perfect squeezing at the self-pulsing frequency Ω_0.
S_{a2} is the following correction[23]:

$$S_{a2} = (16\Omega_0^4/17)\{11016\Omega_0^{10} + 4998\omega^2\Omega_0^8 - 6881\omega^4\Omega_0^6 + 3870\omega^6\Omega_0^4 - 852\omega^8\Omega_0^2 + 89\omega^{10}\}/\{\omega^2(\omega^2 - 4\Omega_0^2)^2(\omega^4 - 2\omega^2\Omega_0^2 + 9\Omega_0^4)^2\} \tag{8}$$

This term diverges for $\omega = 0$ and $\omega = 2\Omega_0$, which are the Fourier frequencies of the mean amplitude in equation (4) and decreases rapidly at higher frequencies.
More interesting than the field *amplitude* is the noise spectrum of the field *intensity* $I_2 =| A_2(t) |^2$, because it is the quantity which is actually measured. In the linear approximation, the fluctuations of this quantity are simply given by :

$$\delta I_2(t) = \bar{A}_2(t)\delta A_2(t)^* + \bar{A}_2(t)^*\delta A_2(t) \tag{9}$$

In the c.w. case, $\bar{A}_2(t)$ is time independent, and the Fourier components of the amplitude and of the intensity are proportional. This is no longer the case when $\bar{A}_2(t)$ oscillates in time. The intensity Fourier spectrum $S_I(\omega)$ is then given by a convolution product and is equal to:

$$S_I(\omega) = S_0 + \epsilon^2 S_{I2} \tag{10}$$

with[23]:

$$S_{I2} = (16/17)\{2244\Omega_0^8 - 3583\omega^2\Omega_0^6 + 1932\omega^4\Omega_0^4 - 342\omega^6\Omega_0^2 + 21\omega^8\}/\{\omega^2 - 4\Omega_0^2)^2(\omega^4 - 2\omega^2\Omega_0^2 + 9\Omega_0^4)^2\} \tag{11}$$

This correction has only a divergence at frequency $2\Omega_0$, corresponding to the evolution frequency of the mean intensity oscillation. Figure 1 displays the variation of the intensity noise spectrum as a function of ω and of the pump intensity $(E_1)^2$ when the self-pulsing threshold $I_{sp} = (E_1)^2_{sp}$ is crossed. Apart from the large noise increase for frequencies close to $2\Omega_0$, the noise variation is smooth around the bifurcation. As the correction S_{I2} is positive, the squeezing is reduced at any frequency when the intensity increases from I_{sp}. But it is interesting to notice that the intensity squeezing is still almost perfect above the bifurcation in the vicinity of the self-pulsing frequency Ω_0 and, in a way similar to the OPO threshold case, remains very good even far from threshold.

4. BEYOND THE PARAMETRIC LIMIT

It is well-known that pure parametric processes are very efficient to reduce quantum fluctuations. In the semi-classical approach, this feature appears very clearly, as there is no noise source associated to such media: the only noise sources are the vacuum fluctuations of the input modes and the pump field fluctuations. For three-wave mixing processes, the non-linear crystals used in the experiments turn out to be almost perfect parametric media, and introduce a negligible amount of excess noise. This is no longer the case for four wave-mixing processes : most experiments use quasi-resonant media to get a non negligible nonlinear coefficient, and therefore the dynamics of the medium plays an important role in the noise properties of the optical system. This problem has already received a lot of attention[24,25,26,27]. In this paper, our purpose will be simply to show how the semi-classical approach must be modified in this respect, and more precisely to calculate the excess noise sources which are introduced by the atomic medium.

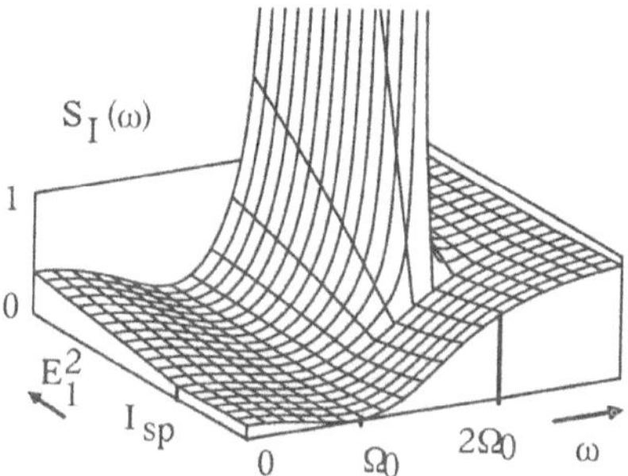

Fig. 1. Intensity noise spectrum for the second harmonic output field in the vicinity of the self-pulsing threshold $(0.9I_{sp} < (E_I)^2 < 1.5I_{sp}\ ;\ 0 < \omega < 3\gamma_2/\tau)$.

Let us consider more specifically the problem of a Kerr medium inserted in an optical cavity[7] (Figure 2). We assume for the moment that the medium is perfect. Then its only effect is to add to the usual cavity round trip phase shift $\Phi = kL$ (k: field wave vector, L: cavity length) an intensity dependent phase shift α_{Kerr} equal to :

$$\alpha_{Kerr} = k\chi^{(3)}L \mid E \mid^2 \tag{12}$$

where E is the intracavity field amplitude. This term may induce bistability in the system. We allow for extra linear losses in the cavity, which are due to a non-perfect mirror M', introducing another coupling port in the optical system. It is then straightforward to calculate the output field A^{out} as a function of the input field A^{in} and of the other input field A'^{in} through mirror M':

$$A^{out} = \frac{\gamma + i\varphi + i\alpha_{Kerr} - \gamma'}{\gamma - i\varphi - i\alpha_{Kerr} + \gamma'} A^{in} + \frac{t}{\gamma - i\varphi - i\alpha_{Kerr} + \gamma'} t' A'^{in} \tag{13}$$

In this expression, $\gamma = 1 - r$, $\gamma' = 1 - r'$, where r, t and r', t' are the amplitude reflexion and transmission coefficients respectively of the coupling mirror M and of the "loss" mirror M'. γ, γ' and the cavity detuning φ are assumed to be small.

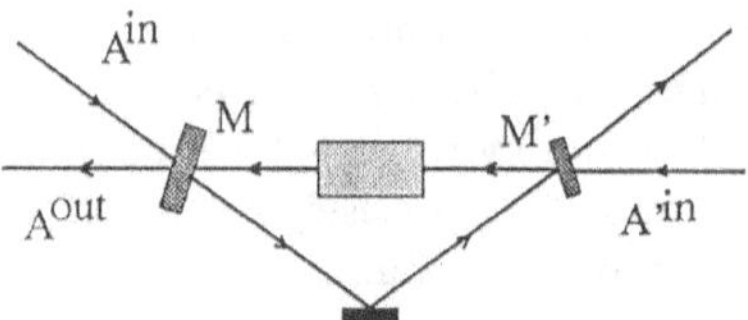

Fig. 2. Scheme of an optical bistable device, made of a nonlinear medium inserted in an optical cavity with two non perfect mirrors M and M'.

From this expression, one first computes the mean output field A^{out}, assuming that at the input $A^{in} \neq 0$ and $A'^{in} = 0$. The output noise is then calculated by linearizing Eq. (13) around its mean value. The Fourier component at frequency ω of the field fluctuations, $\delta A^{out}(\omega)$, is obtained as a function of the input field noise $\delta A^{in}(\omega)$ and of the vacuum fluctuations $\delta A'^{in}(\omega)$. The expression takes a simple form when expressed for the column matrix $[\delta A^{out}(\omega)]$ of elements $\delta A^{out}(\omega)$ and $\delta A^{out*}(-\omega)$:

$$\begin{aligned}[\delta A^{out}(\omega)] &= \frac{\gamma + i\varphi + i\omega\tau + i[\alpha_{Kerr}] - \gamma'}{\gamma - i\varphi - i\omega\tau - i[\alpha_{Kerr}] + \gamma'} [\delta A^{in}(\omega)] \\ &+ \frac{t}{\gamma - i\varphi - i\omega\tau - i[\alpha_{Kerr}] + \gamma'} t' [\delta A'^{in}(\omega)]\end{aligned} \tag{14}$$

(τ: cavity round trip time). $[\alpha_{Kerr}]$ is the 2×2 matrix $k\chi^{(3)}L \begin{bmatrix} \bar{E}^* & \bar{E} \\ \bar{E}^* & \bar{E} \end{bmatrix}$. As this matrix is non-diagonal, there is a possibility of squeezing[28] in such a system, and especially of intensity squeezing in a definite region of the parameter space[7]. The second term in Eq. (14) gives the expression of the excess noise in the output field due to the existence of linear losses.

The problem that we want to solve is to precise how the previous expression is modified when the perfect Kerr medium with linear losses is replaced by a more realistic quantum system, such as an atomic system. For the sake of simplicity, we will restrict ourselves here to the case of a assembly of N two-level atoms. The intracavity field is then coupled to new dynamical variables, namely the dipole moment of each atom i, described by the raising and lowering operators $\mathbf{S}_i^+$ and $\mathbf{S}_i^-$, through the usual electric dipole interaction:

$$\mathbf{H_I} = -\hbar g \Sigma_i (\mathbf{E}_i^+ \mathbf{S}_i^+ + \mathbf{E}_i^- \mathbf{S}_i^-) \tag{15}$$

g being the coupling constant, proportional to the dipole matrix element between the two levels, and $\mathbf{E}_i^+$ (resp. $\mathbf{E}_i^-$) the positive (resp. negative) frequency part of the electric field operator of the cavity mode acting on the ith atom. The evolution of the atomic variables is ruled by the Bloch equations, which take into account the previous atom-cavity mode coupling, but also the relaxation mechanisms, such as spontaneous emission or collisions. To further simplify the problem, we will make the extra assumptions that this atomic relaxation is not strongly modified by the cavity mode (no "cavity QED" effects), and that the interatomic coherences are negligible (no superfluorescence). We will neglect all transverse effects and assume perfect phase matching between the different atoms. We give here the essential steps of the derivation, which is detailed in ref[29,30].

As usual, the first step consists in determining the mean values of the different field and atomic variables for given initial conditions (nonzero mean input field A^{in}, atoms in the ground state). One obtains a bistable behaviour, which has been extensively studied[31]. One then defines atomic and field fluctuation operators $\delta\mathbf{S}_i^{\pm}(t)$, $\delta\mathbf{A}^{\mathrm{in}}(t)$, $\delta\mathbf{A}^{\mathrm{out}}(t)$ and $\delta\mathbf{E}_i^{\pm}(t)$, as:

$$\delta\mathbf{S}_i^{\pm}(t) = \mathbf{S}_i^{\pm}(t) - <\mathbf{S}_i^{\pm}>, \quad \delta\mathbf{A}^{\mathrm{in}}(t) = \mathbf{A}^{\mathrm{in}}(t) - <\mathbf{A}^{\mathrm{in}}>, \ etc... \tag{16}$$

Knowing the expression (15) of the intracavity field-atom interaction and the properties of the cavity, one can calculate the output field fluctuations as a function of the input field fluctuations and of the atomic dipole fluctuations. One finds for the Fourier component at frequency ω:

$$\delta\mathbf{A}^{\mathrm{out}}(\omega) = \frac{\gamma + i\varphi + i\omega\tau}{\gamma - i\varphi - i\omega\tau}\,\delta\mathbf{A}^{\mathrm{in}}(\omega) + \frac{it}{\gamma - i\varphi - i\omega\tau} g \sum_i \delta\mathbf{S}_i^-(\omega) \tag{17}$$

with the same notations as in equation (13). The effect of the atomic fluctuations on the output field appears, as expected, as filtered by the cavity resonance curve. This expression is not sufficient to calculate the variance of such fluctuations, because there is a strong correlation between the atomic noise (last term) and the input noise (first term), that we will know determine.

The equivalent of the linearization procedure of the semi- classical theory is here the use of the *quantum linear response theory* [32], which enables us to compute the response of a quantum system to a small perturbation. Let us stress here that the small perturbation is provided by the field fluctuations around its mean value. The response of the atoms to the mean field has been treated in an exact way in the first step of this derivation. One finds an expression of the following form for the column matrix $[\delta \mathbf{S}_i(\omega)]$ of elements $\delta \mathbf{S}_i^-(\omega)$ and $\delta \mathbf{S}_i^+(\omega)$:

$$[\delta \mathbf{S}_i(\omega)] = [\chi(\omega)][\delta \mathbf{E}_i(\omega)] + [\delta \mathbf{F}_i(\omega)] \tag{18}$$

where $[\chi(\omega)]$ is the atomic susceptibility matrix, and $[\delta \mathbf{F}_i(\omega)]$ the proper atomic noise term. Both $[\chi(\omega)]$ and the atomic noise correlation functions are calculated knowing the dynamics of the atoms around the mean value, i.e. using the Bloch equations of the system and the quantum regression theorem.

One eventually obtains the following input-output quantum relation for the field fluctuations:

$$\begin{aligned}[\delta \mathbf{A}^{\mathbf{out}}(\omega)] &= \frac{\gamma + i\varphi + i\omega\tau + i[\alpha(\omega)]}{\gamma - i\varphi - i\omega\tau - i[\alpha(\omega)]}[\delta \mathbf{A}^{\mathbf{in}}(\omega)] \\ &+ \frac{t}{\gamma - i\varphi - i\omega\tau - i[\alpha(\omega)]}[\delta \mathbf{F}_{\mathbf{at}}(\omega)]\end{aligned} \tag{19}$$

where :

$$[\alpha(\omega)] = Ng^2 \begin{bmatrix} 1 & 0 \\ 0 & -1 \end{bmatrix} [\chi(\omega)] \tag{20}$$

is the total susceptibility of the N atom system, and:

$$[\delta \mathbf{F}_{\mathbf{at}}(\omega)] = i \begin{bmatrix} 1 & 0 \\ -1 & 0 \end{bmatrix} \sum_i [\delta \mathbf{F}_i(\omega)] \tag{21}$$

is the global atomic noise. Expression (19) gives therefore an explicit expression for the output field noise properties knowing the input field noise and the atomic evolution equations. Comparing this expression with Eq. (14), one observes that the real part of $[\alpha(\omega)]$ corresponds in the pure parametric medium case to the matrix $[\alpha_{Kerr}]$, whereas its imaginary part is the generalization of the linear loss parameter γ', the transmitted vacuum noise input $t'\delta A'^{in}$ being replaced by the atomic noise term $[\delta \mathbf{F}_{\mathbf{at}}(\omega)]$.

Let us stress here an important point : $\chi(\omega)$ and $\delta \mathbf{F}_i(\omega)$, originating from the same dynamical equations, are not independent quantities. As a consequence, one can derive a simple relation between the imaginary part of the atomic susceptibility and the atomic noise, which nothing else than a particular case of the fluctuation-dissipation theorem. This relation can be shown to ensure the proper commutation relation for the output field $\delta \mathbf{A}^{\mathbf{out}}$, which provides a serious check of the consistency of the method.

The method presented here has the interest of decoupling the cavity mode evolution from the determination of the atomic evolution and induced noise, without any assumptions about the relative orders of magnitude for the different time constants in the problem (no adiabatic elimination of a variable whatsoever). In the two-level problem that we have presented here, it gives exactly the same analytical expressions as the usual method in the most general case[26].

When the atom field detuning becomes very large compared to the other parameters of the problem, it is easy to show that Eq. (19) gets identical to Eq. (14), using the linear absorption coefficent as the cavity linear loss. This gives strong support to the intuitive picture that linear losses, independently of their precise physical origin, can be viewed as a simple scattering of photons out of the cavity mode, which couples in some amount of vacuum fluctuations.

The method presented here in the two-level atom bistability case is readily extendable to more complex problems, for example a three-level system interacting with two modes of the field[33]. Its generalization to other problems such as semi- conductors or media with gain due to external pumping is presently under consideration.

ACKNOWLEDGMENTS

Thanks are due to P. Grangier and L. Lugiato for stimulating discussions. This work has been done within the frame of the EEC contract ESPRIT BRA 3186.

REFERENCES

1. S. Reynaud, A. Heidmann, *Opt. Commun* 71, 209 (1989).
2. S. Reynaud, *Ann. Phys. Fr.* 15, 63 (1990).
3. C. Fabre, S. Reynaud, "Fundamental systems in Quantum Optics", *Les Houches session 53, J. Dalibard, J.M. Raimond and J. Zinn-Justin eds,* Elsevier Science Publishers B.V. (1991).
4. S. Reynaud, C. Fabre, E. Giacobino, *J. Optical Society Am.* B4, 1520 (1987).
5. C. Fabre, E. Giacobino, A. Heidmann, S. Reynaud, *J. de Phys.* 50, 1209 (1989).
6. C. Fabre, E. Giacobino, A. Heidmann, L. Lugiato, S. Reynaud, M. Vadacchino, Wang Kaige, *Quantum Optics* 2, 159 (1990).
7. S. Reynaud, C. Fabre, E. Giacobino, A. Heidmann, *Phys. Rev.* A40, 1440 (1989).
8. B. Yurke, *Phys. Rev.* A29, 408 (1984), and B. Yurke, J.S. Denker, Phys. Rev. A29, 1419 (1984).
9. C.W. Gardiner, H.J. Collett, *Phys. Rev.* A30, 1386 (1984), and H.J. Collett, C.W. Gardiner, Phys. Rev. A31, 3761 (1985).
10. Y. Yamamoto, N. Imoto, S. Machida, *Phys. Rev.* A33, 3243 (1986), and Y. Yamamoto, S. Machida, O. Nillson, Phys. Rev. A34, 4025 (1986).
11. E. P. Wigner, *Phys. Rev.* 40, 749 (1932).
12. A. Ekert, P. Knight, *to be published in Phys. Rev.* A (1991).
13. T. Marshall, E. Santos, *Phys. Rev.* A41, 1582 (1990).
14. D. Kupiszewska, K. Rzazewski, *Phys. Rev.* A42, 6869 (1990).
15. L. Lugiato, P. Galatola, L.M. Narducci, *Opt. Commun.* 76, 276 (1990).
16. B. Yurke, *Phys. Rev.* A29, 408 (1984).
17. M.J. Collett, D.F. Walls, *Phys. Rev.* f A32, 2887 (1985).

18. L.A. Wu, M. Xiao, H.J. Kimble, *J. Opt. Soc. Am.* B4, 1465 (1987).
19. P.D. Drummond, K.J. McNeil, D.F. Walls, *Optica Acta* 27, 321 (1980).
20. 20. S.F. Pereira, M. Xiao, H.J. Kimble, *J. Hall, Phys. Rev.* A38, 4931 (1988).
21. A. Sizmann, R. Horowicz, G. Wagner, G. Leuchs, *Opt. Commun.* 80, 138 (1990).
22. P. Mandel, T. Erneux in "Laser Physics", *Lecture Notes in Physics 182, 182,* J. Harvey, D. Walls eds, Springer-Verlag, Heidelberg (1983).
23. N. Pettiaux, P. Mandel, C. Fabre, *Phys.Rev.Letters* 66, 1838 (1991).
24. L. Lugiato, G. Strini, *Opt. Commun.* 41, 67 (1982).
25. H.J. Carmichael, *Phys. Rev.* A33, 3262 (1986).
26. M.D. Reid, *Phys. Rev.* A37, 4792 (1988).
27. L. Lugiato, M. Vadacchino, F. Castelli, p. 161 in "Squeezed and non-classical light", *P. Tombesi, E. Pike eds,* Plenum Press (new York) 1989, and refs in.
28. M. Kitagawa, Y. Yamamoto, *Phys. Rev.* A34, 3974 (1986).
29. J.M. Courty, P. Grangier, L. Hilico, S. Reynaud, *to be published in Opt. Commun.* (1991).
30. L. Hilico, C. Fabre, E. Giacobino, S. Reynaud (to be published).
31. L. Lugiato, "Theory of Optical Bistability", *Progress in Optics Vol. XXI,* E. Wolf ed., North-Holland (Amsterdam), 1984, and refs in.
32. H. Mori, *Prog. in Theoretical Phys.* 33, 423 (1965).
33. P. Grangier, J.F. Roch, S. Reynaud (to be published).

QUANTUM SUPERPOSITIONS: SQUEEZING AND DISSIPATIONS

A. Vidiella Barranco, V. Bužek, P. L. Knight, and W. K. Lai

Optics Section, Blackett Laboratory, Imperial College,
London SW7 2BZ, U.K.

1. INTRODUCTION

The linear superposition principle is one of the most fundamental features of quantum mechanics[1]. In particular, the interference of quantum amplitudes results from this principle. It has been realized recently that the interference between various states composing a quantum superposition state gives rise to various non-classical effects[2-6], for instance squeezing, that is a reduction of quadrature fluctuations below the level associated with the vacuum[7-8]. The aim of our talk is to clarify the origin of squeezing in a superposition of quantum states of light and to study the influence of damping on squeezing.

We start our talk with a simple example of a superposition state of the form

$$|\Psi\rangle = C_m|m\rangle + C_n|n\rangle, \tag{1}$$

where $|m\rangle$ and $|n\rangle$ are number states of a single mode of the electromagnetic field. We will show in the talk that the states described by equation (1) can exhibit squeezing which depends on particular values of complex amplitudes C_m and C_n. In the second part of our talk we will discuss the origin of non-classical effects in a superposition of coherent states $|\alpha\rangle$. We will analyze in detail the superposition of two coherent states:

$$|\Psi\rangle = A\left(|\alpha_1\rangle + |\alpha_2\rangle\right); \tag{2}$$

(A is the normalization constant) as well as the one-dimensional continuous superposition of coherent states of the form[4]:

$$|\xi\rangle \equiv C_F \int_{-\infty}^{\infty} F(\alpha,\xi)|\alpha\rangle d\alpha, \tag{3}$$

where α is the real amplitude of a coherent state $|\alpha\rangle$; C_F is the normalization constant defined as

$$C_F^{-2} = \int\int_{-\infty}^{\infty} F(\alpha,\xi)F(\alpha',\xi)\exp[-(\alpha-\alpha')^2/2]d\alpha d\alpha'. \tag{4}$$

and $F(\alpha,\xi)$ is the distribution function.

Quantum Measurements in Optics, Edited by P. Tombesi and
D.F. Walls, Plenum Press, New York, 1992

Due to the fact that physical systems in general are not isolated but are interacting with an environment (heatbath) we will study in the final part of our talk the influence of damping on quantum interference. In particular, we will analyze in detail the way in which squeezing is deteriorated by the presence of a heatbath.

2. SUPERPOSITION OF NUMBER STATES

In this part of the talk we will study how squeezing of quadrature fluctuations can appear in the superposition of two number states given by equation (1). To study light squeezing we introduce two quadrature operators $\hat{a}_1$ and $\hat{a}_2$ corresponding to the creation and annihilation operators $\hat{a}^\dagger$ and $\hat{a}$ of the field mode under consideration:

$$\hat{a}_1 = \frac{\hat{a} + \hat{a}^\dagger}{2}; \qquad \hat{a}_2 = \frac{\hat{a} - \hat{a}^\dagger}{2i}. \tag{5}$$

Operators $\hat{a}^\dagger$ and $\hat{a}$ obey the ordinary bosonic commutation relation $[\hat{a}, \hat{a}^\dagger] = 1$ (we adopt $\hbar = 1$) and hence

$$[\hat{a}_1, \hat{a}_2] = \frac{i}{2}. \tag{6}$$

One of the consequences of the commutation relation (6) is the uncertainty relation for the variances of the quadrature operators :

$$\langle(\Delta\hat{a}_1)^2\rangle\langle(\Delta\hat{a}_2)^2\rangle \geq \frac{1}{16}, \tag{7}$$

where the variance of the operator $\hat{a}_i$ is defined as $\langle(\Delta\hat{a}_i)^2\rangle = \langle\hat{a}_i^2\rangle - \langle\hat{a}_i\rangle^2$ and is related to the normally ordered variance $\langle : (\Delta\hat{a}_i)^2 : \rangle$ as follows:

$$\langle(\Delta\hat{a}_i)^2\rangle = \frac{1}{4} + \langle : (\Delta\hat{a}_i)^2 : \rangle. \tag{8}$$

The state for which the equality in equation (7) holds is called the minimum uncertainty state (MUS). For instance, the vacuum and the coherent states of light are examples of MUS. For these states the variances in both quadratures are equal to 1/4. The state is called squeezed if the variance of the quadrature operator is less than vacuum fluctuation (i.e. less than 1/4). Alternatively, one can say that the state is squeezed if the normally ordered variance is less than zero. It is not necessary for the squeezed state to be the MUS. Nevertheless, the squeezed vacuum state (see definition below) is the MUS.

The variances of the quadrature operators can be written as:

$$\langle(\Delta\hat{a}_1)^2\rangle = \frac{1}{4} + \frac{1}{2}\left[\langle\hat{a}^\dagger\hat{a}\rangle + Re\langle\hat{a}^2\rangle - 2(Re\langle\hat{a}\rangle)^2\right]; \tag{9.a}$$

$$\langle(\Delta\hat{a}_2)^2\rangle = \frac{1}{4} + \frac{1}{2}\left[\langle\hat{a}^\dagger\hat{a}\rangle - Re\langle\hat{a}^2\rangle - 2(Im\langle\hat{a}\rangle)^2\right], \tag{9.a}$$

from which it follows that squeezing can appear only if the expectation values $\langle\hat{a}\rangle$ and/or $\langle\hat{a}^2\rangle$ are non-zero (of course this is not sufficient condition for observation of squeezing). The non-zero values of $\langle\hat{a}\rangle$ and/or $\langle\hat{a}^2\rangle$ are associated with the off-diagonal terms of the density matrix in the number-state basis. In other words they

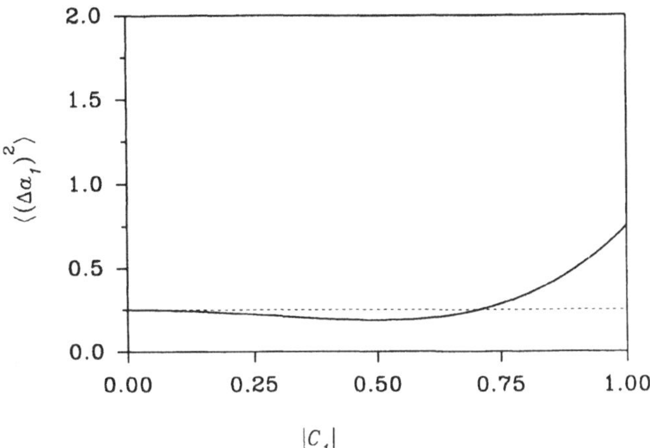

Fig. 1. Variance in the $\hat{a}_1$ quadrature for the vacuum and one-photon superposition states as a function of amplitude $|C_1|$. Squeezing occurs for values of the variance less than 1/4.

appear due to the quantum interference between the number states $|n-1\rangle$ and $|n\rangle$; and $|n-2\rangle$ and $|n\rangle$, respectively.

The variances of the quadrature operators in the superposition state (1) are (we suppose that $m < n$):

$$\langle(\Delta\hat{a}_1)^2\rangle = \frac{1}{4} + \frac{1}{2}\Big[m|C_m|^2 + n|C_n|^2 + \delta_{m,n-2}|C_m||C_n\sqrt{n(n-1)}\cos\Delta\phi$$
$$-2\delta_{m,n-1}|C_m|^2|C_n^2 n\cos^2\Delta\phi\Big]; \tag{10.a}$$

$$\langle(\Delta\hat{a}_2)^2\rangle = \frac{1}{4} + \frac{1}{2}\Big[m|C_m|^2 + n|C_n|^2 - \delta_{m,n-2}|C_m||C_n\sqrt{n(n-1)}\cos\Delta\phi$$
$$-2\delta_{m,n-1}|C_m|^2|C_n^2 n\sin^2\Delta\phi\Big], \tag{10.b}$$

where $\Delta\phi$ is the relative phase between C_n and C_m. From the above it follows that the superposition of number states given by the equation (1) is not the MUS. Nevertheless, if $m = n-1$ or $m = n-2$, then squeezing of the quadrature variances (10) can be observed for certain values of the parameters C_n and C_m. Let for simplicity suppose $m = 0$ and $n = 1$. If we take $\cos^2\Delta\phi = 0$, then in this case we find that the variances $\langle(\Delta\hat{a}_1)^2\rangle \geq 1/4$ for all values of $|C_1|$, whereas $\langle(\Delta\hat{a}_2)^2\rangle$ can be squeezed for certain values of $|C_1|$. On the other hand, for $\cos^2\Delta\phi = 1$ this relationship is reversed and $\langle(\Delta\hat{a}_1)^2\rangle$ is squeezed.

From figure 1 it follows that squeezing in $\langle(\Delta\hat{a}_1)^2\rangle$ is possible for values of $|C_1| < 1/\sqrt{2}$. Moreover, from this figure we see that the variance $\langle(\Delta\hat{a}_1)^2\rangle$ decreases to its minimum value 3/16 at $|C_1| = 1/2$ and then increases to its maximum value of 3/4 at $|C_1| = 1$. We note that one-photon superposition state can exhibit squeezing because,

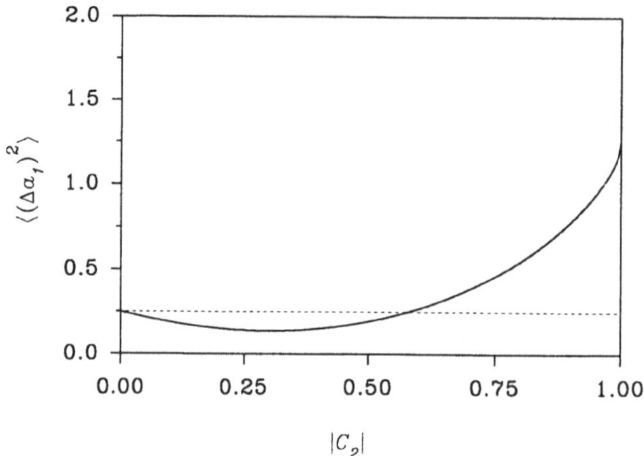

Fig. 2. Variance in the $\hat{a}_1$ quadrature for the vacuum and one-photon superposition states as a function of amplitude $|C_2|$. Squeezing occurs for values of the variance less than 1/4.

although the expectation values of the squares of the quadrature operators are equal, the state has a nonvanishing mean electric field. This electric field provides the required phase sensitivity and is responsible for squeezing. Formally the nonvanishing electric field arises as the consequence of the interference between the states $|0\rangle$ and $|1\rangle$. It should be also said that the superposition state $|\Psi\rangle = C_0|0\rangle + C_1|1\rangle$ is a state which in principle can be generated by a photon interaction Hamiltonian that is linear in $\hat{a}$ and $\hat{a}^\dagger$. For example, the Jaynes-Cummings model with a cavity field initially prepared in the vacuum state and the atom prepared in the coherent superposition of upper and lower states is a good example of a system in which one-photon superposition states can be generated (for details see Ref.[2]).

From equation (10) it follows that superposition of the vacuum state and two-photon state, i.e. $|\Psi\rangle = C_0|0\rangle + C_2|2\rangle$, can exhibit squeezing as well. In fact the degree of squeezing in this case can be even larger then in the previous case. In particular, if we put in equations (10), with $m = 0$ and $n = 2$, $\cos\Delta\phi = -1$, then we can observe reduced fluctuations in the $\hat{a}_1$ quadrature for $|C_2|1/\sqrt{3}$ (see figure 2). The highest degree of squeezing is obtained for $|C_2|^2 = 1/2 - 1/\sqrt{6}$ and is approximately equal to 45% below the standard quantum limit. Superposition of higher numbers of photons do give rise to photon sensitive noise properties. However, the fluctuations in these states are not reduced (squeezed) below the vacuum level. The reason for this is that the noise associated with a number state, other than the vacuum, is too large to be squeezed by the superposition process.

So far we have only discussed pure superposition states. Nevertheless, statistical mixtures can also exhibit squeezing. We will discuss this feature in the last part of our talk in which we will study the effect of damping on squeezing (see also Refs.[9-11]).

3. SUPERPOSITION OF COHERENT STATES

In this part of our talk we will discuss the non-classical effects which arise due to the quantum interference between coherent states. There have recently been several proposals for the generation of optical superposition states in nonlinear processes[12–15], so it is meaningful to analyze the origin of non-classical effects in superposition of coherent states. We will concentrate mainly on quadrature squeezing. This problem is very interesting, because the coherent states are always associated with the "most" classical states one can imagine the in framework of quantum theory[16]. On the other hand as we will see the quantum interference between these states leads to generation of states whose properties are as far as one can imagine from "classical" states[3,17–19], that is the superposition states can exhibit quadrature squeezing[17], higher order squeezing[20], sub-Poissonian photon statistics[21] and other non-classical features.

Let us consider a superposition of two coherent states given by equation (2). The density matrix corresponding to this superposition is given by the following expression

$$\hat{\rho} = A(|\alpha_1\rangle\langle\alpha_1| + |\alpha_2\rangle\langle\alpha_2| + |\alpha_1\rangle\langle\alpha_2| + |\alpha_2\rangle\langle\alpha_1|), \tag{11}$$

while the density matrix describing statistical mixture of two coherent states $|\alpha_1\rangle$ and $|\alpha_2\rangle$ is

$$\hat{\rho}_M = p_1|\alpha_1\rangle\langle\alpha_1| + p_2|\alpha_2\rangle\langle\alpha_2|, \tag{12}$$

where p_i is the probability to find the system in the state $|\alpha_i\rangle$. These probabilities are normalized to 1.

3.1. *Wigner function*

Now we introduce the notion of the Wigner function through the characteristic function $C^{(W)}(\xi)$, which is associated with the symmetrical order of the bosonic (photon) operators and is given by the relation[22]

$$C^{(W)}(\xi) = Tr\left[\hat{\rho}\exp(i\xi\hat{a}^\dagger + i\xi^*\hat{a})\right]. \tag{13}$$

The Wigner function is defined as the Fourier transform of the characteristic function $C^{(W)}(\xi)$:

$$W(\beta) = \pi^{-2}\int d^2\xi \exp[-i(\xi\beta^* + \xi^*\beta)]C^{(W)}(\xi). \tag{14}$$

The Wigner function corresponding to the superposition of two coherent states (2) can be written in the form

$$W(\beta) = A(W_1 + W_2 + W_{12}) \tag{15.a}$$

where

$$W_i = \frac{2}{\pi}\exp(-2|\alpha_i - \beta|^2); \tag{15.b}$$

and

$$W_{12} = \frac{2}{\pi} \exp\left[-\frac{1}{2}\left(|\alpha_1|^2 + |\alpha_2|^2\right)\right] \{\exp\left[\alpha_2\alpha_1^* - 2(\beta - \alpha_2)(\beta^* - \alpha_1^*)\right]$$

$$+ \exp\left[\alpha_1\alpha_2^* - 2(\beta - \alpha_1)(\beta^* - \alpha_2^*)\right]\} \tag{15.c}$$

The terms W_i are the Wigner functions corresponding to the coherent states $|\alpha_i\rangle$, while the term W_{12} arises due to the quantum interference between coherent states under consideration.

The Wigner function for the statistical mixture (12) is given by the relation

$$W_M = p_1 W_1 + p_2 W_2 \tag{16}$$

and it does not contain the term describing the quantum interfence between coherent states.

3.2. *Even coherent states*

To simplify our task we will suppose that $\alpha_1 = -\alpha_2 = \alpha$, where α is a real parameter. In this case we obtain from (2) the following state

$$|\Psi\rangle = A^{1/2}\left\{|\alpha\rangle + |-\alpha\rangle\right\}, \tag{17}$$

with the normalization constant

$$A^{-1} = 2[1 + \exp(-2\alpha^2)].$$

The state (17) is called[23] the even coherent state.

The Wigner function corresponding to this state can be found using the general expression (15) and is presented in Figure 3, where $x = Re\beta$ and $y = Im\beta$. If we compare this function with the Wigner function (see Figure 4) corresponding to the statistical mixture of states $|\alpha\rangle$ and $|-\alpha\rangle$ described by the density matrix

$$\hat{\rho}_M = \frac{1}{2}\left(|\alpha\rangle\langle\alpha| + |-\alpha\rangle\langle-\alpha|\right) \tag{18}$$

we can directly observe that the term W_{12} corresponding to the quantum interference between states $|\alpha\rangle$ and $|-\alpha\rangle$ should play an important role in statistical properties of superpositions of coherent states.

3.3. *Quadrature squeezing*

We can easily find the variances (9) of the quadrature operators (5) in the statistical mixture (18):

$$\langle(\Delta\hat{a}_1)^2\rangle_M = \frac{1}{4} + \alpha^2; \qquad \langle(\Delta\hat{a}_2)^2\rangle_M = \frac{1}{4}. \tag{19}$$

From above it follows that the fluctuations in the $\hat{a}_1$ quadrature are larger in the case of the statistical mixture compared to the vacuum-state (or the coherent-state) value, which is equal to 1/4. Fluctuations in $\hat{a}_2$ remain the same for both the statistical mixture (18) as well as for the coherent state $|\alpha\rangle$ (or $|-\alpha\rangle$).

On the other hand, for the even coherent state (17) we find the reduction of fluctuations in $\hat{a}_2$ quadrature (i.e. in the y-direction in the phase space – see Figure 3):

$$\langle(\Delta\hat{a}_2)^2\rangle = \frac{1}{4} - \frac{\alpha^2\exp(-2\alpha^2)}{1 + \exp(-2\alpha^2)}. \tag{20.a}$$

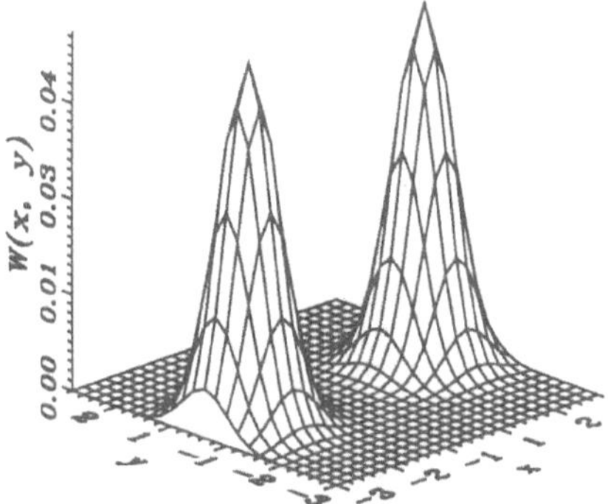

Fig. 3. Wigner function corresponding to the even coherent state (17) with $\alpha = 2$. The role of the interference term is transparent.

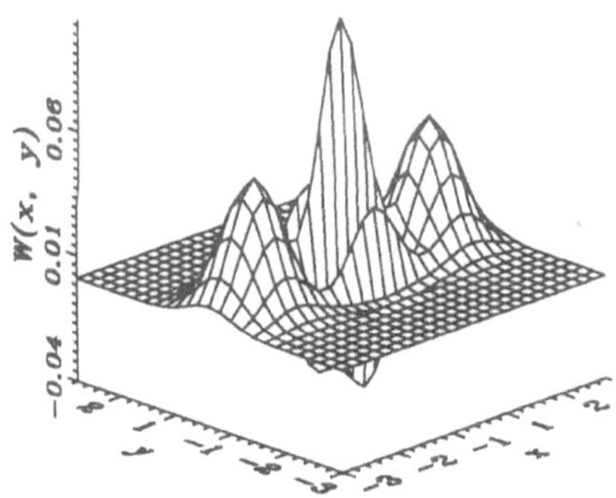

Fig. 4. Wigner function corresponding to the statistical mixture (18) with $\alpha = 2$ and $p_1 = p_2 = 1/2$.

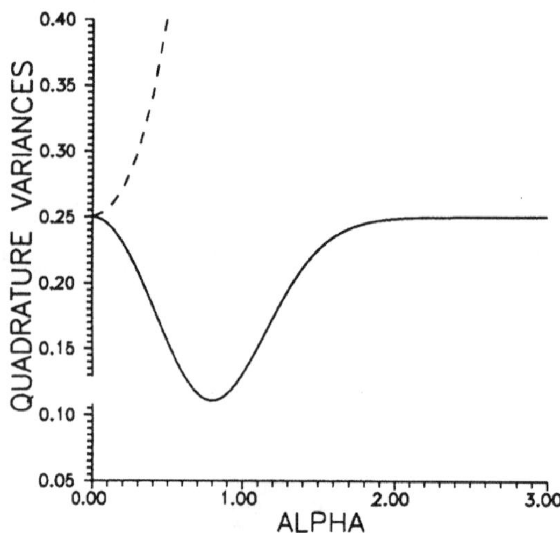

Fig. 5. Quadrature variances given by equations (20) versus parameter α. The dashed line corresponds to $\langle(\Delta\hat{a}_1)^2\rangle$ and the solid line corresponds to $\langle(\Delta\hat{a}_2)^2\rangle$.

Simultaneously fluctuations in $\hat{a}_1$ are enhanced:

$$\langle(\Delta\hat{a}_1)^2\rangle = \frac{1}{4} + \frac{\alpha^2}{1+\exp(-2\alpha^2)}. \tag{20.b}$$

Variances $\langle(\Delta\hat{a}_i)^2\rangle$ versus the parameter α (which is related to the intensity of the even coherent state) are plotted in Figure 5. We see that the maximum reduction in the fluctuations can be obtained for quite small values of α. From the above it follows that quadrature squeezing can emerge as a consequence of the quantum interference between coherent states. We should note that even coherent states (17) exhibit not only quadrature squeezing, but also higher-order squeezing as well as amplitude squared squeezing[17,20].

3.4. *Photon number distribution*

Here we discuss briefly properties of the photon number distribution of the statistical mixture (18) as well as the superposition (17) of two coherent states. The photon number distribution is defined as

$$P_n = \langle n|\hat{\rho}|n\rangle. \tag{21}$$

For the superposition of coherent states (17) we can write P_n in the form:

$$P_n = A\left\{P_n^{(1)} + P_n^{(2)} + P_n^{(12)}\right\}, \tag{22}$$

where $P_n^{(i)}$ is the Poissonian photon number distribution of the coherent state $|\alpha_i\rangle$:

$$P_n^{(i)} = \frac{|\alpha_i|^{2n}}{n!}\exp(-|\alpha_i|^2), \tag{23}$$

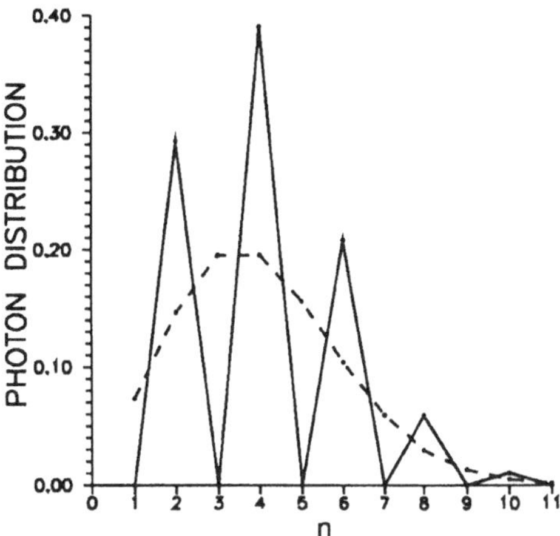

Fig. 6. Photon number distribution for a superposition of two coherent states (17) with amplitude $\alpha = 2$. The dashed line is the distribution of the corresponding statistical mixture (18).

and $P_n^{(12)}$ is the term arising due to the quantum interference between coherent states $|\alpha_1\rangle$ and $|\alpha_2\rangle$:

$$P_n^{(12)} = \frac{1}{n!}\left\{(\alpha_1\alpha_2^*)^n + (\alpha_2\alpha_1^*)^n\right\}\exp\left[\frac{1}{2}(|\alpha_1|^2 + |\alpha_2|^2)\right]. \tag{24}$$

The photon number distribution of the statistical mixture (18) is

$$P_n^M = p_1 P_n^{(1)} + p_2 P_n^{(2)}. \tag{25}$$

We see that in the case of the statistical mixture the photon number distribution (25) is just the sum of two Poissonian distributions corresponding to two independent coherent states $|\alpha_1\rangle$ and $|\alpha_2\rangle$. On the other hand, in the case of a superposition of coherent states $|\alpha_1\rangle$ and $|\alpha_2\rangle$ the term corresponding to the quantum interference plays an important role. To see this clearly we will assume that $\alpha_1 = -\alpha_2 = \alpha$. In this case the statistical mixture (25) has just the Poissonian photon number distribution

$$P_n^M = \frac{|\alpha|^{2n}}{n!}\exp\left(-|\alpha|^2\right). \tag{26}$$

The superposition of coherent states under consideration (i.e. the even coherent state) has the following photon number distribution:

$$P_n = \begin{cases} \frac{2\exp(-|\alpha|^2)}{1+\exp(-|\alpha|^2)}\frac{|\alpha|^{2n}}{n!} & \text{if } n = 2m; \\ 0 & \text{if } n = 2m+1. \end{cases} \tag{27}$$

The oscillations in P_n are very similar to those in the case of the squeezed vacuum discussed by Schleich and Wheeler[24,25]. Generally, these oscillations are due to quantum interference in the phase space.

3.5. *Continuous superposition of coherent states*

Recently Janszky and Vinogradov[4,5] extended the idea of superposition of coherent states and investigated the quadrature variances of a continuous one-dimensional superposition of coherent states (3). They have shown that such superposition of coherent states can lead to significant reduction of fluctuations in one of the quadratures. With the superposition state $|\xi\rangle$ given by equation (3) one can find expressions for the mean values of products of the creation $\hat{a}^\dagger$ and the annihilation $\hat{a}$ operators of the field mode in the following form:

$$\langle(\hat{a}^\dagger)^m \hat{a}^n\rangle = \int\int_{-\infty}^{\infty} F(\alpha,\xi)F(\alpha',\xi)\exp[-(\alpha-\alpha')^2/2]\alpha^m(\alpha')^n d\alpha d\alpha'. \tag{28}$$

In particular, if $F(\alpha,\xi)$ is taken to be the Gaussian function

$$F(\alpha,\xi) = \exp\left[-\frac{(1-\xi)}{2\xi}\alpha^2\right] \tag{29}$$

with $\xi \in (0,1)$ and

$$C_F^{-1} = \pi^{-1/2}\frac{(1-\xi^2)^{1/4}}{(2\xi)^{1/2}}, \tag{30}$$

then one can find for the variances of the quadrature operators $\hat{a}_1$, $\hat{a}_2$ given by the equation (10) the following expressions

$$\langle(\Delta\hat{a}_1)^2\rangle = \left(\frac{1}{4}\right)\frac{(1+\xi)}{(1-\xi)}; \tag{31.a}$$

$$\langle(\Delta\hat{a}_2)^2\rangle = \left(\frac{1}{4}\right)\frac{(1-\xi)}{(1+\xi)}. \tag{31.b}$$

From this one can conclude that the states $|\xi\rangle$: i) belong to the class of the minimum uncertainty states; ii) the fluctuations in the "second" quadrature are reduced below the shot noise limit.

We see, therefore, that the one-dimensional superposition (with Gaussian distribution) of coherent states leads to states exhibiting a large degree of squeezing (in the limit $\xi \to 1$). We now demonstrate by direct calculations that if $F(\alpha,\xi)$ is the Gaussian function (29), then the state $|\xi\rangle$ is equal to the squeezed vacuum state generated by the action of the squeeze operator $\hat{S}(\hat{a}^\dagger,\hat{a},\xi)$:

$$\hat{S}(\hat{a}^\dagger,\hat{a},\xi) = \exp\left[\frac{r}{2}(\hat{a}^\dagger)^2 - \frac{r}{2}\hat{a}^2\right]; \qquad \xi = \tanh r, \tag{32}$$

on the vacuum state $|0\rangle$ of the field mode. To do so we decompose the coherent state $|\alpha\rangle$ in equation (3) into number states:

$$|\alpha\rangle = \exp(-\alpha^2/2)\sum_{n=0}^{\infty}\frac{\alpha^n}{\sqrt{(n!)}}|n\rangle, \tag{33}$$

and exchange the order of integration and summation procedures, i.e. we rewrite equation (3) in the form

$$|\xi\rangle = C_F\sum_{n=0}^{\infty}\frac{1}{\sqrt{(n!)}}|n\rangle\int_{-\infty}^{\infty} d\alpha\alpha^n\exp\left[-\frac{1}{2\xi}\alpha^2\right]. \tag{34}$$

After performing the integration on the r.h.s of equation (34) we find

$$|\xi\rangle = \left(1-\xi^2\right)^{1/4} \sum_{n=0}^{\infty} \frac{[(2n)!]^{1/2}}{2^n n!} \xi^n |2n\rangle, \tag{35}$$

from which it follows that the one-dimensional superposition of coherent states (3) with the distribution function (29) is identical to the squeezed vacuum state:

$$|\xi\rangle = C_F \int_{-\infty}^{\infty} F(\alpha,\xi) d\alpha \hat{D}(\alpha)|0\rangle = \hat{S}(\xi)|0\rangle. \tag{36}$$

We should stress here that the last equation describes the relation between the states, but not between the displacement and the squeeze operators themselves.

Now we turn our attention to the displacement of the superposition state such that there is a mean field amplitude. We show that a one-dimensional superposition of coherent states with the distribution function $F(\alpha,\xi,\beta)$ centered at a non-zero value of α is equal to the squeezed coherent state[6]. We take for our distribution function $F(\alpha,\xi,\beta)$ the displaced form

$$F(\alpha,\xi,\beta) = \exp\left[-\frac{(1-\xi)}{2\xi}(\alpha - x_0)^2\right], \tag{37}$$

with the normalization constant C_F given by the equation (4) and with a displacement

$$x_0 = \left(\frac{1+\xi}{1-\xi}\right)^{1/2} \beta.$$

In this case from equation (3) we obtain for the state $|\xi\rangle$ the following expression:

$$|\xi\rangle = (1-\xi^2)^{1/4} \exp\left[-\frac{(1-\xi)}{2} x_0^2\right] \sum_{n=0}^{\infty} \sqrt{(n!)}\, [(1-\xi)x_0]^n$$

$$\times \left\{ \sum_{k=0}^{\|n/2\|} \frac{1}{(n-2k)!k!} \left(\frac{\xi}{2(1-\xi)^2 x_0^2}\right)^k \right\} |n\rangle, \tag{38}$$

where $\|x\|$ denotes the greatest integer less than or equal to x. Using the new parametrization:

$$\mu = \frac{1}{(1-\xi^2)^{1/2}}; \qquad \nu = \frac{-\xi}{(1-\xi^2)^{1/2}}, \tag{39}$$

with $\mu^2 - \nu^2 = 1$ and $|\mu| > |\nu|$, we can rewrite equation (38) in the form:

$$|\xi\rangle = \mu^{-1/2} \exp\left[-\frac{(1-\nu/\mu)}{2}\beta^2\right] \sum_{n=0}^{\infty} \frac{1}{\sqrt{(n!)}} \left(\frac{\nu}{2\mu}\right)^{n/2} H_n(\beta/\sqrt{2\mu\nu})|n\rangle, \tag{40}$$

where $H_n(x)$ is the Hermite polynomial. It is obvious that the last expression obtained describes precisely the squeezed coherent state as defined by Yuen[26] i.e., we have explicitly proved that

$$\hat{S}(\xi)\hat{D}(\beta)|0\rangle = C_F \int_{-\infty}^{\infty} d\alpha\, F(\alpha,\xi,\beta)\hat{D}(\alpha)|0\rangle. \tag{41}$$

In other words, we can construct, through a one-dimensional superposition of coherent states with a properly chosen distribution function, the squeezed coherent state. Obviously, the physical reason for squeezing is the same as for the case of squeezed vacuum state discussed earlier. It is amusing that a superposition of the most classical of field states, the coherent states, can through the action of quantum interference, generate the archetypal nonclassical field states – the squeezed vacuum and the squeezed coherent state.

It can also be shown that the squeezed number state[27] defined as a result of the action of the squeezing operator $\hat{S}(\xi)$ on the number state $|n\rangle$ can be constructed as a one-dimensional superposition of displaced number states[28], the states obtained through the action of the displacement operator on the number state, that is

$$\hat{S}(\xi)|n\rangle = C_F \int_{-\infty}^{\infty} F(\alpha,\xi) d\alpha \hat{D}(\alpha)|n\rangle, \tag{42}$$

where the function $F(\alpha,\xi)$ and the normalization constant C_F are given by equations (29) and (30), respectively.

4. INFLUENCE OF DAMPING ON QUANTUM INTERFERENCE

Now we will study the decay of a field mode initially prepared either in a superposition of number states or in a superposition of coherent states. To be specific, we shall assume that the density operator $\hat{\rho}$ for the field mode obeys a zero temperature master equation in the Born-Markov approximation. This equation in the interaction picture can be written as (see, for instance, Ref.[22]):

$$\frac{\partial \hat{\rho}}{\partial t} = \frac{\gamma}{2}\left(2\hat{a}\hat{\rho}\hat{a}^{\dagger} - \hat{a}^{\dagger}\hat{a}\hat{\rho} - \hat{\rho}\hat{a}^{\dagger}\hat{a}\right), \tag{43}$$

where γ is the decay constant. Such model was previously studied by Walls and Milburn[9] and Savage and Walls[10]. These authors have shown that the off-diagonal terms in the field density operator, expressed in a coherent-state basis, are weighted with a factor which rapidly supresses theses coherences. The effect of the decay on observable quantities was studied by Phoenix[11] who has shown that mean values of observable quantities, arising from off-diagonal coherence, do not decay on a faster time scale than other terms arising from the diagonal terms. As it will be seen later this is the reason why squeezing in the superposition of coherent states decay on the same time scale as the mean photon number. On the other hand we will show that the photon number distribution P_n, which is related to the diagonal matrix elements in the number-state basis, is very sensitive to the presence of damping. The same is true also for that part of the Wigner function which corresponds to the quantum interference between coherent states.

4.1. *Decay of superposition of number states*

Following Barnett and Knight[29] we can write the solution of the master equation (43) in the form

$$\hat{\rho}(t) = \exp(\hat{L}t)\exp\left[\frac{\hat{J}}{\gamma}\left(1 - e^{-\gamma t}\right)\right]\hat{\rho}(0); \tag{44}$$

where two superoperators $\hat{L}$ and $\hat{J}$ are defined by their action on the density operator, i.e.

$$\hat{J}\hat{\rho} = \gamma \hat{a}\hat{\rho}\hat{a}^{\dagger}; \tag{45.a}$$

$$\hat{L}\hat{\rho} = -\frac{\gamma}{2}\left(\hat{a}^{\dagger}\hat{a}\hat{\rho} + \hat{\rho}\hat{a}^{\dagger}\hat{a}\right). \tag{45.b}$$

Let for simplicity suppose the field to be initially in the superposition of $|0\rangle$ and $|1\rangle$ states (1). In this case we find for the density matrix $\hat{\rho}(t)$ at $t > 0$ the following expression:

$$\hat{\rho}(t) = \left[|C_0|^2 + |C_1|^2(1-\mu)\right]|0\rangle\langle 0| + |C_1|^2\mu|1\rangle\langle 1|$$
$$+C_0C_1^*\mu^{1/2}|0\rangle\langle 1| + C_0^*C_1\mu^{1/2}|1\rangle\langle 0|, \tag{46}$$

where $\mu = \exp(-\gamma t)$. From the above we see that the diagonal and the off-diagonal terms of the density matrix in the number-state basis do not decay with the same rate. Nevertheless, if we evaluate the expression for the mean photon number and for the variance of the quadraure operators we find that:

$$\bar{n}(t) = \mu\bar{n}(0), \tag{47}$$

and

$$\langle(\Delta\hat{a}_1(t))^2\rangle = \frac{1}{4} + \mu\langle:(\Delta\hat{a}_1(0))^2:\rangle, \tag{48}$$

from which it follows that squeezing degrades with the same rate as the mean photon number decays. In other words, we see that in spite of the fact that the superposition of number states decays into the mixture of number states squeezing of the quadrature variances can be observed.

The same result can be obtained for the superposition of the vacuum and two-photon state. The $\hat{\rho}$ matrix in this case is

$$\hat{\rho}(t) = \left[1 - |C_2|^2\mu(2-\mu)\right]|0\rangle\langle 0| + 2\mu(1-\mu)|C_1|^2|1\rangle\langle 1|$$
$$+\mu^2|C_2|^2|2\rangle\langle 2| + C_0C_2^*\mu|0\rangle\langle 2| + C_0^*C_2\mu|2\rangle\langle 0|. \tag{49}$$

In addition to the fact that the mean photon number and the degree of squeezing decay with the same rate, from the above we can also observe that not only off-diagonal terms but diagonal terms of the density matrix in the number-state basis decay with different rates. As we will see later this is the reason why the non-classical effects related to the diagonal terms of the $\hat{\rho}$ matrix in the number-state basis (for instance, oscillations of the photon number distribution) are much more sensitive to the presence of decay than squeezing.

4.2. *Decay of superposition of coherent states*

Using the formal solution (44) for the $\hat{\rho}$ matrix one can find the time-dependent expression for the density matrix of the superposition of two coherent states at $t > 0$ (for details see Ref.[11]):

$$\hat{\rho}(t) = A\sum_{i,j=1}^{2}\langle\alpha_i|\alpha_j\rangle^{1-\mu}|\mu\alpha_j\rangle\langle\mu\alpha_i|. \tag{50}$$

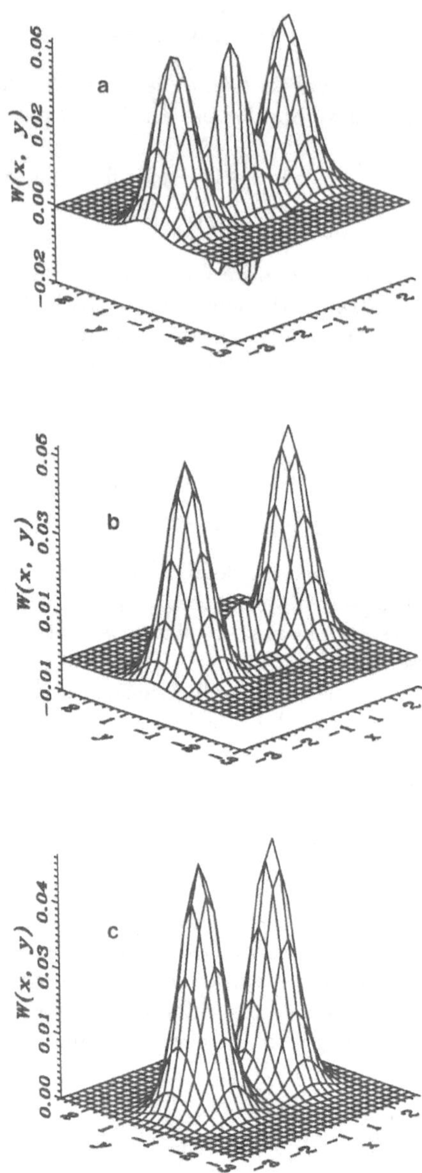

Fig. 7. The Wigner function of the initial even coherent state influenced by damping. The transition from the Wigner function of the superposition state (see Fig.3) towards the Wigner function of the mixture of states (see Fig.4) can be observed: $\alpha = 2$ and $\gamma t = 0.1$ (a); $\gamma t = 0.3$ (b) and $\gamma t = 0.5$ (c).

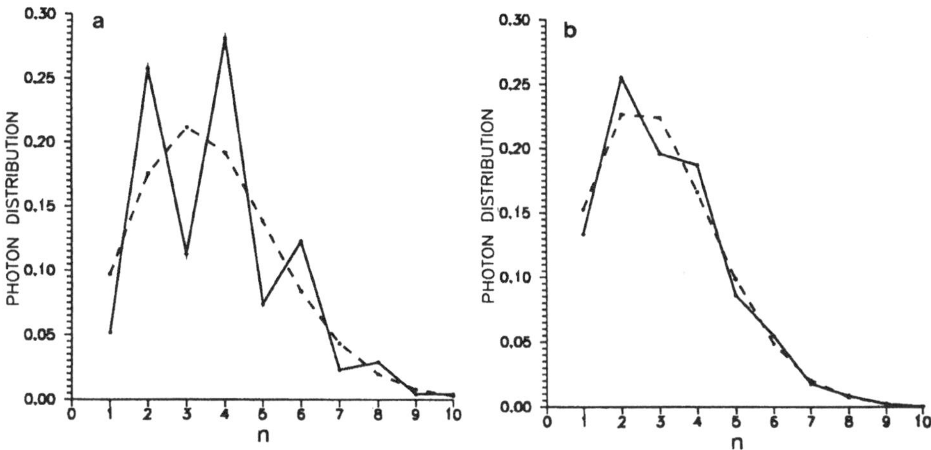

Fig. 8. Photon number distribution of the even coherent state (solid line) and the mixture of corresponding coherent states become more closer to each other under the influence of damping. Oscillatory behaviour of the photon number distribution (see Fig.6) at $t = 0$ diminishes with increase of γt: compare figure (a) with $\gamma t = 0.1$ and (b) with $\gamma t = 0.3$. For $\gamma t = 0.5$ the photon number distribution for the superposition and for the mixture are almost the same ($\alpha = 2$).

The last expression reflects the fact that the off-diagonal terms of the density matrix are rapidly dephased at a rate governed by the separation of the coherent states. The interference terms in the Wigner function (see equation (6.c)) are proportional to the off-diagonal terms of the $\hat{\rho}$ matrix which means that Wigner function of the statistical mixture decays very rapidly towards the Wigner function of the statistical mixture. This can be well seen from figure 7 in which the Wigner function of the initial even coherent state is ploted for various values of μ.

Similar sensitiveness of the quantum interference terms with respect to decay can be also observed in the case of the photon number distribution. As we said earlier the photon number distribution of the even coherent state exhibits oscillations (see Fig.6), while the statistical mixture of states $|\alpha\rangle$ and $-|\alpha\rangle$ has the Poissonian distribution. The oscillations of the photon number distribution have their origin in the quantum interference decsribed by the off-diagonal terms (in the coherent-state basis) of the density matrix (see equation (24)). From here it follows that the oscillations of P_n, i.e. the interference term $P_n^{(12)}$ in equation (24), disappear very rapidly and that only the terms corresponding to the statistical mixture give the dominant contribution to the P_n (see Fig.8).

From above we can conclude that both the Wigner function and the photon number distribution are very sensitive to the rapid destruction of the off-diagonal coherences. On the other hand, one can find (see, for instance Ref.[11]) that the normally-ordered expectation values $\langle(\hat{a}^\dagger)^m\hat{a}^n\rangle$ are

$$\langle(\hat{a}^\dagger)^m\hat{a}^n\rangle = A\sum_{i,j=1}^{2}\langle\alpha_i|\alpha_j\rangle(\mu\alpha_j)^n(\mu\alpha_i)^m, \tag{51}$$

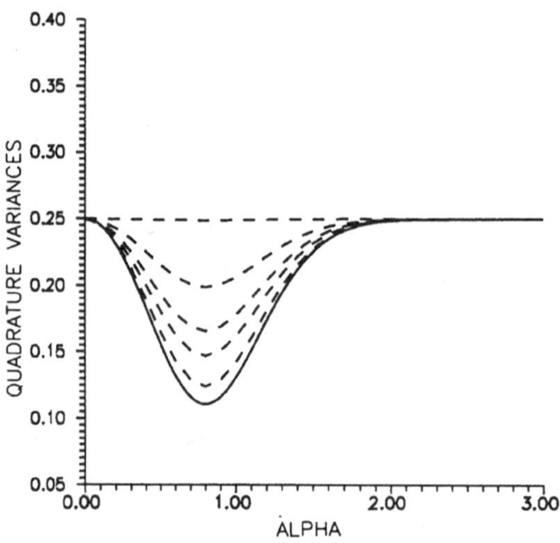

Fig. 9. The quadrature variance $\langle(\Delta\hat{a}_1(t))^2\rangle$ of the even coherent state versus parameter α for various values of γt. Various lines in this figure correspond from bottom to up to the following values of γt: 0.0; 0.1; 0.2; 0.5; 1.0 and 5.0, respectively.

which means that the terms arising from the off-diagonal coherences in the superposition state (2) decay on the same time scale as those arising from the diagonal elements. In fact this result can be easily derived directly from the quantum Langevin equation for the operators $\hat{a}^\dagger$ and $\hat{a}$ (see Ref.[22]). Now we understand that the intensity of the field mode as well as the mean values $Re(\langle\hat{a}^2\rangle)$ and $(Re\langle\hat{a}\rangle)^2$ decay with the same rate which means that for the variances of the quadrature operators we again obtain (compare with equation (48)):

$$\langle(\Delta\hat{a}_i(t))^2\rangle = \frac{1}{4} + \mu\langle: (\Delta\hat{a}_i(0))^2 :\rangle. \tag{52}$$

Comparing figure 8 with figure 9 in which the quadrature variances of the even coherent state for various values of γt are plotted we can easily see that squeezing is much more robust with respect to damping than, for instance the oscillations of the photon number distribution. Finally we should mention that the higher order squeezing or amplitude squared squeezing are much more sensitive with respect to damping than ordinary squeezing, because the non-classical effects in these cases are assiciated with higher order moments of photon operators which decay faster than the intensity of the field.

5. SUMMARY

In this talk we have discussed the origin of non-classical effects in superposition of quantum states of light. In particular we have shown that a superposition of coherent states, i.e. the "most classical" field states, can through the action of quantum interference generate the archetypal non-classical states: the squeezed vacuum and the squeezed coherent state. We have also studied the decay of such superpositions.

We have shown that in spite of the fact that Wigner functions of superposition states are very sensitive with respect to disipations, quadrature squeezing degrade with the rate equal to the system decay rate.

ACKNOWLEDGMENTS

We would like to thank S.M.Barnett, K.Wodkiewicz and S.J.D.Phoenix for many helpful and stimulating discussions. This work was supported in part by the U.K. Science and Engineering Research Council.

REFERENCES

1. P.A.M.Dirac, *The Principles of Quantum Mechanics*, 4th edition (Oxford University Press, Oxford, 1987).
2. K.Wodkiewicz, P.L.Knight, S.J.Buckle, and S.M.Barnett, *Phys. Rev. A* 35: 2567 (1987).
3. L.Mandel, *Physica Scripta* T12: 34 (1986).
4. J.Janszky and An.V.Vinogradov, *Phys. Rev. Lett.* 64: 2771 (1990).
5. P.Adam and J.Janszky, *Phys. Lett. A* 149: 67 (1990).
6. V.Bužek and P.L.Knight, *Opt. Commun.* 81: 331 (1991).
7. R.Loudon and P.L.Knight, *J. Mod. Opt.* 34: 709 (1987).
8. K.Zaheer and M.S.Zubairy, *Advances in Molecular and Optical Physics, vol. 28.*, edited by D.Bates and B.Bederson (Academic Press, New York, 1990), p. 143.
9. D.F.Walls and G.J.Milburn, *Phys. Rev. A* 31: 2403 (1985).
10. C.M.Savage and D.F.Walls, *Phys. Rev. A* 32: 2316 (1985).
11. S.J.D.Phoenix, *Phys. Rev. A* 41: 5132 (1990).
12. B.Yurke and D.Stoler, *Phys. Rev. Lett* 57: 13 (1986).
13. G.J.Milburn and C.A.Holmes, *Phys. Rev. Lett.* 56: 2237 (1986).
14. A.Mecozzi and P.Tombesi, *Phys. Rev. Lett.* 58: 1055 (1987).
15. A.Mecozzi and P.Tombesi, *J. Opt. Soc. Am. B* 4: 1700 (1987).
16. R.Loudon, *Quantum Theory of Light* (Clarendon, Oxford, 1983).
17. M.Hillery, *Phys. Rev. A* 36: 3796 (1987).
18. K.Vogel and H.Risken, *Phys. Rev. A* 40: 2847 (1990).
19. W.Schleich, J.P.Dowling, R.J.Horowicz, and S.Varro, *New Frontiers in Quantum Electrodynamics*, edited by A.Barut (Plenum Press, New York, 1990), p. 31.
20. V.Bužek, I.Jex, and Tran Quang, *J. Mod. Opt.* 37: 159 (1990).
21. V.Bužek, P.L.Knight, and A. Vidiella Barranco, *Squeezed States and Uncertainty Relations*, edited by D.Han and Y.S.Kim, in press.
22. W.H.Louisell, *Quantum Statistical Properties of Radiation* (Wiley, New York, 1973).
23. J.Peřina, *Quantum Statistics of Linear and Nonlinear Optical Phenomena* (Riedel, Dordrecht, 1984).
24. W.P.Schleich and J.A.Wheeler, *J. Opt. Soc. Am. B* 4: 1715 (1987).
25. W.Schleich, D.F.Walls, and J.A.Wheeler, *Phys. Rev. A* 38: 1177 (1988).
26. H.P.Yuen, *Phys. Rev. A* 13: 2226 (1976).
27. M.S.Kim, F.A.M.DeOliveira, and P.L.Knight, *Phys. Rev. A* 40: 2494 (1989).
28. F.A.M.DeOliveira, M.S.Kim, P.L.Knight, and V.Bužek, *Phys. Rev. A* 41: 2645 (1990).
29. S.M.Barnett and P.L.Knight, *Phys. Rev. A* 33: 2444 (1986).

TIME-DEPENDENT INVARIANTS AND NONCLASSICAL LIGHT

V.I.Man'ko

Lebedev Physics Institute, 117924 Moscow, Leninsky pr. 53
USSR

1. UNCERTAINTY RELATIONS

The Heisenberg uncertainty relation[1]

$$\Delta q \Delta p \geq \frac{\hbar}{2} \tag{1}$$

where the dispersions of position Δq and momentum Δp may be calculated for any given state described by the density operator $\hat{\rho}$ as

$$\Delta q = \sqrt{\mathrm{Tr}\hat{\rho}\hat{q}^2 - (\mathrm{Tr}\hat{\rho}\hat{q})^2}\ , \tag{2}$$

$$\Delta p = \sqrt{\mathrm{Tr}\hat{\rho}\hat{p}^2 - (\mathrm{Tr}\hat{\rho}\hat{p})^2}\ . \tag{3}$$

is the base of quantum approach to consider the motion of a particle in any potential. Another uncertainty relation was obtained[2] by Schrödinger

$$\Delta q \Delta p \geq \frac{\hbar}{2} \frac{1}{\sqrt{1 - r^2}}. \tag{4}$$

Here the variable r is the correlation coefficient of the coordinate and momentum. It can be calculated as follows

$$r = (\Delta q \Delta p)^{-1} \left\{ \frac{1}{2} \langle \hat{q}\hat{p} + \hat{p}\hat{q} \rangle - \langle \hat{p} \rangle \langle \hat{q} \rangle \right\} \tag{5}$$

The inequality (4) is the Schrödinger uncertainty relation. For $r=0$ it becomes the Heisenberg uncertainty relation (1). Glauber[3] introduced the notion of coherent states using

Quantum Measurements in Optics, Edited by P. Tombesi and
D.F. Walls, Plenum Press, New York, 1992

the electromagnetic field states. This is the best available approximation for classical states of the field which works in the quantum case. The coherent state of the field oscillator for a given mode may be obtained if we demand the following condition of minimizing the uncertainty product in Heisenberg inequality (1), i.e.

$$\Delta q \Delta p = \frac{\hbar}{2} \tag{6}$$

At the same time the dimensionless dispersions of coordinate Δq^* and momentum Δp^* of the oscillator must be equal. For the usual quantum oscillator with the frequency ω and the mass m it means that

$$\Delta q^* = \frac{\Delta q}{\sqrt{\hbar/m\omega}}, \qquad \Delta p^* = \frac{\Delta p}{\sqrt{\hbar m\omega}} \tag{7}$$

Thus we have instead of (6) the relations

$$\Delta q^* \ \Delta p^* = 1/2 \ , \ \ \Delta q^* = \Delta p^* \ . \tag{8}$$

The normalized coherent state $|\alpha\rangle$ which is the solution for condition (8) is the eigenstate of the oscillator annihilation operator and in the coordinate representation it has the form

$$\langle q|\alpha\rangle = \pi^{-1/4}\, l^{-1/2} \exp\{-\frac{|\alpha|^2}{2} - \frac{\alpha^2}{2} - \frac{q^{*2}}{2} + 2^{1/2}\, \alpha q^*\},$$

$$l = (\hbar/m\omega)^{1/2},\ q^* = q/l\ . \tag{9}$$

Let us now, following Ref.[4], consider a state of minimal uncertainty product in the Schrödinger inequality, instead of minimizing Heisenberg uncertainty relation (1). Thus we look for a state for which

$$\Delta q \Delta p = \frac{\hbar}{2} \frac{1}{\sqrt{1 - r^2}}. \tag{10}$$

This state turned out to be a pure state described by a Gaussian wave packet of the form

$$\langle q|\beta\rangle = \Psi_o(q) \exp\{-\frac{|\beta|^2}{2} + \frac{2^{1/2}\, \beta q}{(u - v)l} + \frac{\beta^2}{2} \frac{u^* - v^*}{u - v}\} \tag{11}$$

where

$$\Psi_o(q) = \pi^{-1/4} \left((u - v)\, l\right)^{-1/2} \exp\{-\frac{u + v}{2(u - v)} \frac{q^2}{l^2}\}\ . \tag{12}$$

The dispersions of coordinate and momentum are determined in terms of two complex variables u and v

$$(\Delta q)^2 = l^2 \left\{\frac{1}{2} + |v|^2 - \mathrm{Re}(uv^*)\right\} , \tag{13}$$

$$(\Delta q)^2 = l^{-2} \hbar^2 \left\{\frac{1}{2} + |v|^2 + \mathrm{Re}(uv^*)\right\} , \tag{14}$$

The correlation coefficient r of position and momentum is also expressed in terms of these complex parameters

$$r = \mathrm{Im}(uv^*) \left[\left(\frac{1}{2} + |v|^2\right)^2 - \mathrm{Re}(uv^*)\right]^{-1/2} . \tag{15}$$

The numbers u and v obey the relation $|u|^2 - |v|^2 = 1$ and may be considered as parameters of the group SU(1.1). The physical meaning of the correlated state[11] is analogous to the physical meaning of the coherent state. The correlated state is as close to the classical state of an oscillator as possible in the frame of quantum mechanics if one has a nonzero correlation coefficient. The coherent state minimizes the Heisenberg uncertainty relation and the correlated state the Schrödinger uncertainty relation, respectively. If we do not demand the equality of dispersions minimizing the Heisenberg uncertainty relation, we will obtain squeezed states, if one of the dispersions is smaller than its value for the ground state. But the correlation coefficient is not taken into account for this state. In fact, all three sorts of status: coherent, squeezed and correlated are true Gaussian wave packets which have been constructed by Schrödinger[5].

2. OSCILLATOR WITH TIME-DEPENDENT FREQUENCY

Let us remind now the problem of the parametric quantum oscillator. The Gaussian packets and Fock states for this problem have been found by Husimi[6]. It turned out that the parametric oscillator serves as a model for studying the correlated and squeezed states. As it was shown in Ref.[7] it has liner integrals of motion, which in Schrödinger representation explicitly depend on time. So, let us assume that a parametric oscillator has the frequency $\omega(t)$, which depends on time. We introduce the dimensionless parameters

$$\Omega(t) = \frac{\omega(t)}{\omega_0} , \quad \hat{Q} = \frac{\hat{q}}{l} , \quad \hat{P} = -il\frac{\partial}{\partial q} , \quad \hat{p} = -i\hbar\frac{\partial}{\partial\partial q} , \tag{16}$$

where $\hat{q}$ and $\hat{p}$ are coordinate and momentum operator, respectively. In such variables the Schrödinger equation for the oscillator has the form

$$i\frac{\partial\Psi}{\partial t} = \left\{\frac{\hat{p}^2}{2} + \Omega^2(t)\frac{\hat{Q}^2}{2}\right\} , \quad \hbar=1 . \tag{17}$$

We assume the frequency $\omega(t)$ to be equal to ω_0 at the initial momentum t=0. The time we will consider to be dimensionless. It means the time unity in this case is ω_0^{-1}. It can be checked that in the Schrödinger representation the operator

$$\hat{A}(t) = i2^{-1/2} (\varepsilon(t)\hat{P} - \dot{\varepsilon}(t)\hat{Q}) \tag{18}$$

is an integral of motion[7] for a parametric oscillator with the Hamiltonian

$$\hat{H} = \frac{\hat{p}^2}{2} + \Omega^2(t)\frac{\hat{Q}^2}{2} \tag{19}$$

The complex function $\varepsilon(t)$ is supposed to be the complex solution of the classical equation of motion

$$\ddot{\varepsilon}(t) + \Omega^2(t)\,\varepsilon(t) = 0 \tag{20}$$

with the initial conditions

$$\varepsilon(0) = 1\,, \qquad \dot{\varepsilon}(0) = i \tag{21}$$

The operator $\hat{A}^+(t)$ is an integral of motion, too, and the commutation relations of these integrals of motion are boson ones

$$[\hat{A}(t), \hat{A}^+(t)] = 1 \tag{22}$$

if the classical complex oscillator trajectory satisfies the condition

$$\dot{\varepsilon}\varepsilon^* - \dot{\varepsilon}^*\varepsilon = 2i\,. \tag{23}$$

The normalized vacuum state $\Psi_o(Q,t)$ may be found from the condition

$$\hat{A}\Psi_o = 0\,, \tag{24}$$

which is solved, for example, by the function

$$\Psi_o(Q,t) = \pi^{-1/4}\,\varepsilon^{-1/2}\exp\left[i\,\frac{\dot{\varepsilon}(t)Q^2}{2\varepsilon(t)}\right]. \tag{25}$$

The importance of the solution (25) is contained in the following properties

$$(\Delta Q)^2 = \frac{|\varepsilon|^2}{2}\,, \qquad (\Delta P)^2 = \frac{|\dot{\varepsilon}|^2}{2} \tag{26}$$

and

$$r = \frac{(|\varepsilon\dot{\varepsilon}|^2 - 1)^{1/2}}{|\varepsilon\dot{\varepsilon}|}\,. \tag{27}$$

Thus, the state (25) is the correlated state. The parameters u and v are expressed in terms of a complex classical trajectory

$$u = 2^{-1/2}(\varepsilon(t) - i\dot{\varepsilon}(t)), \quad v = 2^{-1/2}(\varepsilon(t) + i\dot{\varepsilon}(t)) \tag{28}$$

The normalized coherent state $\Psi_\alpha(Q,t)$, which is the eigenstate of the invariant $\hat{A}(t)$ (18)

$$\hat{A}\Psi_\alpha(Q,t) = \alpha\Psi_\alpha(Q,t) \tag{29}$$

has the form

$$\Psi_\alpha(Q,t) = \Psi_o(Q,t)\exp\left\{-\frac{|\alpha|^2}{2} + \frac{2^{1/2}\alpha Q}{\varepsilon(t)} - \frac{\alpha^2 \varepsilon^*(t)}{2\,\varepsilon(t)}\right\}. \tag{30}$$

The variances of the coordinate $(\Delta Q)^2$ and momentum $(\Delta P)^2$ as well as the correlation coefficient r are given for the state (30) by the same formula (26) and (27). So, we see that the coherent state of a parametric oscillator - with respect to the integral of motion $A(\hat{t})$ - is the correlated state. The squeezing coefficient for a correlated state is determined by the classical trajectory $\varepsilon(t)$

$$\left(\frac{\Delta Q}{\Delta P}\right)^2 = \left|\frac{\varepsilon}{\dot{\varepsilon}}\right|^2. \tag{31}$$

Thus we may generate correlated states using the parametric quantum systems like the oscillator with time-dependent frequency. If we have the frequency time dependence of the form

$$\Omega^2(t) = 1 - 2\kappa\delta(t) \tag{32}$$

the dispersions of coordinate and momenta are expressed as follows

$$(\Delta Q)^2 = \frac{1}{2} + 2\kappa^2\sin^2 t + \kappa\sin 2t\ ; \tag{33}$$

$$(\Delta P)^2 = -(\Delta q)^2 + 1 + 2\kappa^2\ ; \tag{34}$$

Here we demonstrate the distribution function of photons in a correlated state[8]. The distribution function W_n is defined as the square of the coefficients of expansion of the correlated state in terms of Fock states. It has the form

$$W_n = \frac{1}{|u|n!}\exp\left[-|\beta|^2 + \frac{\beta^2 v^*}{2u} + \frac{\beta^{*2} v}{2u^*}\right]\left|H_n\left(\frac{\beta}{\sqrt{2u\,v}}\right)\right|^2. \tag{35}$$

3. NONSTATIONARY CASIMIR EFFECT AND VACUUM QUANTUM GENERATOR

One of possible methods to generate squeezed light is to use the nonstationary Casimir effect when moving resonator walls produce continuous time-dependent reconstruction of electromagnetic vacuum state. The work against the Casimir forces produces two effects. The first is the effect of photons generation from the vacuum. Thus, the resonator with mechanically trembling walls is a quantum vacuum generator of electromagnetic radiation. The second effect is the effect of squeezing of quantized modes in the resonator due to parametric change of vacuum energy. Both effects exist simultaneously.

Let us discuss a general scheme of arising of Casimir effect. For many physical systems their Hamiltonian may be written down as the sum of energies of one-dimensional oscillator-modes of small linear vibrations

$$H = \hbar \sum_n \left[\hbar \omega_n(L) \, a_n^+ a_n + \frac{\hbar \omega_n(L)}{2} \right]$$

The sum over the index n is the sum over all the modes in the system under discussion. The frequencies $\omega_n(L)$ depend on the parameters of classical system which are described by formal vector $L=(L_1,L_2,\ldots L_n)$. The components of these vectors are different physical parameters, for example, the geometrical parameters, such as length of radius of a cylindrical resonator, refraction index of the media in the resonator, magnetic and electric field components for the electron in these fields, etc.. The quantity of these parameters is the dimension of the vector L. If the index n changes in infinite interval the energy of the ground or vacuum state

$$E_{vac} = \sum_n \frac{\hbar \omega_n(L)}{2}$$

is infinite. But the energy may be considered as the term which is infinite and does not depend on the parameters L and the term E(L), depending on the parameters L, which is the reason for Casimir forces. The reason for these forces is formally connected with the dependence of the system modes frequencies $\omega_n(L)$ on the outer parameters L. So,

$$\delta E_{vac} = \left(\frac{\hbar}{2} \sum_n \frac{\partial \omega_n(L)}{\partial L} \right) \delta L \ .$$

If the change of the vacuum energy connected with the change of the parameters δL is negative, the Casimir forces act in such direction in order to increase this change of the parameters. So, the neutral plates in the Casimir effect attract, since the vacuum electromagnetic energy is less for smaller distances between these plates. At the same time the metallic sphere tends to increase its radius since the vacuum electromagnetic energy is less for larger sphere radius. Such behaviour is well known.

For nonstationary quantum systems the system parameters may be time-dependent ($L=L(t)$) due to the existence of external forces. Thus, the plates in the Casimir effect may be moved by external mechanical forces. The refraction index of the media may vary with time, the geometrical dimensions of the resonator may be influenced by external mechanical forces. In all these cases the vacuum or ground state energy must change. This means that the vacuum state is continuosly reconstructed. For each of the time moments the state is no more the vacuum state due to the change of the parameters. If we have the system like photons or other quanta in a box with changing volume it corresponds to the process with the creation of quanta, for example, the photons. The effect of this kind in the case of time-dependent vacuum was discussed as particle creation in Ref.[9]. So, due to the work against the Casimir forces one form of energy may be converted into the other form. Thus, for waving neutral plates (due to mechanical external forces) between the plates the photons must be created and this means that the mechanical energy from outer sources is converted into electromagnetic energy of photons. It is interesting that this effect must create the quanta of all other fields existing in nature. Thus, due to Casimir forces we can have the generator of photons in parametric resonator which may be called as quantum vacuum generator. As was mentioned above, the idea of quanta generation due to parametric vacuum reconstruction is not a new idea. The new element is the suggestion to realize this idea in different types of experiments. Possible experiments with the resonator, the geometrical form and dimensions of which are time-dependent, were discussed. More sensitive measurements in the resonator with stable walls have been made in Ref.[10]. The second type of analogous idea but in different interpretation is the discussion of semiconductor laser given in Ref.[11]. By different kinds of influence it is possible to vary with time the waveguide properties in semiconductor that will produce the same analogue of nonstationary Casimir effect.

It is possible to discuss the third reduction of nonstationary Casimir effect using Josephson junction. If there is no external voltage in Josephson junction, but its parameters are time-dependent, the vibrations of current and voltage will be excited in it. This suggestion[12] is based on the analogy of Josephson junction and a conventional resonant circuit (quantized resonant circuit). In classical resonant circuit it is impossible to excite electrical vibrations without external source of voltage. But for quantum resonant circuit, due to Casimir nonstationary forces, it is possible to transform mechanical energy which may be the reason for the change of the circuit parameters into electrical energy of current vibrations.

If this idea is realized, it will be quantum vacuum generator of electrical vibrations. The current and voltage in this case play the role of conjugate quantum observables and in parametric Josephson junction they may be squeezed. Thus, the quantum noise in Josephson junction may be reduced for current. In this case the voltage will have larger noise. The discussion of Josephson junction as quantum vacuum generator has been presented in Ref.[12]. The squeezed and correlated states of Josephson junction may be also excited, if changing its parameter with time. Berry phase can also be an object of the research for the parametric Josephson junction.

In all three realizations of nonstationary Casimir effect the change of parameters may be organized, for example, by the influence of external sound or external laser light. In the periodic case the modes, which are most sensitive to the external influence, are the modes corresponding to parametric resonance conditions. The chaotic (unstable) behaviour of the systems under investigation may be also checked for special time-periodic kicking corresponding to δ-kicking situation discussed in Ref.[13].

If we have the squeezed and correlated modes light (wave function or density matrix is a general Gaussian function in many dimensions) the photon distribution function in this state is explicitly expressed in terms of Hermite polynomial of several variables. Due to this it has wavy behaviour with respect to photons labels like in one-dimensional case[14]. The possibility to use the resonator with moving walls to generate squeezed light was discussed in Ref.[15].

REFERENCES

1. W.Heisenberg, Z.Phys. 43, 122 (1927).
2. E.Schrödinger, Ber.Kgl.Akad.Wiss. 296 (1930).
3. R.Glauber, Phys.Rev.Lett. 10, 84 (1963).
4. V.V.Dodonov, E.V.Kurmyshev and V.I.Man'ko, Phys.Lett. A 79, 150 (1980).
5. E.Schrödinger, Naturwissenschaften, 14, 664 (1926).
6. K.Husimi, Progr.Theor.Phys. 9, 381 (1953).

7. I.A.Malkin and V.I.Man'ko, Phys.Lett. 32 A, 243 (1970).
8. V.V.Dodonov, A.Klimov and V.I.Man'ko, Journal of Soviet Laser Research 9, 126 (1989b).
9. H.Ezawa, Y.Tomozawa and H.Umezawa, Nuovo Cim. Ser.X, Vol.5, p.810 (1957).
10. H.Walther. In: Proc. of Internal Conference on Quantum Optics, Hyderabad University, 5-10 Jan. 1991.
11. E.Yablonovitch, J.P.Heritage, D.E.Aspnes and Y.Yafel, Phys.Rev.Lett. vol.63, 976 (1989).
12. V.V.Dodonov, V.I.Man'ko and O.V.Man'ko, Journ. of Soviet Laser Research, 10, issue 5, p.43 (1989); Proceedings of Lebedev Physics Institute, Moscow, Nauka Publ., Vol.208 (1991) (in press).
13. G.Karner, V.I.Man'ko and L.Streit, University of Bielefeld preprints n.398/89; 399/89; Proc. of Lebedev Physics Institute, Nauka Publ., Vol.208 (1991) (in press).
14. V.V.Dodonov, A.B.Klimov and V.I.Man'ko, Phys.Lett. Vol.134 A, p.211 (1989).
15. V.V.Dodonov, A.B.Klimov and V.I.Man'ko, Phys.Lett. Vol.142 A, p.511 (1989); Vol.149 A, p.225 (1990); V.I.Man'ko. In: ECOOSA90 Proceed., IOP Publishing, Bristol (1991) (in press);
S.Sarkar. In: Photon and Quantum Fluctuations, Adam Hilger, Bristol (1988), ed. by E.R.Pike, H.Walther, p.151.

A UNIFIED APPROACH TO THE QUANTUM MASTER AND LANGEVIN EQUATIONS

L. Accardi, Y.G. Lu*

Centro Matematico V.Volterra, Dipartimento di Matematica
Universita' di Roma II

1. THE PHYSICAL ORIGINS OF THE QUANTUM BROWNIAN MOTIONS

The **quantum Brownian motions** (QBM) were introduced in quantum-optics in the early 1960's as good approximations of the electro-magnetic (EM) field in certain limiting situations such as the singular coupling, the van Hove limit (weak coupling limit), the low density limit, ... (cf. Haken [16], Lax [18], von Waldenfels [22] and also [8a] for a survey). The goal of these approximations was to explain dissipation in quantum theory.

In the series of papers [2,..., 9, 19], a systematic and detailed analysis of the weak coupling and low density limit had been pursued in a variety of models, both in the Boson and Fermion case (there are even some results for the weak coupling limit for the free noise [12]). The main achievements of the above mentioned series of papers are:

1) to provide a solid physical and mathematical foundation for the use of the Quantum Brownian Motions in physics;
2) to give a rigorous meaning to the statement **...the QBM's are approximations of quantum fields...**;
3) to check the theoretical validity of the various forms of **quantum Langevin equations** proposed in the heuristic physical literature (quantum optics, Nuclear Magnetic Resonance (NMR), ...).

The general scheme of these investigations is the following:

1) One starts with two quantum Hamiltonian systems in interaction. One of these systems is called the *small system* or simply *the system* and the other one is called *reservoir, or heat bath, or field, ...* In many models the small system has discrete spectrum and the field has integrable correlations (which implies continuous spectrum). The latter property must be enjoyed by at least one of the two systems in order for the techniques introduced up to now to be applicable.

2) The interaction or the statistical correlations of the two systems depend on certain parameters and one studies the limiting situation when these parameters

*On leave of absence from Beijing Normal University

Quantum Measurements in Optics, Edited by P. Tombesi and
D.F. Walls, Plenum Press, New York, 1992

(typically: strength of interaction, density of particles,...) tend to zero. The idea is to extract some physically relevant information by **renormalizing**, i.e. **rescaling**, some of the physical parameters of the problem (in our case-time) as functions of the parameters which tend to zero. The basic criteria, to choose the appropriate rescalings, consist in fixing some quantities which, after the rescaling should remain finite.

3) The limit of point 2) is taken in some technical sense, to be described below. In the limit the system remains unaltered in the sense that it is described by the same state space, while the field *converges* to a quantum Brownian motion.

4) The most interesting phenomena take place at the level of evolution. These can be schematically described as follows.

At the Hamiltonian level:

- the state of the composite system satisfies the Schrödinger equation in interaction representation.
- The observables of the small, as well as of the composite system, satisfy the Heisenberg equation.
- The state of the small system satisfies the reduced equation, obtained by *integrating away* (by means of a partial trace) the field degrees of freedom. Even in the simplest (linear) models, this equation is a complicated integro-differential equation which gives very little insight into the physics of the problem.

At the stochastic level (i.e. after the limit):

- the state of the composite system satisfies a (quantum) stochastic differential equation.
- The observables of the small system, satisfy a (quantum) Langevin equation.
- The state of the small system satisfies a (quantum) master equation.

In all cases the term *quantum* is in parenthesis because, with an appropriate choice of the interaction, one can obtain in the limit the classical noises (e.g. the classical Brownian motion or Poisson process) and the evolution equations become classical stochastic differential equations.

2. THE BASIC PHYSICAL EXAMPLE: THE DAMPED HARMONIC OSCILLATOR

Some steps of the abstract scheme illustrated at the end of the previous Section, are known in the physical literature, especially in quantum optics and in the theory of damped quantum systems.

A typical example is the deduction of the quantum Langevin equation for an atom in interaction with an electromagnetic field. After introducing the *dipole* and the *rotating wave* approximations, the Hamiltonian of the composite system is:

$$H = \omega a^+ a + \int \mathrm{d}k f(k) A_k^+ A_k + \int \mathrm{d}k [A_k^+ \otimes a + h.c.] g(k) \tag{2.1}$$

where f, g are suitable cutoff functions, a, a^+ are the annihilation and creation operators of the atom (described as an harmonic oscillator), A_k, A_k^+ are the annihilation and creation operators corresponding to the k-th mode of the field in momentum representation.

Denoting

$$a_t := e^{itH} a e^{-itH} \; ; \; A_k(t) = e^{itH} A_k e^{-itH} \tag{2.2}$$

the Heisenberg equations of motion for a_t and $A_k(t)$ become:

$$\frac{\mathrm{d}}{\mathrm{d}t} a_t = -i\omega a_t - i \int \mathrm{d}k A_k(t) g(k) \tag{2.3}$$

$$\frac{\mathrm{d}}{\mathrm{d}t} A_k(t) = -i f(k) A_k(t) - i g(k) a(t)$$

Solving for $A_k(t)$ one finds

$$A_k(t) = A_k e^{-itf(k)} - i g(k) e^{-itf(k)} \int_0^t \mathrm{d}s e^{isf(k)} a_s$$

and inserting $A_k(t)$ in the equation for a_t:

$$\frac{\mathrm{d}}{\mathrm{d}t} a_t = -i\omega a_t - i \int \mathrm{d}k g(k) A_k e^{-itf(k)} - \int_0^t \mathrm{d}s a_s \int \mathrm{d}k g^2(k) e^{-i(t-s)f(k)}$$

Now, denoting

$$\int \mathrm{d}k g^2(k) e^{-i(t-s)f(k)} =: \phi(t-s) \tag{2.4}$$

$$-i \int \mathrm{d}k g(k) A_k e^{-itf(k)} =: \xi_t \tag{2.5}$$

the reduced equation for a_t becomes

$$\frac{\mathrm{d}}{\mathrm{d}t} a_t = -i\omega a_t - \int_0^t \mathrm{d}s \phi(t-s) a_s + \xi_t \tag{2.6}$$

Notice that up to now **no approximations** have been made.

Equation (2.6) is a **quantum Langevin equation** with ξ_t playing the role of a (quantum) **random force**.

Moreover, from the definition (2.5) of ξ_t, it follows that the random force (quantum noise) acts **exclusively on the field degrees of freedom.**

Now let Ψ denote the Fock vacuum for the $A_k^{\pm}$ (i.e. for the field) and let us consider the Ψ-statistics for the random force ξ_t. Because of Gaussianity this statistics is uniquely determined by the 2-point (correlation) functions

$$\langle \Psi, \xi_t \Psi \rangle = \langle \Psi, \xi_t^* \xi_s^* \Psi \rangle = \langle \Psi, \xi_t \xi_s \Psi \rangle = 0 \; ; \qquad \langle \Psi, \xi_t \xi_s^* \Psi \rangle = \phi(t-s) \tag{2.7}$$

where the function $\phi(t-s)$ is the same as ones appearing in the equation (2.6) and defined by (2.4).

In the usual (classical) Langevin equation, the random force is a **white noise**, i.e. a mean zero, δ-correlated random process. In the physical literature it has been remarked that, with the choices:

$$f(k) = k \tag{2.8}$$

$$g(k) = g_0 \in \mathrm{R} \tag{2.9}$$

the function $\phi(t-s)$, defined by (2.4) becomes a positive multiple of the δ-function, say $\kappa\delta(t-s)$ with

$$\kappa = 2\pi g_0^2 \tag{2.10}$$

and the random force takes the form

$$\alpha_t := -ig_0 \int \mathrm{d}k A_k e^{-itk} . \tag{2.11}$$

From (2.11) it follows that the Ψ-statistics of α_t is Gaussian with mean zero and correlations

$$\langle \Psi, \alpha_t \Psi \rangle = \langle \Psi, \alpha_s \alpha_t \Psi \rangle = \langle \Psi, \alpha_s^+ \alpha_t^+ \Psi \rangle = 0 \tag{2.12}$$

$$\langle \Psi, \alpha_s \alpha_t^+ \Psi \rangle = \kappa \delta(t-s) \tag{2.13}$$

Moreover the commutation relations for the α_s^+, α_t are:

$$[\alpha_s, \alpha_t] = [\alpha_s^+, \alpha_t^+] = 0 \qquad ; \forall s, t$$

$$[\alpha_t^+, \alpha_t] = \int e^{-i(t-s)k} g_0^2 \mathrm{d}k = 2\pi g_0^2 \cdot \delta(t-s)$$

This means that, with respect to the state Ψ, the pair (p_t), (q_t), defined by

$$p_t = \frac{1}{2i}[\alpha_t - \alpha_t^+] \; ; \; q_t = \frac{1}{2}[\alpha_t + \alpha_t^+] \tag{2.15}$$

is a quantum white noise in the sense defined in Section (4.) below.

With the choices (2.8) and (2.9) and in the above notations, the equation for a_t becomes

$$\frac{\mathrm{d}}{\mathrm{d}t} a_t = -i\omega a_t - \kappa a_t + \alpha_t \tag{2.16}$$

which is a **true Langevin equation** in the sense that the random force (α_t) has the correct statistics, given by (2.13). In the right hand side of equation (2.16) one recognizes **three types of terms**:

(i) a purely Hamiltonian term: $-i\omega a_t$

(ii) a damping (dissipation) term: $-\kappa a_t$

(iii) a fluctuation term: the random force α_t.

The *fluctuation* effects compensate the *dissipation* effects and the overall evolution of the composite system is **unitary**; in particular the commutation relations

$$[a_t, a_t^+] = 1$$

are preserved at each time. This fact was remarked by Senitzky ([20]), Lax, Haken (see also the historical survey in [1]).

Notice however that, with the choice $f(k) = k$, the field Hamiltonian becomes:

$$\int_{-\infty}^{+\infty} \mathrm{d}k \cdot k \cdot A_k^+ A_k$$

which is not bounded below! This means that the models defined by the choices (2.8) and (2.9) are not plausible physical models. Such models are called **singular coupling models**.

Strictly related to the singular coupling models is the so called **singular coupling limit**, which can be described as follows: one starts by assuming a priori condition (2.8) (i.e. $f_k = k$) but, instead of assuming a priori condition (2.9), one introduces a family $g^{(\lambda)}$, of cut-off functions in the interaction, (λ is a real parameter) such that $g_k^{(\lambda)}$ tends to g_0 (real constant), in the distribution sense, as $\lambda \to 0$. With such a choice the ϕ-function (here a $\phi^{(\lambda)}$ function), defined by (2.4) converges to a δ-function.

The singular coupling limit was studied by Hepp and Lieb ([17]) and, in a more rigorous and general form by Gorini, Kossakowski, Frigerio, Verri ([13, 14, 15]). Several problems however are left open by the above mentioned approach:

1) First of all it requires from the beginning a field Hamiltonian unbounded below and therefore it brings no contribution to the basic issue of the approximation of (classical and quantum) noises by Hamiltonian systems, namely: how is it possible that, in some physically plausible limiting situations, objects with flat spectrum, such as the quantum Brownian motions, are approximated by usual Hamiltonian systems, with spectrum bounded below?

2) The approximations were *very* nonrigorous. In particular: *different approximations led to different equations* (cf. [10, 11]).

3) Lack of a unifying principle for the various equations obtained: each Langevin equation of motion for each observable required a different ad hoc procedure.

4) To determine all the parameters defining the QBM as functions of the underlying parameters of the microscopic model.

5) The method does not extend to nonlinear interactions.

6) The physical meaning of the *number process* was not recognized.

A satisfactory solution to the problems could be found only within the framework of the weak coupling and low density limit.

3. RANDOM VARIABLES AND STOCHASTIC PROCESSES

A classical real valued random variable $\xi : (\Omega, \mathcal{F}, P) \to R$ is equivalent, via the spectral theorem, to a self-adjoint operator, still denoted ξ, on $L^2_{\mathbf{C}}(\Omega, \mathcal{F}, P)$

$$(\xi f)(\omega) = \xi(\omega) f(\omega)$$

i.e. the operator ξ is the multiplication by the random variable ξ itself and the domain of ξ is the set of all the functions $f \in L^2(\Omega)$ such that

$$\int |\xi(\omega) f(\omega)|^2 \mathrm{d}P < \infty \tag{3.1}$$

in the sense that any classical real valued random variable defines a self-adjoint operator by (3.1) and conversely, any self-adjoint operator can be put in this form by means of a unitary transformation. This suggests the following

DEFINITION (3.1) A R-valued *random variable* is a triple $\{\mathcal{H}, \Phi, \xi\}$ where:

i) $\mathcal{H}$ **C**-Hilbert space; $\Phi \in \mathcal{H}$; $\|\Phi\| = 1$

ii) $\xi : D(\xi) \subseteq$ (dense) $\mathcal{H} \to \mathcal{H}$ is a self adjoint operator.

The spectral measure μ_Φ^ξ of ξ in Φ defined by

$$\langle \Phi, f(\xi)\Phi \rangle =: \int_R f(x) \mathrm{d}\mu_\Phi^\xi(x); \qquad \forall f \in L^\infty(R; \mathrm{Borel})$$

is called the *distribution of the r.v.* ξ (i.e. $\mu^\xi_\Phi(I) = \text{Prob}_\Phi(\xi \in I)$; $I \subset R$ an interval).

Remark 1. If $n \in N$ the condition $\Phi \in D(\xi^n)$ is equivalent to $\langle \Phi, \xi^n \Phi\rangle = \int_R x^n \mathrm{d}\mu^\xi_\Phi(x) < +\infty$ i.e. to the fact that the n-th moment of ξ is finite.

Remark 2. There is no distinction between classical and quantum random variables. *Example* A quantum field referred to the vacuum (or a coherent, or thermal) state.

DEFINITION (3.2) A *real valued stochastic process* is a triple $\{\mathcal{H}, \Phi, (\xi_t)_{t\in R}\}$ ($t \in R$-index set) such that each $\{\mathcal{H}, \Phi, \xi_t\}$ is a random variable. The process is called *classical* if the spectral projections of the ξ_t commute among themselves (i.e. the ξ_t strongly commute).

DEFINITION (3.3) The process $\{\mathcal{H}, \Phi, (\xi_t)\}$ is said to admit moments of order $n(\in N)$ if $\forall t_1, \ldots, t_n \in R$

$$\Phi \in D(\xi_{t_1} \cdot \xi_{t_2} \cdot \ldots \cdot \xi_{t_n}) \tag{3.2}$$

$$|\langle \Phi, \xi_{t_1} \cdot \xi_{t_2} \cdot \ldots \cdot \xi_{t_n} \cdot \Phi\rangle| < +\infty \tag{3.3}$$

DEFINITION (3.4) A process $\{\mathcal{H}, \Phi, (\xi_t)_{t\in R}\}$ is called a mean 0 Gaussian stochastic process if

i) it admits moments of any order

ii) $\langle \Phi, \xi_{t_1} \cdot \ldots \cdot \xi_{t_n} \Phi\rangle = 0$ if n is odd and if $n = 2p$

$$\langle \Phi, \xi_{t_1} \cdot \ldots \cdot \xi_{t_n} \Phi\rangle = \sum \varepsilon(i_1, j_1; \ldots; i_p, j_p) c(t_{i_1}, t_{j_1}) \cdot \ldots \cdot c(t_{i_p}, t_{j_p})$$

where the sum is extended to a subset of pair partitions $(i_1, j_1; \ldots; i_p, j_p) \subseteq$ of $\{1, \ldots, 2p\}$ and with

$$c(t_j, t_k) := \langle \Phi, \xi_{t_j} \xi_{t_k} \Phi\rangle, \qquad \varepsilon(i_1, j_1; \ldots; i_p, j_p) \in C$$

Example Gaussian in the sense of classical probability; Bose or Fermi Gaussian in the usual sense of quantum field theory; Free Gaussian in the sense of Voiculescu (cf. [23]).

DEFINITION (3.5) A classical R-valued white noise is a family (p_t) of random variables such that the stochastic process $\{\mathcal{H}, (p_t), \Phi\}$ is Gaussian with mean 0 and covariance

$$\langle \Phi, p_s p_t \Phi\rangle = \delta(s - t)$$

Defining then

$$P_t := P(\chi_{[0,t]}) := \int_0^t p_s \mathrm{d}s \tag{3.4}$$

one has, for any s, t, $[P_s, P_t] = 0$ hence (P_t) is a classical process in the sense of Definition (3.2) and

$$\langle \Phi, P_s P_t \Phi\rangle = \int_0^s \mathrm{d}u_1 \int_0^t \mathrm{d}u_2 \langle \Phi, p_{u_1} p_{u_2} \Phi\rangle = \min\{s, t\} = \langle \chi_{[0,s]}, \chi_{[0,t]}\rangle_{L^2(R)}$$

where we always use the notation

$$\chi_I(x) = \begin{cases} 0 & \text{if } x \notin I \\ 1 & \text{if } x \in I \end{cases}$$

The process (3.4) is called a real valued classical Brownian motion with zero initial condition. More generally

DEFINITION (3.6) A classical real valued Brownian motion (with 0 initial condition) is a classical R-valued Gaussian (mean 0) process $\{\mathcal{H}, \Phi, (P_t)_{t\in R}\}$ with covariance $\langle \Phi, P_s P_t \Phi \rangle = \min\{s,t\}$.

4. QUANTUM BROWNIAN MOTIONS

The basic idea is that a quantum Brownian motion is a pair of non commuting classical BM. More precisely

DEFINITION (4.1) A **quantum Brownian motion** is a quadruple:

$$\{\mathcal{H}, \Phi; (P_t), (Q_t), t \in R\}$$

such that:
- i) both $\{\mathcal{H}, \Phi, (P_t)\}$ and $\{\mathcal{H}, \Phi, (Q_t)\}$ are classical Brownian motions.
- ii) $[P_s, Q_t] = i\lambda \min\{s,t\}$; $\lambda \in R$.

THEOREM *Quantum Brownian motions exist*
Proof Cf. the example below.

Example 1. the Fock BM over $L^2(R)$. Let

$$\mathcal{H} = \Gamma(L^2(R_+)) = \bigoplus_{n\geq 0} \otimes_{\text{sym.}} L^2(R_+)$$

be the Fock space over $L^2(R_+)$. Denote

$$\Phi := \text{ vacuum vector } \quad f \in L^2(R_+)$$
$$\psi(f) := \Phi + \sum_{n\geq 1} \frac{\otimes^n f}{\sqrt{n!}} ; \tag{4.1}$$

exponential (non-normalized coherent) *vector*. For any test function $f \in L^2(R_+)$, introduce the *annihilation operator*

$$A(f)\psi(g) := \langle f, g\rangle_{L^2(R_+)}\psi(g)$$

and the *creation operator*

$$A^+(f) = \text{ adjoint of } A(f)$$

They satisfy the **canonical commutation relations** (CCR):

$$[A(f), A^+(g)] = \langle f, g\rangle_{L^2(R)} \tag{4.2}$$

A simple calculation now shows that, if we define:

$$P_t := \frac{1}{2i}\{A(\chi_{[0,t]}) - A^+(\chi_{[0,t]})\}$$
$$Q_t := \frac{1}{2}\{A(\chi_{[0,t]}) + A^+(\chi_{[0,t]})\}$$

then

$$\{\mathcal{H}, \Phi, (P_t), (Q_t)\}$$

is a quantum Brownian motion in the sense of Definition (4.1).

In the Physical literature, one often uses the notation:

$$A_t = A(\chi_{[0,t]}) = \int_0^t a_s \mathrm{d}s$$
$$[a_s, a_t^+] = \delta(t-s)$$

The connections between classical and quantum Brownian motions are best understood by recalling some basic rules of the classical Ito calculus.

It is well known that the increments $\mathrm{d}w_t = w_{t+\mathrm{d}t} - w_t$ of the classical Wiener process satisfy the relation:

$$(\mathrm{d}w_t)^2 \equiv \mathrm{d}t \equiv \mathrm{d}w_t \cdot \mathrm{d}w_t \qquad ; \qquad (\mathrm{d}w_t)^3 \equiv 0$$

where $\equiv$ means *equivalent up to terms of order* $o(\mathrm{d}t)$ in an appropriate topology. In general if we introduce the notation

$$\mathrm{d}X(t) := X(t+\mathrm{d}t) - X(t)$$

(this is a finite defference!) then one has

$$\mathrm{d}(XY) = \mathrm{d}X \cdot Y + X \cdot \mathrm{d}Y + \mathrm{d}X \cdot \mathrm{d}Y$$

Example $\mathrm{d}(w_t^2) = \mathrm{d}w_t \cdot w_t + w_t \cdot \mathrm{d}w_t + \mathrm{d}w_t \cdot \mathrm{d}w_t \equiv 2w_t\mathrm{d}w_t + \mathrm{d}t$. This means:

$$w_t^2 - w_{t_0}^2 = \int_{t_0}^t 2w_t \mathrm{d}w_z + (t - t_0)$$

In general one has the identity

$$\mathrm{d}w_t^n \equiv w_t^{n-1}\mathrm{d}w_t + \mathrm{d}w_t^{n-1} \cdot w_t + \mathrm{d}w_t \cdot \mathrm{d}w_t^{n-1}$$

Which has to be understood in a technical sense that shall not be discussed here.

Moreover, introducing the notation we obtain the algebraic identity which is the essence of Ito's formula, namely:

$$A(f)\psi(g) = \langle f, g\rangle\psi(g)$$

where $A(f)$ is the annihilator $\psi(g)$ is a non normalized coherent (exponential) vector (cf. (4.1)) and f, g are test functions in $L^2(R_+)$. Now choose $f = \chi_{[t,t+dt]}$ and denote

$$A_t = A(\chi_{[0,t]}) = \int_0^t a_s \mathrm{d}s \ .$$

Formally:

$$a_s\psi(f) = f(s)\psi(f)$$

i.e. as is an *annihilation density*. Then

$$dA_t = A_{t+dt} - A_t = A(\chi_{[t,t+dt]})$$

$$dA_t\psi(g) = \langle\chi_{[t,t+dt]}, g\rangle\psi(g) = \left(\int_t^{t+dt} g(s)ds\right)\psi(g) \equiv g(t)dt \cdot \psi(g)$$

where the last approximation is true if g is a continuous function.

The quantum Ito tables stems from the above approximation, i.e.

$$dA_t\psi(g) \equiv g(t)dt \cdot \psi(g) .$$

In fact this implies:

$$\begin{aligned}\langle\psi_f, dA_t^+ dA_t\psi_g\rangle &= \langle dA_t\psi_f, dA_t\psi_g\rangle \\ &\equiv (f(t)dt)^*(g(t)dt)\cdot\langle\psi_f,\psi_g\rangle \equiv O(dt^2)\langle\psi_f,\psi_g\rangle .\end{aligned}$$

Hence, in our notations:

$$dA_z^+ dA_t \equiv 0 .$$

Similarly:

$$(dA_t)^2 \equiv (dA_z^+)^2 \equiv 0 .$$

But

$$\begin{aligned}dA_t dA_t^+ &= A(\chi_{[t,t+dt]})\cdot A^+(\chi_{[t,t+dt]}) = [A(\chi_{[t,t+dt]}), A^+(\chi_{[t,t+dt]})]+ \\ &\quad + A^+(\chi_{[t,t+dt]})A(\chi_{[t,t+dt]}) = \langle\chi_{[t,t+dt]},\chi_{[t,t+dt]}\rangle + O(dt^2) = dt + O(dt^2) .\end{aligned}$$

Hence, in our notations:

$$dA_t dA_t^+ \equiv dt .$$

The classical Ito table is deduced, as an application of the above: Define $W_t = A_t + A_t^+$. If $s < t$

$$[W_s, W_t] = [W_s, W_s] + [W_s, W_{[s,t]}] = 0$$

where, in the last identity, we have used the property that *the future commutes with the past* (an immidiate consequence of (4.2)). We conclude that (W_t) is a commutative family and therefore $\{(W_t),\Phi\}$ is a classical stochastic process. Moreover (W_t) is mean 0 and Gaussian since both A_t, A_t^+ are. Finally

$$dW_t^2 = (dA_t + dA_t^+)^2 = dA_t^2 + dA_t dA_t^+ + dA_t^+ dA_t + (dA_t^+)^2 \equiv dt$$

Therefore (W_t) is a classical Wiener process.

Example **The number (gauge) process.** Another important example of stochastic process is the number process.

Let be given a pre-Hilbert space H; denote $\Gamma(H) = \oplus_{n\geq 0}(\otimes\text{sym}.H)^n$ the Fock space over H, Φ the Fock vacuum. For each $X \in B(H)$ with the norm less than 1, define a bounded operator $\Gamma(X)$ on $\Gamma(H)$ by the relation

$$\Gamma(X)\psi(f) := \psi(Xf) \tag{4.3}$$

A special case of (4.3) is $X = e^{ith}$ with the operator h being self-adjoint. By Stone-von Neumann Theorem and (4.3), (e^{ith}), as well as $(\Gamma(e^{ith}))$, are unitary group on H, and $\Gamma(H)$ respectively. Denote $N(h)$ the generator of $(\Gamma(e^{ith}))$, i.e.

$$e^{itN(h)}\psi(f) := \psi(e^{ith}f) = \Gamma(e^{ith})\psi(f)$$

and moreover for an arbitrary operator k, we define $N(k)$ by complex linearity:

$$N(k) = N\left(\left[\frac{k+k^*}{2}\right] + i\left[\frac{k-k^*}{2i}\right]\right) := N\left(\frac{k+k^*}{2}\right) + iN\left(\frac{k-k^*}{2i}\right)$$

The distribution of $N(h)$ in a coherent state (described by normalized exponential vectors–coherent vectors) is (a linear combination of) Poisson, i.e., if $h = h^*$

$$\begin{aligned}
&\langle e^{-\frac{1}{2}\|f\|}\psi(f),\, e^{itN(h)} \cdot e^{-\frac{1}{2}\|f\|}\psi(f)\rangle = \\
&= e^{-\|f\|^2}\langle\psi(f),\, \Gamma(e^{ith})\psi(f)\rangle = e^{-\|f\|^2}\langle\psi(f), \psi(e^{ith}f)\rangle \\
&= \exp\{-\|f\|^2 + \langle f, e^{ith}f\rangle\} = \exp\langle f, (e^{ith} - 1)f\rangle \ .
\end{aligned}$$

Example 1. Let $h = h^2 = h^*$ be an orthogonal projection. Then:

$$e^{ith} - 1 = e^{it} \cdot h + h^{\perp} - 1 = (e^{it} - 1)h$$

therefore the characteristic function of $N(h)$ is $\exp(e^{it} - 1)\|hf\|^2$ which corresponds to a Poisson distribution with intensity

$$\|hf\|^2$$

5. BASIC IDEA OF THE NEW APPROACH TO THE WEAK COUPLING LIMIT

We shall only discuss the Boson case: let H_0 denote the System space; H_1 the 1-particle Reservoir space; H_S the system Hamiltonian; H the 1-particle reservoir Hamiltonian; $\beta > 0$ the inverse temperature; $\mu \in R$ the chemical potential $z = e^{\beta\mu}$ the fugacity; $n(\beta, z)$ the density of particles (per unit volume). It is known that $n(\beta, z)/n \to$ const. as $z \to 0$. Let

$$Q_z = \frac{1 + ze^{-\beta H}}{1 - ze^{-\beta H}} = \coth\left(\frac{1}{2}\beta(H - \mu)\right) \tag{5.1}$$

denote the covariance of the KMS state of the reservoir at inverse temperature β. Notice that

$$z = 0 \iff \beta = +\infty \iff \quad \text{Fock state}$$

Q_z is densely defined on a domain $D(Q) \subseteq H_1$.

Let $S_t^1 : H_1 \to H_1$ denote the 1-particle dynamics and suppose that

$$S_t^1 Q_z = Q_z S_t^1 \tag{5.2}$$

Denote $W(H_1)$ the CCR-C* algebra over H_1 and φ_{Q_z} the free Bose grand canonical equilibrium state at inverse temperature β and fugacity z characterized by

$$\varphi_{Q_z}(W(f)) = \exp(-\frac{1}{2}\langle f, Q_z f\rangle) \tag{5.3}$$

$\{\mathcal{H}_z, \pi_z, \Phi_z\}$ the GNS triple of $\{W(H_1), \varphi_{Q_z}\}$. Because of (5.2), the state φ_{Q_z} is invariant for the free reservoir dynamics

$$u_t^R(W(f)) = W(S_t^1 f) .$$

Denoting u_t^S the free system dynamics we define the free dynamics

$$u_t^0 = u_t^S \otimes u_t^R \tag{5.5}$$

The basic assumption on S_t^1 is that there exists a nonempty subset $K \subset H_1$ such that

$$\int_R |\langle f, S_t^1 g\rangle| dt < +\infty \quad \forall f, g \in K \tag{5.6}$$

For example, in the case, corresponding to the free Bose gas,

$$H_1 = L^2(R^d) ; \quad d \geq 3$$

the $S_t^1 = e^{-it\Delta}$, where Δ is the Laplacian and K can be chosen to be

$$K = L^2 \cap L^\infty \cap L^1(R^d) .$$

In general we shall always assume that

$$QK \subseteq K \tag{5.7}$$

Notice that one can always assume that

$$S_t K \subseteq K \quad ; \qquad \forall t \tag{5.8}$$

A typical interaction, found in quantum optics is:

$$\lambda V_g = \frac{\lambda}{i}(B \otimes A^+(g) - B^+ \otimes A(g)) \tag{5.9}$$

where $g \in K \subseteq H_1$ and one makes the further assumption:

$$u_t^S(B) = e^{-i\omega_0 t} B . \tag{5.10}$$

In laser theory one arrives at such an interaction starting from the usual electromagnetic field and introducing the dipole and the rotating wave approximations.

For example, one can choose (H_S, u_t^S) to be the harmonic oscillator and

$$B = a \quad \text{the annihilation operator}$$

$$u_t^S = \mathrm{Ad} U_t^S = Ade^{it\omega_0 a^+ a}$$

where ω_0 is a privileged frequency: the laser frequency.

Not all the above assumptions are essential. For example:

Boson CCR: for Fermion reservoirs nothing changes (cf. [4, 9]).

Linearity in $A(g)$, $A^+(g)$: $A^p(g)$, $A^+(g))^p$ can be handled with similar techniques (cf. [6, 19]).

Rotating wave approximation (i.e. condition (5.10)): not necessary (cf. [5]).

Considering the Schrödinger equation in interaction representation

$$\partial_t U_t^{(\lambda)} = \frac{1}{i} V^{(\lambda)}(t) U_t^{(\lambda)} = \frac{1}{i} V^{(\lambda)}(t) U_t^{(\lambda)} \tag{6.1}$$

and its solution by the iterated series:

$$U_t^{(\lambda)} = 1 + \sum_{n=1}^{\infty} (-i)^n \int_0^t \mathrm{d}t_1 \int_0^{t_1} \mathrm{d}t_2 \cdots \int_0^{t_{n-1}} \mathrm{d}t_n V^{(\lambda)}(t_1) \cdots V^{(\lambda)}(t_n) \tag{6.2}$$

The first term of the series is

$$\int_0^t \mathrm{d}t_1 V^{(\lambda)}(t_1) = \int_0^t \mathrm{d}t_1 \lambda (D \otimes A^+(S_{t_1} g) - c.c.) = D \otimes A^+ \left(\lambda \int_0^t \mathrm{d}t_1 S_{t_1} g \right) - c.c. \tag{6.3}$$

where $S_t := S_t^1 e^{-i\omega_0 t}$. This suggests to introduce the collective (in time) test functions

$$g_{\lambda,t} := \lambda \int_0^t \mathrm{d}t_1 S_{t_1} g$$

and

$$A(\lambda \int_0^t \mathrm{d}t_1 S_{t_1} g) = A(g_{\lambda,t})$$

as collective fields.

The vacuum statistics of the collective fields $A(g_{\lambda,t})$, $A^+(g'_{\lambda,t})$ is clearly Gaussian, with mean zero

$$\langle \Phi, A(g_{\lambda,t}) \Phi \rangle = 0$$

and variance

$$\langle \Phi, A(g_{\lambda,t}) A^+(g'_{\lambda,t}) \Phi \rangle = \lambda^2 \int_d^t t_1 \int_0^t \mathrm{d}t_2 \langle S_{t_1} g, S_{t_2} g' \rangle = \lambda^2 \int_0^t \mathrm{d}s \int_0^t \mathrm{d}t \langle g, S_{s-r} g' \rangle$$

with the change of variables $s - r = u$ this becomes

$$\lambda^2 \int_0^t \mathrm{d}s \int_{-s}^{t-s} \mathrm{d}u \langle S_u g, g' \rangle$$

and with the further change of variables $\lambda^2 s = v$ we obtain

$$\int_0^{\lambda^2 t} \mathrm{d}v \int_{-v/\lambda^2}^{t-v/\lambda^2} \mathrm{d}u \langle S_u g, g' \rangle \, .$$

These naturally suggests to introduce the **time renormalization** $t \to t/\lambda^2$. After that, we obtain:

$$\int_0^t \mathrm{d}v \int_{-v/\lambda^2}^{(t-v)/\lambda^2} \mathrm{d}u \langle S_u g, g' \rangle \to t \int_{-\infty}^{+\infty} \langle S_u g, g' \rangle \mathrm{d}u \; ; \quad \text{as } \lambda \to 0$$

If we compute the 2-point function at 2 different times t, t', repeating the same arguments as above one finds

$$\langle \Phi, A(g_{\lambda,t/\lambda^2}) \cdot A^+(g'_{\lambda,t'/\lambda^2})\Phi \rangle = \int_0^t dv \int_{-v/\lambda^2}^{(t'-v)/\lambda^2} du \langle S_u g, g' \rangle$$

which, as $\lambda \to 0$ converges to

$$\min\{t, t'\} \cdot \int_{-\infty}^{+\infty} \langle S_u g, g' \rangle du = \langle \chi_{[0,t]}, \chi_{[0,t']} \rangle \cdot (g|g') =: (g_t|g'_{t'})$$

This is precisely the covariance of a **classical** Wiener process with values in the Hilbert space of test functions endowed with the scalar product

$$(g|g') := \int_R \langle S_u g, g' \rangle du \tag{6.4}$$

In order to get a feeling of the scalar product (6.4), it is convenient to consider the example stated before, i.e. $H_1 = L^2(R^d)$, $(d \geq 3)$; $S_t^1 = e^{-it\Delta}$ (Δ-the Laplacian). In this case one has

$$\begin{aligned}
(f|g)_1 &= \int_R e^{-i\omega_0 t} \langle f, e^{-it\Delta} g \rangle_{L^2(R^d)} dt = \int_R dt \int_{R^d} dk e^{-i(\omega_0 - k^2)t} \hat{f}(k)^* \hat{g}(k) \\
&= \int_R dt \int_0^\infty \rho^{d-1} d\rho \int_{S^{(d-1)}} d\sigma_{d-1} E^{-i(\omega_0 - \rho^2)t} \hat{f}(\rho, \sigma)^* \hat{g}(\rho, \sigma) \\
&= \int_0^\infty \rho^{d-1} d\rho \delta(\rho^2 - \omega_0) \int_{S^{(d-1)}} \hat{f}(\rho, \sigma)^* \hat{g}(\rho, \sigma) d\sigma_{d-1}
\end{aligned}$$

where $S^{(d-1)}$ denotes the unit sphere in R^d and $d\sigma_{d-1}$ the uniform measure on it.

Therefore, using the identity

$$\delta(\rho^2 - \omega_0) = \frac{\delta(\rho - \sqrt{\omega_0})}{2\rho}$$

one finds

$$(f|g)_1 = \omega_0^{\frac{d-2}{2}} \int_{S^{(d-1)}} \hat{f}(\sqrt{\omega_0}, \sigma)^* (\sqrt{\omega_0}, \sigma) d\sigma_{d-1} \ . \tag{6.5}$$

Summing up:

$$(f|g)_1 = \omega_0^{\frac{d-2}{2}} \langle \hat{f}(\sqrt{\omega_0}, \cdot), \ \hat{g}(\sqrt{\omega_0}, \cdot) \rangle_{L^2(S^{d-1})} \ .$$

In the non Fock (finite temperature equilibrium) case one is lead to consider the scalar product

$$(f|g)_Q = \int_R e^{-i\omega_0 t} \langle f, e^{-it\Delta} Q g \rangle dt$$

with $Q = \coth \frac{\beta}{2}\Delta$. In this case the calculations are exactly the same and in the result one simply introduces the factor $\coth \frac{\beta}{2}\sqrt{\omega_0}$ replacing the factor $\omega_0^{(d-2)/2}$ in (6.5). Notice that in both cases if $\omega_0 = 0$ then the scalar product vanishes.

A more intuitive idea of the sense in which the collective fields approximate the quantum Brownian motions, can be achieved by means of an approximation which illustrates the analogy with the central limit theorem of classical probability. In the expression for the collective field

$$A(g_{\lambda,t}) = \lambda \int_0^{t/\lambda^2} dt_1 A(S_{t_1} g)$$

putting $\frac{1}{\lambda^2} = N$, so that $\lambda \to 0 \Leftrightarrow N \to \infty$ and using the approximation

$$\int_t^{t+dt} du A(S_n g) \equiv A(\chi_{[t,t+dt]} S_t g)$$

one obtains

$$A(g_{\lambda,t}) = \frac{1}{\sqrt{N}} \int_0^{Nt} du A(S_u g) \equiv \frac{1}{\sqrt{N}} \sum_{j=1}^{[Nt]} A(\chi_{[t_j,t_j+1]} S_{t_j} g) \tag{6.6}$$

where $\{t_1, \ldots, t_{[Nt]}\}$ is a partition of the interval $[0, t/\lambda^2] = [0, Nt]$. Now the random variables $A(\chi_{[t_j,t_j+1]} S_{t_j} g)$ are uniformly infinitesimal (as $\max(t_{j+1} - t_j) \to 0$) and, even if not independent, they are weakly dependent because of condition (5.6) and Gaussianity. If the random variables were classical, then the right hand side of (6.6) would be the typical sum which, in the functional central limit theorems, converges in law to the classical Wiener process. Here the variables do not commute and the effect is a **quantum functional central limit theorem.**

Summing up: starting from the Schrödinger equation in interaction representation:

$$\partial_t U_t^{(\lambda)} = \frac{1}{i} V_t^{(\lambda)} U_t^{(\lambda)}$$

and introducing the time renormalization:

$$U_t^{(\lambda)} \longrightarrow U_{t/\lambda^2}^{(\lambda)}$$

and the collective fields

$$A(f_{\lambda,[S,t]}) = A(\lambda \int_{S/\lambda^2}^{T/\lambda^2} dt S_t f)$$

we expect that, as $\lambda \to 0$:

1) The collective annihilator $A(f_{\lambda,[0,t]})$ converges in some sense to the annihilator of a QBM $A_t(f)$.
2) The renormalized wave operator at time t, $U_{t/\lambda^2}^{(\lambda)}$, converges in some sense to some unitary operator U_t living on the QBM space.
3) Some subspace $\mathcal{H}_{(\lambda)}$ of collective vectors converges, in some sense to the Hilbert space of the QBM.

7. CONVERGENCE OF THE QUANTUM FIELDS TO THE QUANTUM NOISES

In this section we clarify the following questions:

i) In which sense the convergences mentioned at the end of the previous section are to be understood?

ii) Which kind of equation is satisfied by U_t?

iii) What does the Heisenberg equation look like in the limit?

Concerning the first problem, we expect that, for any observable of the small system one has:

$$\lim_{h\to 0}\langle \Phi_\lambda, j^\lambda_{t/\lambda^2}(a_0)\Phi'_\lambda\rangle = \lim_{\lambda\to 0}\langle \Phi_\lambda, U^{\lambda *}_{t/\lambda^2}\cdot a_0\cdot U^\lambda_{t/\lambda^2}\Phi'_\lambda\rangle = \langle \Phi, U^*_t\cdot a_0\cdot U_t\Phi'\rangle$$

for some unitary operator U_t living in the L^2 space of the QBM.

The collective vectors Φ_λ have to be chosen case by case and they depend on the interaction. In the case of the interaction (5.9) one introduces the collective Weyl operators

$$W\left(\lambda\int_{S/\lambda^2}^{T/\lambda^2} S_u f \mathrm{d}u\right) \;;\quad f\in K$$

and the collective coherent vectors

$$W\left(\lambda\int_{S/\lambda^2}^{T/d^2} S_u f \mathrm{d}u\right)\cdot\Phi_Q = \Phi\left(\lambda\int_{S/\lambda^2}^{T/\lambda^2} S_u f \mathrm{d}u\right)$$

The physical idea underlying this choice is that the collective long time behaviour of a system is best described in collective long time states.

THEOREM I The approximating processes

$$\left\{\mathcal{H}_Q, W\left(\lambda\int_{S/\lambda^2}^{T/\lambda^2} S_u f \mathrm{d}u\right),\quad \Phi_Q\right\}$$

converges to the QBM

$$\{\Gamma_Q(L^2(R,\mathrm{d}t;\tilde{K})),\ W(\chi_{[S,T]}\otimes f)\,,\quad \Phi_Q\}$$

with values in the completion $\tilde{K}$ of K for the scalar product

$$(f|g)_1 := \int_R \cdot\langle f, S_t g\rangle \mathrm{d}t \qquad\qquad f,g\in K$$

THEOREM II The limit, as $\lambda\to 0$, of

$$\langle u_1\otimes\Phi_Q\left(\lambda\int_{S_1/\lambda}^{T_1/\lambda^2} S_u f_1 \mathrm{d}u\right), U^{(\lambda)}_{t/\lambda^2}u_2\otimes\Phi_Q\left(\lambda\int_{S_2/\lambda^2}^{T_2/\lambda^2} s_u f_2 \mathrm{d}u\right)\rangle$$

exists $\forall\, u_1, u_2 \in H_0$, $\forall\, f_1, f_2\in K$; $\forall\, S_1, T_1, S_2, T_2$ and is equal to:

$$\langle u_1\otimes\Phi(\chi_{[S_1,T_1]}\otimes f_1), U_t u_2\otimes\Phi(\chi_{[S_2,T_2]}\otimes f_2)\rangle$$

where U_t is the solution of the QSDE

$$dU_t = \Big\{ B \otimes dA_g^+(t) - B^+ \otimes dA_g(t) -$$
$$- \left[(g|g)^-_{\frac{1}{2}(Q+1)} B^+ B \otimes 1 + (g|g)^-_{\frac{1}{2}(Q-1)} BB^+ \otimes 1 \right] dt \Big\} U_t$$
$$U_0 = 1$$

Notice that the dt term is *not* the **Ito correction**, but the Ito correction **plus a shift in the system Hamiltonian.** This shift is an operator generalization of the Lamb shift. To calculate the precise form of it we use the Ito table of the QBM

$$2 \operatorname{Re} (g|g)^-_{\frac{1}{2}(Q+1)} dt = \left(g | \frac{Q+1}{2} g \right)_1 dt = dA_g^Q(t) d(A_g^Q(t))^+$$
$$2 \operatorname{Re} (g|g)^-_{\frac{1}{2}(Q-1)} dt = \left(g | \frac{Q-1}{2} g \right)_1 dt = d(A_g^Q(t))^+ dA_g^Q(t) \ .$$

Thus, writing $(g|g)^-_{\frac{1}{2}(Q\pm1)}$ in terms of real and imaginary part, and introducing the **shift Hamiltonian**

$$\delta H = \operatorname{Im} (g|g)^-_{\frac{1}{2}(Q+1)} B^+ B + \operatorname{Im} (g|g)^-_{\frac{1}{2}(Q-1)} BB^+$$

the stochastic differential equation becomes

$$dU_t = \Big[B \otimes dA_g^+ - B^+ \otimes dA_g - i\delta H dt - \left\{ \left(g | \frac{Q+1}{2} g \right) B^+ B \otimes 1 + \left(g | \frac{Q-1}{2} g \right) BB^+ \otimes 1 \right\} dt \Big] U_t \tag{7.1}$$

Finally, notice that the scalar coefficients of the dt-term in equation (7.1) have the form

$$(g|g)^-_{\frac{1}{2}(Q\pm1)} = \int_{-\infty}^{0} \langle g, S_t \frac{1}{2}(Q \pm 1) g \rangle dt \ .$$

Recalling that $S_t = e^{-i\omega_0 t} S_t^1 = e^{it(H-\omega_0)}$ and changing $t \to -t$ the integral becomes

$$\int_0^{\infty} dt \langle e^{-it(H-\omega_0)} g, \frac{1}{2}(Q \pm 1) g \rangle = \lim_{\varepsilon \to 0} \langle \frac{1}{i(H-\omega_0) + i\varepsilon} g, \frac{Q \pm 1}{2} g \rangle \tag{7.2}$$

where we use the notation (standard in the physical literature):

$$\frac{1}{i(H-\omega_0) + i\varepsilon} \to \delta(H - \omega_0) + i\mathcal{P} \frac{1}{H - \omega_0} \tag{7.3}$$

and $\mathcal{P}$ denotes principal part.

If the 1-particle reservoir Hamiltonian H has an absolutely continuous energy spectrum, i.e. for some (at least continuous) function $J_{g,g}(\omega)$ one has

$$J_{g,g}(\omega) d\omega = \langle g, dE_H(\omega) g \rangle$$

($E_H(\cdot)$ being the spectral measure of H), and if, in energy representation, also the operator Q is the multiplication operator by some function $q(\omega)$, then the integral on the left hand side of (7.2) can be explicitly calculated and gives

$$\begin{aligned}(g|g)^{-}_{\frac{1}{2}(Q\pm1)} &= \int_{-\infty}^{0} \mathrm{d}t \int_{R} \mathrm{d}\omega e^{i(\omega-\omega_0)t}\frac{1}{2}(q(\omega)\pm 1)J_{g,g}(\omega) \\ &= \frac{\pi}{2}(q(\omega_0)\pm 1)J_{g,g}(\omega_0) - i\mathcal{P}\int_{-\infty}^{+\infty}\mathrm{d}\omega\frac{\frac{1}{2}(q(\omega)\pm 1)J_{g,g}(\omega)}{(\omega-\omega_0)}\end{aligned} \tag{7.4}$$

where the principal value integral is purely imaginary since the spectral density $J_{g,g}(\omega)$ is positive and $q(\omega) \geq 1$. The meaning of the spectral density $J_{f,g}(\omega)$ is best clarified by the identities

$$(f|g)_1 = 2\pi J_{f,g}(\omega_0) \quad ; \qquad (f|g)_Q = 2\pi q(\omega_0)J_{f,g}(\omega_0)$$

where the index 1 labeles the Fock case ($Q = 1$).

Formula (7.2) and its particularization (7.4) establishes an interesting connection between the results discussed here and the known results on (non relativistic) renormalization theory. This connection shall be made more precise by further investigations from which we expect that some interesting results will emerge.

REFERENCES

1. L. Accardi: *Noise and dissipation in quantum theory.* Rev. math. Phys. Vol. 2, No 2, (1990) 127–176.
2. L.Accardi , A.Frigerio, Y.G.Lu: *The weak coupling limit as a quantum functional central limit.* Commun. Math. Phys. 131, (1990), 537–570.
3. L.Accardi , A.Frigerio and Y.G.Lu: *The Weak Coupling Limit (II) : The Langevin equation and finite temperature case,* submitted to Publications RIMS Oct. 1989.
4. L.Accardi , A.Frigerio and Y.G.Lu: *The Weak Coupling Limit For Fermions,* to appear in Journ. Math. Phys.
5. L.Accardi , Y.G.Lu: *The Weak Coupling Limit Without Rotating Wave Approximation,* to appear in Anna. Ins. H.Poincaré Phys. Th.
6. L.Accardi , Y.G.Lu: *The Weak Coupling Limit For Non-Linear interactions. Ascona* Stochastic Processes, Physics and Geometry, 1–26.
7. L.Accardi , Y.G.Lu: *The Number Process as Low Density Limit of Hamiltonian Models,* to appear in Commun. Math. Phys.
8. L.Accardi , Y.G.Lu: *The Low Density Limit of Quantum Systems,* to appear in J. Physics A.

8a. L.Accardi, A.Frigerio, R. Alicki, Y.G.Lu: An invitation to the weak coupling and the low density limits, in Quantum Probability and related fields.Vol. VI. World Scientific (1991).

9. L.Accardi , Y.G.Lu: *The Fermion number processes as a functional central limit of quantum Hamiltonian models,* to appear in Anna. Ins. H.Poincaré.
10. G.S. Agarwal: Quantum Optics. Springer 1974.
11. N. Cartwright: *How the laws of Physics lie.* Clarendon Press 1988.
12. F.Fagnola, Y.G.Lu: *The weak coupling limit for free noise,* submitted to J. Theo. Proba.

13. A. Frigerio, V. Gorini: *N-level systems in contact with a singular reservoir. II.* J. Math. Phys. 17, 2123-2127.
14. A. Frigerio, M. Verri: *Long-time asymptotic properties of dynamical semi-groups on W^*-algebras,* Math. Jap. 24, 53-63 (1979).
15. V. Gorini, A. Kossakowski: *N-level systems in contact with a singular reservoir.* J. Math. Phys. 1298–1305.
16. H. Haken: *Laser theory.* Springer 1984.
17. K. Hepp, E.H. Lieb: *Phase transitions in resevoir-driven open systems, with applications to superconductors and lasers.* Helv. Phys. Acta 46, 575-603.
18. M. Lax: *Phys. Rev.* 145(1965)111-129.
19. Y.G.Lu: *The weak coupling limit for quadratic interaction,* submitted to J. Math. Phys.
20. I.R. Senitzky: *Dissipation in Quantum Mechanics. The harmonic oscillator.* Phys. Rev.119(1960)670.
21. L. van Hove: *Quantum-mechanical perturbation giving rise to a statistical transport equation.* Physica (Utricht) 21, 517- 540.
22. W.von Waldenfels: *Ito Solution of the linear quantum stochastic differential equation describing light emission and absorption.* SLNM 1055 (ed) Accardi L., Frigerio A., Gorini V.
23. D. Voiculescu: *Free noncommutative random variables, random matrices and the II_1 factors of free groups, in* Quantum Probability and related fields. Vol. VI. World Scientific (1991).

RATE EQUATION APPROACH TO SQUEEZING

S. Smart and S. Swain

Department of Applied Mathematics and Theoretical Physics
The Queen's University of Belfast
Belfast BT7 1NN
The United Kingdom

Rate equations have been frequently used in quantum optics in the past - often as a first approximation to some more complicated set of equations [1]. The advantage of this form of equation is that it enables the various contributing terms to be easily interpreted physically: the disadvantage is that they almost always involve some drastic approximation in their construction. In previous publications [2], we have shown how *exact* equations of motion for the diagonal elements of the density matrix may be derived which possess a rate-equation structure. In that paper, it was essential that the field be quantised and spontaneous decay was taken into account using the conventional master equation formalism; here we modify the treatment to deal with a classical external field in the presence of a squeezed (quantum) field.

For simplicity, we consider an atomic system with nondegenerate energy levels E_i, which are labelled in such a way that $E_i > E_j$ if $i > j$, and we use a system of units in which $\hbar = 1$. The hamiltonian for the interacting system of atom and fields may be written

$$H = \sum_i E_i |i\rangle\langle i| + V; \qquad V(t) = \sum_{i,j} \mu_{ij} |i\rangle\langle j|\, e^{-i\omega_{ij}t} \tag{1}$$

where ω_{ij} denotes the frequency of the laser that couples the atomic levels $|i\rangle$ and $|j\rangle$. (For simplicity, we assume that there is only one such laser field coupling a given pair of levels.) The master equation for a system interacting with squeezed white light is of the form [3]

$$\dot{\rho}_{ab} = -(iE_{a,b} + \Gamma_{ab})\,\rho_{ab} - i\sum_c [V_{ac}(t)\,\rho_{cb} - \rho_{ac}V_{cb}(t)] + \delta_{ab}\sum_c \gamma_{ca}\,\rho_{cc} - \xi_{ab}\,\rho_{ba}\, e^{-2i\omega_{ab}t} \tag{2}$$

where we have assumed the centre frequency of the squeezed mode corresponding to the laser frequency ω_{ab} to be equal to that frequency. Also, $E_{i,j} \equiv E_i - E_j$,

$\gamma_{ij} = \chi_{ij} \times \begin{cases} (N_{ij}+1) & \text{if } i > j \\ N_{ij} & \text{if } i < j \end{cases}$, $\Gamma_{ij} \equiv 1/2 \sum_n(\gamma_{in} + \gamma_{jn}) = \Gamma_{ji}$ and $\xi_{ij} = \chi_{ij}M_{ij}$ where χ_{ij} is a damping constant and N_{ij} and M_{ij} are squeezing parameters [3]. Denoting the Laplace transform of $\rho(t)$ by $\tilde{\rho}(z)$, and defining

$$\sigma_{ab} \equiv \tilde{\rho}_{ab}(z + i\omega_{ba}) \tag{3}$$

we obtain from the Laplace transform of (2):

$$\varepsilon_{ab}\,\sigma_{ab} + i\sum_{c\neq b} \mu_{ac}\,\sigma_{cb} - i\sum_{c\neq a} \mu_{cb}\sigma_{ac} + \xi_{ab}\sigma_{ba} = B_{ab} \quad (b \neq a) \tag{4}$$

Quantum Measurements in Optics, Edited by P. Tombesi and D.F. Walls, Plenum Press, New York, 1992

where

$$\varepsilon_{ab}(z) = z + i(E_{a,b} - \omega_{ab}) + \Gamma_{ab} \quad (5)$$

$$B_{ab}(z) = \rho_{ab}(0) - i\mu_{ab}[\tilde{\rho}_{bb}(z) - \tilde{\rho}_{aa}(z)]. \quad (6)$$

Formally, this is equivalent to the corresponding equation in the absence of squeezing, except for the last term, which has the interesting effect of connecting an off-diagonal element of the density matrix with its transpose. Squeezing also shows its effects in the expression for γ_{ij}, where the squeezing parameter N_{ij} plays the role of a photon number, and γ_{ij} describes emission for $i > j$ and absorption for $i < j$.

We have shown [2] that the solution of this set of equations is

$$\sigma_{ab} = \frac{B_{ab}}{(ab)} - \frac{B_{ba}\xi_{ab}}{(ab,ba)} - \sum_{c}^{*}\left[\frac{i\mu_{ac}B_{cb}}{(ab,cb)} - \frac{i\mu_{cb}B_{ac}}{(ab,ac)}\right] +$$
$$+ \sum_{c}^{*}\left[\frac{i\xi_{ab}\mu_{bc}B_{ca}}{(ca,ba,ab)} - \frac{i\xi_{ab}B_{bc}\mu_{ca}}{(bc,ba,ab)}\right] +$$
$$+ \sum_{c,d}^{*}\left[-\frac{\mu_{ac}\mu_{cd}B_{db}}{(ab,cb,db)} + \frac{\mu_{ac}\mu_{db}B_{cd}}{(ab,cb,cd)} + \frac{\mu_{cb}\mu_{ad}B_{dc}}{(ab,ac,dc)} - \frac{\mu_{cb}\mu_{dc}B_{ad}}{(ab,ac,ad)}\right] + \cdots\cdots \quad (7)$$

where $(ab,cd) = (ab)(cd)_{ab}$, etc. and

$$(ab) = \varepsilon_{ab} + \sum_{c}^{*}\left[\frac{\mu_{ac}\mu_{ca}}{(cb)_{ab}} + \frac{\mu_{bc}\mu_{cb}}{(ac)_{ab}}\right] - \frac{\xi_{ab}\xi_{ba}}{(ba)} -$$
$$- \sum_{c,d}^{*}\left[\frac{i\mu_{ac}\mu_{cd}\mu_{da}}{(cb,db)_{ab}} - \frac{i\mu_{bc}\mu_{cd}\mu_{db}}{(ac,ad)_{ab}}\right] + \cdots\cdot \quad (8)$$

The starred summation sign means that terms are excluded from the sum over $c,d,\cdots$ which would make any of the ordered pairs in the $(ab,cd,ef,\cdots)$ functions equal to ab. (Detailed definitions of the various quantities involved in (7) and (8) are to be found in [2].) The solution (7) is exact if the system of equations is finite. The quantities $(cd,ef,\cdots)_{ij,kl,\ldots}$ are defined in a similar way to eqs. (7) - (8), except that in addition to the restrictions given previously, the variables summed over in the definition may not be equal to any of the subscripts pairs ij, kl, $\cdots$.

Equation (7) gives the solution for the off-diagonal elements in terms of the diagonal elements and the initial coherences $\rho_{ab}(0)$ through the factors B_{ab}. We substitute these solutions into the equation for $\sigma_{aa} \equiv \tilde{\rho}_{aa}(z) \equiv \tilde{P}_a(z)$:

$$(z + \Gamma_{aa})\tilde{P}_a = \rho_{aa}(0) - i\sum_{c\neq a}\mu_{ac}\,\sigma_{ca} + i\sum_{c\neq a}\mu_{ca}\,\sigma_{ac} + \sum_{c\neq a}\gamma_{ca}\,\sigma_{cc} \quad (9)$$

For simplicity, we assume $\rho_{ab}(0) = 0$ for all a, b with $a \neq b$. That is, there are no initial coherences. (This assumption is not essential). Substituting into eq. (9), we find after some rearrangement that the equation may be written in rate equation form (in Laplace space):

$$z\,\tilde{P}_a(z) - P_a(0) = \sum_{c\neq a}\{[\gamma_{ca} + \tilde{W}_{ca}(z)]\,\tilde{P}_c(z) - [\gamma_{ac} + \tilde{W}_{ac}(z)]\,\tilde{P}_a(z)\} \quad (10)$$

where, to third order in μ:

$$\tilde{W}_{ca}(z) = 2\,\mathcal{Re}\left\{\frac{\mu_{ac}\mu_{ca}}{(ca)} + \frac{\mu_{ca}\xi_{ac}\mu_{ca}}{(ca,ac)} -\right.$$
$$\left. - i\sum_{d}^{*}\left[\frac{\mu_{ca}\mu_{ad}\mu_{dc}}{(da,dc)} + \frac{\mu_{ca}\mu_{ad}\mu_{dc}}{(ac,dc)} + \frac{\mu_{ca}\mu_{ad}\mu_{dc}}{(ca,da)}\right] + \cdots\right\} \quad (11)$$

The left hand side of eq. (10) is the Laplace transform of the derivative $\dot{P}_a(t)$: the first term on the right hand side is the rate of transitions (both *incoherent* - the γs and *coherent* - the W's) into level a from all other levels c. The final term is the rate of all transitions out of level a. $\tilde{W}_{ac}(z)$ is the *generalised* transition rate from level a to level c, due to the classical external fields (e.g. second and third terms) and the interplay between the external and squeezed fields (e.g. second term). We use the term "generalized" because, whilst the W's are real, they are not necessarily positive - a property which reflects the fact that quantum interference effects are included.

It is usually simpler to solve the system of equations (10) in Laplace space, and then to perform the Laplace inverse of the solutions to obtain $P_a(t)$, rather than inverting the system of equations to begin with. If we do adopt the latter approach, we obtain a non-Markoffian system:

$$\dot{P}_a(t) = \sum_{c \neq a} \left\{ \left[\gamma_{ca} P_c(t) + \int^t W_{ca}(t-\tau) P_c(\tau)d\tau \right] - \left[\gamma_{ac} P_a(t) + \int^t W_{ac}(t-\tau)] \tilde{P}_a(\tau)d\tau \right] \right\} \tag{12}$$

However, we have a simpler situation if, as often is the case, we are only interested in the steady state. This is given by the Laplace identity

$$\underset{t \to \infty}{\mathrm{Lim.}}\, P(t) = \underset{z \to 0}{\mathrm{Lim.}}\, z\tilde{P}(z) \tag{13}$$

which, applied to eq. (10), yields

$$\dot{P}_a(t) = \sum_{c \neq a} \{ [\gamma_{ca} + \tilde{W}_{ca}(0)]\, P_c(t) - [\gamma_{ac} + \tilde{W}_{ac}(0)]\, P_a(t) \}, \tag{14}$$

a Markoffian system which is valid for times $t \gg T_c$, where T_c is the largest decay time for the transients of the system.

As a particular illustration, we consider a two-level atom undergoing resonance fluorescence in the presence of a squeezed vacuum [4]. Our approach leads to the following picture in terms of the coherent (W) and incoherent decay rates (γ) between the levels.

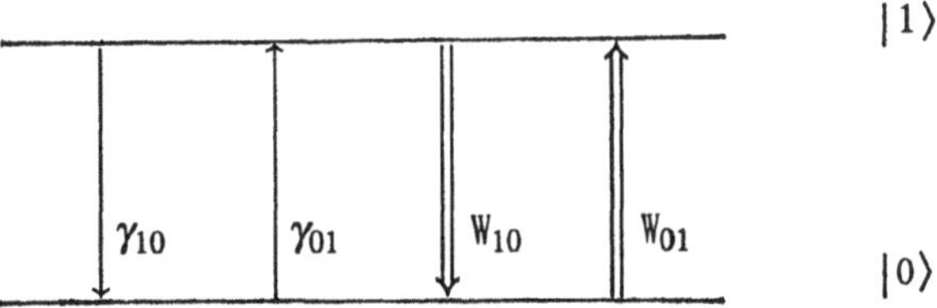

Figure 1. The transition rates in the two level system.

By inspection of the Figure 1, we obtain the following rate equations for the atomic level occupation probabilities (after decay of transients):

$$\dot{P}_0 = (W_{10} + \gamma_{10})P_1 - (W_{01} + \gamma_{01})P_0 \tag{15}$$

$$\dot{P}_1 = (W_{01} + \gamma_{01})P_0 - (W_{10} + \gamma_{10})P_1 \tag{16}$$

where $\gamma_{10} = \gamma(N+1)$, $\gamma_{01} = \gamma N$, $\xi = \gamma M e^{i\varphi}$, $g = |g| e^{i\varphi_L}$, $\Phi = 2\varphi_L - \varphi$, $\Delta = E_{1,0} - \omega_L$ is the detuning and φ_L the laser phase. Here $W_{10} = W_{01} = W$ with W being obtained from (11) (on setting $z = 0$) as

$$W = \frac{2|g|^2\gamma(N+1/2+M\cos\Phi)}{\gamma^2[(N+1/2)^2-M^2] + \Delta^2} \tag{17}$$

(We stick to the notation of [4] as much as possible.) The steady state solution of the above equations

$$P_1(\infty) = \frac{\gamma_{01}+W_{01}}{W_{01}+\gamma_{01}+W_{10}+\gamma_{10}} = \frac{1}{2}\frac{\gamma N + W}{\gamma(N + 1/2) + W} \tag{18}$$

may be shown to agree with that obtained in [4].

In the absence of squeezing, we have $N = M = 0 \Rightarrow \gamma_{01} = 0$, and

$$W \longrightarrow W(0) = \frac{|g|^2\gamma}{\gamma^2/4 + \Delta^2}. \tag{19}$$

In the limit $\gamma \to 0$, this gives $W(0) = 2\pi|g|^2\delta(E_{1,0} - \omega_L)$ – the Fermi "Golden Rule".

With squeezing, we have the inequality $M^2 \leq N(N+1)$. Within this picture, the effect of squeezing is to (i) introduce an upward incoherent rate γ_{01}, (ii) modify the downward incoherent rate γ_{10} and (iii) modify the coherent transition rate W by a phase dependent term in the numerator and a phase-independent term in the denominator. The latter arise from the terms involving ξ in (11) and (8). For simplicity, we restrict our subsequent attention to the case where $\Delta = 0$ and M is maximal. We see that when $N \gg 1$, the rate W becomes very large for $\Phi = 0$ [$W = 8NW(0)$] and very small for $\Phi = \pi$ [$W = W(0)/2N$], compared with its value in the absence of squeezing. These features are illustrated in Figure 2, where we take $\gamma = 1$, $|g|^2 = 2$ (which gives $W(0) = 4$). This shows W for various values of Φ from 0 to π as a function of the squeezing parameter N for small values of N.

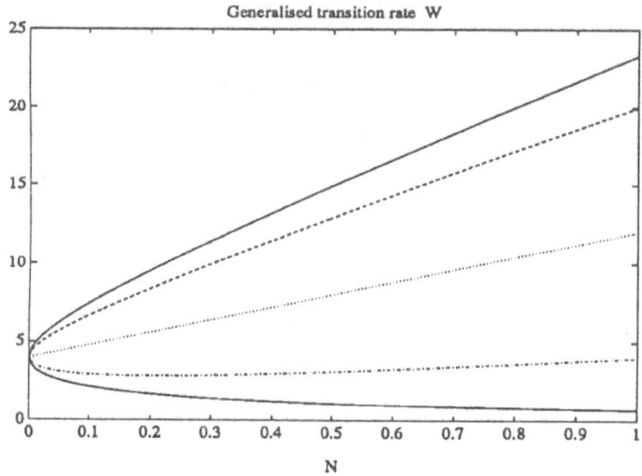

Figure 2. The coherent rate W as a function of N for $\Phi = 0, \pi/4, \pi/2, 3\pi/4$ and π, $\Phi = 0$ being the top graph and $\Phi = \pi$ the bottom. ($\gamma = 1$.)

Turning now to $P_1(\infty)$, we see from (18) that there are two ways in which this quantity can approach its asymptotic value of 1/2. We limit our discussion to the situation $|g| \sim 1$. The first is if $W \gg \gamma$, which as we have seen from the previous paragraph, occurs if $\Phi = 0$, even for small values of N. However, if $\Phi = \pi$, $W \ll \gamma$ for all N. The second is if $N \gg 1$, no matter what the value of W. These features are clearly seen in Figure 3, which shows the behaviour for small values of N, and in Figure 4, which gives the global picture. In the absence of squeezing, the intensity of the applied classical field is such as to produce a steady state value of P_1 of about 0.444. For $\Phi = \pi$ and $3\pi/4$, $P_1(\infty)$ decreases at first from this value as the squeezing is increased. This is due to the decrease in the value of W. However, for N sufficiently large, the incoherent rates begin to dominate W, and $P_1(\infty)$ begins to increase.

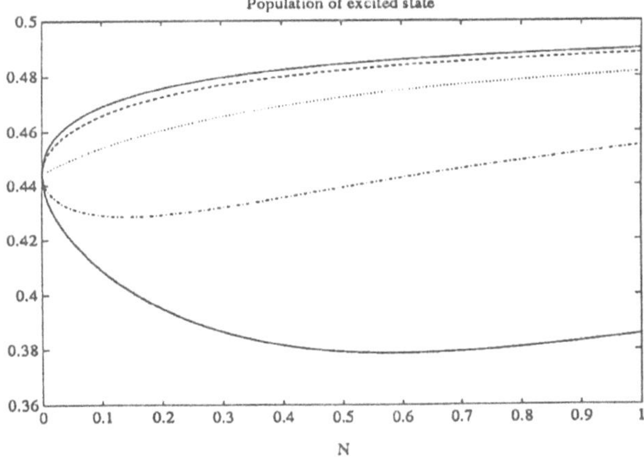

Figure 3. The excited state population $P_1(\infty)$ as a function of N for $\Phi = 0, \pi/4, \pi/2, 3\pi/4$ and π, $\Phi = 0$ being the top graph and $\Phi = \pi$ the bottom. ($\gamma = 1$.)

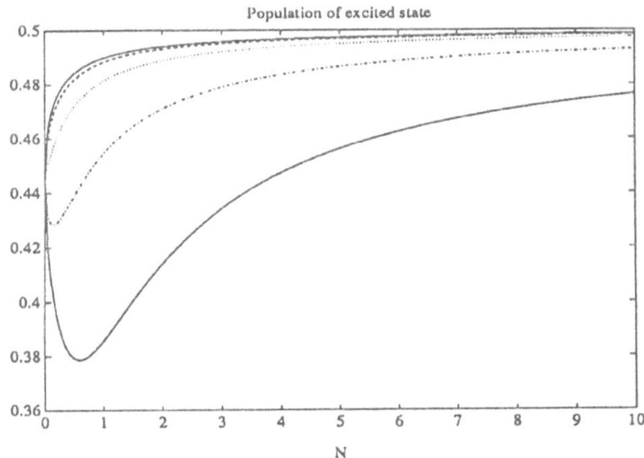

Figure 4. As Figure 3 but with a larger range of N values.

Thus the behaviour of $P_1(\infty)$ – its rapid increase with N for small values of Φ, and for $\Phi \sim \pi$, its fall to a minimum followed by a steady rise – may be understood in terms of the behaviour of the coherent and incoherent transition rates.

Acknowledgement. S. Smart wishes to thank the Department of Education for Northern Ireland for financial support.

References

[1] See, for example, Ackerhalt J. R. and Shore B. W., Phys. Rev. A16, 277 (1977); R. J. Cook and H. J. Kimble, Phys. Rev. Lett. **54**, 1023 (1985).

[2] S. Swain, J. Phys. B13, 2375, (1980); Adv. At. Mol. Phys. **22**, 387 (1986).

[3] C W Gardiner and M J Collett, Phys. Rev. A31, 3761 (1985)

[4] H J Carmichael, A S Lane and D F Walls, Phys. Rev. Lett. **58**, 2539 (1987).

QUANTUM NOISE REDUCTION IN LASERS BY DYNAMIC PUMP NOISE SUPPRESSION

H. Ritsch[1] and P. Zoller[2]

1. Institute for Theoretical Physics, University of Innsbruck
A-6020 Innsbruck, Austria

2. Institute for Theoretical Physics, University of Innsbruck
A-6020 Innsbruck, Austria and
Joint Institute for Laboratory Astrophysics and
Department of Physics, University of Colorado, Boulder, Co 80309

1. INTRODUCTION

The standard theories for a single mode laser predict that far above threshold the laser generates a coherent state (with a phase randomly varying in time). Hence the counting statistics for the output photons is Poissonian and the intensity fluctuation spectrum is shot noise limited (this defines the so-called standard quantum limit (SQL)). Recently there have been several suggestions for developing a laser which produces states with reduced quantum fluctuations. Examples are a laser with sub-Poissonian pump[1-4] (pumping with amplitude squeezed light or a sequence of regularly spaced short pump pulses) or a laser with an intracavity nonlinear absorber[5-8]. In this work we identify a new mechanism of a dynamcial pump noise suppression in multilevel laser systems which leads (at least in principle) to a complete reduction of the low frequency intensity noise[9].

2. LASER WITH DYNAMIC PUMP NOISE REDUCTION

Let us assume that the laser active medium consists of a fixed large number N of m- level atoms, with one pair of levels ($|1\rangle - |2\rangle$) resonantly coupled to a single lasing mode. From the lower lasing level the electron is subsequently repumped into the upper lasing level via the (m–2) remaining levels. The corresponding transitions from $|j\rangle \rightarrow |i\rangle$ are described by transition rates w_{ij}. For simplicity we treat the whole problem in the good cavity limit, where one can adiabatically eliminate the atoms. The corresponding Langevin equation for electric field amplitude is[10]

Quantum Measurements in Optics, Edited by P. Tombesi and
D.F. Walls, Plenum Press, New York, 1992

$$\dot{\alpha} = -\frac{1}{2}\,\gamma\alpha\,[\,1 - \frac{C\,d_o}{1 + \alpha\alpha^+/n_s}\,] + \zeta_\alpha(t) \tag{1}$$

A similar equation holds for α^+ obtained by the replacement $\alpha \leftrightarrow \alpha^+$. In (1) the first term is the cavity damping while the second term describes gain. γ is the damping constant, $C=2\,\mu^2\,N/\Gamma_{12}\gamma$ the cooperativity parameter, with μ the dipolmoment matrix element of the lasing transition and Γ_{12} the transverse damping rate. Furthermore d_o denotes the zero- cavity-field population inversion and n_s the saturation photon number. $\zeta_\alpha(t)$ is a Langevin force with diffusion coefficients

$$D_{\alpha\alpha^+} = N\,\mu^2 \int_0^\infty d\tau\,(\langle\sigma^+,\sigma^-(\tau)\rangle + \langle\sigma^+(\tau),\sigma^-\rangle)\ , \tag{2}$$

$$D_{\alpha\alpha} = N\,\mu^2 \int_0^\infty d\tau\,\langle\sigma^-(\tau),\sigma^-\rangle\ ,\ D_{\alpha^+\alpha^+} = N\,\mu^2 \int_0^\infty d\tau\,\langle\sigma^+,\sigma^+(\tau)\rangle\ , \tag{3}$$

Here $\sigma^\pm$ refers to the atomic raising (lowering) operator on the laser transition. The atomic correlation functions in Eq.(2) have to be evaluated with the help of the Quantum Regression Theorem from the Optical Bloch Equations for the multilevel system, where the interaction of the atom with the cavity mode is represented by c-number amplitudes α, α^+.

A simple, but rather realistic, example for a laser atom is a four level system (m=4) as depicted in Fig. 1. We have solved Eq.(1) for this case and derived the following analytical expression for the Mandel Q-Parameter defined by $(\langle\hat{n}\hat{n}\rangle - \langle\hat{n}\rangle^2)/\langle\hat{n}\rangle = 1 + Q$ (with $\hat{n}$ the photon number operator of the cavity mode):

$$Q = -\,\frac{w_{23}w_{34}w_{41}\ (2w_{23} + 2w_{34} + w_{41} + 2w_{43})}{(2w_{23}2w_{34} + w_{23}w_{41} + w_{34}{}^{\backprime}v_{41} + w_{41}w_{43})} \tag{4}$$

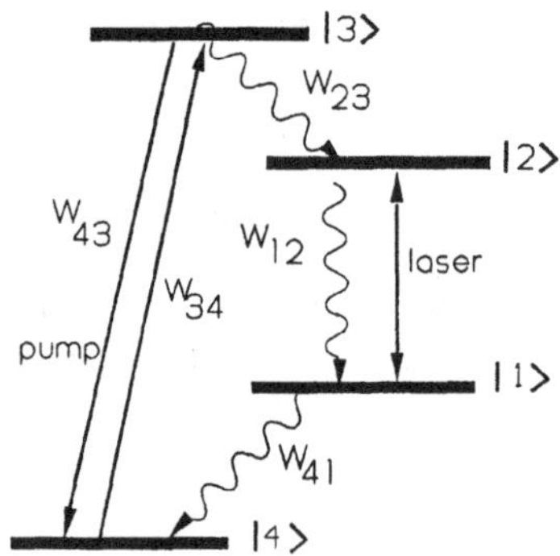

Fig. 1. Four level laser scheme. $|2\rangle - |1\rangle$ and $|4\rangle - |3\rangle$ are the laser and pump transition, respectively.

Fig. 2 shows Q as a function of the pump rate w_{34} for the laser operated well above threshold $Cd_o << 1$ and various combinations of the atomic decay rates. Note that Q is negative, exhibiting nonclassical sub-Poissonian statistics and shows a minimum for some indermediate value of the pump rate w_{34}. Neglecting spontaneous emission, a short calculation yields an optimum Q value of $Q = -2/7$ in the case of $w_{42}=(4/3)w_{34}$ and $w_{13}=w_{34}$. Including the spontaneous emission rate w_{21} on the lasing transition to lowest order in w_{21}/w_{42} we find:

$$Q = -\frac{2}{7} + \frac{5}{7}\frac{w_{21}}{w_{42}} , \tag{5}$$

which for a typical laser, where we have a fast decay out of the lower lasing level, is very small.

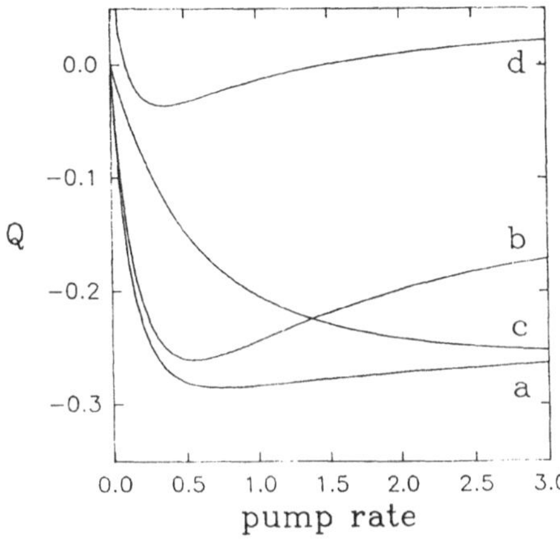

Fig. 2. Mandel Q parameter as function of the pump rate $w_{34} \equiv B$. The parameters for the various curves are $w_{23}=3/2$, $A=w_{12}=0$ (solid line), $w_{23}=10$, $A=w_{12}=0$ (dashed line), $w_{23}=3/2$, $A=5$ (dot-dashed line), $w_{23}=3/2$, $A=0$, $w_{12}=1/4$ (dotted line). All rates are in units of w_{14} (compare Fig. 1). A and B are the Einstein coefficients on the pump transition.

The physical origin of this sub-Poissonian behaviour is closely related to the well known antibunching in the atomic resonance fluorescence. Once one of the laser active atoms has emitted a photon, its wavefunction is projected into the lower of the two lasing levels. Hence it cannot reemit a second photon until it is reexcited to the upper lasing level. This delay between two sucessive emissions introduces a regularity in the pump process, which leads to the negative Q as calculated above. By increasing the number of intermediate steps in this recycling process this noise reduction can be enhanced even further. To understand the dependence of the Mandel Q parameter on the number of atomic levels involved in this recycling process of the active electron, we consider the m-level scheme shown in Fig. 3. Again $|2\rangle \rightarrow |1\rangle$ is the laser transition and electrons are recycled back from $|1\rangle$ to the upper state $|2\rangle$ via the levels $|m\rangle$, $|m-1\rangle$, ...$|3\rangle$.

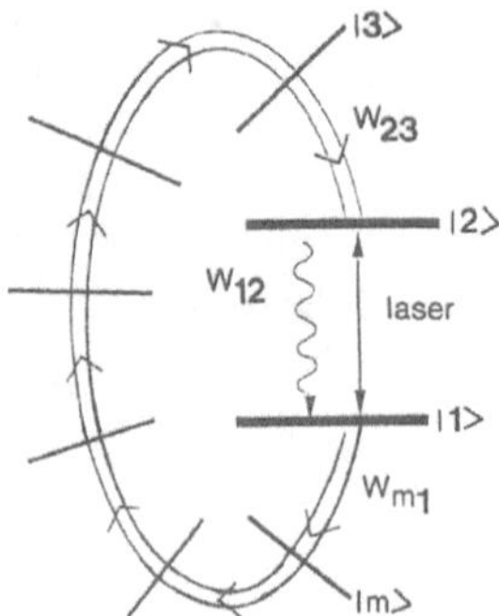

Fig. 3. Symbolic laser scheme with m levels.

Each of these steps is modelled by a unidirectional incoherent rate process. To identify the mechanism of noise suppression we note that for a single step $|j\rangle \rightarrow |j-1\rangle$ with rate r the conditional probability for the electron jumping to $|i\rangle$ in the time interval [t,t+dt], provided it was prepared in $|j\rangle$ at time t=0, is given by an exponential decay law $\tilde{c}(t) = r\, e^{-rt}$. On the other hand $m-1$ consecutive rate steps lead to

$$\tilde{c}(t) = r(r\,t)^{m-1} \quad e^{-rt} / m! \tag{6}$$

Thus the recycling of the electron through these steps leads to anticorrelation in the sense that $\tilde{c}(t=0) = 0$ for $m > 2$. For fixed mean jump time $\tilde{c}(t)$ approaches a delta function when m tends to infinity. In this limit here are no more fluctuations and the stochastic process becomes deterministic. Fig. 4 shows the Mandel Q parameter as a function of the cooperativity parameter C for m=3,4,...,10 assuming optimum conditions of matched rates $w_{23} = \ldots = w_{m-1\,m} = 1/2\, w_{m1}$ and no decay on the laser transition, $w_{12} = 0$. We see that for large C and increasing number of atomic levels Q decreases. We have been able to show that for C >> 1 the Mandel Q parameter is given by

$$Q = \frac{1}{2}\,\frac{m-2}{m-1} \qquad (C \gg 1) \tag{7}$$

which predicts Q = 0 for the two-level system, $Q = -1/4$ and $Q = -1/3$ for the three and four-level system, respectively, and $Q \rightarrow -1/2$ for m >> 1 which corresponds to the ideal limit of complete noise suppression. In this limit the results are identical to the case of a laser pumped by a regular injection of excited atoms. However, in our case no external regularty is assumed from the outside, but it is internally generated by the nonlinear dynamics of the laser atoms themselves.

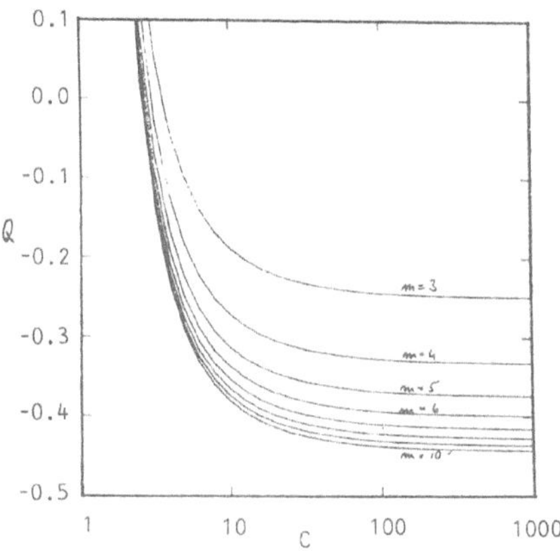

Fig. 4. Mandel Q parameters as a function of the cooperativity parameter C for m=3,4,...,10 levels (compare Fig. 1).

ACKNOWLEDGEMENTS

This work was supported by the Fonds zur Foerderung der wissenschaftlichen Forschung under grant Nr. P7295.

REFERENCES

1. M.A.M.Marte and P.Zoller, Phys. Rev. A 40, 5774 (1989).
2. M.A.M.Marte and P.Zoller, Quantum Opt. 2, 229 (1990).
3. C.Benkert, M.O.Scully, J.Bergou, L.Davidovich, M. Hillary and M. Orzag, Phys. Rev. A 41, 2756 (1990).
4. F.Haake, S.M.Tan and D.F.Walls, Phys. Rev. A 40, 7121 (1989); ibid. 41, 2808 (1990).
5. H.Ritsch, Quantum Opt. 1, (1990).
6. M.J.Collett, A.S.Lane and D.F.Walls, Phys. Rev. A, to be published.
7. Gorbachev, Opt. Comm. (1990).
8. T.A.B.Kennedy and D.F.Walls, Phys. Rev. A 40, 6366 (1989).
9. H.Ritsch, P.Zoller, C.W.Gardiner and D.F. Walls, Phys. Rev. A , submitted.
10. C.W.Gardiner, Quantum Noise (Springer Verlag, Berlin,1991).

IV. OPTICAL PROPAGATION, COMMUNICATION AND IMAGING

WIDEBAND QUANTUM COMMUNICATION: A NEW FRONTIER?

P. D. Drummond[1] and Carlton M. Caves[2]

1. Department of Physics, University of Queensland
St. Lucia, Queensland 4072, Australia
2. National Center for Integrated Photonic Technology
University of Southern California
Los Angeles, California 90089, U.S.A.

1. INTRODUCTION

An important question in communication theory is the channel capacity[1] C—the maximum rate of information transfer (maximum mutual information[2] per second)—obtained with a given input power P. In this paper we consider channels in which information is encoded in a linear bosonic field, such as the electromagnetic field propagating in vacuum. For a *single* channel of this sort—for example, a single transverse mode of an electromagnetic transmission line—we show that there is a physically sensible upper limit on the channel capacity, imposed by quantum mechanics.[3-10] To achieve this limit requires radiation whose bandwidth approaches its maximum frequency. Throughout this paper we assume lossless channels, on the grounds that losses introduce noise that degrades performance; we further assume that the channels have no gain.

In Sec. II we review a theorem due to Holevo,[11] which establishes the maximum information transfer of a quantum channel in terms of quantum entropies. Because of its generality, Holevo's theorem deserves to be called the fundamental theorem of quantum communication theory. In Sec. III we use this theorem to obtain a universal upper bound on the wideband capacity of *any* physically realizable single linear bosonic channel with a bounded communication energy. In the long-time limit, this results in a universal limit C_{WB} (in bits/s) on the channel capacity in terms of the maximum time-averaged power P:

$$C_{\mathrm{WB}} = \frac{\pi}{\ln 2}\sqrt{\frac{2P}{3h}}\,. \tag{1.1}$$

Our proof frees the upper bound (1.1) from dependence on frequency multiplexing or the use of number states,[3-10] and shows that it is a general entropy-based upper limit that applies to *any* technique for encoding information on a single-channel linear bosonic field. This new proof is especially significant in view of claims in the literature[12-14] that infinite capacities are achievable. We point out that these claims

Quantum Measurements in Optics, Edited by P. Tombesi and
D.F. Walls, Plenum Press, New York, 1992

of infinite capacity rely on an assumption that photon energies are approximately independent of frequency, which cannot hold over large bandwidths.

Recent advances in nonclassical state production, together with the development of high T_c superconductors, might make accessible the *physical* conditions for approaching the $(P/h)^{1/2}$ limit. The recent advances in nonclassical state production stem from the pioneering work of Slusher *et al.*,[15] who first generated and detected quadrature-squeezed light[16] with noise below the shot-noise level. This development was followed quickly by observations of quadrature squeezing by other groups.[17-18]

The availability of quadrature-squeezed states makes it interesting to investigate the capacity of such channels. In Sec. IV we discuss the capacity limit C_{QS} for a single wideband frequency-multiplexed quadrature-squeezed channel:

$$C_{\text{QS}} = \frac{2}{\ln 2}\sqrt{\frac{P}{h}} \, . \tag{1.2}$$

Thus coherent quadrature communication can achieve $\sqrt{6}/\pi \simeq 78\%$ of the capacity maximum (1.1).

The direct observation of photon-number squeezing (or antibunching) in light from a diode laser[19] demonstrates the potential for using number states in a communication system. Another potential source of number-squeezed light is photon twinning, with feedback from one twin, in a nondegenerate parametric oscillator.[20] We use the term "number-state channel" to refer to a channel that uses number-state inputs with ideal direct detection of the output quanta. In Sec. V we review the derivation[4,6-10] of the channel capacity of a single wideband frequency-multiplexed number-state channel; in this case it is possible to reach the fundamental limits given by quantum theory. However, there are as yet no techniques that allow efficient photon transmission and detection over bandwidths that approach the maximum frequency.

Efficient transmission of information at the wideband quantum limit is an extremely ambitious goal. It is characterized by a wideband spectrum, similar to blackbody radiation, making it necessary to detect quanta on time scales of their period. In this wideband limit virtual quantum effects are extremely important.[21] The region of ultra-wideband quantum noise is an unexplored area of quantum noise theory and quantum measurement. In Sec. VI we consider the possibility that a high T_c superconducting waveguide, excited with quadrature-squeezed states, might meet the *physical* conditions for approaching the wideband capacity (1.2). In this regard the recent report of microwave-frequency squeezing by a Josephson parametric amplifier is a promising start.[22]

2. QUANTUM COMMUNICATION AND CHANNEL CAPACITIES

2.1. *Shannon's theorem*

Shannon's theorem is the fundamental theorem of classical communication theory, and provides a quick entry into thinking about quantum-limited capacities. The theorem[1,2] states that a single (typically narrow-band) channel with additive, white,

Gaussian noise has a capacity

$$C = \delta \log_2\left(1 + \frac{P}{S\delta}\right) , \tag{2.1}$$

where P is the signal power, δ is the channel bandwidth (in Hz), and S is the (white) power spectrum of channel noise (i.e., the noise power per Hz). For a massless bosonic channel operating at frequency f, one might expect quantum mechanics to place a lower bound $S \geq hf$ on the noise power spectrum. With this supposition Shannon's theorem yields a quantum-limited capacity

$$C = \delta \log_2\left(1 + \frac{P}{hf\delta}\right) . \tag{2.2}$$

In describing a bosonic channel, one distinguishes a wave-like regime $P/hf\delta \gg 1$ and a particle-like regime $P/hf\delta \ll 1$. Recalling that Shannon's theorem arises from a classical wave analysis, one might guess that the Shannon capacity (2.2) has general, but approximate validity as the optimum capacity for narrowband channels in the wave-like regime. What about wideband capacities in this wave-like regime? For present purposes it is sufficient to estimate the wideband capacity by letting $\delta \sim f$— i.e., by letting the bandwidth be as large as possible consistent, at least crudely, with energy hf for a quantum. The resulting optimum wideband single-channel capacity in the wave-like regime is

$$C \sim f \log_2\left(1 + \frac{P}{hf^2}\right) = \sqrt{\frac{P}{h}}\sqrt{\frac{hf^2}{P}} \log_2\left(1 + \frac{P}{hf^2}\right) < 1.2\sqrt{\frac{P}{h}} , \tag{2.3}$$

where the maximum occurs for $f \simeq \frac{1}{2}\sqrt{P/h}$. This analysis indicates that in the wave-like regime the frequency dependence of the energy of a quantum is responsible for the $\sqrt{P/h}$ form of the capacity maximum.

A different heuristic argument yields an optimum narrowband capacity in the particle-like regime $P/hf\delta \ll 1$. Quanta are transmitted at a rate P/hf, so each quantum can occupy, roughly speaking, one of $hf\delta/P$ wave-packet modes. The resulting optimum narrowband single-channel capacity in the particle-like regime is

$$C = \frac{P}{hf} \log_2\left(\frac{hf\delta}{P}\right) . \tag{2.4}$$

What about wideband capacities in this particle-like regime? We can estimate this as before by letting $\delta \sim f$, thus finding again an optimum wideband single-channel capacity $C \sim \sqrt{P/h}$.

It is useful to summarize here the lesson that we can draw from these qualitative arguments: for wideband channels the frequency dependence of the energy of a quantum is a vitally important factor, and leads in either regime to a wideband capacity maximum $\sim \sqrt{P/h}$, this capacity being achieved at the transition zone between the two regimes. Results obtained *neglecting* the Einstein relation $E = hf$, that seem to imply no wideband capacity limit,[12–14] are in fact only valid for $\delta << f$.

We now turn to more rigorous proofs, using Holevo's[11] fundamental theorem in quantum communication theory. This theorem establishes an upper bound on the information transfer (mutual information) in terms of the entropy of the channel density operator and the entropies of the input states.

In order to state Holevo's theorem, it is necessary first to define what we mean by the use of a quantum channel for communication over some finite time interval. The input alphabet for the channel is encoded as one of the quantum states or density operators $\hat{\rho}_a$, where $a = 1, \ldots, \mathcal{A}$. We emphasize that we are not talking about transmitting a sequence of the states $\hat{\rho}_a$; each state $\hat{\rho}_a$ constitutes the *entire* message transmitted during the time interval $\mathcal{T}$. The input states $\hat{\rho}_a$ typically describe quantum excitations of arbitrary wavepackets of longitudinal modes, each wavepacket having a finite temporal extension in the communication channel.

The input is described statistically by probabilities $p_A(a)$ to transmit state $\hat{\rho}_a$. The total information transmitted down the channel is given by

$$H(A) = -\sum_a p_A(a) \log_2 p_A(a) , \tag{2.5}$$

and the (unconditioned) channel density operator is

$$\hat{\rho} = \sum_a p_A(a) \hat{\rho}_a . \tag{2.6}$$

At the output one attempts to retrieve the transmitted information by making some sort of measurement. The most general quantum-mechanical measurement[22] can be described in terms of a complete set of bounded non-negative Hermitian operators $\hat{F}_b$, where $b = 1, \ldots, \mathcal{B}$. These operators are called "effects;" they must be complete in the sense that they provide a resolution of the unit operator as

$$\sum_b \hat{F}_b = \hat{1} . \tag{2.7}$$

Such a generalized measurement can be regarded as a measurement of a quantity B, a "generalized observable," whose possible values are regarded as an output alphabet, labeled for convenience by $b = 1, \ldots, \mathcal{B}$.

To describe the output statistically, we need the conditional probability to obtain output b, given that state $\hat{\rho}_a$ was transmitted:[22]

$$p_{B|A}(b|a) = \mathrm{tr}\,(\hat{\rho}_a \hat{F}_b) . \tag{2.8}$$

This probability is normalized to unity because of the completeness property (2.7). The unconditioned probability for output b is

$$p_B(b) = \sum_a p_{B|A}(b|a) p_A(a) = \mathrm{tr}\,(\hat{\rho} \hat{F}_b) . \tag{2.9}$$

The generalized measurement reduces to an ideal quantum measurement if the effects are a complete set of projection operators $\hat{\Pi}_a$ that are orthogonal, i.e.,

$$\hat{\Pi}_a \hat{\Pi}_{a'} = \delta_{aa'} \hat{\Pi}_a , \tag{2.10}$$

but this is by no means the most general choice. For example, the effects can be multiples of non-orthogonal projection operators, which make up an (over)complete set. The coherent-state projectors of a single mode provide an example of this, which can be approximately realized[23,24] by heterodyne detection.

The information transmitted from input to output is quantified by the mutual information,[2]

$$H(B;A) = H(B) - H(B|A) = \sum_{a,b} p_{B|A}(b|a)p_A(a)\log_2\left(\frac{p_{B|A}(b|a)}{p_B(b)}\right) , \tag{2.11}$$

where

$$H(B) = -\sum_{b} p_B(b)\log_2 p_B(b) \tag{2.12}$$

is the total information available at the output and

$$H(B|A) = -\sum_{a,b} p_{B|A}(b|a)p_A(a)\log_2 p_{B|A}(b|a) \tag{2.13}$$

is a conditional output information that characterizes the channel noise.

In terms of these definitions, Holevo's theorem states that

$$\max_{\{\hat{F}_b\}} H(B;A) \leq S(\hat{\rho}) - \sum_{a} p_A(a)S(\hat{\rho}_a) , \tag{2.14}$$

where the maximum is taken over all possible generalized measurements—i.e., over all complete sets of effects $\hat{F}_b$—and where

$$S(\hat{\rho}) \equiv -\mathrm{tr}\,(\hat{\rho}\log_2\hat{\rho}) \tag{2.15}$$

is the quantum entropy of the density operator $\hat{\rho}$ (in bits).

What makes Holevo's theorem nontrivial and interesting is that it applies to completely general situations—i.e., any type of input states, and any type of detection at the channel output. The input states might be, for example, an (over)complete set of one-dimensional non-orthogonal projectors (non-orthogonal pure states, of which coherent states are an example), in which case one could attempt to transmit information $H(A) \gg S(\hat{\rho})$. Equally, one could measure an (over)complete set of effects (for example, single-mode coherent states measured by heterodyne detection[23,24]) in an attempt to receive information $H(B) \gg S(\hat{\rho})$. Physically, the reason one cannot transmit information in excess of $S(\hat{\rho})$ is that quantum mechanics limits the number of "distinguishable" signals that can be transmitted down the channel. Should one try to transmit and detect information in excess of $S(\hat{\rho})$, the channel becomes noisy enough that the information transmitted does not exceed $S(\hat{\rho})$. Holevo's theorem makes this physical intuition precise and establishes it for general inputs and outputs.

For a given set of input states the channel capacity is obtained by maximizing the mutual information over the input probabilities $p_A(a)$ and over the generalized measurement, to get the maximum information that can be transmitted, and then dividing by $\mathcal{T}$ to get an information rate:

$$C \equiv \frac{1}{\mathcal{T}} \max_{\{p_A(a)\}} \max_{\{\hat{F}_b\}} H(B;A) . \tag{2.16}$$

Holevo's theorem (2.14) establishes a capacity upper bound in terms of the maximum entropy of the channel density operator,

$$C \leq \frac{1}{T} \max_{\{\hat{\rho}_a\}} \max_{\{p_A(a)\}} \max_{\{\hat{F}_b\}} H(B;A) \leq \frac{1}{T} \max_{\hat{\rho}} S(\hat{\rho}) = \frac{S_{\max}}{T} . \tag{2.17}$$

where the maximum with respect to $\hat{\rho}$ is taken subject to any channel constraints.

The capacity upper bound, $S_{\max}/T$, can be achieved in the following way. Find a complete orthonormal set of basis states $|a\rangle$ in which the density operator $\hat{\rho}$ that maximizes $S(\hat{\rho})$ is diagonal, i.e.,

$$\hat{\rho} = \sum_a p(a)|a\rangle\langle a| , \tag{2.18}$$

and then choose $\hat{\rho}_a = |a\rangle\langle a| = \hat{F}_a$ and $p_A(a) = p(a)$. Holevo's theorem shows that such a choice of input states and input probabilities is the only way to achieve the capacity maximum.

Holevo[11] proved his theorem for a finite-dimensional channel Hilbert space, a finite input alphabet, and a finite output alphabet. Thus the theorem does not appear at first to apply to the coherent-state channel mentioned several times in this section, if a continuous alphabet is used. We argue that because the transmitter and detector are physical devices with finite energy and resolution, an idealized continuous alphabet is not physically possible, although of mathematical interest. The requirements of Holevo's theorem are in fact always satisfied by *physically realizable* channels.

3. CAPACITY UPPER BOUNDS FOR A WIDEBAND CHANNEL

We now consider a single linear bosonic communication channel, which we think of as single transverse mode of an electromagnetic transmission line, and we derive a general upper bound on the channel's wideband capacity.

3.1. *Description of channel*

In this section we assume a *finite* transmission time T. If the channel is dispersionless, then the assumption of a finite transmission time allows us to use the standard procedure of quantizing in terms of discrete frequency-domain modes defined by periodic boundary conditions on a length $\mathcal{L} = cT$, where c is the propagation speed down the channel. This procedure yields *one* mode at each of a set of allowed frequencies

$$f_j = j/T , \quad j = 1, 2, \ldots , \tag{3.1}$$

which are multiples of a minimum frequency $1/T$. (There are actually two modes at each frequency, propagating in opposite directions; we use the mode that propagates in the direction of interest.) We emphasize that *a single channel can be defined operationally by saying that there is one mode at each of the frequencies (3.1).*

It might seem that our work in this section is restricted to dispersionless channels. However, for dispersive channels, we can also argue from a standard procedure using wave-packet techniques; the same upper bound is obtained.

The Hilbert space for the channel is spanned by the photon-number eigenstates (Fock states) $|\{n_j\}\rangle$ of the frequency-domain modes, where n_j is the number of photons in mode j. The Fock state $|\{n_j\}\rangle$ is an energy eigenstate with energy

$$E_N = \sum_j hf_j n_j = \frac{h}{T}\sum_j jn_j = \frac{Nh}{T} . \tag{3.2}$$

All the energy eigenvalues E_N are multiples of the quantum h/T for the lowest mode and can be labeled by the integer

$$N = \sum_{j=1}^{\infty} jn_j . \tag{3.3}$$

If we let $\mathcal{N}_N$ denote the number of states with energy E_N, then $\mathcal{N}_N$ is the number of ways of writing the positive integer N as a sum (3.3). Determining $\mathcal{N}_N$ is precisely the elementary number-theoretic problem of partitioning N—i.e., of determining the number of ways that up to N positive integers can be added together to give N.

3.2. *Channel with energy maximum*

We now consider the case in which a time-averaged power constraint is implemented by saying that there is a *maximum* energy $E_N = PT = Nh/T$ (where we can assume again that N is an integer). During the transmission time one is willing to transmit any energy up to and including E_N, but no energy higher than E_N. Such a maximum-energy constraint allows large fluctuations in instantaneous power, provided the overall time-average power limit is not exceeded.

The maximum-energy constraint restricts the Hilbert space to the finite-dimensional bounded energy subspace that includes all the subspaces with energy $E_n \leq E_N$, i.e., $n = 0, 1, \ldots, N$. We can apply Holevo's theorem directly to this subspace. We find the capacity upper bound for this maximum-energy channel is just

$$C_{\max} = \frac{S_{\max}}{T} = \frac{1}{T}\log_2\left(\sum_{n=0}^{N} \mathcal{N}_n\right) . \tag{3.4}$$

An asymptotic expansion of the sum in (3.4) is known,[25] based on the asymptotic expansion of the partition function. We derive the leading correction here to illustrate how it arises. The number of states $\mathcal{N}_n$ rises so rapidly with n that the sum is dominated by the last few terms, so we write it as

$$\sum_{n=0}^{N} \mathcal{N}_n = \frac{1}{4\sqrt{3}N} \exp\left(\pi\sqrt{\frac{2N}{3}}\right) \sum_{k=0}^{N} \exp\left(-\frac{\pi}{\sqrt{6N}}k\right) \left[1 + O\left(\frac{k}{N}\right) + O\left(\frac{k^2}{N^{3/2}}\right)\right] . \tag{3.5}$$

The resulting capacity upper bound for a maximum-energy channel is

$$C_{\max} = \frac{\pi}{\ln 2}\sqrt{\frac{2P}{3h}} - \frac{1}{2T}\log_2\left(\frac{8\pi^2 PT^2}{h}\right) + \frac{1}{T}O\left(\frac{1}{\sqrt{P/hT}}\right) . \tag{3.6}$$

The $T \to \infty$ limit then gives the capacity upper bound C_{WB} of (1.1).

The power constraint is often imposed in a somewhat less restrictive way than we have done above, as a restriction just on the ensemble average power. This allows some messages to exceed the time-average power limit, but yields an identical result for capacity in the long-time limit. In the examples given in the remainder of the paper we shall use this looser restriction, which allows for a close analogy with canonical ensemble averages in statistical mechanics. First we demonstrate explicitly that the capacities of idealized single wideband frequency-multiplexed zero-temperature coherent-state and quadrature-squeezed channels lie below the capacity bound (1.1), although the wideband quadrature-squeezed capacity (1.2) can approach this limit rather closely. We note that these examples use continuous alphabets.

4.1. *Narrowband channels*

As a preliminary we consider a bandwidth limited channel with frequency f and bandwidth $\delta \ll f$, generalizing below to the wideband case. The channel is excited by successive longitudinal wave-packet modes, which are emitted at a rate $\delta \geq 1/\mathcal{T}$.[26] The emission of each wave packet constitutes a "use" of the channel, so δ is the "use rate" (number of uses per second). We let $\bar{n}$ denote the average number of signal photons transmitted per use—i.e., the average number of photons per second per Hz. The average photon transmission rate is $\delta\bar{n}$, and hence the average input power to the channel is given by

$$P = \delta h f \bar{n} \, . \tag{4.1}$$

a. Coherent-state channel. By a "coherent-state channel" we mean the following: at each use the transmitter emits a coherent state, which is read out at the output by ideal heterodyne detection—i.e., by detection of both field quadratures. The input is described statistically by the probabilities $p_A(\alpha)d^2\alpha$ at each use to transmit a coherent state

$$\hat{\rho}_\alpha = |\alpha\rangle\langle\alpha| \, , \tag{4.2}$$

where α lies within $d^2\alpha$. The two-dimensional probability density $p_A(\alpha)$ is defined with respect to the integration measure $d^2\alpha = d\alpha_R \, d\alpha_I$ and turns out to be the Glauber-Sudarshan P function,[26] as is evident from writing the unconditioned channel density operator (per use):

$$\hat{\rho} = \int d^2\alpha \, p_A(\alpha)\hat{\rho}_\alpha \, . \tag{4.3}$$

The average-power constraint becomes a constraint on the mean number of photons per use,

$$\bar{n} = \mathrm{tr}(\hat{\rho}\hat{n}) = \int d^2\alpha \, |\alpha|^2 p_A(\alpha) \, . \tag{4.4}$$

Ideal heterodyne detection is described[23,24] by an overcomplete set of effect densities

$$\hat{F}_\beta = \frac{1}{\pi}|\beta\rangle\langle\beta| \, , \tag{4.5}$$

which are multiples of coherent-state projectors. They give the conditional probability density to read out β at the output, given that α was transmitted:

$$p_{B|A}(\beta|\alpha) = \mathrm{tr}(\hat{\rho}_\alpha \hat{F}_\beta) = \frac{1}{\pi}|\langle\beta|\alpha\rangle|^2 = \frac{1}{\pi}\exp\left(-|\beta-\alpha|^2\right) . \tag{4.6}$$

Since the channel noise is additive and Gaussian, we can invoke Shannon's theorem to find the capacity.[1,2] The mutual information per use is maximized by a Gaussian input probability density,

$$p_A(\alpha) = \frac{1}{\pi\bar{n}}\exp\left(-\frac{|\alpha|^2}{\bar{n}}\right) , \tag{4.7}$$

which makes the unconditioned channel density operator a thermal state, and which, together with the conditional probability density, leads to an output probability density

$$p_B(\beta) = \int d^2\alpha\, p_{B|A}(\beta|\alpha)p_A(\alpha) = \frac{1}{\pi(1+\bar{n})}\exp\left(-\frac{|\beta|^2}{1+\bar{n}}\right) . \tag{4.8}$$

If one calculates a total output information and a conditional output information with respect to the measures $d^2\alpha$ and $d^2\beta$, one finds

$$H(B) = -\int d^2\beta\, p_B(\beta)\log_2 p_B(\beta) = \log_2[e\pi(1+\bar{n})] , \tag{4.9}$$

and

$$H(B|A) = \int d^2\alpha\, p_A(\alpha)\left(-\int d^2\beta\, p_{B|A}(\beta|\alpha)\log_2 p_{B|A}(\beta|\alpha)\right) = \log_2(e\pi) \tag{4.10}$$

Although neither $H(B)$ nor $H(B|A)$ is invariant under changes in integration measure—thus both are ambiguous—their difference, the mutual information, is invariant, as it must be, since it leads to the capacity of a coherent-state channel[27]

$$C = \delta\log_2(1+\bar{n}) . \tag{4.11}$$

The mathematical description of a coherent-state channel has a compelling physical interpretation. There is one quantum's worth of channel noise, half of which is zero-point noise intrinsic to a coherent state and the other half of which is additional noise that arises in any attempt to measure both field quadratures.[27,28] The total output information is ambiguous because any continuous variable, viewed as a limit of a discrete variable, carries an infinite amount of information. The quantum-mechanical noise, however, sets a scale for distinguishing nearby coherent states, thus making the mutual information—and hence the channel capacity (4.11)—finite and perfectly well defined.

b. Quadrature-squeezed channel. By a "quadrature-squeezed channel" we mean the following: at each use the channel is excited into a quadrature-squeezed state[16] that has reduced (squeezed) noise in the field quadrature that carries information in its coherent excitation; at the output the information is read out by ideal homodyne detection of the squeezed quadrature. The squeezed states can be denoted explicitly by $|x_1, 0\rangle_{(r,0)}$.[29] Here x_1 denotes the information-bearing expectation value (coherent excitation) of field quadrature $\hat{x}_1$, and r is the squeeze parameter, which we choose to be the same for all transmitted states. The 0 in the squeezed-state ket denotes the vanishing expectation value of the orthogonal field quadrature $\hat{x}_2$—vanishing because excitation of the orthogonal quadrature wastes energy; the 0 in the subscript means that the phase of the squeezing is such that the first quadrature has reduced noise. Suppressing superfluous notation and writing the input states as density operators, we define

$$\hat{\rho}_{x_1} = |x_1, 0\rangle_{(r,0)}\,{}_{(r,0)}\langle x_1, 0| \; . \tag{4.12}$$

The input is described statistically by the probability density $p_{X_1}(x_1)$ (defined with respect to the standard measure dx_1) at each use to transmit the state $\hat{\rho}_{x_1}$.The unconditioned channel density operator (per use) is

$$\hat{\rho} = \int dx_1\, p_{X_1}(x_1)\hat{\rho}_{x_1} \; , \tag{4.13}$$

and the average-power constraint becomes

$$\bar{n} = \mathrm{tr}(\hat{\rho}\hat{n}) = \sigma^2 + \sinh^2 r \; , \tag{4.14}$$

where

$$\sigma^2 = \int dx_1\, x_1^2 p_{X_1}(x_1) \tag{4.15}$$

is the second moment of x_1 with respect to the input probability density $p_{X_1}(x_1)$.

Ideal homodyne detection is described[23,24] by effect densities

$$\hat{F}_{y_1} = |y_1\rangle\langle y_1| \; , \tag{4.16}$$

which are projectors onto the eigenstates of field quadrature $\hat{x}_1$ (i.e., $\hat{x}_1|y_1\rangle = y_1|y_1\rangle$). The effect densities give the conditional probability density to read out value y_1, given that x_1 was transmitted,

$$p_{Y_1|X_1}(y_1|x_1) = \mathrm{tr}(\hat{\rho}_{x_1}\hat{F}_{y_1}) = |\langle y_1|x_1, 0\rangle_{(r,0)}|^2 = \frac{1}{\sqrt{2\pi(e^{-2r}/4)}} \exp\left(-\frac{(y_1 - x_1)^2)}{2(e^{-2r}/4)}\right) \; . \tag{4.17}$$

The variance of this distribution, $e^{-2r}/4$, describes quadrature noise that is reduced by e^{-2r} from the one-quarter quantum's worth of zero-point quadrature noise.

Again the channel noise is additive and Gaussian, so Shannon's theorem[1,2] (modified for a channel where only one quadrature is detected) tells us that the mutual information is maximized by a Gaussian input probability density,

$$p_{X_1}(x_1) = \frac{1}{\sqrt{2\pi\sigma^2}} \exp\left(-\frac{x_1^2}{2\sigma^2}\right) \; , \tag{4.18}$$

and that the channel capacity at fixed r is given by

$$C = (\delta/2)\log_2\left(1 + \frac{4\sigma^2}{e^{-2r}}\right) = (\delta/2)\log_2[1 + 4e^{2r}(\bar{n} - \sinh^2 r)] \; . \tag{4.19}$$

Maximizing with respect to the squeeze parameter r yields the capacity of a quadrature-squeezed channel,[27]

$$C = \delta\log_2(1 + 2\bar{n}) \; , \tag{4.20}$$

which is achieved when[23,30]

$$e^{2r} = 2\bar{n} + 1 \,. \tag{4.21}$$

The capacities of zero-temperature narrowband coherent-state and quadrature-squeezed channels can be summarized by

$$C = \delta \log_2(1 + \gamma\bar{n}) \,, \tag{4.22}$$

where $\gamma = 1$ applies to a coherent-state channel and $\gamma = 2$ applies to a quadrature-squeezed channel.

These coherent-state and quadrature-squeezed capacities are examples of the narrowband wave-like capacity (2.2). We note that the qualitative arguments arising from Shannon's theorem in Sec. II give the exact capacity of a coherent-state channel, and are approximately valid for a quadrature-squeezed channel in the wave-like regime. Quantum mechanics limits these capacities by restricting the number of phase-space cells that are consistent with the power constraint. Although squeezed states can have very low variances in one quadrature, this advantage is partially offset by the fact that both quadratures can be used for coherent state communication.

4.2. *Wideband channels*

Consider now a single zero-temperature frequency-multiplexed channel, in which each frequency bin is a narrowband coherent-state or quadrature-squeezed channel of the sort just considered. The capacity is given by

$$C = \frac{\delta}{\ln 2} \sum_i \ln(1 + \gamma\bar{n}_i) \,. \tag{4.23}$$

The maximum capacity is obtained by varying $\bar{n}_i$ as a function of frequency, with P held constant. Carrying out this procedure and replacing sums by integrals, one finds that there is a natural upper cutoff frequency

$$f_c = \sqrt{\frac{2\gamma P}{h}} \,, \tag{4.24}$$

beyond which communication is too inefficient to be useful. The optimal photon-number distribution is nonthermal,

$$\gamma\bar{n} = \begin{cases} \dfrac{f_c}{f} - 1, & 0 \le f \le f_c \,, \\ 0, & f \ge f_c \,. \end{cases} \tag{4.25}$$

and the channel capacity for wideband frequency-multiplexed coherent-state and quadrature-squeezed channels is given by

$$C_{\mathrm{CQ}} = \frac{f_c}{\ln 2} = \frac{1}{\ln 2}\sqrt{\frac{2\gamma P}{h}} \,, \tag{4.26}$$

where the subscript CQ is supposed to remind one that this capacity applies both to Coherent-state channels and to Quadrature-squeezed channels. For a coherent-state channel ($\gamma = 1$) the results (4.24)–(4.26) were obtained by Gordon[3] and by Marko.[5] For a quadrature-squeezed channel ($\gamma = 2$) the capacity (4.26) reduces to the capacity C_{QS} given in (1.2).

5. CAPACITIES OF A NUMBER-STATE CHANNEL

5.1. *Narrowband channel*

As before, we consider first a narrowband number-state channel with frequency f and bandwidth $\delta \ll f$, generalizing below to the wideband case. The channel is excited by successive longitudinal "wave-packet" modes, which are emitted at a rate δ. The inputs to the channel are photon-number states of the wave-packet modes, so the input alphabet is the set of non-negative integers and a transmitted message is a sequence of integers. The input is described statistically by probabilities $p_N(n)$ to transmit n photons down the channel at each use. The total information transmitted down the channel is given by the Shannon information

$$H(N) = -\sum_n p_N(n) \log_2 p_N(n) . \tag{5.1}$$

At the channel output the photons are counted by an ideal photocounter. Thus the output alphabet is also the set of non-negative integers, and a received message is a sequence of integers. Channel noise, should there be any, is characterized by a conditional probability $p_{M|N}(m|n)$, the probability to count m photons at the output when n photons are transmitted at the input. The unconditioned probability to count m photons at the output is

$$p_M(m) = \sum_n p_{M|N}(m|n) p_N(n) . \tag{5.2}$$

The conditional probability that n photons were transmitted, given m counted at the output, is obtained from Bayes's theorem:

$$p_{N|M}(n|m) = \frac{p_{M|N}(m|n) p_N(n)}{p_M(m)} . \tag{5.3}$$

The information transmitted per use from input to output is quantified by the mutual information,

$$H(M;N) = H(M) - H(M|N) = \sum_{n,m} p_{M|N}(m|n) p_N(n) \log_2\left(\frac{p_{M|N}(m|n)}{p_M(m)}\right) , \tag{5.4}$$

where

$$H(M) = -\sum_m p_M(m) \log_2 p_M(m) \tag{5.5}$$

is the total information available at the output and $H(M|N)$ is a conditional output information that characterizes the channel noise. In the case of an ideal, noiseless channel at zero temperature, the mutual information equals the input information, since

$$H(M) = H(N) , \qquad H(M|N) = 0 . \tag{5.6}$$

For simplicity, we shall focus on the noiseless case here. Although the results are easily generalised to noisy channels, it is clear from (5.4) that the introduction of noise or finite temperatures can only degrade channel performance.

The channel capacity is the bandwidth times the maximum mutual information, where the maximum is taken over the possible input probabilities $p_N(n)$, subject to the constraint on average photon number $\bar{n}$ (i.e., subject to the constraint on average input power $P = \delta h f \bar{n}$). In symbols C is defined by

$$C \equiv \delta \max_{\{p_N(n)\}} H(M;N)\ , \tag{5.7}$$

where the maximum is taken subject to the constraints

$$\sum_{n=0}^{\infty} p_N(n) = 1\ , \qquad \sum_{n=0}^{\infty} n p_N(n) = \bar{n}\ . \tag{5.8}$$

This problem is the same as finding the canonical distribution for a harmonic oscillator and has the standard result,

$$p_N(n) = \frac{1}{1+\bar{n}} \left(\frac{\bar{n}}{1+\bar{n}} \right)^n\ , \tag{5.9}$$

This gives the result for the capacity of a single narrowband number-state channel as[5–7,26,31,32]

$$C = \delta[\bar{n} \log_2(1+\bar{n}^{-1}) + \log_2(1+\bar{n})]\ . \tag{5.10}$$

The number-state capacity is a combination of the wave-like and particle-like capacities (2.2) and (2.4). We emphasize that even though the capacity (5.10) is derived here as the capacity of a number-state channel, the techniques developed in Sec. III can be used to show that it is an upper bound on the capacity of *any* single narrowband linear bosonic channel with an ensemble average power constraint.

5.2. *Wideband channel*

We now consider a single zero-temperature frequency-multiplexed channel, in which each frequency bin is a narrowband number-state channel of the sort just considered; i.e., the inputs to each bin are number states which are detected at the output by an ideal photocounter. Let δ be the bandwidth of each frequency bin, let f_i be the frequency of the ith bin, and let $\delta \bar{n}_i$ be the average photon transmission rate in the ith bin. The total input power and the channel capacity

$$C = \frac{\delta}{\ln 2} \sum_i [\bar{n}_i \ln(1+\bar{n}_i^{-1}) + \ln(1+\bar{n}_i)] \tag{5.11}$$

are evaluated by summing over the frequency bins.

One maximizes C by varying $\bar{n}_i$ as a function of frequency, with P held constant. This process is precisely equivalent to maximizing the entropy in a channel of photons with positive momentum, and the resulting photon-number distribution is that of a one-dimensional black body,

$$\bar{n}_i = \frac{1}{e^{\beta h f_i} - 1}\ , \tag{5.12}$$

where β is an "inverse temperature." Replacing the sum (5.11) by an integral over all positive frequencies, one finds the standard results that $P = \pi^2/6h\beta^2$ and

$C = 2\beta P/\ln 2 = \pi^2/3h\beta \ln 2$, which lead to the wideband capacity C_{WB} of (1.1).[4,6–10] Previous derivations of the wideband capacity (1.1) apply only to a frequency-multiplexed number-state channel[4,6,7] or *assume* a connection between information and quantum entropy $-\mathrm{tr}(\hat{\rho}\log\hat{\rho})$.[8–10] The upper bound C_{WB} is independent of the medium used and clearly has a general entropic origin. As we demonstrated in Sec. III, it is an upper bound on the capacity of any single linear bosonic channel, regardless of how the information is encoded in the bosonic field.

6. PHYSICAL REALIZATIONS OF A WIDE-BAND CHANNEL

We now wish to consider the question of how to realize the type of communication channel considered here. This question might have direct technological applications, for example, in constructing communication channels in high-speed computers with minimal power expenditure. We note that the communication input power doesn't necessarily have to be dissipated, although schemes to recycle this power could be complex. Since efficient photon-number detection is somewhat impractical at the low frequencies where ultra-wideband channels exist physically, it is most likely that coherent communication would have to be used. Obviously, the use of squeezed states would give the optimum performance, especially if low bit-error rates are required.

The required transmitting medium must then be nearly lossless over a wide bandwidth from f_{min} to f_c. As we wish to neglect thermal effects, we must require $kT \ll hf_{\mathrm{min}}$, so that the thermal occupation number $\bar{n}_T \ll 1$ even at the minimum frequency. A superconducting waveguide[33] is an excellent example of this type of wideband medium, when it is operated at a temperature well below its transition temperature T_c. Such operation is necessary so that the cutoff frequency f_c remains below the absorption edge at $f = 2\Delta/h$, where 2Δ is the energy at which electrons are excited over the superconducting bandgap and the waveguide is no longer lossless ($2\Delta = 3.5\,kT_c$ in a BCS superconductor[33]).

The advantage of the recently developed high T_c superconductors[34] is that much higher photon energies can be used than in low T_c superconductors. The bandgap is larger, mainly due to the larger T_c, but also because $2\Delta \simeq 8\,kT_c$ in these materials.[35] Suppose one chooses $f_{\mathrm{min}} = 0.02\,f_c$. With the readily obtainable transition temperature $T_c \simeq 90°$ K, an upper cutoff frequency $f_c \simeq 5$ THz is still a factor of three below the absorption edge. For $T \simeq 1°$ K the thermal occupation number at f_{min} is $\bar{n}_T \simeq 0.001$. The channel capacity with quadrature-squeezed inputs would then be $C \simeq 6$ Tbits/s, with input power $P \simeq 4$ nW. We note that thin film experiments have shown very high reflectivities and low infrared absorption in Y-Ba-Cu-O films.[36] We suggest that microwave squeezed states[18] propagating down cryogenic waveguides, with Josephson junctions[37] or picosecond pulse sampling techniques,[36] might meet the physical requirements for quantum-limited wideband communication.

ACKNOWLEDGMENTS

CMC thanks Ben Schumacher for bringing Holevo's theorem to his attention, and PDD thanks Anne Streat and Brendan McKay for discussions on summing partition functions. PDD acknowledges support from the University of Waikato, New Zealand. CMC's work was supported in part by the U.S. Defense Advanced Research Projects Agency through the National Center for Integrated Photonic Technology and in part by the U.S. Office of Naval Research (Contract No. N00014-91-J-1167).

REFERENCES

1. C. E. Shannon, *Bell Syst. Tech. J.* **27**: 379 (1948); **27**: 623 (1948).
2. For a general treatment of information theory and its application to communication, see R. G. Gallager, *Information Theory and Reliable Communication*, Wiley, New York, (1968).
3. J. P. Gordon, in: *Advances in Quantum Electronics*, J. R. Singer, ed., Columbia University, New York, (1961).
4. D. S. Lebedev and L. B. Levitin, *Dokl. Akad. Nauk SSSR* **149**: 1299 (1963) [English translation: *Sov. Phys. Dokl.* **8**: 377 (1963)].
5. H. Marko, *Kybernetik* **2**: 274 (1965).
6. D. S. Lebedev and L. B. Levitin, *Information and Control* **9**: 1 (1966).
7. J. I. Bowen, *IEEE Trans. Inf. Theory* **IT–13**: 230 (1967).
8. J. B. Pendry, *J. Phys. A* **16**: 2161 (1983).
9. J. D. Bekenstein, *Phys. Rev. A* **37**: 3437 (1988).
10. J. D. Bekenstein and M. Schiffer, *Int. J. Mod. Phys. C*, to be published.
11. A. S. Kholevo (Holevo), *Problemy Peredachi Informatsii* **9**(3): 3 (1973) [English translation: *Problems of Information Transmission* **9**: 177 (1973)].
12. C. W. Helstrom, *Proc. IEEE (Lett.)* **62**: 139 (1974).
13. H. P. Yuen, R. S. Kennedy, and M. Lax, *IEEE Trans. Inf. Theory* **IT–21**: 125 (1975).
14. J. R. Pierce, E. C. Posner, and E. R. Rodemich, *IEEE Trans. Inf. Theory* **IT–27**: 61 (1981).
15. R. E. Slusher, L. W. Hollberg, B. Yurke, J. C. Mertz, and J. F. Valley, *Phys. Rev. Lett.* **55**: 2409 (1985).
16. For reviews of squeezed light, see R. Loudon and P. L. Knight, *J. Mod. Opt.* **34**: 709 (1987) and M. C. Teich and B. E. A. Saleh, *Quantum Opt.* **1**: 153 (1989); see also *Squeezed States of the Electromagnetic Field*, edited by H. J. Kimble and D. F. Walls, special issue of *J. Opt. Soc. Am. B* **4**: 1453–1741 (1987).
17. R. M. Shelby, M. D. Levenson, S. H. Perlmutter, R. G. DeVoe, and D. F. Walls, *Phys. Rev. Lett.* **57**: 691 (1986); L.-A. Wu, H. J. Kimble, J. L. Hall, and H. Wu, *ibid.* **57**: 2520 (1986); M. W. Maeda, P. Kumar, and J. H. Shapiro, *Opt. Lett.* **12**: 161 (1987); B. L. Schumaker, S. H. Perlmutter, R. M. Shelby, and M. D. Levenson, *Phys. Rev. Lett.* **58**: 357 (1987); M. G. Raizen, L. A. Orozco, M. Xiao, T. L. Boyd, and H. J. Kimble, *ibid.* **59**: 198 (1987); M. Xiao, L.-A. Wu, and H. J. Kimble, *ibid.* **59**: 278 (1987); P. Grangier, R. E. Slusher, B. Yurke, and A. LaPorta, *ibid.* **59**: 2153 (1987); R. E. Slusher, P. Grangier, A. LaPorta, B. Yurke, and M. J. Potasek, *ibid.* **59**: 2566 (1987); S. Pereira, M. Xiao, H. J. Kimble, and J. L. Hall, *Phys. Rev. A* **38**: 4931 (1988); P. Kumar, O. Aytür, and J. Huang, *Phys. Rev. Lett.* **64**: 1015 (1990).
18. R. Movshovich, B. Yurke, P. G. Kaminsky, A. D. Smith, A. H. Silver, R. W. Simon, and M. V. Schneider, *Phys. Rev. Lett.* **65**: 1419 (1990).
19. S. Machida, Y. Yamamoto, and Y. Itaya, *Phys. Rev. Lett.* **58**: 1000 (1987); S. Machida and Y. Yamamoto, *ibid.* **60**: 792 (1988); W. H. Richardson and R. M. Shelby, *ibid.* **64**. 400 (1990).

20. A. Heidmann, R. J. Horowicz, S. Reynaud, E. Giacobino, C. Fabre, and G. Camy, *Phys. Rev. Lett.* **59**: 2555 (1987); P. R. Tapster, J. G. Rarity, and J. S. Satchell, *Phys. Rev. A* **37**: 2963 (1988); T. Debuisschert, S. Reynaud, A. Heidmann, E. Giacobino, and C. Fabre, *Quantum Opt.* **1**: 3 (1989); O. Aytür and P. Kumar, *Phys. Rev. Lett.* **65**: 1551 (1990).
21. P. D. Drummond, *Phys. Rev. A* **35**: 4253 (1987); *Physica Scripta* **T21**: 70 (1988).
22. For a discussion of generalized measurements, see K. Kraus, *States, Effects, and Operations: Fundamental Notions of Quantum Theory*, Springer, Berlin (1983) or E. B. Davies, *Quantum Theory of Open Systems*, Academic, London (1976).
23. H. P. Yuen and J. H. Shapiro, *IEEE Trans. Inf. Theory* **IT–24**: 657 (1978); J. H. Shapiro, H. P. Yuen, and J. A. Machado Mata, *ibid.* **IT–25**: 179 (1979); H. P. Yuen and J. H. Shapiro, *ibid.* **IT–26**: 78 (1980).
24. J. H. Shapiro and S. S. Wagner, *IEEE J. Quantum Electron.* **QE–20**: 803 (1984).
25. G. H. Hardy and S. Ramanujan, *Proc. Lon. Math. Soc. (Series 2)* **17**: 75 (1918); H. Rademacher, *Proc. Lon. Math. Soc. (Series 2)* **43**: 241 (1937); see also Sec. 24.2.1 of *Handbook of Mathematical Functions*, M. Abramowitz and I. A. Stegun, eds., U.S. GPO, Washington, D.C. (1964).
26. R. J. Glauber, *Phys. Rev. Lett.* **10**: 84 (1963); E. C. G. Sudarshan, *ibid.* **10**: 277 (1963).
27. J. P. Gordon, *Proc. IRE* **50**: 1898 (1962); Y. Yamamoto and H. A. Haus, *Rev. Mod. Phys.* **58**: 1001 (1986).
28. E. Arthurs and J. L. Kelly, Jr., *Bell. Syst. Tech. J.* **44**: 725 (1965); C. M. Caves, *Phys. Rev. D* **26**: 1817 (1982); E. Arthurs and M. S. Goodman, *Phys. Rev. Lett.* **60**: 2447 (1988); S. L. Braunstein, C. M. Caves, and G. J. Milburn, *Phys. Rev. A* **43**: 1153 (1991).
29. C. M. Caves and B. L. Schumaker, *Phys. Rev. A* **31**: 3068 (1985); B. L. Schumaker and C. M. Caves, *ibid.* **31**: 3093 (1985).
30. H. P. Yuen, *Phys. Lett. A* **56**: 105 (1976).
31. T. E. Stern, *IRE Trans. Inf. Theory* **IT–6**: 435 (1960).
32. H. Takahasi, in *Advances in Communications Systems*, Vol. 1, A. V. Balakrishnan, ed., Academic, New York (1965).
33. See, for example, J. R. Schrieffer, *Theory of Superconductivity* , Benjamin, Reading, PA (1983).
34. J. G. Bednorz and K. A. Müller, *Z. Phys. B* **64**: 189 (1986); C. W. Chu, P. H. Hor, R. L. Meng, L. Gao, Z. J. Huang, and Y. Q. Wang, *Phys. Rev. Lett.* **58**: 405 (1987).
35. Z. Schlesinger, R. T. Collins, D. L. Kaiser, and F. Holtzberg, *Phys. Rev. Lett.* **59**: 1958 (1987).
36. I. Bozovic, D. Kirillov, A. Kapitulnik, K. Char, M. R. Hahn, M. R. Beasley, T. H. Geballe, Y. H. Kim, and A. J. Heeger, *Phys. Rev. Lett.* **59**: 2219 (1987).
37. R. Sobolewski, D. P. Butler, T. Y. Hsiang, C. V. Stancampiano, G. A. Mourou, *Phys. Rev. B* **33**: 4604 (1986).

QUANTUM MECHANICAL TREATMENT OF A PROPAGATING OPTICAL BEAM

Nobuyuki Imoto[1], John R. Jeffers and Rodney Loudon

Department of Physics, University of Essex
Wivenhoe Park, Colchester CO4 3SQ (U.K.)

1. INTRODUCTION

Quantum mechanics has been established on the basis of the Hamiltonian formula which describes the time evolution of the system. In any textbook, the quantization procedure starts from the box-quantization, in which spatial modes of a cavity are first defined, and then the time evolution of the modes is described. The Hamiltonian $\hat{H}$ has a role of "time evolution generator," which governs the time evolution of an operator $\hat{a}$ as

$$\frac{d}{dt}\hat{a} = \frac{1}{i\hbar}[\hat{a}, \hat{H}] \tag{1}$$

in the Heisenberg picture and governs the time evolution of the density operator $\hat{\rho}$ as

$$\frac{d}{dt}\hat{\rho} = \frac{1}{i\hbar}[\hat{H}, \hat{\rho}] \tag{2}$$

in the Schrödinger picture. Either standing wave modes or travelling wave modes can be used by choosing a proper boundary condition. A Fabry-Perot cavity can be used for determining standing wave modes, while a periodic boundary condition (a ring cavity) can be used for determining travelling wave modes. Once a spatial mode is defined, its evolution is purely time evolution which occurs simultaneously throughout the cavity. If we consider, for simplicity, beam dependence on time t and propagation direction z only, the usual quantum theory can be characterized by the spatial Fourier transform of the field

$$E(z,t) = \sqrt{\frac{1}{2\pi}} \int dk_z E(k_z, t) e^{ik_z z} + \text{c.c.} \tag{3}$$

The Fourier coefficient $E(k_z, t)$ is then regarded as a quantum mechanical operator. $E(k_z, t)$ is factorized as $E_0(k_z)\hat{a}(k_z, t)$, where $E_0(k_z)$ is the single-photon electric field and $\hat{a}(k_z, t)$ is the annihilation operator of mode k_z normalized so that $\hat{a}^\dagger(k_z, t)\hat{a}(k_z, t)$ becomes the photon number (of mode k_z at time t in quantization volume V.)

In many experiments in quantum optics, however, there usually is no particular cavity that defines the modes, and the optical beam evolves as it propagates through the medium. In a typical experiment of squeezing, for example, a cw beam propagates along a nonlinear

Quantum Measurements in Optics, Edited by P. Tombesi and
D.F. Walls, Plenum Press, New York, 1992

optical medium, changing its quantum mechanical property as a function of position in the medium. In this case, there is no time evolution but a spatial evolution as the beam is stationary. Other examples are beam attenuation in a lossy medium and amplification in a gain medium. In all of these cases, it may be difficult to use non-evolving travelling wave modes for a periodic boundary condition or standing wave modes for a Fabry-Perot cavity.

The above problem has been known and there have been many attempts to describe the propagation of the beam.[2-11] In particular, a usual way is to replace t by $-z/c$ simply in Eq. (1) or (2). One can find its justification in Shen's theory,[2] in which a localized Hamiltonian is defined in each chopped section of an optical beam. In this theory, however, it may be difficult to treat continuous (or nearly continuous) modes because the mode spacing will be large if the chopped beam volume is made small for the purpose of describing a substantial spatial evolution of the beam.

Another method is to expand the spatially evolving beam by spatial modes defined in a large cavity that includes all relevant optical devices.[3,4] In this treatment, however, an interaction involving a mode is assumed to occur all over the cavity at the same time. Accordingly, there is a difficulty in describing localized phenomena such as beam generation at a light emitter, beam absorption at a detector, and beam interaction at a beamsplitter.

In the classical treatment of an optical beam or a microwave beam, however, there is a standard method to treat a spatially evolving beam, namely, to look at the spatial change of the amplitude of an oscillating field at a frequency ω. This method usually leads to coupled mode equations,[12] which describe energy transfer along propagation direction between two (or more) frequency modes. In this treatment, the field is expanded as

$$E(z,t) = \sqrt{\frac{1}{2\pi}} \int d\omega E(\omega, z) e^{-i\omega t} + \text{c.c}, \tag{4}$$

which is in contrast to Eq. (3). It is then expected that a novel quantum mechanical theory is possible by regarding the Fourier coefficient $E(\omega, z)$ as a quantum mechanical operator. This treatment is attempted by Caves and Crouch[7] in terms of "spatial Langevin equation" in a rather phenomenological way.

The purpose of this paper is to give a quantum theory for a spatially evolving optical beam, in which the spatial evolution is directly written as

$$\frac{d}{dz}\hat{a} = \frac{1}{i\hbar}[\hat{a}, \hat{I}_z] \tag{5}$$

where $\hat{I}_z$ is a spatial evolution generator. The present theory also gives a justification for the treatment to regard the Fourier coefficient $E(\omega, z)$ as a quantum mechanical operator. $E(\omega, z)$ is factorized as $E_0(\omega)\hat{a}(\omega, z)$, where $E_0(\omega)$ is the single-photon electric field and $\hat{a}(\omega, z)$ is the annihilation operator normalized so that $\hat{a}^\dagger(\omega, z)\hat{a}(\omega, z)$ is the photon number (of mode ω passing through point z during quantization time T).

Some fundamental examples are given as beam propagation problems such as a directional coupler, parametric coupling, linear loss and linear gain. The systematic usage of the present theory will be brought into relief through these examples. Also, an expression for the noise power spectrum of the detected photocurrent is given in the context of the present theory.

The expression of $\hat{I}_z$ for the electromagnetic wave is particularly different from the Hamiltonian or the momentum operator when there is z-component of the field. Such a case occurs for wave propagation in a waveguide. We show that there are two kinds of orthogonal relationships for transverse modes according to two different cases, in one of which a wavenumber (along the propagation direction) is given and the transverse modes have different frequencies, and in the other of which a frequency is given and the transverse modes have different wavenumbers. In the former case, the orthogonality is generally expressed by the fact that mode cross-terms vanish in the Hamiltonian, i.e., the Hamiltonian is distributed to transverse modes while in the latter case, the Hamiltonian or the momentum operator has mode cross-terms but $\hat{I}_z$ does not. This shows that $\hat{I}_z$ is naturally connected to the mode orthogonality when frequency ω is fixed while $\hat{H}$ is naturally connected to the mode orthogonality when k is fixed.

2. QUANTIZATION PROCEDURE

2.1. *General theory*

It is difficult to derive the present quantum theory characterized by Eq. (5) from the usual quantum mechanics characterized by Eq. (1) or vice versa because both the present theory and the usual quantum theory have no time-space symmetry. The reason that time t is treated specially as a parameter in the usual quantum mechanics is that t is already treated specially as a parameter in the classical Hamiltonian formulation. Similarly, coordinate z should be treated specially in a classical principle which leads to a quantum theory of spatial evolution. Therefore, if we require that the present theory and the usual theory have the same roots, the roots should be a principle which treats time and space equally. One of such a principle is the general minimum-action principle:

$$\delta \iiiint \mathcal{L} dt dx dy dz = 0, \tag{6}$$

where $\mathcal{L}$ is the Lagrangian density of the field, which is a function of the field amplitude, Φ, and its temporal and spatial derivatives. We start from this principle.

There is complete parallelism between the quantization procedure described here and the usual quantization procedure, as is summarized in Fig. 1. In this paper, we describe the quantization procedure for spatial evolution, i.e., we describe the theory in the right half side of Fig. 1. It is of course useful to compare it with the left half side of the figure to see the parallelism.

We define a *spatial* Lagrange function, $L'(z)$, as $L'(z) \equiv \iint_A \int_0^T \mathcal{L} dt dx dy$. Here, A is the cross-sectional area in x-y plane, and T is an arbitrary time duration introduced for quantization. The volume AT in t-x-y space corresponds to the quantization volume V in the usual theory. The general minimum-action principle, Eq. (6), is then reduced to a minimum-action principle with respect to z as

$$\delta \int_{z_0}^{z_1} L'(z) dz = 0. \tag{7}$$

The Euler equation derived from this variance problem becomes

$$\frac{\partial}{\partial z}\left[\frac{\partial L'}{\partial(\partial\Phi/\partial z)}\right] = \frac{\partial L'}{\partial \Phi}. \tag{8}$$

Variable Π, which is canonically conjugate to Φ, is defined as $\Pi \equiv \partial\mathcal{L}/\partial(\frac{\partial\Phi}{\partial z})$. We define a Legendre transform $\mathcal{I}_z$ of $\mathcal{L}$ and its integral I_z as

$$\delta \iiiint \mathcal{L} dt dx dy dz = 0$$

($\mathcal{L}$: Lagrangian density)
Minimum action principle

Present theory

$$\delta \int_{z_0}^{z_1} L(z) dz = 0$$

where $L(z) \equiv \iint_A \int_0^T \mathcal{L} dt dx dy$

$$\Downarrow$$

$$\frac{\partial}{\partial z}\left[\frac{\partial L}{(\partial \Phi/\partial z)}\right] = \frac{\partial L}{\partial \Phi}$$

$\Downarrow$ (Legendre transform)

$$\frac{dA}{dz} = \{A, I_z\}$$

where $I_z \equiv \iint_A \int_0^T \mathcal{I}_z dt dx dy,$

$\mathcal{I}_z \equiv \Pi \cdot \dfrac{\partial \Phi}{\partial z} - \mathcal{L},$ and

$\Pi \equiv \dfrac{\partial \mathcal{L}}{\partial(\partial \Phi/\partial z)}$ (conjugate observable)

$$\Rightarrow \{\Phi(t,x,y,z), \Pi(t',x',y',z)\} = \delta(t'-t)\delta(x'-x)\delta(y'-y)$$

Quantization :

$$\Rightarrow [\Phi(t,x,y,z), \Pi(t',x',y',z)] = i\hbar\delta(t'-t)\delta(x'-x)\delta(y'-y)$$

$$\frac{d\hat{A}}{dz} = \frac{1}{i\hbar}[\hat{A}, \hat{I}_z]$$

Usual theory

$$\delta \int_{t_0}^{t_1} L(t) dt = 0$$

where $L(t) \equiv \iiint_V \mathcal{L} dx dy dz$ (Lagrangian)

$$\Downarrow$$

$$\frac{\partial}{\partial t}\left[\frac{\partial L}{(\partial \Phi/\partial t)}\right] = \frac{\partial L}{\partial \Phi} \quad \text{(Lagrange equation)}$$

$\Downarrow$ (Legendre transform)

$$\frac{dA}{dt} = \{A, H\}$$

where $H \equiv \iiint_V \mathcal{H} dx dy dz$ (Hamiltonian),

$\mathcal{H} \equiv \Pi \cdot \dfrac{\partial \Phi}{\partial t} - \mathcal{L}$ (Hamiltonian density), and

$\Pi \equiv \dfrac{\partial \mathcal{L}}{\partial(\partial \Phi/\partial t)}$ (conjugate observable)

$$\Rightarrow \{\Phi(t,x,y,z), \Pi(t,x',y',z')\} = \delta(x'-x)\delta(y'-y)\delta(z'-z)$$

Quantization :

$$\Rightarrow [\Phi(t,x,y,z), \Pi(t,x',y',z')] = i\hbar\delta(x'-x)\delta(y'-y)\delta(z'-z)$$

$$\frac{d\hat{A}}{dt} = \frac{1}{i\hbar}[\hat{A}, \hat{H}] \quad \text{(Heisenberg)}$$

Figure 1. Comparison between the present and usual theories

$$\mathcal{I}_z \equiv \Pi \cdot \frac{\partial \Phi}{\partial z} - \mathcal{L} \tag{9}$$

and

$$I_z \equiv \int\int_A \int_0^T \mathcal{I}_z dt dx dy. \tag{10}$$

Then, similar to the usual analytical mechanics, the Lagrange's equation of propagation, Eq. (8), is rewritten as

$$\frac{da}{dz} = \{a, I_z\}, \tag{11}$$

where a is an amplitude of a time-domain mode, and $\{a, I_z\}$ is the Poisson bracket defined as

$$\{a, I_z\} \equiv \sum_j \frac{\partial a}{\partial q_j}\frac{\partial I_z}{\partial p_j} - \frac{\partial a}{\partial p_j}\frac{\partial I_z}{\partial q_j}. \tag{12}$$

Here, q_j and p_j respectively are the amplitudes of the normal mode of the field and its conjugate variable. The classical commutation relationship at a point z is written as

$$\Big\{\Phi(t, x, y, z), \Pi(t', x', y', z)\Big\} = \delta(t' - t)\delta(x' - x)\delta(y' - y). \tag{13}$$

We will refer to I_z as the *spatial evolution generator* as it governs the spatial evolution of the system. In this sense, I_z is similar to the Hamiltonian, which governs the temporal evolution in the usual theory. The spatial evolution generator has the dimension of momentum, but it should be noted that I_z is not equal to the z-component of the usual momentum operator, as will be clarified later.

So far the theory is classical. Now, we set a quantization postulate that the classical Poisson bracket is replaced by the commutator bracket as

$$\Big[\hat{\Phi}(t, x, y, z), \hat{\Pi}(t', x', y', z)\Big] = \text{Constant} \cdot \delta(t' - t)\delta(x' - x)\delta(y' - y). \tag{14}$$

The constant in this equation has a dimension of action, and is set to be equal to $i\hbar$. Classical equation of propagation (11) is then replaced by a quantum equation of *propagation* as

$$\frac{d\hat{a}}{dz} = \frac{1}{i\hbar}[\hat{a}, \hat{I}_z]. \tag{15}$$

2.2. *Quantization of an electromagnetic beam*

In this section, we apply the general theory described in the previous section to the electromagnetic field. Lagrangian density $\mathcal{L}$ of the electromagnetic field is written as[13,14]

$$\mathcal{L} = -\frac{1}{2\mu_0}\sum_{n,m} g^{mm} g^{nn} \left(\frac{\partial \Phi_m}{\partial x^n}\right)^2, \quad (n, m = 0, 1, 2, 3) \tag{16}$$

where g^{nn} is Minkowski's tensor, Φ_m is defined by 4-dimensional vector potential as $(\Phi_0, \Phi_1, \Phi_2, \Phi_3) \equiv (\frac{\phi}{c}, A_x, A_y, A_z)$, and x^n is defined as $(x^0, x^1, x^2, x^3) \equiv (-ct, x, y, z)$. Conjugate variables then become

$$\begin{aligned}(\Pi_0, \Pi_1, \Pi_2, \Pi_3) &= \left(\frac{\partial \mathcal{L}}{(\partial \Phi_0/\partial z)}, \frac{\partial \mathcal{L}}{(\partial \Phi_1/\partial z)}, \frac{\partial \mathcal{L}}{(\partial \Phi_2/\partial z)}, \frac{\partial \mathcal{L}}{(\partial \Phi_3/\partial z)}\right) \\ &= \left(\varepsilon_0 \frac{\partial \phi}{\partial z}, -\frac{1}{\mu_0}\frac{\partial A_x}{\partial z}, -\frac{1}{\mu_0}\frac{\partial A_y}{\partial z}, -\frac{1}{\mu_0}\frac{\partial A_z}{\partial z}\right).\end{aligned} \tag{17}$$

It is easily shown, using Eqs. (9), (16), and (17), that energy density $\mathcal{I}_z$ is equal to the minus of the (z,z) component of Maxwell's energy-momentum tensor, $-T_{zz}$. Thus we obtain

$$\hat{I}_z = \iint_A \int_T (-T_{zz}) dt dx dy. \tag{18}$$

It should be noted here that the usual Hamiltonian density is equal to the (t,t) component of Maxwell's tensor, that is, Hamiltonian density u. The usual Hamiltonian is equal to $\iint\int_V u dx dy dz$. It should also be noted that I_z differs from the z component of the usual momentum, P_z, which is defined as $\iint\int_V (S_z) dx dy dz / c$. Here, vector S_j $(j = x, y, z)$ is the Poynting vector.

Using the field components, the spatial evolution generator for a linear medium is expressed as

$$I_z = \iint_A \int_0^T \left[E_z D_z + H_z B_z - \frac{1}{2}(\boldsymbol{E} \cdot \boldsymbol{D} + \boldsymbol{H} \cdot \boldsymbol{B}) \right] dt dx dy. \tag{19}$$

This is compared to the usual expression for the linear medium Hamiltonian

$$H = \iint_A \int_0^L \frac{1}{2} (\boldsymbol{E} \cdot \boldsymbol{D} + \boldsymbol{H} \cdot \boldsymbol{B}) dx dy dz. \tag{20}$$

Comparing Eq. (19) and Eq. (20), it is known that I_z is equal to $-H/c$ if (A) $E_z = H_z = 0$, and if (B) I_z and H do not depend on t and z so that $\int_0^T dt$ can be replaced by $\int_0^L$ with $L = cT$. Conditions (A) and (B) are satisfied if a plane wave having no z-component of the field is assumed and if an optical medium is spatially homogeneous and temporally constant. This explains why a conventional method simply replacing $t \rightarrow -\frac{z}{c}$ is valid in many cases. In general, however, it is needless to say that I_z should be used in the spatial evolution picture and H should be used in time evolution picture. Later, we will see a difference between I_z and H for waveguide modes which have nonzero E_z or H_z component.

3. APPLICATIONS

3.1. *Directional coupler*

A directional coupler has two waveguides which are coupled to each other as is shown in Fig. 2.

Considering a y-polarized wave (TE wave) for simplicity, the electric field contains only y components. The electric field is a summation of field $\hat{E}_{\mathrm{A}}$ for waveguide A and $\hat{E}_{\mathrm{B}}$ for waveguide B as

$$\hat{E}_y = \hat{E}_{\mathrm{A}} + \hat{E}_{\mathrm{B}} \tag{21}$$

Using slowly varying annihilation operator $A(z)$ and transverse mode function $\psi_{\mathrm{A}}(x,y)$ for waveguide A and operator $B(z)$ and mode function $\psi_{\mathrm{B}}(x,y)$ for waveguide B, each electric field is expressed as

$$\hat{E}_{\mathrm{A}} = \sqrt{\frac{\hbar k}{2\varepsilon_{\mathrm{A}} T}}\, e^{-i(\omega t - kz)} [\hat{A}(z) \psi_{\mathrm{A}}(x,y) + \mathrm{H.c.}] \tag{22}$$

and

$$\hat{E}_{\mathrm{B}} = \sqrt{\frac{\hbar k}{2\varepsilon_{\mathrm{B}} T}}\, e^{-i(\omega t - kz)} [\hat{A}(z) \psi_{\mathrm{B}}(x,y) + \mathrm{H.c.}] \tag{23}$$

Here, wavenumber k, which is determined by ω via the dispersion relationship, is assumed to

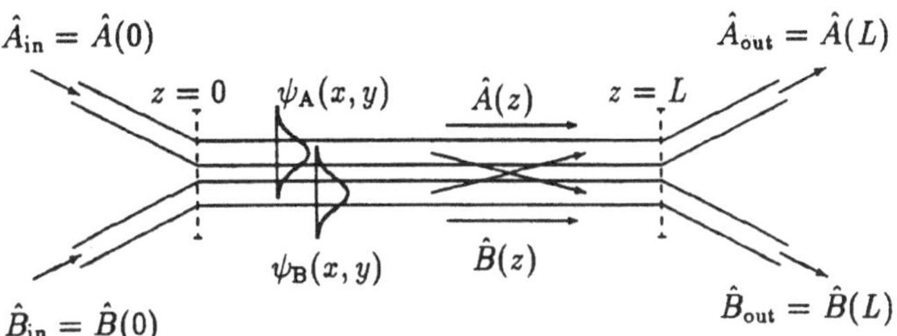

Figure 2. Directional coupler

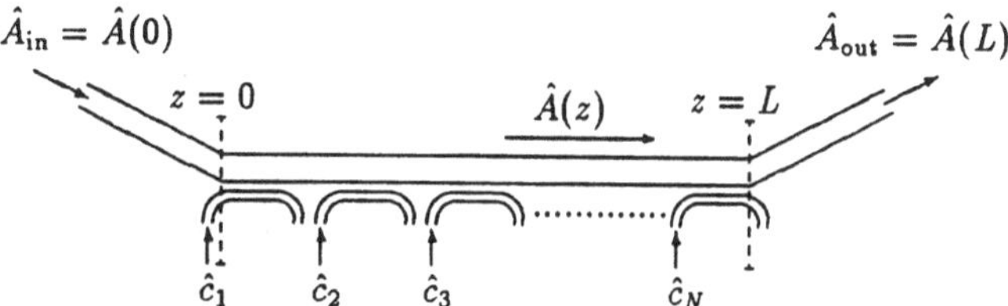

Figure 3. Distributed linear loss. A distributed loss is modeled by successive tiny directional couplers. The spacing Δz of couplers will be made infinitesimally small and the number N of the couplers will be made infinite with $N\Delta z = L = \text{const}$.

be equal for the two waveguides. The dielectric constant ε_A is a function of x and y, which expresses the index profile of waveguide A, and ε_B is for waveguide B.

The perturbation polarization P_{pert} is expressed as $E_A \Delta\varepsilon_B(x,y) + E_B \Delta\varepsilon_A(x,y)$, where $\Delta\varepsilon_A \equiv \varepsilon_{\text{total}} - \varepsilon_A$ is the dielectric constant perturbation for waveguide B, and $\Delta\varepsilon_B \equiv \varepsilon_{\text{total}} - \varepsilon_B$ is that for waveguide A. Thus we obtain

$$P_{\text{pert}} = e^{-i(\omega t - kz)} \left[\sqrt{\frac{\hbar k}{2\varepsilon_A T}}\, \hat{A}(z)\psi_A(x,y)\Delta\varepsilon_B(x,y) + \sqrt{\frac{\hbar k}{2\varepsilon_B T}}\, \hat{B}(z)\psi_B(x,y)\Delta\varepsilon_A(x,y) + \text{H.c.} \right] \tag{24}$$

The interaction part of the spatial evolution generator is obtained as

$$\begin{aligned} \hat{I}_{\text{int}} &= \iint_A \int_0^T \left(E_z P_z - \frac{1}{2} \boldsymbol{E} \cdot \boldsymbol{P} \right) dt\,dx\,dy \\ &= \hbar\kappa \hat{A}^\dagger \hat{B} + \hbar\kappa^* \hat{A} \hat{B}^\dagger, \end{aligned} \tag{25}$$

where κ is the coupling constant defined as

$$\kappa \equiv k \iint \left[\psi_A^*(x,y)\Delta\varepsilon_B(x,y)\psi_B(x,y) + \psi_B^*(x,y)\Delta\varepsilon_A(x,y)\psi_A(x,y) \right] dx\,dy. \tag{26}$$

Here, $\varepsilon_{A,B}/\varepsilon_{\text{total}} \simeq 1$ is used, and small terms $k \iint \psi_A^* \Delta\varepsilon_B \psi_A\, dx\,dy$ and $k \iint \psi_B^* \Delta\varepsilon_A \psi_B\, dx\,dy$ are neglected. The equation of propagation then yields a pair of coupled mode equations

$$\frac{d}{dz}\hat{A} = -i\kappa \hat{B} \quad \text{and} \quad \frac{d}{dz}\hat{B} = -i\kappa^* \hat{A}, \tag{27}$$

The solution is obtained as

$$\begin{cases} \hat{A}(z) = \hat{A}(0)\cos(|\kappa| z) - \frac{i\kappa}{|\kappa|}\hat{B}(0)\sin(|\kappa| z), \\ \hat{B}(z) = \hat{B}(0)\cos(|\kappa| z) - \frac{i\kappa^*}{|\kappa|}\hat{A}(0)\cos(|\kappa| z). \end{cases} \tag{28}$$

This leads to the input-output relationship of a beamsplitter

$$\hat{A}_{\text{out}} = \sqrt{T}\, \hat{A}_{\text{in}} + e^{i\theta}\sqrt{R}\, \hat{B}_{\text{in}} \tag{29}$$

$$\hat{B}_{\text{out}} = \sqrt{T}\, \hat{B}_{\text{in}} - e^{-i\theta}\sqrt{R}\, \hat{A}_{\text{in}}, \tag{30}$$

where $T \equiv \cos^2(|\kappa| L)$, $R \equiv \sin^2(|\kappa| L)$, and $\theta \equiv \arg(\kappa) - \pi/2$.

The above treatment is similar to classical treatment of coupled mode equations in a sense that it describes the spatial evolution of the amplitude of mode $e^{-i\omega t}$. There are, however, also some different points. The coupled mode equations are derived directly from the Maxwell's equation in the classical theory, while, in the present treatment, they are derived from the equation of propagation with the spatial evolution generator. Also, when there is no substantial input for waveguide B, the amplitude B is just zero in the classical theory, while, in the quantum treatment, $\hat{B}$ should not be assumed to be zero but is treated as a Hilbert space operator which is in a vacuum state. This vacuum state is referred to as "vacuum noise" for mode A which is induced by branching beam A. Using vacuum noise operator $\hat{F} \equiv -\frac{i\kappa}{|\kappa|}\hat{B}_{\text{in}}$, Eq. (29) is written as

$$\hat{A}_{\text{out}} = \sqrt{T}\, \hat{A}_{\text{in}} + \sqrt{1-T}\, \hat{F} \tag{31}$$

If $\hat{F}$ is a radiation from a heat bath, it is regarded as being in a thermal state with a finite temperature.

3.2. *Distributed linear loss*

Distributed linear loss is modeled by successive couplings with distributed directional couplers which are weakly coupled. As is shown in Fig. 3, N directional couplers are supposed to be coupled to the main waveguide with an equal interval δz in the coupling region of length $L(=N\delta z)$. The vacuum noise operator for each coupler is denoted by $\hat{f}_j(1 \leq j \leq N)$. We assume for simplicity that all couplers have the same coupling constant κ, the same coupling length δz, and thus the same loss loss value $\delta\eta[=\sin^2(|\kappa|\delta z)]$. The input-output relation for jth coupler leads to a difference equation of $\hat{A}$ at z as

$$\hat{A}(z+\delta z) = \sqrt{1-\delta\eta}\,\hat{A}(z) + \sqrt{\delta\eta}\,\hat{c}_j \simeq \left(1-\frac{\delta\eta}{2}\right)\hat{A}(z) + \sqrt{\delta\eta}\,\hat{c}_j \tag{32}$$

Setting $\delta z \to 0$, we obtain a differential equation

$$\frac{d}{dz}\hat{A}(z) = -\frac{\alpha}{2}\hat{A}(z) + \sqrt{\alpha}\hat{f}(z) \tag{33}$$

where $\alpha \equiv \frac{\delta\eta}{\delta z}$ is loss per unit length, and $\hat{f}(z)$ is defined as

$$\hat{f}(z) \equiv \lim_{\delta z\to 0}\frac{\hat{c}_j}{\sqrt{\delta\eta}} = \frac{1}{|\kappa|}\lim_{\delta z\to 0}\frac{\hat{c}_j}{\sqrt{\delta z}}. \tag{34}$$

It should be noted that $\hat{f}(z)$ expresses noise sources distributed at all z's. It is easy to show a commutation relation

$$[\hat{f}(z), \hat{f}^\dagger(z')] = \delta(z'-z). \tag{35}$$

This means that noise sources at all z's are independent.

The solution of Eq. (33) is obtained as

$$\hat{A}(z) = \sqrt{T}\,\hat{A}(0) + \sqrt{1-T}\,\hat{F} \tag{36}$$

where T is the total transmission defined as

$$T \equiv e^{-\alpha z} \tag{37}$$

$$\hat{F} \equiv \sqrt{\frac{\alpha}{1-e^{-\alpha z}}}\int_0^z \exp\left[-\frac{\alpha(z-z')}{2}\right]\hat{f}(z')dz' \tag{38}$$

$\hat{F}$ is regarded as a vacuum noise as it satisfies the commutation relation $[\hat{F}, \hat{F}^\dagger] = 1$. This means that distributed linear loss is equivallent to a concentrated linear loss with a total loss value of $1-T$. Above calculation procedure, however, will be used for non-trivial cases such as nonlinear effects with distributed loss.

3.3. *Parametric amplification*

We consider the parametric amplification of a propagating beam within pump-nondepletion approximation. As is shown in Fig. 4, the signal beam of frequency ω_s and the idler beam of frequency ω_i couple with each other via the second-order nonlinear effect pumped by the pump beam of frequency $\omega_p = \omega_s + \omega_i$. Thus the relevant modes are signal (s), idler (i) and the pump (p). Considering y-polarized TE plane wave,[15] the electric field is expressed as

$$\hat{E}_y = \sqrt{\frac{\hbar k_s}{2\varepsilon TA}}\, e^{-i(\omega_s t - k_s z)}\hat{A} + \sqrt{\frac{\hbar k_i}{2\varepsilon TA}}\, e^{-i(\omega_i t - k_i z)}\hat{B} + \sqrt{\frac{\hbar k_p}{2\varepsilon TA}}\, e^{-i(\omega_p t - k_p z)}\alpha_p + \text{H.c.} \quad (39)$$

Here the effective cross-sectional area, denoted by A, is assumed to be the same for all the three modes. $\hat{A}$ and $\hat{B}$ respectively are the slowly varing annihilation operators for the signal and idler modes. The pump amplitude, denoted as α_p, is regarded as a c-number amplitude and is not depleted.

Since the nonlinear perturbation polarization, P_{pert} is expressed as $\chi^{(2)}\hat{E}^2$, the interaction part of the spatial evolution generator is expressed as $\int\int_A \int_0^T \chi^{(2)}\hat{E}_y^3 dt dx dy$. If Eq. (39) is substituted into this expression, $6^3(=216)$ terms will appear. Most terms, however, vanish and only terms of the form $e^{\pm i(\omega_p - \omega_s - \omega_i)t}$ remain through the integration $\int_0^T dt$. The spatial evolution generator then becomes

$$\hat{I}_{int} = \hbar\kappa\hat{A}^\dagger\hat{B}^\dagger\, e^{i\Delta k z} + \hbar\kappa^*\hat{A}\hat{B}\, e^{-i\Delta k z}, \quad (40)$$

where $\Delta k \equiv k_p - k_s - k_i$ is the phase mismatch, and the coupling constant κ is defined as

$$\kappa \equiv \int\int \left(\frac{\chi^{(2)}}{2\varepsilon}\right)\sqrt{\frac{k_p k_s k_i}{2\varepsilon AT}}\,\alpha_p dx dy = \left(\frac{\chi^{(2)}}{2\varepsilon}\right)\sqrt{\frac{k_p k_s k_i}{2\varepsilon T}}\,\alpha_p. \quad (41)$$

Assuming $\Delta k = 0$ for simplicity, coupled mode equations are obtained as

$$\frac{d}{dz}\hat{A} = -i\kappa\hat{B}^\dagger \quad \text{and} \quad \frac{d}{dz}\hat{B}^\dagger = i\kappa^*\hat{A}. \quad (42)$$

The solution is written as

$$\begin{cases} \hat{A}(z) = \hat{A}(0)\cosh(|\kappa|z) - \frac{i\kappa}{|\kappa|}\hat{B}^\dagger(0)\sinh(|\kappa|z) \\ \hat{B}^\dagger(z) = \hat{B}^\dagger(0)\cosh(|\kappa|z) + \frac{i\kappa^*}{|\kappa|}\hat{A}(0)\sinh(|\kappa|z). \end{cases} \quad (43)$$

Introducing parameters $\xi \equiv \cosh(|\kappa|z)$ and $\nu \equiv -\frac{i\kappa}{|\kappa|}\sinh(|\kappa|z)$, Eq. (43) is written as

$$\hat{A}_{out} = \xi\hat{A}_{in} + \nu\hat{B}_{in}^\dagger \quad (44)$$

and

$$\hat{B}_{out}^\dagger = \xi\hat{B}_{in}^\dagger + \nu^*\hat{A}_{in}, \quad (45)$$

with $|\xi|^2 - |\nu|^2 = 1$. This means that the output beam can be squeezed when the input beams are in coherent states. This input-output relation conserves the commutation relation, but does not conserves the energy (total photon number) because the signal and idler beams are amplified (or deamplified) gaining energy from the pump beam.

When mode B is in a vacuum state, Eq. (45) describes a linear gain with spontaneous emission noise. Using noise operator $\hat{F}^\dagger \equiv \frac{\kappa^*}{|\kappa|}\hat{B}_{in}^\dagger$ and gain $G \equiv |\xi|^2$, Eq. (45) is written as

$$\hat{A}_{out} = \sqrt{G}\,\hat{A}_{in} + \sqrt{G-1}\,\hat{F}^\dagger. \quad (46)$$

If $\hat{F}^\dagger$ is associated with a gain medium having a certain temperature, it is regarded as being in a thermal state.

3.4. *Distributed linear gain*

Distributed linear loss is modeled by a number of independent parametric amplification

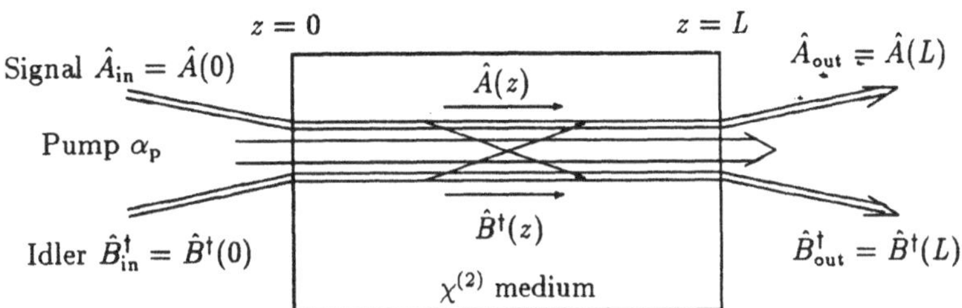

Figure 4. Parametric coupling between the signal and idler beams.

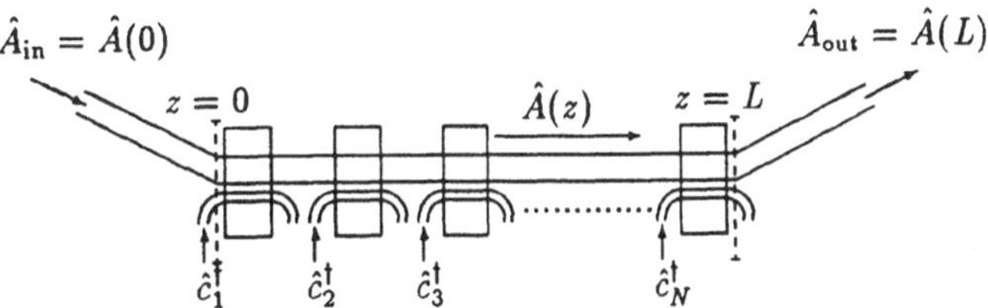

Figure 5. Distributed linear gain. A distributed gain is modeled by successive parametric amplification processes. The spacing Δz of the nonlinear crystals will be made infinitesimally small and the number N of the couplers will be made infinite with $N\Delta z = L = \text{const}$.

devices which are distributed over a certain coupling region. As is shown in Fig. 5, N parametric amplification devices are supposed to be lined up with an equal interval δz in the coupling region of length $L(= N\delta z)$. The spontaneous emission noise operator for each is denoted by $\hat{f}_j^\dagger (1 \leq j \leq N)$. We assume for simplicity that all parametric devices have the same coupling constant κ, the same coupling length δz, and thus the same gain value $G'[= \cosh^2(|\kappa|\delta z)]$. The input-output relation for the jth device leads to a difference equation of $\hat{A}$ at z as

$$\hat{A}(z+\delta z) = \sqrt{G'}\,\hat{A}(z) + \sqrt{G'-1}\,\hat{f}_j \simeq \left(1 + \frac{G'-1}{2}\right)\hat{A}(z) + \sqrt{G'-1}\,\hat{f}_j^\dagger \tag{47}$$

Setting $\delta z \to 0$, we obtain a differential equation

$$\frac{d}{dz}\hat{A}(z) = \frac{g}{2}\hat{A}(z) + \sqrt{g}\hat{f}^\dagger(z) \tag{48}$$

where $g \equiv \frac{G'-1}{\delta z}$ is gain exponent per unit length, and $\hat{f}^\dagger(z)$ is defined as

$$\hat{f}^\dagger(z) \equiv \lim_{\delta z \to 0} \frac{\hat{f}_j^\dagger}{\sqrt{G'-1}} = \frac{1}{|\kappa|}\lim_{\delta z \to 0} \frac{\hat{f}_j^\dagger}{\sqrt{\delta z}}. \tag{49}$$

The commutation relation, Eq. (35), also holds for $\hat{f}$.

The solution of Eq. 48 is obtained as

$$\hat{A}(z) = \sqrt{G}\,\hat{A}(0) + \sqrt{G-1}\,\hat{F}^\dagger \tag{50}$$

where G is the total gain defined as

$$G \equiv e^{gz} \tag{51}$$

$$\hat{F}^\dagger \equiv \sqrt{\frac{g}{e^{gz}-1}}\int_0^z \exp\left[\frac{g(z-z')}{2}\right]\hat{f}^\dagger(z')dz' \tag{52}$$

$\hat{F}$ is regarded as the total spontaneous emission noise operator, which satisfies the commutation relation $[\hat{F}, \hat{F}^\dagger] = 1$. This means that distributed linear gain is equivallent to a concentrated linear gain with a total gain value of G.

4. DIFFERENCE BETWEEN $\hat{H}$ AND $\hat{I}_z$ IN A WAVEGUIDE

As was described in section 2.2, the Hamiltonian and the spatial evolution generator $\hat{I}_z$ are different when there is a z-component of the field. Such an example can be found in guided-wave optics.

Let us consider first TE mode orthogonality in a dielectric slab waveguide. As is shown in Fig. 6, the coordinates are chosen so that x denotes the direction of thickness of the slab waveguide. Taking z as the direction of propagation, no y dependence appears in any physical quantity. Assuming $E_y(t,x,z) = \tilde{E}_y(x)e^{-i(\omega t - kz)}$, etc., the Maxwell's equation leads to

$$\frac{\partial^2 \tilde{E}_y}{\partial x^2} + \left[\omega^2\mu_0\varepsilon(x) - k^2\right]\tilde{E}_y = 0, \tag{53}$$

$$\tilde{H}_x = \frac{k}{\omega\mu_0}\tilde{E}_y, \tag{54}$$

and

$$\tilde{H}_z = \frac{i}{\omega\mu_0}\frac{\partial \tilde{E}_y}{\partial x}. \tag{55}$$

E_x, E_z and H_y are zero for a TE wave. Eq. (53) is a Sturm-Liouville eigenvalue equation, whose solution leads to a dispersion relationship between ω and k. Fig. 7 shows a schematic picture of the ω-k dispersion relationship. In the figure, different curves correspond to different transverse modes. The curves approach the line $\omega = (c/n_{\rm core})k$ as the frequency ω increases because the field is more concentrated in the core, and the curves approach the line $\omega = (c/n_{\rm clad})k$ as the frequency ω decreases because the field spreads over the cladding of the waveguide.

As is shown in Fig. 7, the transverse modes have different ω's when k is fixed (Case 1), while they have different k's when ω is fixed (Case 2). A typical situation for Case 1 is a waveguide laser with Fabry-Perot or a ring resonator, in which the wavenumber along z is fixed by the boundary condition and the gain line shape. The laser may oscillate in several transverse modes having slightly different frequencies. A typical situation for Case 2 is just a laser beam launching into a multimode waveguide. The laser beam with a fixed frequency splits into several transverse modes, which propagate with different propagation constants (wavenumbers). In the following, we consider the orthogonality relationships for Case 1 and 2.

4.1. *Waveguide modes for a fixed value of k*

The wave equation (53) becomes

$$\frac{\partial^2 \tilde{E}_m^y}{\partial x^2} + \left[\omega_m^2 \mu_0 \varepsilon(x) - k^2\right] \tilde{E}_m^y = 0, \tag{56}$$

$$\tilde{H}_m^x = \frac{k}{\omega_m \mu_0} \tilde{E}_m^y, \tag{57}$$

and

$$\tilde{H}_z = \frac{i}{\omega_m \mu_0}\frac{\partial \tilde{E}_m^y}{\partial x}, \tag{58}$$

where m is the transverse-mode suffix and the component suffixes x, y and z are put into superscripts. Eq. (56) is a Sturm-Liouville problem with x-dependent function $\varepsilon(x)$ multiplied by the eigenvalue ω_m. In this case, the orthogonality is written, from the general theory of the Sturm-Liouville problem, as

$$\int \varepsilon \left(\tilde{E}_m^y\right)^* \tilde{E}_n^y dx = 0 \quad \text{for} \quad n \neq m. \tag{59}$$

Using Eqs. (56)–(59), it is easily shown that the orthogonality is also written as

$$\int \left[\varepsilon \left(\tilde{E}_m^y\right)^* \tilde{E}_n^y + \mu_0 \left(\tilde{H}_m^x\right)^* \tilde{H}_n^x + \mu_0 \left(\tilde{H}_m^z\right)^* \tilde{H}_n^z\right] dx = 0. \tag{60}$$

This leads to the orthogonality expression in terms of the Hamiltonian as follows. The Hamiltonian $\hat{H}$ of the field is calculated as

$$\begin{aligned}
\hat{H} &= L\sum_n \int \left[\varepsilon|\tilde{E}_n^y|^2 + \mu_0|\tilde{H}_n^x|^2 + \mu_0|\tilde{H}_n^z|^2\right] dx \\
&\quad + L\sum_{n\neq m} e^{i(\omega_m-\omega_n)t} \int \left[\varepsilon \left(\tilde{E}_m^y\right)^* \tilde{E}_n^y + \mu_0 \left(\tilde{H}_m^z\right)^* \tilde{H}_n^x + \mu_0 \left(\tilde{H}_m^z\right)^* \tilde{H}_n^z\right] dx \\
&= \sum_n \hbar\omega_n \left(\hat{a}_n^\dagger \hat{a}_n + \frac{1}{2}\right),
\end{aligned} \tag{61}$$

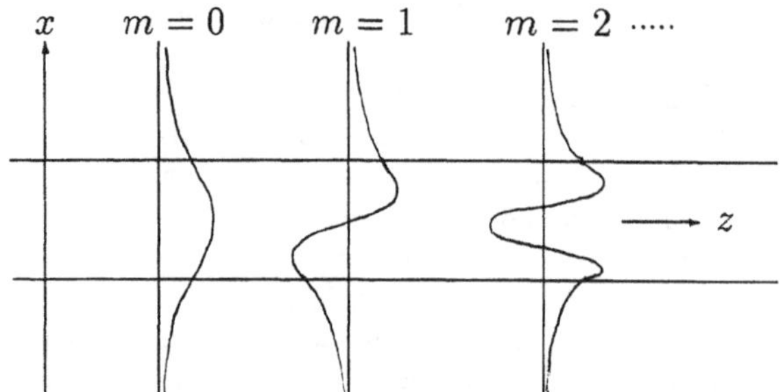

Figure 6. Slab waveguide and the transverse modes

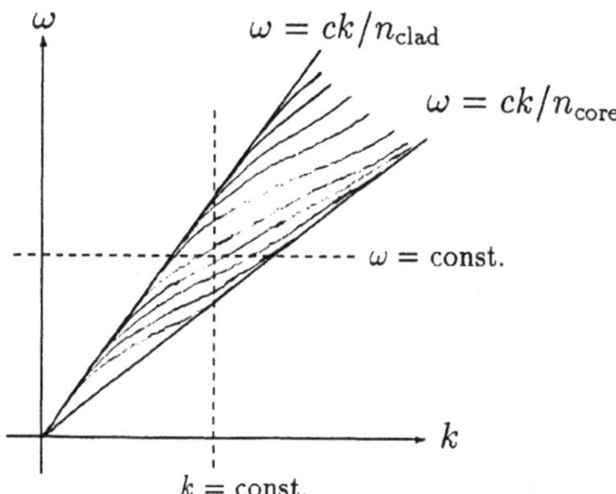

Figure 7. Schematic graph of ω-k dispersion relationship for an optical waveguide.

where L is the cavity length along the z-axis. The cross terms of this equation vanish due to Eq. (60). This means that the Hamiltonian is diagonal in transverse modes.

For TM modes, in which H_y, E_x and E_z are the only nonzero components, the orthogonality relation becomes

$$\int \varepsilon^2 \left(\tilde{E}_m^x\right)^* \tilde{E}_n^x dx = 0, \tag{62}$$

instead of Eq. (59). If, however, we calculate the Hamiltonian, we will see that the cross-terms vanish in the Hamiltonian and thus the Hamiltonian is diagonal in transverse modes. In this sense, the apparent difference between TE orthogonality equation (59) and TM orthogonality equation (62) is superficial and the orthogonality appears in the Hamiltonian in a more general way. This is the case of different transverse modes for a fixed wavenumber k.

4.2. *Waveguide modes for a fixed value of ω*

The wave equation (53) becomes

$$\frac{\partial^2 \tilde{E}_m^y}{\partial x^2} + \left[\omega^2 \mu_0 \varepsilon(x) - k_m^2\right] \tilde{E}_m^y = 0, \tag{63}$$

$$\tilde{H}_m^x = \frac{k_m}{\omega \mu_0} \tilde{E}_m^y, \tag{64}$$

and

$$\tilde{H}_z = \frac{i}{\omega \mu_0} \frac{\partial \tilde{E}_m^y}{\partial x}, \tag{65}$$

Now, the transverse-mode suffix m is put to k. This time, the eigenvalue k_m is separated from the x-dependent function $\varepsilon(x)$. In this case, the orthogonality is written, from the general theory of the Sturm-Liouville problem, as

$$\int \left(\tilde{E}_m^y\right)^* \tilde{E}_n^y dx = 0 \quad \text{for} \quad n \neq m. \tag{66}$$

Using Eqs. (63)–(66), it is easily shown that the orthogonality is also written as

$$\int \left[-\varepsilon \left(\tilde{E}_m^y\right)^* \tilde{E}_n^y - \mu_0 \left(\tilde{H}_m^x\right)^* \tilde{H}_n^x + \mu_0 \left(\tilde{H}_m^z\right)^* \tilde{H}_n^z\right] dx = 0. \tag{67}$$

This leads to the orthogonality expression in terms of the spatial evolution generator $\hat{I}_z$ as follows. The spatial evolution generator $\hat{I}_z$ of the field is calculated as

$$\begin{aligned}
\hat{I}_z &= T \sum_n \int \left[-\varepsilon |\tilde{E}_n^y|^2 - \mu_0 |\tilde{H}_n^x|^2 + \mu_0 |\tilde{H}_n^z|^2\right] dx \\
&\quad + T \sum_{n \neq m} e^{-i(k_m - k_n)z} \int \left[-\varepsilon \left(\tilde{E}_m^y\right)^* \tilde{E}_n^y - \mu_0 \left(\tilde{H}_m^x\right)^* \tilde{H}_n^x + \mu_0 \left(\tilde{H}_m^z\right)^* \tilde{H}_n^z\right] dx \\
&= \sum_n \hbar k_n \left(\hat{a}_n^\dagger \hat{a}_n + \frac{1}{2}\right),
\end{aligned} \tag{68}$$

where T is the quantization time. The cross terms of this equation vanish due to Eq. (67). This means that the spatial evolution generator is diagonal in transverse modes.

For TM modes, in which H_y, E_x and E_z are the only nonzero components, the orthogonality relation becomes

$$\int \varepsilon \left(\tilde{E}_m^x\right)^* \tilde{E}_n^x dx = 0, \tag{69}$$

instead of Eq. (66). If, however, we calculate the spatial evolution generator, we will see that the cross-terms vanish in the spatial evolution generator and it is diagonal in transverse modes. In this sense, the apparent difference between TE orthogonality equation (66) and TM orthogonality equation (69) is superficial and the orthogonality appears in the spatial evolution generator in a more general way. This is the case of different transverse modes for a fixed frequency ω.

5. CONCLUSION

A quantum theory is given for treating spatial evolution of a propagating optical beam. The role of time and space are interchanged in compararison with the usual theory. A spatial evolution generator, which plays a similar role to the usual Hamiltonian is obtained in general and for the electromagnetic field. Several beam propagation problems such as directional coupler, distributed loss, parametric amplification, and distributed gain are solved using the present method. The spatial evolution generator is different from the Hamiltonian particularly when there is z-component of the field. Slab waveguide modes, which usually have nonzero E_z or H_z are investigated to see the difference. There are two different orthogonality relationships according to whether the wavenumber is fixed or the frequency is fixed. For the former case, mode cross terms vanish in the Hamiltonian, while for the latter case, mode cross terms vanish in the spatial evolution generator. This fact also shows that the Hamiltonian is fundamental for the time evolution picture with a fixed k but the spatial evolution generator is fundamental for the spatial evolution picture with a fixed ω.

ACKNOWLEDGMENT

This work is partially supported by Science and Engineering Research Council, U. K.

REFERENCES

1. Present and permanent address: NTT Basic Research Labs., 3-9-11 Midori-cho, Musashino-shi, Tokyo 180 (Japan).
2. Y. R. Shen, Quantum theory of nonlinear optics, in *International School of Physics ≪ Enrico Fermi ≫* "Quantum Optics", R. J. Glauber, ed., page 489, Academic Press, New York (1969).
3. R. Lang, M. O. Scully and W. E. Lamb, Jr., Phys. Rev. A**7**:1788(1973).
4. Xiao-Ping Feng and K. Ujihara, Phys. Rev. A**41**:2668(1990).
5. B. Yurke and J. S. Denker, Phys. Rev., A**29**:1419(1984).
6. M. J. Collett and R. Loudon, J. Opt. Soc. Amer., B**4**:1525(1987).
7. C. M. Caves and D. D. Crouch, *ibid.* 1535(1987).
8. I. Abram, Phys. Rev., A**35**:4661(1987).

9. N. Imoto, in Proc. 3rd Int. Symp. Foundations of Quantum Mechanics (ISQM-Tokyo '89), pages 306-314(1990).

10. B. Huttner, S. Serulnik and Y. Ben-Aryeh, Phys. Rev., **A42**:5594(1990).

11. K. J. Blow, R. Loudon, S. J. D. Phoenix and T. J. Shepherd, Phys. Rev., **A42**: 4102(1990).

12. A large number of coupled mode equations are treated, although the optical field is not quantized, in A. Yariv, *Quantum Electronics* (2nd Edition), John Wiley & Sons, New York (1975).

13. N. N. Bogoliubov and C. V. Shirkov, *Introduction to the theory of quantized fields*, Interscience Publishers, Inc., New York (1959).

14. A. Messiah, *Quantum Mechanics*, McGraw-Hill, New York (1961).

15. In practice, a proper polarization is chosen for each mode so that the birefringence of the crystal is utilized for the phase matching condition such as Type-I or II phase matching condition. Here, however, we assume all modes are y-polarized for simplicity.

DIRECT SPACE QUANTUM OPTICS:

LIGHT PROPAGATION IN A NONLINEAR MEDIUM

I. Abram[1] and E. Cohen[2]

1. Centre National d'Etudes des Telecommunications
   196 Avenue Henri Ravera, 92220 Bagneux, France
2. Service de Physique Theorique, CEA - Saclay
   91191 Gif-sur-Yvette Cedex, France

1. INTRODUCTION

How can we describe the propagation of a short pulse of nonclassical light through a transparent linear or nonlinear medium? For example, how is a short pulse of squeezed light generated when an intense laser pulse undergoes parametric downconversion in a traveling-wave configuration inside a nonlinear crystal? How does such a squeezed pulse undergo self-phase modulation as it propagates in a nonlinear optical fiber? What are its quantum statistics at the exit of the fiber?

The standard theory of Nonlinear Optics[1] describes successfully the propagation of light through a nonlinear medium by relying directly on the macroscopic Maxwell equations. However, in this theory the electric and magnetic fields are represented by c-numbers and are thus treated classically. The standard theory of nonlinear optics, therefore, cannot address problems that involve the quantum statistics of light such as spontaneously-initiated nonlinear propagative processes or the propagation of pulses of nonclassical light.

Within the conventional approach to Quantum Optics, the propagation of a short pulse of light has been traditionally addressed through modal analysis[2]. This technique consists of constructing conceptually a cavity around the medium and defining the normal modes of the electromagnetic field inside that cavity, so that a short pulse can then be obtained by "mode-locking", as is said in the laser parlance. That is, the pulse is expressed as a linear superposition of the modes of the cavity of quantization and thus moves in space as the relative phases of the modes evolve in time. When the pulse propagates in a nonlinear medium, in addition to its translation it also undergoes a distortion that, in the modal approach, can be described by the excitation of additional modes: The new modes become excited as the pulse evolves in time under the modal interaction terms that describe the optical response of the nonlinear medium in the field Hamiltonian.

Quantum Measurements in Optics, Edited by P. Tombesi and D.F. Walls, Plenum Press, New York, 1992

The modal formulation of Quantum Optics can successfully describe experiments that involve a real cavity containing a nonlinear medium such that only a few modes of the cavity are actually excited. In principle, the methods developed to treat the few-modes situation should be able to be extended to account for all optical phenomena, since the modes constitute a complete basis set that can describe all possible configurations of the electromagnetic field. In practice, however, many difficulties arise when the modal approach is applied to situations that would require a very large number (or a continuum) of modes, such as cavity-less, traveling-wave experiments in which a short pulse traverses one or more nonlinear optical elements. These difficulties can be traced to the fact that the modal decomposition of a propagating pulse is essentially a description of the pulse in the reciprocal-space representation, and is thus equivalent to a Fourier transformation of the spatial distribution of the field. Fourier techniques are very successful in solving linear differential equations, since they convert these equations into algebraic equations, which are much easier to solve. However, the use of Fourier techniques in nonlinear differential equations, most often, does not simplify the solution of the original differential equations.

The difficulties of modal analysis in the description of nonlinear propagative phenomena can be seen easily in the case of a Gaussian beam that propagates through a medium with a nonlinear refractive index and undergoes catastrophic self-focusing down to a point. A description of this process through the modal approach starts with a single Gaussian mode but, as the beam propagates, it requires the use of an increasingly larger number of modes, all interacting with each other. In addition, the structure of each mode will have to depend on the degree of focusing of the beam, since the modal structure of the quantization cavity depends on the refractive index of the medium it contains which, in turn, is a local function of the beam intensity. The complexity of the problem of self-focusing when viewed in the reciprocal-space representation, in terms of modes, contrasts with the relative simplicity with which this same problem can be formulated in the direct-space representation, when it is expressed in terms of the local values of the electric field, $E(x,y,z)$, as is often done classically. In the standard theory of Nonlinear Optics, within the Slowly-Varying Amplitude approximation, a relatively simple-looking spatial differential equation can be obtained[1] for a beam propagating along z in a medium characterized by a linear dielectric function ϵ and a third-order susceptibility $\chi^{(3)}$:

$$\frac{\partial E}{\partial z} = i \frac{k\chi^{(3)}}{2\epsilon} |E|^2 E + \frac{i}{2k} \nabla_\perp^2 E \qquad (1)$$

In this equation, the spatial progression along z of a beam with wavevector k is related to the local value of the nonlinear refractive index $(\chi^{(3)} |E|^2/(2\sqrt{\epsilon}))$ and to the transverse structure of the beam, calculated through $\nabla_\perp^2$ the Laplacian along the x and y coordinates. While this differential equation cannot be solved exactly analytically, it is often given as a textbook example, since approximate perturbative solutions are quite easy. Looking at eq. (1), we note that the nonlinear interaction of the electric field with itself is local in the direct-space representation. However, if this equation is Fourier-tranformed so that it is expressed in the de-localized modal basis, a convolution product is obtained and, thus, the resulting equations actually becomes more difficult to solve.

Clearly, for the treatment of propagation of nonclassical light in a nonlinear medium, it would be advantageous if an alternate approach to Quantum Optics could be developed in the direct-space representation. In such a theory the electromagnetic field would not be treated in terms of the conventional delocalized modal (photon) operators, but rather in terms of local field operators, a viewpoint similar to that used in the standard (classical) theory of Nonlinear Optics. Within this direct-space formulation of Quantum Optics, all the methods that have been developed to describe propagative phenomena in classical Nonlinear Optics should be directly translatable into an operatorial language, so they can be applied to the treatment of propagation of nonclassical light.

In what follows, we shall sketch the highlights of such a theory. A more detailed treatment of the direct-space formulation of Quantum Optics can be found elsewhere[3]. In Section 2 we discuss the quantization of the electromagnetic field in direct space and we derive the operatorial equivalent of the macroscopic Maxwell equations. In Section 3 we apply the direct-space operatorial methods to the description of propagation in a linear medium. In Section 4 we examine the description of light propagation in a nonlinear medium, within the context of the direct-space formulation of Quantum Optics. Finally, in Section 5, we summarize our conclusions.

2. THE OPERATORIAL MAXWELL EQUATIONS

In order to develop a direct-space approach to Quantum Optics, we have to go back to the canonical quantization procedure for the electromagnetic field inside an effective nonlinear medium, introduced by Hillery and Mlodinow[3]. To simplify the notation, we examine a geometry in which plane waves propagate along the z-axis with the electric field E polarized along the x-axis and the magnetic field B along the y-axis. E and B can then be treated as scalars. The Lagrangian density that gives equations of motion for the field that are identical to the macroscopic Maxwell equations is,

$$\mathcal{L} = \frac{E^2 - B^2}{2} + \frac{1}{2}\chi^{(1)}E^2 + \frac{1}{3}\chi^{(2)}E^3 + \frac{1}{4}\chi^{(3)}E^4 + \ldots \qquad (2)$$

where the medium is introduced through its linear and nonlinear optical susceptibilities $\chi^{(n)}$, as in the classical theory of Nonlinear Optics. In our simplified treatment, the spatial dependance (inhomogeneity), the frequency dependance (dispersion) and the tensorial properties of $\chi^{(n)}$ are neglected. The quantization of the complete theory (including vectorial aspects, inhomogeneity and dispersion) has been discussed recently by Drummond[5] and by Glauber and Lewenstein[6]. The Lagrangian density (2) permits us to obtain a pair of canonical conjugate variables of the field. The most commonly used pair consists of the vector potential A and the displacement field D. This latter variable is given by

$$D = E + P = E + \chi^{(1)} E + \chi^{(2)} E^2 + \chi^{(3)} E^3 + \ldots \qquad (3)$$

and thus includes the response of the medium in the form of the induced (linear and nonlinear) polarization P.

The field can be quantized, by postulating that the two canonical variables A and D are non-commutinopertos equal-time commutator is given by

$$[D(z,t),A(z',t)] = ic\delta(z-z') \qquad (4)$$

where, in our simplifield geometry, δ is the ordinaty Dirac delta-function, and only the z coordinate need be considered explicitly. c is the speed of light.

From this point on, the field is usually expanded into modes and photon (modal) creation and annihilation operators are defined as linear combinations of the modal components of A and D, such that they follow Bose commutation relations. This passage to the reciprocal-space representation is implemented essentially in order to eliminate the spatial aspects of the quantized field, a feature that facilitates the Hamiltonian formulation of problem, focusing mainly on the temporal evolution of the field. As discussed in the introduction, however, the reciprocal space representation is quite convenient for the free field, or the field in a linear medium, but not necessarily for the field in a nonlinear medium. In this respect, it may be pointed out that the use of the reciprocal-space representation and the definition of photon creation and annihilation operators are not indispensable ingredients in the quantization procedure. Quantization is complete when the equal-time commutator (4) is postulated. This commutator involves the field operator in the direct-space representation, and not in reciprocal space in terms of modes. Thus, starting with the commutator (4), it is possible to develop a direct-space approach to Quantum Optics, without using the modal decomposition of the field, or the photon creation and annihilation operators.

The Lagrangian formalism addresses the spatial and temporal coordinates of the field on the same footing and gives equations of motion of the field relating its temporal evolution to its spatial progression: these are the macroscopic Maxwell equations. Within the Hamiltonian formalism, all the temporal and spatial information for the field is contained in the electromagnetic energy-momentum tensor, which can be calculated directly from the Lagrangian density[7]. Use of the full energy-momentum tensor should, therefore, permit us to address both the spatial and temporal aspects of propagation within the Hamiltonian formalism.

One of the components of the energy-momentum tensor is the canonical Hamiltonian which consists of the energy-density of the field integrated over the volume V of the (infinite) cavity of quantization under periodic boundary conditions, as in the conventional theory of Quantum Optics:

$$H = \int_V \frac{E^2 + B^2}{2} + \frac{1}{2}\chi^{(1)}E^2 + \frac{2}{3}\chi^{(2)}E^3 + \frac{3}{4}\chi^{(3)}E^4 + \ldots \qquad (5)$$

In eq. (5) the Hamiltonian is expressed in terms of E rather than in terms of D, which is the canonical variable. Throughout this paper, we shall use this E-field representation in order to facilitate discussion by making contact with the ideas of classical Nonlinear Optics. All calculations should be based, of course, on the canonical commutator (4) which is in terms of the D field.

For any field operator Q, the Hamiltonian (5) generates infinitesimal displacements along the temporal coordinate through the Heisenberg equation:

$$\frac{dQ}{dt} = i\ [H,Q] = i\ H^{\times}\ Q \qquad (6)$$

We shall use hereafter the Kubo notation for the commutator[8]: the superscript × in eq. (6) denotes the commutator of H with all the operators that follow. This notation permits us to write equations that involve many commutator brackets very compactly. In particular, it permits us to solve the Heisenberg equation (6) formally in a very simple way, as

$$Q(t) = e^{itH^{\times}}\ Q(0) \qquad (7a)$$

as can be verified by expanding the exponential

$$e^{itH^{\times}} Q(0) = Q(0) + (it)\, H^{\times} Q(0) + \frac{1}{2!} H^{\times} H^{\times} Q(0) + \ldots = \quad (7b)$$

$$= Q(0) + (it)\, [H,Q(0)] + \frac{1}{2!} [H,[H,Q(0)]] + \ldots = \quad (7c)$$

$$= e^{iHt}\, Q(0)\, e^{-iHt} \quad (7d)$$

Eq. (7), thus, gives the temporal variation of Q, referred to some initial time t=0, in terms of the time-evolution operator $e^{itH^{\times}}$ which is an exponential of the Hamilonian-commutator operator.

Another component of the energy-momentum tensor that is of interest here is the canonical momentum operator

$$G = \int_V BD = \int_V BE + \chi^{(1)}BE + \chi^{(2)}BE^2 + \chi^{(3)}BE^3 + \ldots \quad (8)$$

which permits us to generate infinitessimal displacements along the spatial coordinate of the field through a Heisenberg-like equation

$$\frac{dQ}{dz} = -i\, G^{\times} Q \quad (9)$$

This equation can be solved in a way analogous to eq. (6), to give the spatial dependance of any field operator as

$$Q(z) = e^{-i(z-z_0)G^{\times}} Q(z_0) \quad (10)$$

in terms of the exponential spatial progression operator $e^{-izG^{\times}}$. Within the reciprocal space representation, explicit use of the momentum operator G is superfluous, since the modes are actually its eigenstates. Thus, evaluation of the spatial progression equation (10) for a modal creation or annihilation operator simply gives the spatially-dependant phase factor e^{ikz} that accompanies the modal operator. In the direct-space representation, on the other hand, the operator Q(z) corresponds to a short pulse, that is to an excitation of the field in the neighbohood of point z, so that eq. (10) actually describes the translation of the excitation that corresponds to the operator Q from point z_0 to point z.

Clearly, in order to describe the propagation of a short pulse of light within the direct-space representation, we have to establish a relationship between the Heisenberg equations for the Hamiltonian (6) and momentum operators (9), such that it can permit us to convert the exponential time-evolution operator (7) into a function of the exponential spatial progression operator (10). This relationship will thus express the time evolution of a short pulse in terms of its motion through space.

It is relatively easy to obtain the relationship between the Hamiltonian and momentum operators, simply by using their definitions (5) and (8) and the basic commutator (4). A little algebra, then, yields this relationship in the form of a set of two commutatorial equations

$$G^{\times} E = \frac{1}{c} H^{\times} B \quad (11a)$$

and

$$G^{\times} B = \frac{1}{c} H^{\times} D \tag{11b}$$

Using the Heisenberg equations (6) and (9) we can see that these commutatorial equations are equivalent to the Maxwell equations

$$\frac{dE}{dz} = - \frac{1}{c} \frac{dB}{dt} \tag{12a}$$

and

$$\frac{dB}{dz} = - \frac{1}{c} \frac{dD}{dt} \tag{12b}$$

Thus, the two commutatorial equations (11) contain all the information on the propagative dynamics of the field that is contained in the Maxwell equations (12) and can, therefore, be used to describe the propagation of a short pulse of light, much in the same way that the classical Maxwell equations are used in classical linear or nonlinear optics. However, in addition to this, the operatorial nature of eqs. (11) implies that these equations can treat at the same time the quantum statistics of light, since the operator algebra that they involve can keep track of the commutation and ordering of the field operators in the course of propagation.

3. PROPAGATION IN A LINEAR MEDIUM

We shall examine here the use of the operatorial Maxwell equations (11) in the description of the propagation of a short pulse of light through a linear dielectric medium. In a linear dielectric these equations reduce to

$$H^{\times} B = c\, G^{\times} E \quad ; \quad \epsilon\, H^{\times} E = c\, G^{\times} B \tag{13}$$

We consider a forward-going electromagnetic pulse which consists of the superposition of the electric and magnetic field operators

$$W^{+}(z,t) = E(z,t) + v\, B(z,t) \tag{14}$$

where $v = c/\sqrt{\epsilon}$ is the speed of light in the dielectric medium. For this electromagnetic wave, the two operatorial Maxwell equations (13) can be combined into a single operatorial wave equation

$$H^{\times} W^{+} = v\, G^{\times} W^{+} \tag{15}$$

which provides a simple substitutional relationship between the commutators of the Hamiltonian and momentum operators. The time evolution of the electromagnetic wave W^{+} can then be written as,

$$W^{+}(z,t) = e^{itH^{\times}} W^{+}(z,0) = \tag{16a}$$

$$= e^{ivtG^{\times}} W^{+}(z,0) = \tag{16b}$$

$$= W^{+}(z-vt,0) \tag{16c}$$

where the substitutional relationship of eq. (15) was used in going from (16a) to (16b), while the spatial progression equation (10) was used in deducing eq. (16c) from (16b).

For a backward-going wave we would have

$$W^{-}(z,t) = E(z,t) - v\,B(z,t) \tag{17}$$

giving for its temporal evolution

$$W^{-}(z,t) = W^{-}(z+vt,0) \tag{18}$$

Eqs. (16) and (18) indicate that the time-evolution of a short electromagnetic pulse in a linear medium consists of the translation of the pulse in space, at a steady speed $\pm v$, as expected from classical optics. In treating this propagation, we did not use the modal description of the field, as is done in the conventional approach to Quantum Optics, but formulated the problem completely in direct-space, by use of local field operators. The direct-space treatment of reflection, refraction, and beamsplitting is relatively straight-forward[3].

4. PROPAGATION IN A NONLINEAR MEDIUM

In a nonlinear medium, the operatorial Maxwell equations (1.11) can be re-written as

$$H^{\times} D = c\, G^{\times} B \quad ; \quad H^{\times} B = c\, G^{\times} \left(\frac{D - P_{NL}}{\epsilon}\right) \tag{19}$$

where we have introduced the nonlinear polarization P_{NL} as in classical Nonlinear Optics. We define a forward-going displacement wave as superposition of the displacement and magnetic field operators

$$V(z,t) = D(z,t) + \sqrt{\epsilon}\, c\, B(z,t) \tag{20}$$

By applying the Maxwell equations (19) onto this displacement wave (20) we can easily show that in our one-dimensional geometry the wave equation is

$$H^{\times} V = v\, G^{\times} V - v\, G^{\times} P_{NL} \tag{21}$$

This equation is reminiscent of the wave equation for the linear medium (15), however, with a very important difference: the presence of the nonlinear polarization term. It indicates that the temporal evolution of a short pulse of light in a nonlinear medium ($H^{\times} V$) involves its movement through space ($v\, G^{\times} V$) and, in addition to this, a distortion that the pulse undergoes as it propagates because of the nonlinear response of the medium ($v\, G^{\times} P_{NL}$). Unlike the case of propagation in a linear medium, this equation does not provide a simple substitutional rule between the Hamiltonian and momentum commutators, which means that propagation cannot be treated for the case of a general nonlinear medium. This is also true in classical Nonlinear Optics: the wave equation cannot be solved in the general case. However, for most nonlinear optical phenomena, the nonlinear response of the medium is much weaker than the propagative effects due to its linear response or to the free-field Hamiltonian. Thus, the nonlinear wave equation (21) can be treated within the framework of a perturbative expansion in powers of the nonlinear polarization.

To order zero in the optical nonlinearity, the Hamiltonian and momentum operators for the field in the medium, reduce to the familiar linear medium Hamiltonian

$$H_0 = \int_V \frac{\epsilon E_0^2 + B_0^2}{2} \tag{22}$$

and linear momentum operators

$$G_0 = \int_V \epsilon\ E_0\ B_0 = \tag{23a}$$

$$= G_{free} + \int_V \chi^{(1)}\ (E_0\ B_0) \tag{23b}$$

while the zero-th order displacement wave reduces (to within the multiplicative constant ϵ) to the forward-going electromagnetic wave W^+ discussed in Section 3:

$$V_0(z,t) = D_0(z,t) + \sqrt{\epsilon}\ c\ B_0(z,t) = \epsilon\ W^+(z,t) \tag{24}$$

To zero-th order, then, the nonlinear wave equation (21) reduces to the corresponding linear equation

$$H_0^\times\ W^+ = v\ G_0^\times\ W^+ \tag{25}$$

indicating that for a short pulse of light propagating inside a nonlinear medium, the dominant effect is the movement of the pulse at a steady speed v, as if the medium were linear.

Using the standard techniques of perturbation theory[7], we can now obtain the time evolution of the displacement wave up to first order in the perturbation as

$$V(z,t) = e^{itH_{1S}^\times}\ \{\epsilon W^+(z,t)\} + V_1(z,t) \tag{26a}$$

$$= e^{itH_{1S}^\times}\ \{\epsilon\ e^{itH_0^\times}\ W^+(z,0)\} + V_1(z,t) \tag{26b}$$

where H_{1S} is the first-order correction to the field Hamiltonian. Let us now focus our attention on the first term in eq. (26a) or (26b). The operator e^{itH_0}, which describes the time evolution of the wave W^+, corresponds to temporal variations in time scales of the order of the optical period, while the operator $e^{itH_{1S}}$ describes variations occurring in a much slower scale. Thus, eq. (26) can be interpreted as representing a fast-varying carrier wave W^+ that propagates in the nonlinear medium and onto which the nonlinear medium imprints a much slower modulation envelope described by the operator $e^{itH_{1S}}$. This is precisely the assumption that is made in classical Nonlinear Optics when the Slowly Varying Amplitude Approximation (SVA) is invoked and, thus, eq. (26) is the operatorial equivalent of the SVA approximation. This means that a theory of nonlinear propagation based on the first-order time-evolution equation (26) is equivalent to the standard theory of Nonlinear Optics but can, in addition, deal with phenomena that involve the quantum statistics of light and nonclassical states.

The first-order wave equation can be obtained after some algebra[3] as

$$H_{1S}^\times\ V = G_{SVA}^\times\ V \tag{27}$$

This equation relates the commutator of the first-order correction to the Hamiltonian H_{1S} to the commutator of a nonlinear effective momentum operator G_{SVA}. This effective momentum operator is precisely the one that describes the spatial progression of a short pulse of light in a nonlinear medium within the SVA. In fact, its Heisenberg-like equation is exactly the same as the spatial differential equation that is obtained in classical nonlinear optics within the SVA. Eq. (27) provides a simple substitutional rule that permits us to convert the slow temporal modulation envelope imprinted on the carrier wave by the nonlinear medium into a spatial progression and describe thus nonlinear traveling wave experiments. Using eq. (27), then, the distortion that a short pulse undergoes when it propagates in a nonlinear medium over a distance vt can be written as

$$e^{itH_{1S}^{\times}} W^{+}(z,t) = e^{ivtG_{SVA}^{\times}} W^{+}(z,t) \tag{28}$$

Within this formalism, then, propagation of a short pulse in a nonlinear medium does not require the decomposition of the pulse into modes, but involves simply the calculation of multiple commutators of the effective momentum operator G_{SVA} (which contains all the information on the nonlinear medium) with operator W^{+} which describes the propagating carrier wave. For the simplest cases, such as parametric downconversion of a strong laser pulse, multiple commutators can be obtained in closed form through iterative formulas[3]. For more complicated situations, numerical calculations or approximations may be necessary.

We shall now examine briefly the structure of G_{SVA}. The expressions for G_{SVA} which are calculated when obtaining eq. (27) can be readily arranged in a way that is evocative of the nonlinear optical process we wish to describe. For example, the effective momentum operator for a medium with quadratic nonlinearity, characterized by the second-order susceptibility $\chi^{(2)}$, can be written in the form

$$G_{SVA}^{(2)} = \int_V \{\frac{\chi^{(2)}}{6} E_0\} (E_0 B_0) + \int_V \{\frac{\chi^{(2)}}{6\epsilon} B_0\} (\frac{\epsilon E_0 + B_0}{2}) \tag{29}$$

The first term in eq. (29) is composed of the product $(E_0 B_0)$ which corresponds to the free-field momentum density, and a pre-factor that is proportional to E_0. Comparing this first term with the expression for the linear momentum operator (23b) we see that the pre-factor is essentially the term that gives rise to the refractive index. Thus, the effective refractive index experienced by a propagating beam in a medium with a quadratic nonlinearity has a contribution that is proportional to the applied external electric field. This is, of course, the electro-optic or Pockels effect. Similarly, the second term in eq. (29) describes magneto-optic effects.

For the case of a medium with cubic nonlinearity, characterized by the third-order susceptibility $\chi^{(3)}$, the effective momentum operator can be written as

$$G_{SVA}^{(3)} = \int_V \{ \frac{\chi^{(3)}}{4\epsilon} \frac{\epsilon E_0 + B_0}{2}\} (E_0 B_0) \tag{30}$$

This operator is the spatial integral of the momentum density $(E_0 B_0)$ multiplied by a pre-factor proportional to the local (linear) energy density $\epsilon E_0 + B_0/2$. Clearly, then, this expression indicates that a light beam propagating in a $\chi^{(3)}$ medium experiences an effective refractive index that has a component proportional to the local energy density (i.e. the local intensity) of the propagating beam, as is observed in the Kerr effect. The effective momentum operator (30) can thus treat readily phenomena that require a local description of the refractive index change, such as self-focusing or self-phase-modulation. The momentum operator (30) can also be expressed in the conventional modal representation of the field in terms of the well-known four-fold products of photon creation and annihilation operators. In this reciprocal-space representation, however, all modes are coupled among each other, and thus one of the main simplifying features of the Kerr nonlinearity, its locality, is lost: When calculating multiple commutators of $G^{(3)}_{SVA}$ with a photon operator, an increasingly larger number of modes are introduced because of the nonlinear coupling among them.

5. CONCLUSIONS

We have presented a direct-space approach to the theory of Quantum Optics, based on the canonical quantization procedure for the electromagnetic field in an effective nonlinear medium and the traditional Hamiltonian formulation of quantum mechanics. This formalism does not use the conventional modal decomposition of the electromagnetic field and the photon operators, but treats the field in terms of local operators. Within the direct-space representation of Quantum Optics, light propagation is treated through the simultaneous use of the Hamiltonian and momentum operators, which address respectively the temporal evolution and the spatial progression of the field on an equal footing.

The Hamiltonian and momentum operators are related to each other through two commutator equations, which are the operatorial equivalent of the macroscopic Maxwell equations. These operatorial Maxwell equations contain all the information on the propagative dynamics of the field inside the effective medium, while at the same time their operatorial algebra permits us to account for the quantum statistics of the field in the course of its propagation. The use of the operatorial Maxwell equations and the local representation for the electromagnetic field permit us to treat the propagation of a short pulse of nonclassical light in a (linear or nonlinear) dielectric through methods analogous to those developed in classical optics.

ACKNOWLEDGEMENT

This work was supported in part by an ESPRIT-Basic Research Action (No. 3186) from the Commission of the European Communities.

REFERENCES

[1] Y.R. Shen, The Principles of Nonlinear Optics, Wiley-Interscience (New York, 1984).
[2] J. Tucker and D.F. Walls, Phys. Rev. 178, 2036 (1969).
[3] I. Abram and E. Cohen, Phys. Rev. A (to be published).
[4] M. Hillery and L.D. Mlodinow, Phys. Rev. A 30, 1860 (1984).
[5] P.D. Drummond, Phys. Rev. A 42, 6845 (1990).
[6] R.J. Glauber and M. Lewenstein, Phys. Rev. A 43, 467 (1991).
[7] C. Itzykson and J.B. Zuber, Quantum Field Theory, McGraw-Hill (New York, 1980).
[8] R. Kubo, J. Phys. Soc. Japan 17, 1100 (1962).

QUANTUM HOLOGRAPHY, SYNTHETIC APERTURE RADAR IMAGING AND COMPUTED TOMOGRAPHIC IMAGING

Walter Schempp

Lehrstuhl für Mathematik I
University of Siegen
D-5900 Siegen, Germany

I think I can safely say that nobody understands quantum mechanics.

Richard P. Feynman (1918-1988)

1. Basic Concepts of Quantum Holography

Quantum holography is a field of research concerned with interference phenomena arising from the quantum nature of coherent radiation. While it was initially dominated by theoretical studies, it is now increasingly an applied area of research with strong influence on photonic computing.

The primary assumption of quantum or photon holography is that the energy transmitted by a beam of coherent radiation is divided into discrete wave packets, or photons, much as an electric current is made up of a flow of electrons. Detailed analysis shows that the arrival times at a photodetector of photons from a classical coherent radiation source such as a laser, exhibit the same Poissonian statistics as does the thermionic emission of electrons from the hot cathode of a vacuum tube. Thus, the photocurrent exhibits fluctuations which resemble the shot noise of the current in the vacuum tube. The quantum noise produced by a photoelectric detector is therefore an intrinsic property of the radiation itself, rather than of the photon detector.

The concept of photon arises from the quantization of the electromagnetic field. The spatial part of Maxwell's equations in vacuo

$$dF = 0 \text{ and } d * F = 4\pi j$$

(differential 2-form $F = B + E \wedge dt$ on the Lorentz space $\mathbb{R}^4$, $*$ = Hodge star operator on $\Lambda^2(\mathbb{R}^4)$) and the boundary conditions ([39]) determine the

Quantum Measurements in Optics, Edited by P. Tombesi and D.F. Walls, Plenum Press, New York, 1992

possible frequencies ν, direction and spatial wavefront profile of the field, collectively referred to as the mode. The field modes are described by wave vectors $\mathbb{k} \neq 0$ in $\mathbb{R}^3$.

In spite of warnings, sometimes the suggestion occurs that the Maxwell waves do form the wave functions of the photon. This suggestion, however, is a false idea. Although a single photon can have a definite position at one time, it is impossible to construct a position operator for the photon. Therefore it cannot have definite positions at all times in a specified time interval, i.e., it cannot have a definite trajectory (Bohr's indeterminacy principle; see [36]). Since in the presence of sources, photons can be absorbed or emitted, one cannot introduce a linear Schrödinger evolution equation for a single photon. In fact, the modes of the electromagnetic field must be quantized and photons of energy $h\nu$ then occur as elementary excitations driven by the quantized Maxwell field modes of frequency ν and label $\mathbb{k}$.

In quantum optics or photonics ([22]), the quantization of the Maxwell field modes is done by expressing the time dependent part of Maxwell's equations in the form of the equation of motion for a classical harmonic oscillator and then replacing the classical harmonic oscillator by its quantum-mechanical counterpart. In this way, the electromagnetic field is considered as an assemblage of driven harmonic oscillators. As a consequence, the energy of the radiation field is quantized and the quanta are referred to as photons. The state of the field can be expressed in terms of number states $|n_{\mathbb{k}}\rangle$ which are states with $n_{\mathbb{k}}$ quanta occupying the mode $\mathbb{k}$. These number states are eigenstates of the Hamiltonian of the quantum-mechanical harmonic oscillator. However, they are not a realistic description of a coherent radiation field, as emitted by a laser. One formal series of number states, the so called coherent state $|\alpha_{\mathbb{k}}\rangle$, is used to represent the coherent radiation field produced by an ideal source such as an ideal laser operating well above threshold. In the Dirac notation, the Glauber formal series expansion

$$|\alpha_{\mathbb{k}}\rangle = \exp(-\tfrac{1}{2}|\alpha_{\mathbb{k}}|^2).\sum_{n\geq 0} \frac{(\alpha_{\mathbb{k}})^n}{\sqrt{n!}} \; |n_{\mathbb{k}}\rangle$$

describes a state $|\alpha_{\mathbb{k}}\rangle$ where the probability of finding the mode occupied by $n_{\mathbb{k}}$ photons exhibits a Poissonian distribution about a mean of $|\alpha_{\mathbb{k}}|^2$ with a width $|\alpha_{\mathbb{k}}|$. Ignoring the slowly time varying phase diffusion (cf. Section 2 infra), the coherent state is considered to be a good approximation to the field produced by a coherent radiation source such as a laser.

Switching to a rigorous formulation, the concept of coherent state leads to the holomorphic quantization procedure, i.e., to the quantization of the complex vector space $\mathrm{Exp}(\mathbb{C}^n \times \mathbb{C}^n)$ of entire holomorphic functions of exponential type on complex phase space $\mathbb{C}^n \times \mathbb{C}^n$. It turns out that the Bargmann-Fock-Segal space of symplectic spinors inside the symplectic Clifford algebra (or Weyl algebra of linear differential operators with constant coefficients) on the underlying 2n-dimensional real symplectic vector space $\mathbb{R}^n \oplus \mathbb{R}^n$ is the natural space in which to carry out this quantization procedure ([1]).

In 1960 a classic paper by Cutrona, Leith, Palermo, and Porcello [6] exposed in the open literature for the first time new ideas previously developed at the University of Michigan regarding applications of coherent optical systems to the processing of radar data. This work was to have enormous impact, lasting even to today.

Joseph W. Goodman (1989)

2. Synthetic Aperture Radar Imaging

Synthetic aperture radar (SAR), first demonstrated in the early 1950s, provides a coherent, high resolution, ground imaging technique implemented from aircraft or spacecraft. It is attractive for reconnaissance purposes since it offers an all-weather capability because it operates at microwave frequencies. In the SAR imaging technique, an effective antenna aperture is synthesized that is much longer than the real antenna in order to generate high azimuth resolution images of the terrain ([3], [5], [9], [11], [12], [16], [20], [21], [24], [29], [30], [38]).

Because ground data collected from many different perspective views of the target area are coherently superposed, the radar platform motion along a straight line-path and a highly stable local oscillator (STALO) reference carried by the platform are prerequisites of the SAR imaging technique. Figure 1 emphasizes the difference between real aperture radar imaging (A) and SAR imaging (B) by X-band radar images covering an area of 18 by 18 km.

Airborne and spaceborne SAR imaging systems form active remote sensing systems which coherently illuminate the terrain with electromagnetic energy at relatively long wavelengths (radar L-band center wavelength $\lambda = 23$ cm, C-band center wavelength $\lambda = 5.7$ cm, X-band center wavelength $\lambda = 3.1$ cm) as the platform moves with respect to the ground being mapped and simultaneously record the coherent ground returns in both amplitude and phase along the path. In SAR remote sensing systems a target at distance r_0 with velocity v relative to the moving platform induces a relativistic chirp amplitude density modulation $\sigma(g^u)$ of the received echo signal of chirp rate

$$u = \frac{4\pi v^2}{\lambda r_0}$$

where $\lambda = c/|\nu|$ denotes the center wavelength of the coherent radar and

$$g^u = \begin{pmatrix} 1 & 0 \\ u & 1 \end{pmatrix} \in \mathbf{Sp}(1, \mathbb{R})$$

maps the symplectic hologram plane $\mathbb{R} \oplus \mathbb{R}$ onto itself. The unipotent matrix g^u represents a geometrization of the chirp wavelet transform. The dependence of the chirp rate u of the range r_0 is called the range-azimuth coupling.

Chronologically, the holographic optical systems for processing SAR

A. REAL-APERTURE IMAGE.

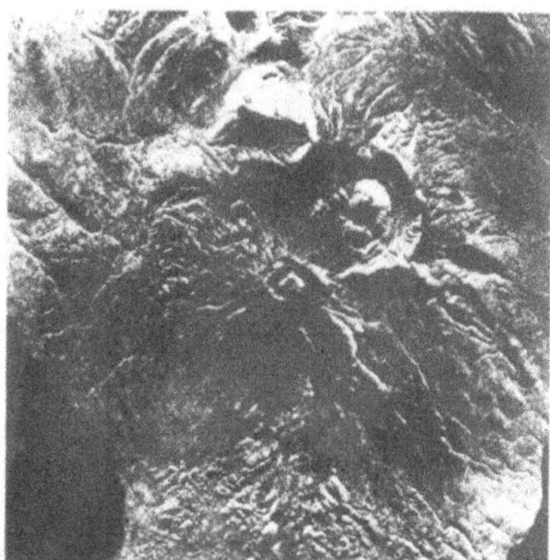

B. SYNTHETIC-APERTURE IMAGE.

Figure 1

imaging data as conceived and successfully developed at the University of Michigan at Ann Arbor in the mid-1950s ([6], [7]) can be regarded as the first analog photonic computers ([14]). Quantum holography rigorously establishes that optical holograms define multi-layer wavelet fan-in/fan-out neural networks ([33], [34]). For an optical visualization of such multi-layer neural networks in crystals we refer to Figures 2 and 3. Figure 2 processed by R. Trebino illustrates the second harmonic generation in a photorefractive crystal, and Figure 3 by D.M. Pepper, J. Feinberg, and N.V. Kukhtarev displays one component of an optical neural network generated a few seconds later by a refractive index volume hologram inside a photorefractive crystal ([40]). By geometrization, the time-scaling (or zooming) wavelet transform $\sigma(g_a)$ generated by the photorefractive effect is associated to the diagonal matrix

$$g_a = \begin{pmatrix} a & 0 \\ 0 & a^{-1} \end{pmatrix} \in \mathbf{Sp}(1,\mathbb{R}) \qquad (a > 0).$$

Basically by an analogous free-space configuration, optical SAR processors use holographic memories to form photonic neurocomputers ([33], [34]). Although today much more efficient digital electronic computers are available for the implementation of advanced digital signal processing algorithms ([2], [3], [28]) than thirty years ago, holographic optical processing is still a valuable tool to deal with the tremendous amount of SAR imaging data. In fact, photonic signal processing technology is so well matched to the SAR data processing requirements that holographic optical processors still account for a major portion of the SAR images produced.

It is a well known fact that the phase information contained in the Fourier transform of an image is generally more important than the amplitude information in determining the image characteristics. Indeed, computer simulations have shown that it is possible to get very satisfactory correlation results with high optical efficiency by using only the phase information carried by the Fourier-plane filter. Optical holography implements the phase information and is therefore completely different from the photographic procedure. It is a two-step coherent imaging process which is based on the principles of inverse scattering. In the first processing step, the writing step, the three-dimensional image is spatially encoded in both amplitude and phase by a quantum interference pattern in the symplectic hologram plane $\mathbb{R} \oplus \mathbb{R}$. The quantum interference pattern as a photonic excitation pattern is driven by coherently mixing the reference beam and the beam scattered by the object, hence by a linear superposition of phase-shifted beam splitter interferometer experiments. Due to the wavefront division performed by the beam splitter device, the relative phase of the reference and object beams remains fixed even if the phase of the input beam of the interferometer fluctuates. The independence of the cross-correlation from the phase diffusion allows to spatially record the quantum interference pattern by a square law photodetector.

In the second processing step of optical holography, the readout step, the raw optical data encoded in the symplectic hologram plane $\mathbb{R} \oplus \mathbb{R}$ are spatially decoded by coherent light illumination of the hologram in order to reconstruct the three-dimensional image of the object in both amplitude and phase.

Because SAR imagery is generated by simultaneously cross-correlating the raw radar data in both amplitude and phase with a reference function

Figure 2

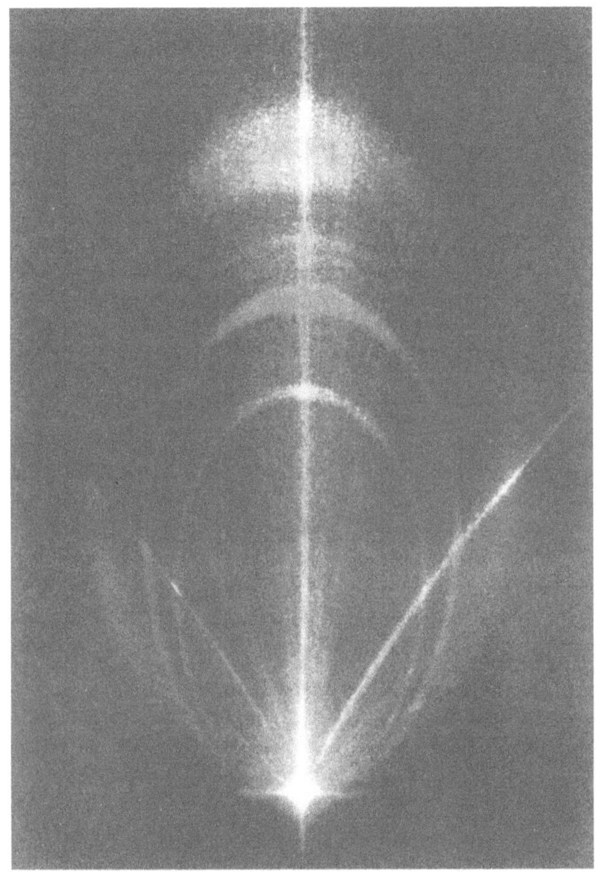

Figure 3

over an integration time which is less than or equal to the time that the radar illuminates a fixed point of the ground, the application of quantum holographic concepts to SAR has been extremely fruitful. Although SAR technology was conceived and successful developed without the aid of quantum holographic concepts, the holographic viewpoint is helpful in understanding the SAR technique in a basic way and the insight into analog photonic computing thus provided has led to ideas that otherwise would perhaps never been conceived.

Everybody involved in medical imaging shares one common dream: To be able to distinguish the structures of the object examined with such sharpness and accuracy that there is no room for speculation. Shortly after the introduction of MRI *as new imaging modality, it was hailed as a technique which did not suffer from the common beam hardening artifacts which destroy the images of the posterior fossa and the spine in X-ray computed tomography. However, it was soon realized that an unfortunate side effect of the complex nature of* MRI *was a whole new set of artifacts.*

Peter A. Rinck (1990)

3. Computed Tomographic Imaging

The magnetic resonance medical imaging group of Siemens Erlangen, the producer of the MAGNETOM imagers, has emphasized the analogy between magnetic resonance imaging (MRI) and optical holography ([13], [15]). As a first example of the quantum holographic approach, the relation between computed tomographic imaging (CTI), MRI, and SAR imaging ([3], [25], [35]) are considered. All of these techniques are high resolution imaging methods based on inverse scattering. Notice that both CTI and MRI are governed, after slice selection, by the Radon transform $\mathcal{R}$ and its back-projector $\mathcal{R}^{\#}$ ([8]). Let $\mathbf{S}_{n-1} = \mathbf{O}(n,\mathbb{R})/\mathbf{O}(n-1,\mathbb{R})$ denote the compact unit sphere oriented toward the outside of the real Euclidean vector space $\mathbb{R}^n$ of dimension $n \geq 2$. The group theoretical link between the transform

$$\mathcal{R}f(\omega,r)d\sigma^{(n-1)}(\omega) \wedge (1 - r^2)^{(1-n)/2}dr$$

with coefficient function

$$\mathcal{R}f(\omega,r) = \int_{\mathbb{R}^n} f(\xi)\varepsilon_{(r-\langle\omega|\xi\rangle)}d\xi$$

(ε = Dirac measure on $\mathbb{R}^n$) of the solid angle differential n-form $d\sigma^{(n-1)}(\omega) \wedge (1 - r^2)^{(1-n)/2}dr$ oriented toward the outside of the compact unit cylinder $\mathbf{S}_{n-1} \times [-1,+1]$ in $\mathbb{R}^n$, and the holographic transform is formed by the reductive dual pair $(\mathbf{O}(n,\mathbb{R}),\mathbf{Sp}(1,\mathbb{R}))$ inside the symplectic group $\mathbf{Sp}(n,\mathbb{R})$; see [32]. The most common imaging method in MRI is based on the Fourier transform on the real Euclidean vector space $\mathbb{R}^n$. As a consequence of the projection-slice theorem ([23]) which is of basic importance for the direct Fourier image reconstruction method for $\mathcal{R}$, one has to phase correct the Fourier transform $\sigma(g_0)$ along the axial direction and then to decompose it into tensor products according to the pair $(\mathbf{O}(n,\mathbb{R}),\mathbf{Sp}(1,\mathbb{R}))$ of subgroups centralizing each other inside $\mathbf{Sp}(n,\mathbb{R})$. The group element g_0 in $\mathbf{Sp}(1,\mathbb{R}) = \mathbf{SL}(2,\mathbb{R})$ is given by the symplectic matrix

$$g_0 = \begin{pmatrix} 0 & -1 \\ 1 & 0 \end{pmatrix} \in \mathbf{Sp}(1,\mathbb{R})$$

which represents a geometric implementation of the Fourier transform $\sigma(g_0)$ on the complex Hilbert space $L^2(\mathbb{R}^n)$ by a 90° rotation. Notice that the group $\mathbf{SL}(2,\mathbb{R})$ is generated by the set of matrices

$$\{g_0\} \cup \{g_a | a > 0\} \cup \{g^u | u \in \mathbb{R}\}$$

which represent the Fourier transform, the time-scaling (or zooming) wavelet transform, and the chirp wavelet transform. Consider the corresponding Iwasawa decomposition

$$\mathbf{SL}(2,\mathbb{R}) = KAN$$

into compact, diagonal, and unipotent subgroups. The one-parameter subgroup K generated by the Weyl element g_0 is isomorphic to the circle group $S_1 = \mathbf{SO}(2,\mathbb{R})$, A normalizes N and both groups are isomorphic to the additive group $\mathbb{R}$. The maximal compact subgroup K of $\mathbf{SL}(2,\mathbb{R})$ gives rise via the restriction $\sigma|K$ to the discrete series $(T_{n/2+l})_{l\geq 0}$ of holomorphic representations of $\mathbf{SL}(2,\mathbb{R})$ with lowest weight $n/2 + 1$. Starting with the lowest weight vector (ground state) for $T_{n/2+l}$, the eigenvectors with eigenvalues $i(n/2 + l + k)$ of the infinitesimal generator $\frac{\partial}{\partial t}$ of K at the Lie algebra level are iteratively computed by means of the Rodrigues' formula for the Gegenbauer (or ultraspherical, or symmetric Jacobi) polynomials

$$P_k^{(p)} = \Gamma(p + 1/2)\Gamma(k + 2p)/\Gamma(2p)\Gamma(k + p + 1/2).P_k^{(p-1/2,p-1/2)}$$

of degree $k \geq 0$ and order $p > -1/2$:

$$(1 - t^2)^{p-1/2}P_k^{(p)}(t) = \frac{(-2)^k}{k!}\Gamma(k + p)\Gamma(k + 2p)/\Gamma(p)\Gamma(2k + 2p)\frac{d^k}{dt^k}(1 - t^2)^{k+p-1/2}$$

In this way, an orthogonal basis of eigenvectors

$$\mathbf{S}_{n-1} \times [-1,+1] \ni (\omega,r) \mapsto Y_{l,m}(\omega).(1 - r^2)^{(n-1)/2}P_k^{(n/2)}(r)$$

obtains ($k \equiv l \pmod 2$) spanning the range of $\mathcal{R}$ in the complex Hilbert space $L^2(\mathbf{S}_{n-1} \times [-1,+1];\ d\sigma^{(n-1)}(\omega) \wedge (1 - r^2)^{(1-n)/2}dr)$. In the preceding tensor product, $Y_{l,m}$ are the surface spherical harmonics on $\mathbb{R}^n$ ($n \geq 3$) of degree $l \geq 0$ such that $m = 1$ for $l = 0$, $1 \leq m \leq n$ for $l = 1$, and according to the representation theory of the compact group $\mathbf{O}(n,\mathbb{R})$,

$$1 \leq m \leq \binom{n + l - 1}{l} - \binom{n + l - 3}{l - 2} = \frac{(n + l - 3)!\,(n + 2l - 2)}{(n - 2)!\,l!}$$

for $l \geq 2$; see [37]. In the case $n = 2$, however, the characters Y_l of the rotation group $\mathbf{S}_1 = \mathbf{O}(2,\mathbb{R})$ are parametrized by the integers $l \in \mathbb{Z}$. The holomorphic representations $(T_{n/2+l})_{l\geq 0}$ of $\mathbf{SL}(2,\mathbb{R})$ with lowest weight $n/2 + l$ lead to Bochner's formula for the Fourier transform $\sigma(g_0)$ in terms of the Hankel transform of order $n/2 + l - 1$. Due to the projection-slice theorem and the self-reciprocal property of Hankel transforms, a singular value decomposition of the Radon transform $\mathcal{R}$ follows by the Hankel transform of order $n/2 + l - 1$ of the Fourier transformed Gegenbauer polynomials $P_k^{(n/2)}$. Moreover, by an application of the classical Funk-Hecke theorem ([4]), a reproducing kernel for the range of $\mathcal{R}$ follows in terms of Gegenbauer polynomials $P_k^{(n/2)}$. Thus the reproducing kernel of $(1 - r^2)^{(1-n)/2}\mathcal{R}$ takes after truncation the tensor product form

$$K_m : ((\omega', r'), (\omega, r)) \mapsto \sum_{0\leq k\leq m} c_{k,m,n} . P_k^{(n/2)}(\langle\omega'|\omega\rangle) . P_k^{(n/2)}(r')P_k^{(n/2)}(r),$$

where $c_{k,m,n} \geq 0$ denote normalization constants. By an application of interpolatory quadrature formulae to K_m with respect to finite sets of nodes in the compact unit cylinder $\mathbf{S}_{n-1} \times [-1,+1]$ inside $\mathbb{R}^n$, effective reconstruction algorithms arise for CTI and MRI, respectively, which reflect the inverse scattering nature of these high resolution imaging methods. As an example, Figure 4 displays sagittal MR images of the human neck spine computed by the direct Fourier image method, one of these high resolution images being localized by a removable Helmholtz coil assembly.

Example 1. In the simplest case $n = 2$, the singular value decomposition of $\mathcal{R}$ in terms of Zernike polynomials of order $|l|$ reads

$$\mathbb{R}^2 \ni z \mapsto \begin{cases} 0 & \text{if } |z| > 1, \\ \left(\dfrac{k+1}{\pi}\right)^{1/2} |z|^{|l|} P^{(0,|l|)}_{(k-|l|)/2}(2|z|^2 - 1) . Y_l\left(\dfrac{z}{|z|}\right) & \text{if } |z| \leq 1 \end{cases}$$

and

$$\mathbf{S}_1 \times \mathbb{R} \ni (\omega, r) \mapsto \begin{cases} 0 & \text{if } |r| > 1, \\ \frac{1}{\pi} Y_l(\omega) . (1 - r^2)^{1/2} P_k^{(1)}(r) & \text{if } |r| \leq 1 \end{cases}$$

where $k \equiv l \pmod 2$, $|l| \leq k$, $Y_l(\omega) = \omega^l$, and $P_k^{(1)} = U_k$ denotes the Čebyšev polynomial of the second kind of degree $k \geq 0$.

The same procedure of constructing a singular value decomposition of the transform

$$\mathcal{R}(\omega, r) d\sigma^{(n-1)}(\omega) \wedge e^{r^2} dr$$

by the theory of surface spherical harmonics with extra tensor product structure provided by the reductive dual pair $(\mathbf{O}(n,\mathbb{R}), \mathbf{Sp}(1,\mathbb{R}))$ inside the symplectic group $\mathbf{Sp}(n,\mathbb{R})$, also works in the case of the non-compact cylinder $\mathbf{S}_{n-1} \times \mathbb{R}$ inside $\mathbb{R}^n$. In this case, the Gegenbauer polynomials $P_k^{(p)}$ have to be replaced by the Hermite polynomials H_k of degree $k \geq 0$ ([32]). The Rodrigues' formula of the Hermite functions (quantum harmonic oscillator wave functions) takes the form

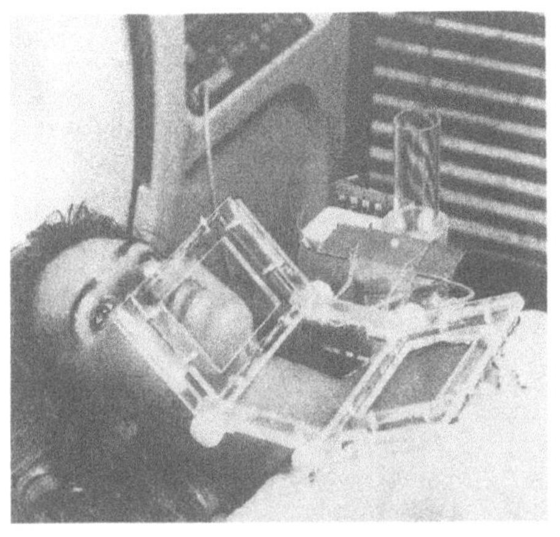

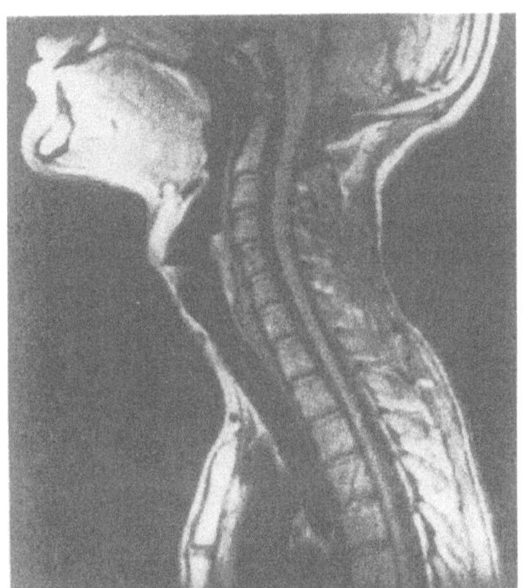

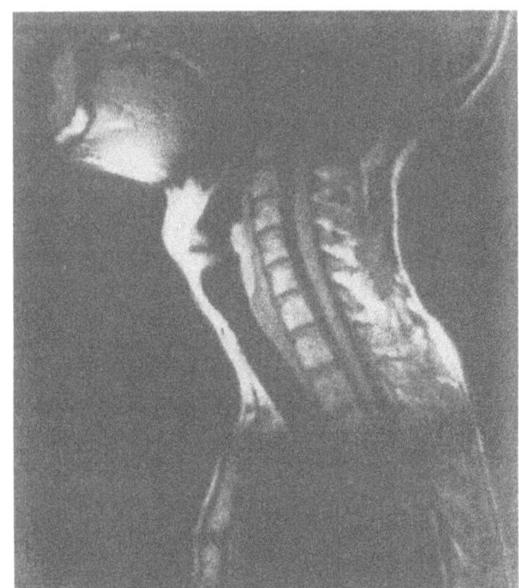

Figure 4

$$e^{-t^2} H_k(t) = (-1)^k . \frac{d^k}{dt^k}(e^{-t^2}).$$

It allows to iteratively compute an orthogonal basis of eigenvectors

$$\mathbf{S}_1 \times \mathbb{R} \ni (\omega, r) \mapsto Y_{1,m}(\omega).e^{-r^2} H_k(r)$$

with $k \equiv 1 \pmod 2$ spanning the range of $\mathcal{R}$ in the complex Hilbert space $L^2(\mathbf{S}_1 \times \mathbb{R}; d\sigma^{(n-1)}(\omega) \wedge e^{r^2} dr)$ and then to combine the projection-slice theorem with Bochner's theorem. In this case, the truncated reproducing kernel of $e^{r^2} \mathcal{R}$ reads by the Funk-Hecke theorem

$$K'_m: ((\omega', r'), (\omega, r)) \mapsto \sum_{0 \le k \le m} c'_{k,m,n} . H_k(\langle \omega' | \omega \rangle) . H_k(r') H_k(r),$$

where $c'_{k,m,n} \ge 0$.

Example 2. For $n = 2$, the singular value decomposition of $\mathcal{R}$ in terms of Laguerre polynomials of order $|l|$ reads

$$\mathbb{R}^2 \ni z \mapsto e^{-|z|^2} |z|^{|l|} L^{(|l|)}_{(k-|l|)/2}(|z|^2) . Y_l(\frac{z}{|z|}),$$

$$\mathbf{S}_1 \times \mathbb{R} \ni (\omega, r) \mapsto C_{k,l} Y_l(\omega).e^{-r^2} H_k(r)$$

where $k \equiv 1 \pmod 2$, $|l| \le k$, and

$$C_{k,l} = (-1)^{(k-|l|)/2} (\Gamma((k+|l|+2)/2))^{1/2} / \pi^{1/4} 2^{k/2} (k!)^{1/2} ((k-|l|)/2)!$$

denote normalization constants.

Notice that the Radon transform $\mathcal{R}$ also allows optical implementations ([10]). The ill-posedness of the CTI reconstruction problem may produce a set of artifacts, a phenomenon which is expected to be overcome by change detection between pairs of CT images. By comparing the two CT images and detecting significant changes the experienced physician can identify those features which are likely to have the greatest medical relevence. Collateral information provided by neurocomputers seems to be very helpful in the high-level comparison procedure ([26]).

We have found that when the notion of what the physicists mean by quantizing a function is suitably generalized and made rigorous, one may develop a theory which goes a long way towards constructing all the irreducible unitary representations of a connected Lie group. The theory is founded in differential geometry.

Bertram Kostant (1970)

It is possible to describe time evolution in an alternative way by introducing a time-dependent unitary transformation affecting both states (vectors) and observables (operators) in such a way as to kill the time dependence of the state vectors. The net result of this manoeuvre is to transfer time dependence from states to observables. The resulting description of time evolution in quantum mechanics is known as the Heisenberg picture.

Michael L.G. Redhead (1987)

4. Holographic Optical Processor

A second example of the quantum holographic approach to SAR are the tilted plane holographic optical processors ([17], [18], [19]): The symplectic hologram plane $\mathbb{R} \oplus \mathbb{R}$ carrying the canonical skew-symmetric nonsingular bilinear form

$$\omega_\nu : \; ((x_1, y_1), (x_2, y_2)) \mapsto \nu . \det \begin{pmatrix} x_1 & x_2 \\ y_1 & y_2 \end{pmatrix}$$

is identified with the polarized cross-section G/C_G of the real three-dimensional Heisenberg nilpotent Lie group

$$G = \left\{ \begin{pmatrix} 1 & x & z \\ 0 & 1 & y \\ 0 & 0 & 1 \end{pmatrix} \mid x, \; y, \; z \in \mathbb{R} \right\}$$

to its one-dimensional center

$$C_G = \left\{ \begin{pmatrix} 1 & 0 & z \\ 0 & 1 & 0 \\ 0 & 0 & 1 \end{pmatrix} \mid z \in \mathbb{R} \right\}$$

and σ denotes the projective metaplectic representation of $\mathbf{Sp}(1,\mathbb{R})$ acting on the standard complex Hilbert space $L^2(\mathbb{R})$ by keeping the line C_G pointwise fixed ([31]). Notice that the transition by a gauge transformation from the polarized cross-section G/C_G to the isotropic cross-section of G yields the Bargmann-Fock-Segal space of symplectic spinors on the symplectic hologram plane $\mathbb{R} \oplus \mathbb{R}$. The metaplectic representation σ actually forms the mathematical expression of the symmetries of the bosonic canonical commutation relations of quantum electrodynamics (QED) and therefore acts as an automorphism of the holographic transform

$$H_\nu(\psi, \varphi; x, y)\nu . dx \wedge dy = \langle U_\nu (\begin{pmatrix} 1 & x & 0 \\ 0 & 1 & y \\ 0 & 0 & 1 \end{pmatrix})\psi | \varphi \rangle . \omega_\nu .$$

The coefficient of the canonical differential 2-form $\omega_\nu = \nu . dx \wedge dy$ on

$\mathbb{R} \oplus \mathbb{R}$ is defined according to the driving action of the irreducible unitary linear representation U_ν $(\nu \neq 0)$ of G on the complex Hilbert space $L^2(\mathbb{R})$ at frequency ν by the two-time average

$$H_\nu(\psi,\varphi;x,y) = \int_{\mathbb{R}} \psi(t-x)\bar{\varphi}(t)e^{2\pi i\nu yt}dt.$$

Because it is not the photons which interfere but rather the transition amplitudes, and the populated modes of these amplitudes can involve many photons, the holographic transform $H_\nu(\psi,\varphi;x,y)\nu.dx \wedge dy$ displays in the Heisenberg picture, after slice selection, the photonic excitation pattern at the point $(x,y) \in \mathbb{R} \oplus \mathbb{R}$ of the symplectic hologram plane driven at frequency ν by the interferring quantized electromagnetic fields ψ, $\varphi \in L^2(\mathbb{R})$. The recording slices correspond to the planar symplectic co-adjoint orbits $(\mathcal{O}_\nu,\omega_\nu)$ in the Kirillov picture ([31]) of the unitary dual $\hat{G}$ of G, and the action of $\mathbf{Sp}(1,\mathbb{R})$ by the metaplectic representation σ is given by the covariance formula

$$H_\nu(\sigma(g)\psi,\sigma(g)\varphi;x,y) = H_\nu(\psi,\varphi;(x,y)g) \quad (g \in \mathbf{Sp}(1,\mathbb{R})).$$

It would be quite difficult to explain the function of the tilted plane holographic optical processor and the STALO reference without Dennis Gabor's fundamental concept of using coherent light illumination for the holographic optical wavefront reconstruction. Mathematically, the wave-front reconstruction is given by the spatial decoding formula which performs an integration over the relative path lengths and view angles according to the filtered backpropagation identity

$$\iint_{\mathbb{R}\oplus\mathbb{R}} H_\nu(\psi,\varphi;x,y)e^{-2\pi i\nu yt}\bar{\psi}(t-x)dxdy = \bar{\varphi}(t) \quad (\|\psi\|_2 = 1).$$

Thus the fundamental aspect of the preceding filtered backpropagation formula is that it transfers in the readout step the information spatially encoded in the symplectic hologram plane $\mathbb{R} \oplus \mathbb{R}$ into temporal amplitude and phase information. It implies the basic principle of holographic interferometry according to which the waves reconstructed from the hologram can interfere with phase-related waves derived from the illuminating coherent light source.

In the SAR holographic optical signal processor, let λ' denote the wavelength of the coherent light scanning the holographic film of transport velocity v'. If

$$v_0 = v/v', \quad \lambda_0 = \lambda/\lambda'$$

are the relative SAR platform velocity and the relative radar wavelength, respectively, then the radar return focus a distance

$$f = \frac{\lambda_0 r_0}{4\pi v_0^2}$$

from the symplectic hologram plane $\mathbb{R} \oplus \mathbb{R}$. It follows that the relativistic effect of the platform motion generates an axial astigmatism. To compensate the linear range variation of the focal length f, a wide-screen equalizer is introduced in the hologram plane. Such an equalizer takes the form of a conical lens or a tilted cylindrical lens which are components of a correcting anamorphic optical system. The recent developments of spatial light modulators (SLMs) have supplied an attractive replacement for the holographic film as an input absorbing medium. Moreover, two-dimensional optical data processors using laser diode illumination, acousto-optic (AO) cell data input, and a charge-couple device (CCD) detector array for the output have been designed ([27]). For each realization of the SAR data processor, however, it is important to notice that the spatial resolution of SAR imaging systems is independent of the range r_0 to the target and the velocity v of the radar platform.

Außer von der Erde oder anderen Planeten her bekannten Erscheinungen finden sich auf der Venus aber auch recht ungewöhnliche Strukturen, die uns Planetenwissenschaftlern einiges Kopfzerbrechen bereiten. Eine unerwartet hohe Zahl asymmetrischer Einschlagkrater beispielsweise läßt darauf schließen, daß die dichte Venusatmosphäre entweder die Flugbahn niedergehender großer Meteorite beeinflußt oder aber die Art und Weise, in der sich die hochgeschleuderte Materie verteilt.

R. Stephen Saunders (1991)

5. Remote Sensing

In the imaging mode, SEASAT SAR operated at a center frequency of 1275 MHz (L-band, λ = 23.5 cm) with pulse duration T = 34 μsec and PRF selections of 1464, 1537, 1580, and 1647 Hz admitting a spatial resolution of 25 m. The depression angle ranged between 67° to 73° and produced an image-swath width of 100 km. The antenna was a 10.74 m by 2.16 m phased array system deployed after orbit insertion. The holographic microwave data for each 100 km wide image-swath have been optically processed to produce four film strips each of which covered a width of 25 km and a length of several thousand kilometers.

The first Shuttle imaging radar (SIR-A) experiment was launched on the second flight of Columbia in November 1981. The satellite orbital altitude was 250 km and the image-swath width 50 km in order to cover a surface area of about 10 million km^2. The SAR antenna of 9.44 m by 2.09 m radiating area was stowed inside the Shuttle cargo bay and operated when the Shuttle was in an inverted attitude. As in the SEASAT SAR, the transmitted pulse was a chirp pulse of 1275 MHz center frequency admitting a spatial resolution of 38 m. The image data were recorded as holographic film on board the Shuttle. The data photographic film was developed and then processed at the laboratory by coherent laser light to generate the original image photographic film at a scale of 1 : 500.000.

A second Shuttle imaging radar (SIR-B) experiment was conducted in October 1984. For SIR-B the SAR antenna was modified, however, to permit the depression angle to be changed during the mission within a range of 30° to 75°. The center wavelength λ was the same as in the earlier missions. The satellite orbital altitude was 225 km and the spatial resolution improved to 25 m.

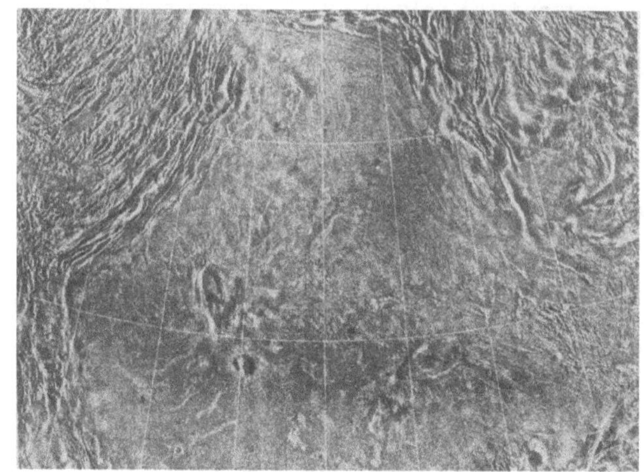

Figure 5

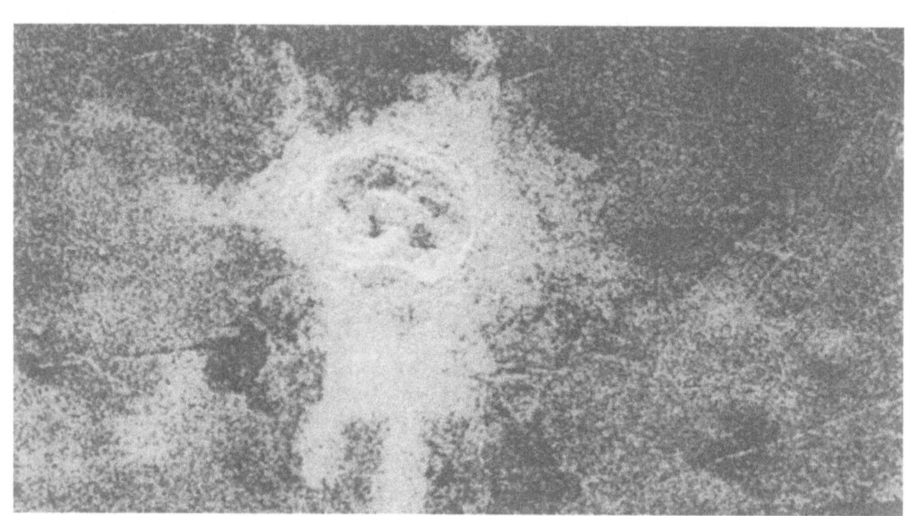

Figure 6

Figure 7

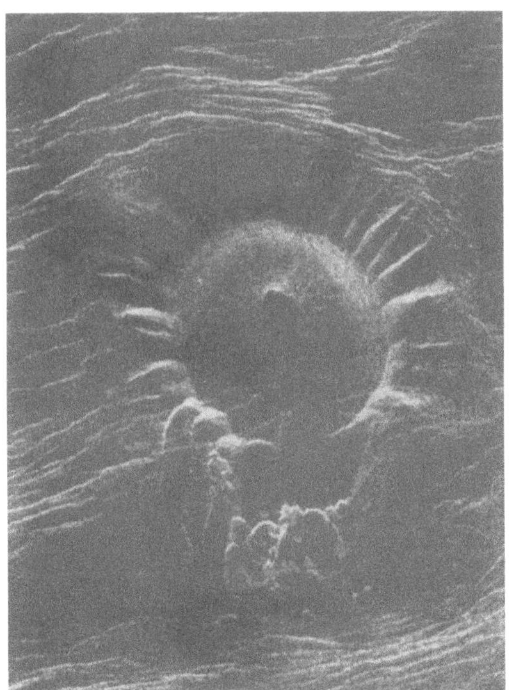

Figure 8

Figure 9

As an example, Figure 5 displays a SAR image of the Lakshmi region of the planet Venus showing details of size down to 1 km. It has been generated by the Soviet Union VENERA 15 and 16 orbiters through the cloud-shrouding atmosphere of Venus which is practically impenetrable for visible light because it reflects about 75 % of the incident sun light. On September 15, 1990 NASA has launched a SAR around Venus for the MAGELLAN mission. The radar is operating at a center frequency of 2385 MHz and provides a resolution down to 120 m which is about 10 % higher than the resolution of the previously processed SAR images of the planet Venus; see Figures 6, 7 and 8. In November 1990 the MAGELLAN SAR started to explore the higher equatorial regions of Venus such as Aphrodite Terra.

By the late 1990s the Cassini spacecraft may be put in an orbit around Saturn and image its moon, Titan, at L-band and K-band on flybys. It will be interesting to compare its resolution with the LACROSSE SAR launched by the Atlantis mission in December 1988. The resolution of the LACROSSE SAR imager being more than 10 times higher than the MAGELLAN SAR is about 1 to 3 meters; see Figure 9 displaying the San Diego Bay.

Acknowledgements

The author would like to thank the Director of the Department of Computer Science of the University of Bonn, Professor Dr. Michael Clausen, whose invitation to deliver lectures on quantum holography and its applications provided the occasion for writing a preliminary version of this article. Moreover, he wishes to gratefully acknowledge a series of illuminating discussions with Director Dr.med. Manfred Crone and Dr.med. Wilhelm Maqua, Department of Radiology, Evangelisches Jung-Stilling Krankenhaus Siegen, concerning their experiences with the various kinds of artifacts generated by the MAGNETOM 1.5 T imager in clinical MRI.

References

1. Babbitt, D., Certain Hilbert spaces of analytic functions associated with the Heisenberg group, *Studies in Mathematical Physics, Essays in Honor of Valentine Bargmann,* E.H. Lieb, B. Simon, and A.S. Wightman, Editors, 19-82, Princeton University Press, Princeton, NJ 1976.
2. Barber, B.C., Theory of digital imaging from orbital synthetic-aperture radar, Int. J. Remote Sensing 6, 1009-1057 (1985).
3. Blahut, R.E., *Theory of Remote Surveillance Algorithms,* Lecture Notes in Radar/Sonar, IMA Preprint Series # 659, Institute for Mathematics and Its Applications, University of Minnesota, Minneapolis, MN 1990.
4. Coifman, R.R., Weiss, G., Representations of compact groups and spherical harmonics, L'Enseignement Math. 14, 121-173 (1968).
5. Cutrona, L.J., Synthetic aperture radar, *Radar Handbook,* M.I. Skolnik, Editor, 23-1-23-25, McGraw-Hill Book, New York, St. Louis, San Francisco, Düsseldorf, Kuala Lumpur, London, Mexico, Montreal, New Delhi, Panama, Rió de Janeiro, Singapore, Sydney, Toronto 1970.
6. Cutrona, L.J., Leith, E.N., Palermo, C.J., and Porcello, L.J., Optical data processing and filtering systems, Trans. IRE IT-6, 386-400 (1960).
7. Cutrona, L.J., Leith, E.N., Porcello, L.J., and Vivian, W.E., On the application of coherent optical processing techniques to synthetic-aperture radar, Proc. IEEE 54, 1026-1032 (1966).
8. Deans, S.R., The Radon transform, *Mathematical Analysis of Physical Systems,* R.E. Mickens, Editor, 81-133, Van Nostrand Reinhold Company, New York 1985.

9. Detlefsen, J., *Radartechnik,* Nachrichtentechnik, Band 18, Springer-Verlag, Berlin, Heidelberg, New York, London, Paris, Tokyo, Hong Kong 1989.
10. Easton, Jr., R.L., Barrett, H.H., Tomographic transformations in optical signal processing, *Optical Signal Processing,* J.L. Horner, Editor, 335-386, Academic Press, San Diego, New York, Berkeley, Boston, London, Sydney, Tokyo, Toronto 1987.
11. Elachi, C., Bicknell, T., Jordan, R.L., and Wu, C., Spaceborne synthetic-aperture imaging radars: Applications, techniques, and technology, Proc. IEEE 70, 1174-1209 (1982).
12. Fitch, J.P., *Synthetic Aperture Radar,* Springer-Verlag, New York, Berlin, Heidelberg, London, Paris, Tokyo 1988.
13. Ganssen, A., Physikalische Grundlagen, *Kernspintomographie,* E. Zeitler, Herausgeber, 31-84, Deutscher Ärzte-Verlag, Köln 1984.
14. Goodman, J.W., A short history of the field of optical computing, *Optical Computing,* B.S. Wherrett, F.A.P. Tooley, Editors, 7-21, Scottish Universities Summer School in Physics Publications, Edinburgh 1989.
15. Gudden, F., Kernspintomographie, *Physik der bildgebenden Verfahren in der Medizin,* H.-J. Maurer, E. Zieler, Herausgeber, 181-196, Springer-Verlag, Berlin, Heidelberg, New York, Tokyo 1984.
16. Harger, R.O., *Synthetic Aperture Radar Systems,* Academic Press, New York and London 1970.
17. Kozma, A., Leith, E.N., and Massey, N.G., Tilted-plane optical processor, Appl. Opt. 11, 1766-1777 (1972).
18. Leith, E.N., Optical processing techniques for simultaneous pulse compression and beam sharpening, IEEE Trans. Aerospace and Electronic Syst. AES-4, 879-885 (1968).
19. Leith, E.N., Quasi-holographic techniques in the microwave region, Proc. IEEE 59, 1305-1318 (1971).
20. Leith, E.N., Synthetic aperture radar, *Optical Data Processing,* D. Casasent, Editor, 89-117, Springer-Verlag, Berlin, Heidelberg, New York 1978.
21. Levanon, N., *Radar Principles,* J. Wiley & Sons, New York, Chichester, Brisbane, Toronto, Singapore 1988.
22. Loudon, R., *The quantum theory of light,* Second Edition, Clarendon Press, Oxford 1983.
23. Ludwig, D., The Radon transform on Euclidean space, Comm. Pure Appl. Math. 19, 49-81 (1966).
24. Moore, R.K., Chastant, L.J., Porcello, L., and Stevenson, J., Imaging radar systems, *Manual of Remote Sensing,* Vol. I, Second Edition, R.N. Colwell, Editor, 429-474, Amer. Soc. of Photogrammetry, Falls Church, VA 1983.
25. Munson, D.C., O'Brien, J.D., and Jenkins, W.K., A tomographic formulation of spotlight-mode synthetic aperture radar, Proc. IEEE 71, 917-925 (1983).
26. Oliver, C.J., White, R.G., Real-time SAR change-detection using neural networks, *Advanced Signal-Processing Algorithms, Architectures, and Implementations,* F.T. Luk, Editor, Proc. SPIE 1348, 40-50 (1990).
27. Psaltis, D., Haney, W., Acoustooptic synthetic aperture radar processors, *Optical Signal Processing,* J.L. Horner, Editor, 191-216, Academic Press, San Diego, New York, Berkeley, Boston, London, Sydney, Tokyo, Toronto 1987.
28. Richards, J.A., *Remote Sensing Digital Image Analysis*, Springer-Verlag, Berlin, Heidelberg, New York, London, Paris, Tokyo 1986.
29. Rihaczek, A.W., *Principles of High-Resolution Radar,* McGraw-Hill, New York, Toronto, London, Sydney 1969
30. Sabins, F.F., *Remote Sensing: Principles and Interpretation,* Second Edition, W.H. Freeman, New York 1987.

31. Schempp, W., *Harmonic Analysis on the Heisenberg Nilpotent Lie Group, with Applications to Signal Theory,* Pitman Research Notes in Math., Vol. 147, Longman Scientific and Technical, Harlow, Essex, and J. Wiley & Sons, New York 1986. Second edition under preparation.
32. Schempp, W., The oscillator representation of the metaplectic group applied to quantum electronics and computerized tomography, *Stochastic Processes in Physics and Engineering,* S. Albeverio, Ph. Blanchard, M. Hazewinkel, and L. Streit, Editors, 305-344, D. Reidel, Dordrecht, Boston, Lancaster, Tokyo 1988.
33. Schempp, W., Neurocomputer architectures, Results in Math. 16, 345-382 (1989).
34. Schempp, W., *Quantum Holography and Neurocomputer Architectures,* Pitman Research Notes in Math., Longman Scientific and Technical, Harlow, Essex, and J. Wiley & Sons, New York (to appear).
35. Schempp, W., High Resolution Imaging: Quantum Holography and Computed Tomography (to appear).
36. Schempp, W., Chen, H.-L., The Uncertainty Inequality in Quantum Holography and Bohr's Indeterminacy Principle, *General Inequalities* 6, W. Walter, Editor, ISNM Series, Birkhäuser Verlag, Basel, Boston, Berlin 1991 (to appear).
37. Schempp, W., Dreseler, B., *Einführung in die harmonische Analyse,* B.G. Teubner, Stuttgart 1980.
38. Wehner, D.R., *High Resolution Radar,* Artech House, Norwood, MA 1987.
39. Weyl, H., Die natürlichen Randwertaufgaben im Außenraum für Strahlungsfelder beliebiger Dimension und beliebigen Ranges, Math. Z. 56, 105-119 (1952).
40. White, J.O., Kwong, S.-K., Cronin-Golomb, M., Fischer, B., and Yariv, A., Wave propagation in photorefractive media, *Photorefractive Materials and Their Applications* II, P. Günter, J.-P. Huignard, Editors, 101-150, Springer-Verlag, Berlin, Heidelberg, New York, London, Paris, Tokyo 1989.

PHOTONIC BAND STRUCTURE

E. Yablonovitch

Bell Communications Research, Navesink Research Center
Red Bank, N.J. 07701-7040

There is an interesting analogy between electron waves in a crystal and light waves in a three dimensionally periodic dielectric structure. Both should be described by band theory. The idea of photonic band structure[1,2] is gaining rapid[3-6] acceptance. The concepts of Reciprocal Space, Brillouin Zones, Dispersion Relations, Bloch Wave functions, Van Hove singularities, etc., are now being applied to optical waves.

A search is underway for three dimensional topologies in which a "photonic bandgap" can open up. This would be an energy band in which optical modes, spontaneous emission, and zero point fluctuations are all absent. Indeed, a photonic bandgap would be essentially ideal since optical dielectric response can be real and dissipationless.

In addition to the obvious applications in atomic physics, inhibited spontaneous emission can now begin to play a role in semiconductor and solid state electronics. If the photonic bandgap overlaps the electronic band edge, spontaneous electron-hole recombination is rigorously forbidden. In a semiconductor laser, this would lead to near unity quantum efficiency into the lasing mode. Photon number state squeezing[7] into that mode would be greatly enhanced.

There have been two main challenges in this field. The first was to show that a full 3-dimensional "photonic bandgap" could actually exist in some type of dielectric structure. The second was to show that such a forbidden gap could be created in a micro-structure amenable to practical micro-fabrication.

At the outset it was realized[1] that a Face-Centered-Cubic (fcc) array in real space would produce the "roundest" Brillouin Zone in reciprocal space. Such a sphere-like Brillouin Zone improves the prospects for a forbidden gap to overlap all the way around its surface. But it was unclear what should be the real-space shape of the atoms in the fcc array. The original suggestion[1] called for cubic atoms. Later, the first experimental effort[8] concentrated on dielectric spheres and on spherical voids in a dielectric background. The spherical void structure appeared to perform particularly well. The history of this field has been a search for that optimal 3-dimensional dielectric geometry, favored by nature and by Maxwell's equations.

During this same period, electronic band theorists began calculating photonic band structure. It rapidly became apparent that the familiar scalar wave band theory, so

Quantum Measurements in Optics, Edited by P. Tombesi and
D.F. Walls, Plenum Press, New York, 1992

frequently used for electrons in solids, was in utter disagreement with experiment on photons[9-12]. Recently a full vector-wave band theory[3-5] became available, which not only agreed with experiment, it high-lighted some discrepancies in experiment. Vector-wave band theory showed that spherical atomic symmetry produced a degeneracy between valence and conduction bands at the W-point of the Brillouin Zone, allowing only a pseudogap, rather than a full photonic bandgap. The solution to this symmetry-induced degeneracy problem is to make the atoms non-spherical.

Therefore we introduce a practical, new, face-centered-cubic (fcc) structure which simultaneously solves two of the outstanding problems in photonic band structure: (1) In this new geometry the atoms are non-spherical, lifting the degeneracy at the W-point of the Brillouin Zone, and permitting a full photonic bandgap rather than a pseudogap. (2) Furthermore, this fully 3-dimensional fcc structure lends itself readily to micro-fabrication on the scale of optical wavelengths. It is created by simply drilling 3 sets of holes 35.26^{o} off vertical into the top surface of a solid slab or wafer, as can be done for example by reactive ion etching. At refractive index $n\sim3.6$, typical of semiconductors, the 3-d forbidden photonic bandgap width in this new structure is ~20% of its center frequency. Calculations indicate that the gap remains open for refractive indices $n\geq2$.

The Wigner-Seitz (W-S) unit cell of the fcc lattice is a rhombic dodecahedron as shown in Fig. 1. The problem of creating an arbitrary fcc dielectric structure reduces to the problem of filling the fcc W-S real-space-unit-cell with an arbitrary spatial distribution of dielectric material. Real space is then filled by repeated translation and close-packing of the W-S unit cells. Scientific progress in this field has been marked by improved choices of how to fill the W-S unit cell with dielectric material. As already mentioned, early proposals called for cubic[1] atoms, then spherical atoms and spherical voids[8] to be inscribed inside the W-S unit cell.

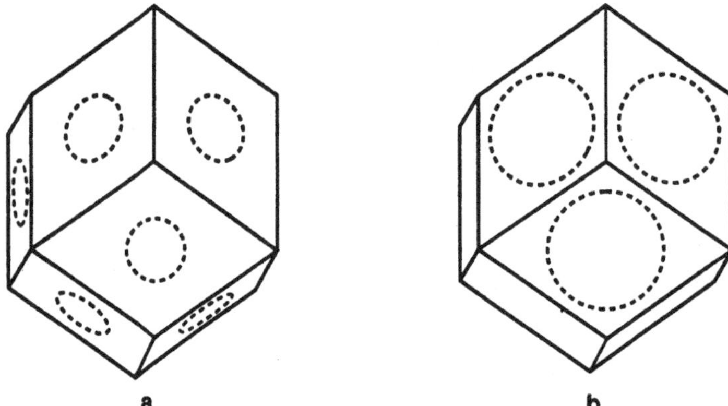

Fig. 1. The Wigner-Seitz real-space-unit-cell of the fcc lattice is a rhombic dodecahedron. In ref.~8, slightly oversized spherical voids were inscribed into the unit cell, breaking through the faces, as illustrated by the dashed lines in (a). The current structure, shown in (b), is non-spherical. Cylindrical holes are drilled through the top 3 facets of the rhombic dodecahedron and exit through the bottom 3 facets. The resulting atoms are roughly cylindrical, and have a preferred axis in the vertical direction.

Fig. 1(a) shows a W-S unit cell filled by an over-sized spherical void, a structure which performed rather well in ref.8. Since the spheres were slightly larger than close-packed, the voids broke through the surfaces of the W-S unit cell as indicated by the dashed circles on the faces of the rhombic dodecahedron in Fig. 1(a). In ref.8 it was already pointed out that there was a symmetry-induced degeneracy at the W-point of the Brillouin Zone in fcc structures. There was a danger that the valence and conduction bands could touch at the degeneracy, closing the photonic bandgap. Based on the weight of experimental evidence however, ref. 8 argued that the degeneracy had only caused adjacent conduction band levels to touch, permitting the gap to remain open. Vector-wave band theory, which has become quite successful recently[3-5], contradicted this. It showed that the degeneracy did indeed cause valence and conduction bands to touch at W, permitting only a pseudogap rather than a full photonic bandgap. Unfortunately, the finite-sized experimental sample in ref. 8 allowed inadequate resolution to detect touching at isolated points on the Brillouin Zone.

The degeneracy at W can be lifted by lowering the spherical symmetry of the atoms inside the W-S unit cell. We have made a close examination[13] of the degenerate wave-functions at W in the nearly-free-photon model, and learned that a distortion of the spherical atoms along the <111>-direction will lift the degeneracy. The W-S unit cell in Fig. 1(b) has great merit for this purpose. Holes are drilled through the top 3 facets of the rhombic dodecahedron and exit through the bottom 3 facets. The beauty of the structure in Fig.1(b) is that a stacking of W-S unit cells results in straight holes which pass clear through the entire "crystal"! The "atoms" are odd-shaped, roughly cylindrical voids centered in the W-S unit cell, with a preferred axis pointing to the top vertex.

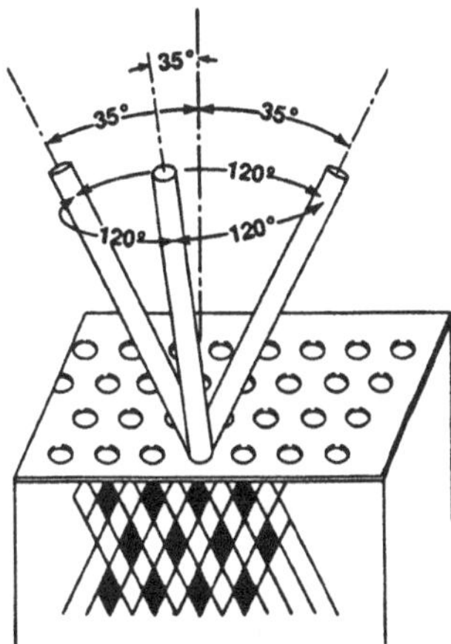

Fig. 2. The method of constructing an fcc lattice of the Wigner-Seitz cells as shown in Fig. 1(b). A slab of material is covered by a mask consisting of a triangular array of holes. Each hole is drilled through 3 times, at an angle 35.26° away from normal, and spread out 120° on the azimuth. The resulting criss-cross of holes below the surface of the slab, suggested by the cross-hatching shown here, produces a fully 3-dimensionally periodic fcc structure, with unit cells as given by Fig. 1(b). The drilling can be done by a real drill bit for microwave work, or by reactive ion etching to create an fcc structure at optical wavelengths.

An operational illustration of the construction which produces an fcc "crystal" of such W-S unit cells is shown in Fig. 2. A slab of material is covered by a mask containing a triangular array of holes. Three drilling operations are conducted through each hole, 35.26° off normal incidence and spread out 120° on the azimuth. The resulting criss-cross of holes below the surface of the slab produces a fully 3-dimensionally periodic fcc structure, with W-S unit cells given by Fig. 1(b)! The drilling can be done by a real drill bit for microwave work, or by reactive ion etching to create an fcc structure at optical wavelengths. We have fabricated such "crystals" in the microwave region by direct drilling into a commercial, low-loss, dielectric material, Emerson & Cumming Stycast-12. Its microwave refractive index, n~3.6, is meant to correspond to that of the common semiconductors, Si, GaAs, etc. By simply scaling down the dimensions, this structure can be employed equally well at optical wavelengths. In this paper we will present the measured and calculated, ω-vs-k, dispersion relations for this new photonic crystal.

There are other ways to break the spherical symmetry of Fig. 1(a). In an important new proposal[5], Ho, Chan and Soukoulis placed two atoms in the W-S unit cell, forming diamond structure and allowing a very wide photonic bandgap. Diamond symmetry can be created by supplementing the operations of Fig. 2 with 3 additional drilling operations, making a total of 6 drilling directions. These 3 new drilling directions, 120° apart, would all lie within the plane of the slab. Therefore they are somewhat difficult to implement experimentally. The 6 drilling directions correspond to the 6 inequivalent <110> channeling holes in diamond structure.

We have experimentally surveyed 3 fcc structures, drilled in accordance with Fig. 2, to different ratios d/a of hole diameter d to fcc unit cube length a: d/a = 0.361, 0.433, and 0.469. The removed volume fraction was approximately 62%, 70%, and 78% in the 3 cases, respectively. The 78% empty structure had the largest forbidden gap in this set and in this Letter we will present results on that structure only. We believe 78% is near the optimal volume fraction for this fcc geometry.

Our procedure is similar to the one we used in ref. 8, except that our dynamic range was improved by using an HP-8510 Network Analyzer for all the measurements. The experimental aim is to fully explore all 4π steradians in reciprocal space, while scanning frequency. The valence band edge frequency is defined by a sudden drop in microwave transmission, while the conduction band edge is defined by the frequency at which the transmitted signal recovers. Conservation of wave vector momentum parallel to the slab entry face determines the band edge position along the surface of the Brillouin Zone. Since there are two polarizations, we can usually determine the 2 valence band edges and 2 of the conduction bands.

Sometimes the coupling of external plane waves to internal Bloch waves is poor, and the experiment can miss one of the conduction band edges, as happened in ref. 8. Finite sample size limits the usable dynamic range, exacerbating the weak coupling problem. Therefore it is important to back up the measurements with numerical calculations, as we have done here. The scalar dielectric constant distribution in Fig. 1 is expanded as a Fourier series in reciprocal space, while Maxwell's equations are expanded[3] as vector plane waves. The eigenvalues converge after a few hundred plane waves are summed in the expansion.

In spite of the non-spherical atoms of Fig. 1(b), the Brillouin Zone (BZ) is identical to the standard fcc BZ shown in textbooks. Nevertheless, we have chosen an unusual perspective from which to view the Brillouin Zone in Fig. 3(a). Instead of having the fcc BZ resting on one of its diamond-shaped facets as is usually done, we have chosen in Fig. 3(a), to present it resting on a hexagonal face. Since there is a preferred axis for the atoms, the distinctive L-points centered in the top and bottom hexagons are 3-fold symmetry axes, and are labeled L_3. The L-points centered in the other 6 hexagons are symmetric only under a 360° rotation, and are labeled L_1. It is helpful to know that the U_3-K_3 points are equivalent since they are a reciprocal lattice vector apart. Likewise the U_1-K_1 points are equivalent.

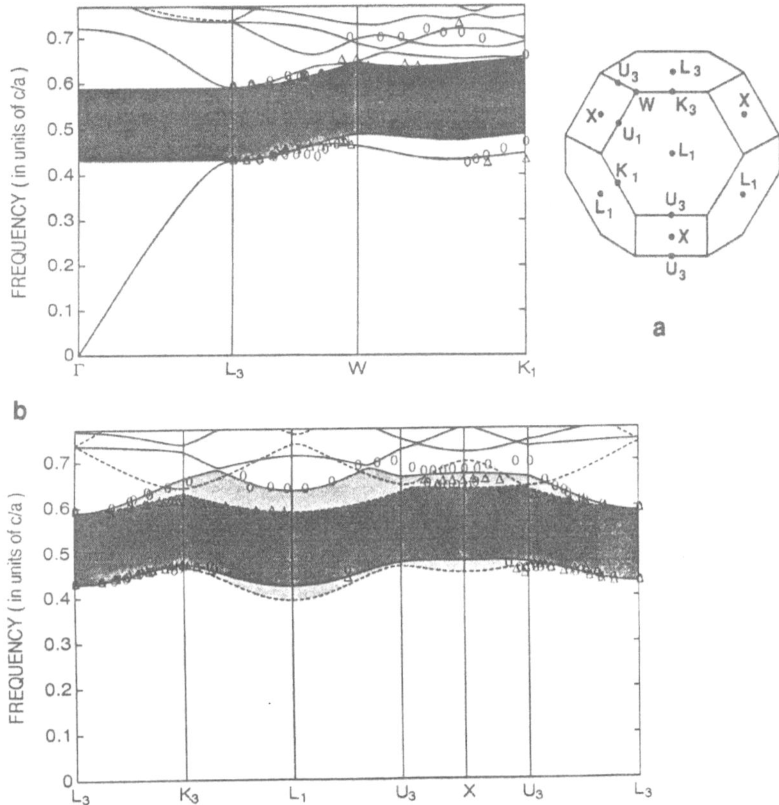

Fig. 3. (a) The Brillouin Zone of an fcc structure incorporatingnon-spherical atoms, as in Fig. 1(b). Since the space lattice is not distorted, this is simply the standard fcc Brillouin Zone lying on a hexagonal face rather than the usual cubic face. Only the L-points on the top and bottom hexagons are 3-fold symmetry axes. Therefore they are labeled L_3. The L-points on the other 6 hexagons are labeled L_1. The U_3-K_3 points are equivalent since they are a reciprocal lattice vector apart. Likewise the U_1-K_1 points are equivalent.

(b) Frequency versus wave-vector, ω-vs-k, dispersion along the surface of the Brillouin Zone shown in 3(a), where c/a is the speed of light divided by the fcc cube length. The ovals and triangles are the experimental points for s~and~p polarization respectively. The solid and dashed lines are the calculations for s and p polarization respectively. The dark shaded band is the totally forbidden bandgap. The lighter shaded stripes above and below the dark band are forbidden only for s and p polarization respectively.

Normal incidence on the slab of Fig. 2 sends the propagation vector directly toward L_3 in reciprocal space ("the North pole"). Tilting the angle of incidence moves the propagation vector along a "meridian" toward the "equator". By choosing different azimuthal angles toward which to tilt, the full reciprocal space can be explored. Fig. 3(b) shows the dispersion relations along different meridians for our primary experimental sample of normalized hole diameter d/a = 0.469 and 78% volume fraction removed. The oval points represent experimental data with s-polarization ($\perp$ to the plane of incidence, $\parallel$ to the slab surface), while the triangular points represent p-polarization ($\parallel$ to the plane of incidence, partially $\perp$ to the slab surface). The horizontal abscissa in the lower graph of Fig. 3(b), L_3 - K_3 - L_1 - U_3 - X - U_3 - L_3 represents a full meridian from the North pole to the South pole of the BZ. Along this meridian the Bloch wave functions separate neatly into s and p polarizations. The s and p polarized theory curves are the solid and dashed lines respectively. The dark shaded band is the totally forbidden photonic bandgap. The lighter shaded stripes above and below the dark band are forbidden only for s and p polarization respectively.

Along the meridian L_3 - W - K_1, the polarizations don't separate neatly, and only the totally forbidden photonic bandgap is shaded. The top of the valence band is at W and is mostly s-polarized, but the valence band peaks at U_3, X, and U_1 are almost as high. The bottom of the conduction band is at L_1, purely p-polarized, is only marginally lower than the valley at L_3. We have also measured the imaginary wave vector dispersion within the forbidden gap. At mid-gap we find an attenuation of 10dB per unit cube length a. Therefore the photonic crystal need not be very many layers thick to effectively expel the zero-point electromagnetic field.

At a typical semiconductor refractive index, n=3.6, the 3-d forbidden gap width is 19% of its center frequency. We have repeated the calculation at lower refractive indices, re-optimizing the hole diameter. Our calculations indicate that the gap remains open for refractive indices as low as n=2.1 using circular holes as in Fig. 2. In reactive ion etching, the projection of circular mask openings at 35° leaves oval holes in the material, which might not perform as well. Fortunately we found, defying Murphy's Law, that the forbidden gap width for oval holes is actually improved, fully 21.7% of its center frequency.

In the visible region, there are many transparent optical materials available with a refractive index above 2.1. Furthermore, state-of-the-art reactive ion etching[14] can produce holes that are $\geq$20 times deeper than their diameter, deep enough to produce an fcc photonic crystal with substantial inhibition in the forbidden gap. It appears that the application of photonic bandgaps to semiconductor physics, optical, and atomic physics may soon be practical.

ACKNOWLEDGEMENTS

EY would like to thank the authors of ref. 5 for telling about their diamond structure calculations before publication, and for intensive discussions of the degenerate Bloch wave functions at the W-point. John Gural deserves special thanks for his patience, dedication, and skilled machining of tens of thousands of holes, which made this project possible.

REFERENCES

1. Yablonovitch E., Phys. Rev. Lett. **58**, 2059 (1987).
2. John S., Phys. Rev. Lett. **58**, 2486 (1987).
3. Leung K. M. and Liu Y.F., Phys. Rev. Lett. **65**, 2646 (1990).

4. Zhang Z. and Satpathy S., Phys.Rev.Lett. **65**, 2650 (1990).
5. Ho K.M., Chan C.T. and Soukoulis C.M., Phys. Rev. Lett. **65**, 3152 (1990).
6. John S. and Wang J., Phys. Rev. Lett. **64**, 2418 (1990).
7. Yamamoto Y. and Machida S., Phys. Rev. A**35**, 5114 (1987).
8. Yablonovitch E. and Gmitter T.J., Phys. Rev. Lett. **63**, 1950 (1989).
9. Satpathy S., Zhang Z. and Salehpour M.R., Phys. Rev. Lett. **64**, 1239 (1990).
10. Leung K.M. and Liu Y.F., Phys. Rev. B**41**, 10188 (1990).
11. John S. and Rangarajan R., Phys. Rev. B**38**, 10101 (1988).
12. Economu E.N. and Zdetsis A., Phys. Rev. B**40**, 1334 (1989).
13. Yablonovitch E., unpublished.
14. Scherer A., van der Gaag B.P., Beebe E.D. and Lin P.S.D., J. Vac. Sci. Technol. B**8**, 28 (1990).

INFORMATION, QUANTUM CORRELATIONS AND COMMUNICATION

Stephen M. Barnett[(1)], D.T. Pegg[(2)] and Simon J.D. Phoenix[(3)]

[(1)] Physics Dept., King's College, Strand, London. WC2R 2LS. UK
[(2)] Division of Science and Technology, Griffith University
Nathan, Brisbane, Queensland 4111. Australia
[(3)] British Telecom Research Laboratories, Martlesham Heath
Ipswich. IPS 7RE. United Kingdom

1. INTRODUCTION

It has been known for more than fifty years[1] that there is something special about quantum correlations. The strength of these correlations leads to consequences that are incompatible with local reality[2] and would appear to suggest the possibility of superluminal communication. However, it has been proven that the statistical properties of one of a pair of correlated systems are *unaltered* by observation of its partner[3]. No information is *transmitted* by the action of the measurement. Nonetheless, this does not imply that quantum correlations have no relevance in the discussion of communications.

2. FORMATION AND CORRELATION STRENGTH

It is true that we cannot transfer information to a quantum system by manipulation of a correlated partner system. However, this does not imply that quantum correlations contain no information. Indeed, we can use the information tied up in the correlations as an absolute measure of the strength of the correlations. We define the index of correlation,[4,5] I_c, to be the quantity of information lost by measuring the properties of the correlated systems *independently*. The quantitative expression of this index is

$$I_c = S_a + S_b - S ,$$

where S is the entropy associated with the correlated state and $S_{a(b)}$ is the entropy derived from the reduced density operator for the *a(b)* system alone. If the two systems are decorrelated then $I_c = 0$, but if they are correlated then $I_c > 0$. Maximising I_c corresponds to optimising the strength of the correlations.

The index of correlation has its maximum value when two conditions are satisfied[5]: (i) $S = 0$, that is the combined system is in a pure state, (ii) $S_a = S_b = S_{a(b)}^{max}$. These conditions are equivalent to the requirement that there should be a *minimum* amount of information associated with the *a* and *b* systems independently *but* the combined state is completely specifiable. For a pair of spin-1/2 particles, optimally correlated states are pure states with random spin properties. One example of such a maximally correlated state is the singlet state often discussed in association with EPR correlations[1,2]

$$|\psi^{max}> = \frac{1}{\sqrt{2}}\left(|-{>}_a\,|+{>}_b - |+{>}_a\,|-{>}_b\right),$$

where $|+(-)>_a$ is the state corresponding to spin *a* aligned up(down) along a chosen direction and $|+(-)>_b$ is the state corresponding to spin *b* aligned up(down) along the same direction. This maximally correlated state is a pure state but has random single spin properties; the probability that either spin will be aligned in a given direction is one half.

If there is an energy constraint then maximally entropic states are thermal. In this case, the states with the strongest correlations will be pure states with thermal singlesystem properties. One important example of this type is well-known in quantum optics. A two-mode squeezed vacuum,

$$|re^{i\phi}> = \sum_{n=0}^{\infty} \frac{(\tanh r\; e^{i\phi})^n}{\cosh r}\,|n>_a\,|n>_b\,,$$

is a pure state with single-mode thermal (Bose-Einstein) properties[6]. As such it is the most strongly correlated of all two-mode states of light[5]. Analysis of pairs of observables shows that half the information contained in the correlations may be associated with the correlation between the photon numbers[7] Other less strongly correlated observables reveal less of the information contained in the correlations.

3. CORRELATIONS AND COMMUNICATION

If a message is transmitted by using two channels, then the correlations between the channels may be used to encode the information. This is true both classically and quantum mechanically and it has the advantage that an eavesdropper observing only one channel will learn nothing from his observations. However, the strength of the quantum correlations allows more information to be stored in the correlations than is possible in an analogous classical system[7]. We will not discuss the details of this principle here but rather illustrate it with an elementary example.

The most elementary correlated state of light is formed when a single photon falls on a 50:50 beamsplitter. The output state is an entangled superposition of the form

$$|\psi^{(\pm)}> = \frac{1}{\sqrt{2}}\left(|1>_a\,|0>_b \pm i\,|0>_a\,|1>_b\right),$$

where the subscripts a and *b* correspond to the a and *b* output modes and the ± sign is determined by which of the two input modes is excited. Such a state is an entangled superposition state and will exhibit the non-locality that is characteristic of correlated quantum states[8].

The index of correlation reveals that the correlations in the states $|\psi^{(\pm)}>$ contain two bits of information. The first corresponds to the photon number and the second to the phase of the superposition. This second piece of information is a consequence of the fact that the state $|\psi^{(\pm)}>$ is a *superposition* of two-mode states rather than a statistical mixture. In this sense, the second piece of information is intrinsically quantum mechanical. It may be read by the receiver with the use of a second beamsplitter. If the path lengths of the resulting Mach-Zehnder interferometer are suitably adjusted the phase of the superposition is revealed by the direction of the photon on leaving the interferometer.

The scheme presented here is only one elementary example of the potential of quantum correlations for communications. More sophisticated techniques are to be found in the newly-developing science of quantum cryptography. An introduction to this field and an example of the role of quantum correlations has been presented by Ekert[9].

4. CONCLUSIONS

In 1928 Eddington[10] wrote : "We often think that when we have completed our study of *one* we know all about two, because 'two' is 'one and one'. We forget that we still have to make a study of 'and'." His point was that features of organisation, arrangement or correlation are crucial to our understanding of the physical world. He also noted that entropy plays a key role in interpretin8 the significance of correlations. More than sixty later the same ideas are bringing to light new subtleties in the quantum theory of "and".

ACKNOWLEDGEMENTS

S.M.B. thanks Paolo Tombesi and Elisabeth Giacobino for a wonderful ten days in Cortina. This work was supported, in part, by G.E.C. Research Ltd., the U.K. Fellowship of Engineering and The Royal Society.

REFERENCES

1. A.Einstein, B.Podolsky and N.Rosen, Can a Quantum Mechanical Description of Reality be Considered Complete ?, Phys.Rev. A **47**; 777 (1935).
2. J.S.Bell, Speakable and Unspeakable in Quantum Mechanics, Cambridge University Press, Cambridge (1987).
3. G.C.Ghirardi, A.Rimini and T.Weber, A General Argument Against Superluminal Transmission Through the Quantum-Mechanical Measurement Process, Lett. Nuovo Cim., **27**; 293 (1980).
4. W.H.Zurek, Information Transfer in Quantum Measurements: Irreversibility and Amplification, in Quantum Optics, Experimental Gravity and Measurement Theory, P.Meystre and M.0.Scully, eds., Plenum Press, New York (1983).
5. S.M.Barnett and S.J.D.Phoenix, Entropy as a Measure of Quantum Optical Correlation, Phys.Rev. A **40**; 2404 (1989).
6. S.M.Barnett and P.L.Knight, Thermofileld Analysis of Squeezing and Statistical Mixtures in Quantum Optics, J. Opt. Soc. Am. B **2**; 467 (1985).
7. S.M.Barnett and S.J.D.Phoenix, Information Theory, Squeezing and Quantum Correlations, submitted to Phys.Rev. A.
8. S.M.Tan, D.F.Walls and M.J.Collett, Nonlocality of a Single Photon, Phys.Rev.Lett., **66**, 252 (1991).
9. A.K.Ekert, Quantum Cryptography Based on Bell's Theorem, published in this volume.
10. A.S.Eddington, The Nature of the Physical World, Cambridge University Press, Cambridge (1928).

V. PHOTON INTERFERENCE AND BELL'S THEOREM

WHAT IS TWO PHOTONS?

INTERFERENCE AND NON-LOCAL EFFECTS USING PHOTON PAIRS.

J. G. Rarity and P. R. Tapster

Royal Signals and Radar Establishment
St Andrews Rd, Malvern
Worcs., UK, WR14 3PS

1. INTRODUCTION

To address correctly the question "What is two photons?" we first have to satisfactorily describe one photon. One possible answer, that preferred by the experimentalists, is "the photon was the quanta that caused the click in our photon counting detector". In other words the smallest quantity of energy extractable from an electromagnetic field by a detector is a photon. These answers are certainly safe in the sense that they allow for interference effects to occur when more than one route to the detector exists. They do not however address non-classical light sources, the simplest of which is the one photon state. Recent experiments have produced approximations to such states using photon pair sources[1–3]. Such states also produce interference effects when input into interferometers[1] supporting Dirac's statement "A photon only interferes with itself"[4]. We must conclude therefore that a photon can be described as a "click" after detection but in order to describe fully the propogation of a quantum of light from a source we need to use a model where probability amplitudes (or field operators) are propagated down all possible paths to the detector where the square modulus of their vector sum provides a probability of detection.

In recent experiments non-degenerate parametric downconversion has been used as a source of photon pairs. In this process a pump beam of short wavelength light illuminates a non-linear crystal. A small amount of the pump light photons are down-converted to pairs of longer wavelength photons emitted simultaneously in directions satisfying phase matching conditions and energy conservation. Pair photons can be selectively detected using suitably placed single photon counting detectors and coincidence counting techniques. Experimenters have exploited this quantum correlation to demonstrate one photon states[2], photon anti-bunching[3], sub-poissonian and sub-shot noise light[5]. More recently, interference effects occuring when the two photons are superposed have been investigated. As discrete photons have random relative phase we do not see classical second order interference effects in such experiments. Interfer-

Quantum Measurements in Optics, Edited by P. Tombesi and
D.F. Walls, Plenum Press, New York, 1992

ence effects are however seen in the coicidence rate, the rate of detection of photon pairs. The first experimental demonstration of these fourth order interference effects was obtained by superposing pair photons on a two element detector[6]. The coincidence rate between elements oscillated with element separation. Although the detected effect was weak due to the finite detector size compared to oscillation period it is clearly non-classical. It is also non-local in the sense that detection of a photon at one position strongly affects the possible positions of detection of its partner. Much clearer effects occur when the pair photons are superposed at beamsplitters[7–11]. Here we will discuss two experiments which clearly show that the interference effects are caused by there being more than one equivalent path for the two photon to propogate from the source to the detector. In a sense the experiments can be described in terms of self interference of two-photons in the spirit of Dirac's statement reproduced above, hence the title "What *is* two photons".

The first experiment exploits the momentum conservation inherent in parametric downconversion to create an entangled state involving two different possible propogation paths for the pair photon to travel from the crystal to the coincidence detectors[10,11,12]. The second experiment exploits the energy conservation inherent in the creation of two downconverted photons from one pump photon. This leads to a phase coherence between pairs emitted at different times. Pair photons emitted at different times can be superposed using identical out of balance Mach-Zehnder interferometers[13–18]. The coincidence rate can be modulated by small changes of path length difference in either interferometer. In the version described here we show that these effects are retained when one of the photons propogates 170m down an optical fibre[19].

2. EXPERIMENTS

The two experimental arrangements are shown in figures 1 and 2. A parametric downconversion crystal cut for non-degenerate operation is pumped by a short wavelength laser. Signal and idler photons are emitted in a broad band cone behind the crystal with pairs satisfying energy conservation

$$\omega_i + \omega_s = \omega_0 \tag{1}$$

where $\omega_{i,s,0}$ are signal, idler and pump angular frequencies respectively. Pairs are selected by placing apertures in the downconverted cone satisfying the phase matching conditions in the crystal (in terms of wave vectors $\vec{k}_i + \vec{k}_s = \vec{k}_0$).

2.1. *Superposition of momentum states*

In the first arrangement (figure 1) double apertures are used and the photon pairs consist of one photon of slightly longer wavelength in the upper outer aperture matched to one of slightly shorter wavelength passing through the lower inner aperture. Alternatively the longer wavelength photon could appear in the outer lower aperture paired with a shorter wavelength photon in the upper inner aperture. The two pairs are arranged to be indistinguishable by placing the apertures symmetrically with respect to the pump beam. Mirrors are then used to superpose the long wavelength (signal) photons at one beamsplitter and the short wavelength (idler) photons at another (in

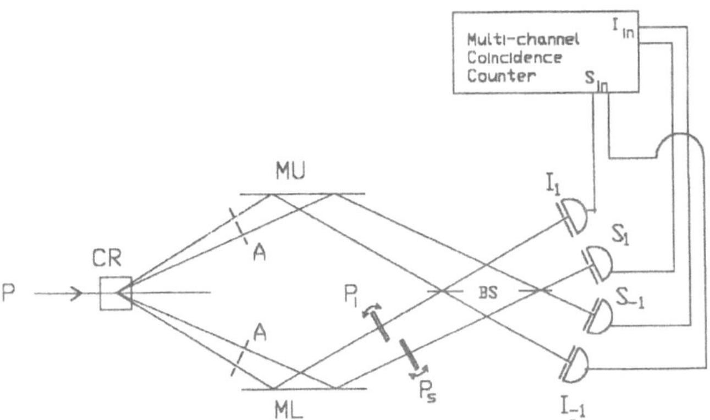

Fig. 1.Schematic of the momentum state superposition apparatus. A 413nm pump beam (P) illuminates a KD*P non-linear crystal (CR). Double apertures (A) are used to select upper and lower signal (dashed) and idler (dot-dash) beams. Phase plates P_i and P_s are placed in the lower idler and signal beams respectively. Signal beams are recombined at a beamsplitter and detected in photon counting detectors S_1 and S_{-1}. Idler beams are recombined at a physically separate point on the beamsplitter and detected in I_1, I_{-1}. Coincidences between signal and idler photodetections are measured in a multi-channel coincidence counter.

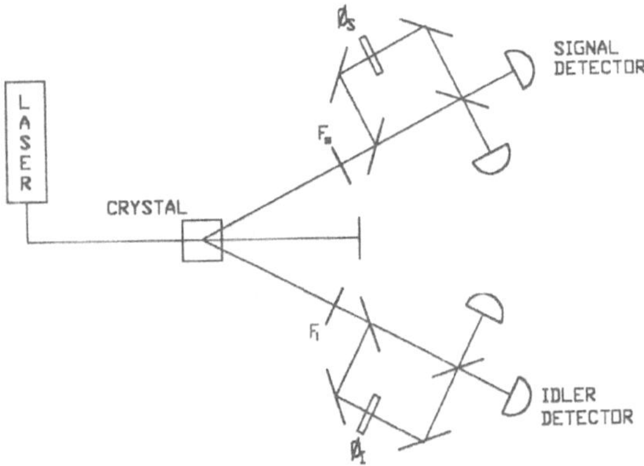

Fig. 2. Schematic of the emission time superposition apparatus. Filters F_s and F_i select energy (and phase) matched photon pairs. Identical out of balance Mach-Zehnder interferometers are placed in each arm of the apparatus. Coincidences between signal and idler detectors are measured as a function of phase shift ϕ_i and ϕ_s.

the actual experiment this was a physically separate point on the same beamsplitter). Signal photon detectors labelled S_1 and S_{-1} and idler detectors labelled I_1 and I_{-1} view the four beamsplitter outputs. When the photon pairs are selectively detected by coincidence counting between signal and idler channels we cannot distiguish which route the pair photon has taken and hence interference effects can occur. A simplified version of the wavefunction for the pair photon selected by the pair of double apertures is

$$\Psi_{si} = \frac{1}{\sqrt{2}}\left(|1_{upper}>_i |1_{lower}>_s e^{i\phi_s} + |1_{lower}>_i |1_{upper}>_s e^{i\phi_i}\right) \tag{2}$$

where $|1>_{s,i}$ denotes a one-photon state in the signal/idler arm of the apparatus and we include here phases $\phi_{i,s}$ introduced by phaseplates in the lower signal and idler beams. The probability $P_{si}(j,k)$ $(j,k=\pm1)$ of seeing a signal photodetection in detector S_j coincident with an idler photodetection in I_k is then given by

$$P_{si}(j,k) = \eta_s\eta_i < \Psi|\hat{a}^\dagger_{ik}\hat{a}^\dagger_{sj}\hat{a}_{sj}\hat{a}_{ik}|\Psi > \tag{3}$$

where η_s and η_i are detector efficiencies and $\hat{a}_{sj}$ and $\hat{a}_{ik}$ are annihilation operators for photons at the signal and idler detectors respectively. The operators at the detectors can be expressed in terms of the operators in the upper and lower branches $\hat{a}_{su,sl,iu,il}$ by projecting back through the beamsplitter. For example

$$\hat{a}_{s1} = i\sqrt{R}\hat{a}_{su} + \sqrt{T}\hat{a}_{sl} \tag{4}$$

where R and T are intensity reflection and transmission coefficients at the beamsplitter and a 90 degree phase change on reflection ensures energy conservation in the beamsplitter.

Substitution of equations 2 and 4 into 3 and applying annihilation operators to their respective states only we obtain the pair photon detection probability,

$$P_{si}(j,j) = RT\eta_s\eta_i[1+\cos(\phi_i-\phi_s)] \tag{5}$$

$$P_{si}(j,k) = \frac{R^2+T^2}{2}\eta_s\eta_i[1-2\frac{RT}{R^2+T^2}\cos(\phi_i-\phi_s)] \quad j\neq k$$

which reduces to

$$P_{si}(j,k) = \frac{1}{4}\eta_s\eta_i[1+jk\cos(\phi_i-\phi_s)]. \tag{6}$$

for 50/50 beamsplitters. The striking fact about this result is that the probability of coincident detection of two apparently separate photons can be modulated with 100% certainty using either of the widely separated phase plates. These apparent non-local effects can be used to demonstrate violations of local realistic theories or Bell inequalities[20]. This experiments is novel in that it does not use polarisation as the locally measured variable. In the experiment all four possible coincidence rates $C(j,k)$ $(j,k=\pm1)$ were measured as a function of phase-plate settings and combined to form the normalised correlation coefficient $E(\phi_s,\phi_i)$.

$$E(\phi_s,\phi_i) = \frac{C(1,1)+C(-1,-1)-C(1,-1)-C(-1,1)}{\Sigma_{jk}C(j,k)} \tag{7}$$

$$= P(1,1)+P(-1,-1)-P(1,-1)-P(-1,1) = \cos(\phi_i-\phi_s)$$

for 50/50 beamsplitters. A plot of the measured correlation coefficient as a function of phase plate setting is shown in figure 2. The measured visibility is less than 100%

because of incomplete spatial mode matching and frequiency matching at the beamsplitters which were close to 50/50 for the wavelengths used. The visibility is however high enough for a demonstration of a violation of Bell's inequality which is described in detail elsewhere[11]. Also described elsewhere is the full multimode theory of this effect including detail of the phase matching which leads to the creation of these two color two-photons[10].

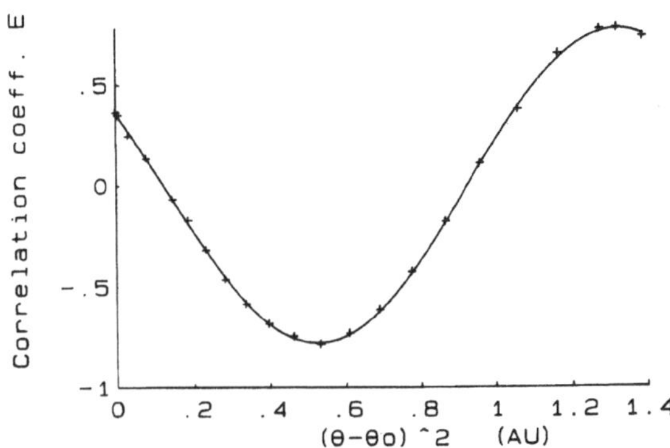

Fig. 3. Correlation coefficient $E(\phi_s, \phi_i)$ defined in equation 7 measured as a function of phase plate setting ϕ_s. Phase plates thickness was adjusted by tilting an angle θ away from nominal perpendicular (θ_0) hence ϕ_s varies linearly as a function of $(\theta - \theta_0)^2$.

2.2. *Superposition of times of emission*

A schematic of the apparatus is shown in figure 2. Out of balance Mach-Zehnder interferometers are placed in the signal and idler arms of the apparatus. There are thus two equivalent routes from the crystal to the detector for each photon of the pair. A simplistic wavefunction for the two photons viewed from the detectors (but normalised at the crystal) assuming 50/50 beamsplitters is thus

$$\Psi_{si} = \frac{1}{4}[|1_{long}>_i |1_{long}>_s e^{i(\phi_i+\phi_s)} e^{i\Delta T(\omega_i+\omega_s)} + |1_{short}>_i |1_{short}>_s \tag{8}$$

$$+|1_{short}>_i |1_{long}>_s e^{i\phi_s} e^{i\Delta T\omega_s} + |1_{long}>_i |1_{short}>_s e^{i\phi_i} e^{i\Delta T\omega_i}]$$

where again $|1>_{s,i}$ denotes a one-photon state in the signal/idler arm of the apparatus, the subscripts *long* and *short* denote propogation via the long and short paths through the interferometer and the small phase shifts $\phi_{i,s}$ incorporate any phase changes occuring on reflection while ΔT represents the mean time delay incurred on taking the long path in either interferometer. Containing a superposition of a *long, long* two-photon with a *short, short* two-photon this wavefunction directly expresses the uncertainty in time of emission of the photon pair. Coherence between the two possible emission times is ensured when the pump laser has a constant frequency ω_0 (ie a long coherence length as compared to ΔT) and when energy conservation is strictly applied in the relationship $\omega_i + \omega_s \equiv \omega_0$. In this situation we can evaluate the probability of coincidenct detection using equations 3 and 6.

$$P_{si}(j,k) = \frac{1}{8}\eta_s\eta_i[2 + \cos(\phi_i + \phi_s + \omega_0 \Delta T)]. \tag{9}$$

The interference here has a maximum visibility of 50% and there has been some discussion of whether this can be truly proved non-classical[14]. The interference is diluted by detection of long-short events where one photon of the pair takes the long path while the other travels via the short path. This introduces a time delay ΔT in their detection times which can in principle be resolved by using a short enough coincidence gate. When these events are discriminated against the factor of two above is reduced to unity and theoretical visibility of the coincidece fringes rises to 100%. This has recently been experimentally demonstrated[18]

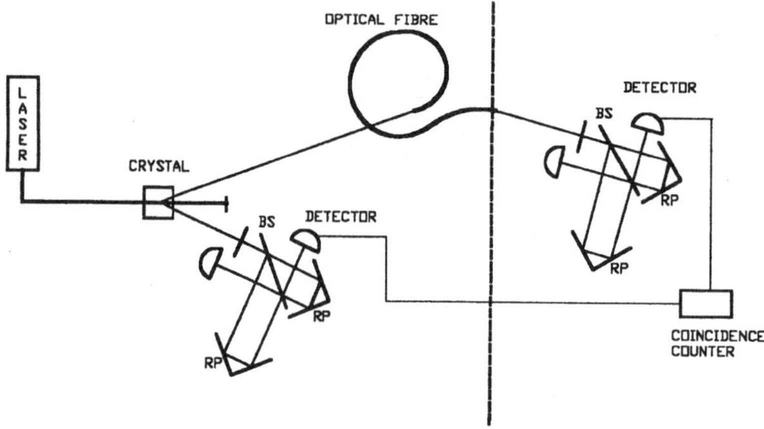

Fig. 4. Experimental demonstration of two-photon interference at large distances. A pump laser illuminates the non-linear crystal and signal and idler photons so created are launched into the two interferometers. A simplified form of Mach-Zehnder is used involving a single beamsplitter (BS) and retroreflecting prisms (RP). One prism is mounted on a 0.1μm resolution stepper motor driven translation stage to allow variation of phase $\phi_{i,s}$ Interferometers are optically separated by 170m of optical fibre. Furthur details of the apparatus appear in the text.

In our experiment (figure 4) a Lithium Iodate crystal pumped by a Helium Cadmium laser operating at 441.6nm wavelength was used as a source of photon pairs. The crystal axis was tilted to produce 883.2nm wavelength pair beams separated by an angle of about 30°. Small ($\sim$ 3mm diameter at 600mm from the crystal) apertures were used to select the beams. The pump beam was unfocussed ($\sim$ 2mm diameter) hence these apertures were multimode. A Mach-Zehnder inerferometer was placed in the idler arm of the apparatus while in the signal arm a microscope objective (times 10, 0.25 numerical aperture (NA)) was used launch the light into a multimode fibre (0.23NA) of length 170m. Light from the fibre end was collimated by another microscope lens and launched into a second Mach-Zehnder interferometer. Photon counting detectors covered by 10nm bandwidth 883nm interference filters were placed to detect the light output from each interferometer. The interferometers were both calibrated to zero pathlength differences by finding white light (10nm bandwidth filtered tungsten bulb) fringes then carefully adjusted to identical 2 millimetre path length differences. At this point no white light fringes could be seen. Scanning the interferometer around the 2 millimetre pathlegth difference position did not produce any oscillations in the photoelectron count rate which was about 70 kilo-counts per second in the idler channel and 30 kilo-counts per second in the signal channel. Fibre losses were about a factor of three. The coincidence rate was measured by delaying the idler channel counts in a pulse generator and using a 10ns gate time multichannel coincidence counter. The background coincidence rate due to random overlap of pulses in the gate time was thus about 20 counts per second while the coincidence peak after background subtraction showed some 600 of coincidences per second (on average).

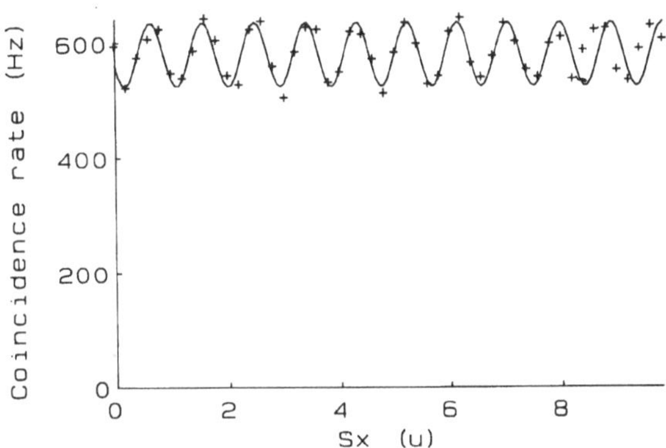

Fig. 5. Coincidence rate as path length difference δX in one interferometer is changed close to the equivalent 2mm path length difference point.

The pathlength difference in one of the Mach-Zehnders was changed in 0.2μm steps using a computer controlled translation stage and 5 second measurements of the coincidence rate were made over a 100μm range of travel around the equal 2mm pathlength difference position.

A section of the collected data is shown in figure 5. Clearly the oscillation has a period close to 883nm but the visibility (11.3%) is somewhat below the expected value of 0.5. The full result is shown in figure 6 along with a computer fit to the data. The disappearance of the effect away from the equal 2mm pathlength difference position reflects the bandwidth (10nm) of individual beams. A full theory taking into account the finite bandwidth of signal and idler beams appears elsewhere[17–19]. The low visibility could be interpreted in terms of a short coherence length pump beam but we presume it is more associated with spatial averaging in the interferometers and mode scrambling in the fibre. Individually measured interferometers showed interference fringes with about 75% visibility when illuminated with both white light and Helium-Neon laser light. We expect the low visibility will be improved by use of better optics in the interferometers but with the present detector resolutions it is still limited to 50%. Recent experiments [18] have shown that much greater visibilities can be acheived using detectors and coincidence electronics with time resolutions better than ΔT

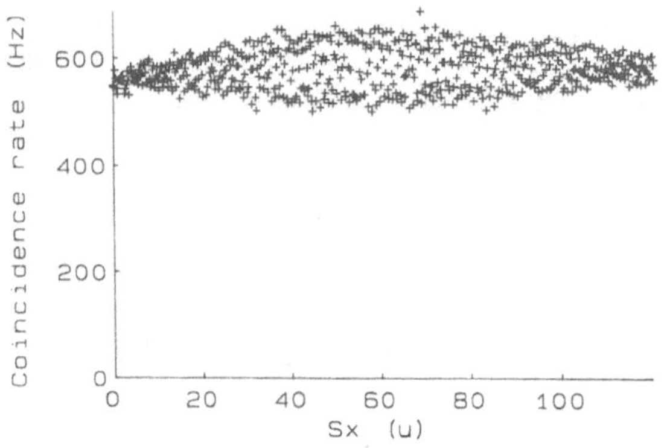

Fig. 6. Full set of coincidence data as path length difference δX in one interferometer is changed over a 100μm scan around the equivalent 2mm path length difference point.

3. DISCUSSION

We have demonstrated using two simple examples that two-photon interference is primarily a phenomenon associated with two indistinguishable paths for the two-photon to travel from source to detector. At no time in this discussion have we proposed interference between different photons of a pair. In fact in both these examples signal and idler photons are recombined at spatially separated points. However, because of the entangled wavefunction, apparent non-local effects are seen validating the need for a fully quantum description of the two-photon.

Uses for these effects are beginning to emerge. For instance the apparatus shown in figure 4 could be used as a prototype secure communication system [21]. The data bits would be encoded as high or low coincidence rates by altering the path length difference in the local (idler) interferometer. For modulation of coincidences between the remote signal detector and a time stamp sent by wire from the idler detector, interferometer path length differences have to be matched to within a few microns (the coherence length set by the single photon bandwidth) while actual path length differences can be up to a few centimetres. The remote interferometer can be thought of as a key which has to be set to the correct value (path length difference) before information (modulation of the coincidence rate) can be received. Intercepting the system at the dashed line in figure 4 and performing singles counting and coincidence measurements would not detect and signal apart from Poisson fluctuations of detection rates.

The delay between emission and arrival at the signal beam interferometer is about $1\mu s$. Rapid adjustment of the phase plate in this arm (on timescales shorter than $1\mu s$) would allow delayed choice tests of local realistic theories[22]. It is also of academic interest to verify whether non-classical correlations are retained over large distances and whether these effects can be used to test the assumption of linearity in quantum mechanics.

3. REFERENCES

1 P Grangier, G Roger and A Aspect, Europhys. Letts. **1**, 173, (1986)

2 C K Hong and L Mandel, Phys. Rev. Lett., **56**, 58 (1986)

3 J G Rarity, P R Tapster and E Jakeman, Opt. Commun. **62**, 201 (1987)

4 P A M Dirac, *The Principles of Quantum Mechanics*, Oxford University Press, Oxford (1930), p9

5 P R Tapster, J G Rarity, and J S Satchell, Phys. Rev. A **37**, 2963 (1988)

6 R Ghosh and L Mandel, Phys. Rev. Lett. **59**, 1903 (1987)

7 C K Hong, Z Y Ou and L Mandel, Phys. Rev. Lett., **59**, 2044 (1987)

8 J G Rarity and P R Tapster, in 'Photons and Quantum Flutuations', E R Pike and H Walther eds., Adam Hilger, p122. (1988); J Opt Soc Am B, **6**, 1221 (1989)

9 Z Y Ou and L Mandel, Phys. Rev. Lett. **61**, 54-57 (1988)

10 J G Rarity and P R Tapster, Phys. Rev. A **41**, 5139 (1990)

11 J G Rarity and P R Tapster, Phys. Rev. Letts. **64**, 2495 (1990)

12 M A Horne, A Shimony and A Zeilinger, Phys. Rev. Lett. **62**, 2209-2212 (1989); M A Horne and A Zeilinger, in 'Symposium on the Foundations of Modern Physics' eds P Lahti and P Mittelstaedt, World Scientific p435-439 (1985)

13 J D Franson, Phys. Rev. Lett. **62**, 2205 (1989)

14 P G Kwiat, W A Vareka, C K Hong, H Nathel and R Y Chiao, Phys. Rev. A **41**, 2910 (1990).

15 Z Y Ou, X Y Zou, L J Wang and L Mandel, Phys. Rev. Lett. **65**, 321 (1990)

16 J G Rarity, P R Tapster, E Jakeman, T Larchuk, R A Campos, M C Teich and B E A Saleh, Phys. Rev. Letts. **65**, 1348 (1990)

17 Z Y Ou, X Y Zou, L J Wang and L Mandel, Phys. Rev. A **42**, 2957 (1990)

18 J Brendel, E Mohler and W Martienssen, Phys. Rev. Letts. **66**, 1142 (1991)

19 J G Rarity and P R Tapster, to be published

20 J S Bell, Physics **1**, 19, (1964)

21 UK patent application 9018973.9

22 A Aspect, P Grangier and G Roger , Phys. Rev. Lett. **49**, 91 (1981)

BELL'S THEOREM WITHOUT INEQUALITIES AND BEYOND

Anton Zeilinger[1,2], Daniel M. Greenberger[3], and Michael A. Horne[4]

1. Atominstitut der Österreichischen Universitäten Schüttelstraße 115, A-1020 Wien (Austria)
2. Institut für Experimentalphysik Universität Innsbruck, A-6020 Innsbruck (Austria)
3. City College of the City University of New York New York, NY 10013 (USA)
4. Stonehill College North Easton, MA 02357 (USA)

1. INTRODUCTION

The investigation of the foundations of quantum mechanics has received momentous stimulation when, in 1964, John Bell established his famous theorem[1]. It is specifically remarkable that Bell's work not only triggered an enormous amount of theoretical work, it also gave rise to an extensive series of experiments. Both the discussion and the experiments focussed on the quantum mechanical correlations in two-particle systems. Recently, an extension to systems with more than two particles was found with rather interesting novel properties[2]. In the present paper we shall firstly point out the similarities and the differences between the two situations and secondly we shall briefly discuss some new aspects concerning entangled states consisting of more than two particles partly by further generalizing the quantum states considered.

We remark in the beginning that in particular the more philosophical aspects of our paper might not be of immediate importance for the practicing quantum mechanic, be she an experimentalist or a theorist.

Quantum Measurements in Optics, Edited by P. Tombesi and D.F. Walls, Plenum Press, New York, 1992

Therefore anybody just interested in the application of quantum mechanics may safely ignore these aspects of our paper. That some of the more technical aspects of multiparticle correlations might have even practical applications is signified by the recent discussions in quantum cryptology[3].

2. TWO PARTICLES VERSUS THREE PARTICLES

Here we would like to give a brief analysis of the two-particle situations where Bell's inequality applies and compare it to the new three- and four-particle situation. The purpose of this chapter is to give an operational introduction for the newcomer to the field. For technical details we refer the reader to the papers in ref. 2.

An important point we wish to stress once more is that it was not the motive of John Bell to introduce some hidden variables into physics in order to go beyond quantum mechanics. All he did was to accept the original arguments of Einstein-Podolsky-Rosen (EPR)[4] at face value and - continuing their line of reasoning - he arrived at a conflict with quantum mechanics. The EPR line of argumentation starts from the observation that in two-particle systems exhibiting *perfect correlation* for some properties one can predict from the results of measurements on one particle exactly the results of certain measurements on the other particle. Thus EPR did explicitly accept predictions of quantum mechanics. EPR then employ their idea of *locality* that the measurement performed on one particle can cause no real change in the other particle. This because the measurement events on the two particles can very well be spacelike separated and this would not change the correlations. A keystone for the EPR argument is then the famous *reality* principle which states that an element of reality must correspond to a physical quantity which can be predicted with certainty. Thus the measured quantity on particle 2 must correspond to an element of reality. But because of the locality condition this must hold for any of the observables for which the quantum state prescribes perfect correlation. The important point now is that states exist which predict perfect correlations for noncommuting observables. Thus, according to EPR, elements of reality must exist corresponding to each of these noncommuting observables. Hence, finally, quantum mechanics cannot be a complete physical theory because there is no quantum state which predicts definite values for noncommuting observables.

Thus EPR arrived at the conclusion that quantum mechanics is an incomplete description of physical reality. They explicitly left open whether or not a complete theory is possible. Yet they expressed their believe in the possibility of constructing such a theory.

Bell, continuing along the lines of reasoning of EPR, explicitly assumed that the elements of reality as demanded by EPR on the basis of the existence of perfect correlations really do exist. Bell then demonstrated that this assumption leads to contradiction with quantum mechanics in cases where measurements on not perfectly correlated observables are performed. The formal expression of this statement are Bell's inequalities.

Operationally, one needs a source emitting particle pairs in correlated states, either one or two detectors for either particle and some variable parameter on both particles. The observables have to be dichotomic. These dichotomic variables usually are spin variables, either the two spin states of a spin-1/2 particle or the polarization states of a photon. Yet, properly chosen momentum-position variables might also lead to violations of Bell's inequalities. That explicit proposal[5] has recently led to the first non-spin verification[6] of violations of Bell's inequalities.

It is rather interesting to note that more than 20 years had to pass after Bell's seminal work until it was found that the application of the EPR program to systems consisting of more than 2 particles can result in even more striking disagreements with quantum mechanics[7]. The spirit of reasoning is the same as the one leading to the Bell inequalities. Here too one starts from the prediction of quantum mechanics of *perfect correlations* between some observables of the three or more particles. Through the application of *locality* and *reality* one arrives again at new theoretical predictions. In contrast to the two-particle case these predictions now are not statistical bounds anymore but they are predictions of perfect correlations for other combinations of variables. Quantum mechanics, on the other hand, also makes definite predictions for the combination of variables but with completely opposite eigenvalues.

Operationally, the three-particle situation is very similar to the two-particle one. One also has dichotomic variables for each particle and a variable parameter. Here too, experiments exploiting spin correlations and experiments exploiting momentum correlations have been proposed.

The most interesting aspect of the new situation is that reasoning using EPR *locality* and *reality* leads to wrong predictions even for perfect correlations. This is an intrinsic inconsistency within the EPR line of argument not present in the two-particle correlations. It shall be interesting to exploit this intrinsic inconsistency further in particular with the aim at identifying the possible sets of assumptions about the physical world consistent with quantum mechanics.

3. ENTANGLED ENTANGLEMENT

The central feature of all EPR-type situations is that the properties of the particles described by the quantum state considered are *more strongly correlated* than any correlations in classical physics. Schrödinger[8] has called this property *entanglement* (in the German original "Verschränkung"). Entanglement means that the properties of a particle in a correlated quantum state cannot be described only upon information obtained from within the particles's backward light cone but that the *results* of measurements on one or more of the other particles of the system are also required. It seems that the great majority of physicists ascribes this to some breakdown of locality or separability (evidently these two terms have to be specified very carefully) in quantum mechanics. Here we do not wish to enter that segment of the discussion but refer the reader to a very thorough presentation of the issues involved[9].

In the present paragraph we rather want to analyze the property of entanglement itself. Consider for example the three-particle correlated state proposed by Mermin[7]

$$|\psi\rangle = \sqrt{1/2}\ [|z+\rangle_1|z+\rangle_2|z+\rangle_3 + |z-\rangle_1|z-\rangle_2|z-\rangle_3]. \quad (1)$$

where $|z+\rangle_i$ and $|z-\rangle_i$ mean states of particle i with spin up or down respectively along the z-direction. The coordinate system defined for each particle need not be the same. For example, in the case of photons propagating in different directions the spin is assumed to be defined along the direction of momentum of each photon. We now consider measurements of a given particle, say particle 1, along its z-direction. According to the reduction of the state vector the state of the remaining particles 2 and 3 after the measurement of particle 1 is either

$$|\psi>' = |z+>_2|z+>_3 \tag{2}$$

or

$$|\psi>' = |z->_2|z->_3 \tag{3}$$

depending on whether the result of the measurement on the first particle is + or -. In any case, after measurement of one particle along its z-direction the other two particles are in a factorizable state. In other words, each of the two remaining particles enjoys a well-defined state on its own after a z-measurement of the first particle, the two particles are not entangled anymore.

On the other hand, suppose we decide to measure particle 1 along some direction within the x-y-plane, say, along the x-axis. Then the state of the other two particles is not factorizable, independent of the specific result of the measurement on particle 1. In other words, the decision of experimenter 1 to measure the spin of particle i along a direction within the x-y-plane or along the z-direction decides also whether the other two particles are left in an entangled state or not after the first measurement. Therefore, in this case, entanglement is an entangled property dependent on the experimenters choice.

Another rather intriguing situation arises if we generalize the EPR discussion even further. After the GHZ generalization to 3 and 4 particles we now consider states consisting of more than two nonfactorizable terms. This, we note, is not possible within the standard Bell-inequality discussions since it is easy to see that any sum of product states of two two-state particles can always be written as a sum of just two terms. For more than two two-state particles this can only be done in limited cases. For our discussion we just consider the rather innocent looking state

$$|\psi> = \sqrt{1/3}\ [\,|z+>_1|z+>_2|z->_3 + |z+>_1|z->_2|z+>_3 + |z->_1|z+>_2|z+>_3 \tag{4}$$

For this state we again contemplate measurements along the z-direction of, say, particle 1. If the result of that measurement is + then the remaining two particles are found in the state

$$|\psi\rangle' = \sqrt{1/2}\,[|z+\rangle_2|z-\rangle_3 + |z-\rangle_2|z+\rangle_3]. \quad (5)$$

and if the result of the measurement on particle 1 is - the remaining two particles are found in the state

$$|\psi\rangle' = |z+\rangle_2|z+\rangle_3. \quad (6)$$

Therefore, if the result of the measurement on particle 1 was +, the other two particles are left in an entangled state and if the result of the first measurement was - the two remaining particles are left in a factorizable state, they each enjoy their own pure state. In this example now it is not the experimenters choice but Nature's choice which decides whether or not the other two particles are in a pure state after the first measurement. We point out that before the first measurement all 3 particles are entangled with each other and there is nothing in state (4) which would decide whether or not after the first measurement the other two particles will be entangled or not.

We therefore have shown in the present paragraph that entanglement itself can be an entangled property and that whether or not the remaining particles are left in an entangled state after a measurement on one particle might depend either on the experimenters choice or on the specific statistical outcome of the first experiment.

4. CONCLUDING COMMENTS

Entanglement is becoming a more and more common phenomenon in experiments. As a specific example of possible future work we might note the possible entanglement in atomic interferometry. There an atom enjoying both beam paths could be entangled with a spontaneously emitted photon. Evidently. our considerations also apply if the base states are not spin states and if other particles than photons are involved.

ACKNOWLEDGMENTS

This work was supported by the Austrian Fonds zur Förderung der wissenschaftlichen Forschung under project No. P 6635 and by the U.S. National Science Foundation (grants No. DMR 87-13559 and INT 87-13341).

REFERENCES

1. J.S. Bell, "On the Einstein-Podolsky-Rosen paradox," Physics 1, 195-200 (1964), reprinted in J.S. Bell, *Speakable and Unspeakable in Quantum Mechanics* (Cambridge U.P., Cambridge, 1987).
2. D.M. Greenberger, M. Horne, and A. Zeilinger, "Going beyond Bell's theorem," in *Bell's Theorem, Quantum Theory, and Conceptions of the Universe*, edited by M. Kafatos, Kluwer Academic, Dordrecht, The Netherlands (1989), pp. 73-76. D.M. Greenberger, M.A. Horne, A. Shimony, and A. Zeilinger, "Bell's theorem without inequalities", Am. J. Phys. **58**, 1131-1143 (1990).
3. S.M. Barnett, S.J.D. Phoenix and D.T. Pegg, "Entropy, information and quantum optical correlations", this conference. A. Ekert, "Quantum Cryptography based on Bell's theorem", this conference.
4. A. Einstein, B. Podolsky, and N. Rosen, "Can quantum-mechanical description of physical reality be considered complete?" Phys. Rev. **47**, 777-780 (1935).
5. M. A. Horne and A. Zeilinger, "A Bell-type EPR experiment using linear momenta", in *Symposium on the Foundations of Modern Physics*, edited by P. Lahti and P. Mittelstaedt (World Scientific, Singapore, 1985), pp. 435-439. M.A. Horne, A. Shimony, and A. Zeilinger, "Two-particle interferometry," Phys. Rev. Lett. **62**, 2209-2212 (1989).
6. J.G. Rarity and P.R. Tapster, "Experimental violation of Bell's inequality based on phase and momentum", Phys. Rev. Lett. **64**, 2495-2498 (1990).
7. N.D. Mermin, "Quantum mysteries revisited", Am. J. Phys. **58**, 731-734 (1990); and "What's wrong with these elements of reality?" Phys Today **43** (6), 9-11 (1990); "Extreme quantum entanglement in a superposition of macroscopically distinct states", Phys. Rev. Lett. **65**, 1838-1840 (1990).
8. E. Schrödinger, "Die gegenwärtige Situation in der Quantenmechanik", Naturwissenschaften **23**, 807-812, 823-828, 844-849 (1935). English translation in "*Quantum Theory and Measurement*", J.A. Wheeler and W.H. Zurek, Eds., Princeton Univ. Press, Princeton (1983).
9. M. Redhead, "*Incompleteness, Nonlocality, and Realism*", Clarendon, Oxford (1987).

THE STOCHASTIC INTERFEROMETRY AND THE FOUNDATIONS OF THE BOSE-EINSTEIN STATISTICS

Francesco De Martini

Dipartimento di Fisica, Università "La Sapienza"
00185 Roma, Italy

1. INTRODUCTION

In the present work we report the realization of the *Stochastic* (or *Non Stationary*) *Interferometer* (referred to as StO-IF) i.e. of an optical interferometer (IF) whose physical properties are determined by the use of the "Stochastic-Beam-Splitter" (StO-BS) as the field-partition device in place of the usual, "stationary" (viz. time t-constant) beam-splitter (StA-BS). The StO-BS has been first introduced and adopted in a previous "2nd-order-coherence" experiment as a relevant part of the *Stochastic-Intensity-Interferometer* [1]. In the first part of the work we shall devote our attention to field-interferometry (if) by which the 1st-order coherence properties of the beam are investigated[2]. As we shall see, the relevant observable of this process, the fringe "Visibility", is strongly determined by the stochastic properties of the noise source exciting the StO-BS. A general QED theory of the stochastic splitting process accounting for any shape of the noise probability distribution p(x) is given. Furthermore a simple information-theory approach will elucidate the behaviour of StO-IF with respect to basic quantum Complementarity arguments. The theoretical investigation is then substantiated by the results of an experiment involving a Mach-Zehnder Stochastic-Interferometer. In the second part of the work the related Stochastic-Intensity-Interferometry will be presented by two different albeit closely connected, "macroscopic" and "microscopic" experimental setups. We shall show that when the photon splitting is interpreted as a two- or multi-channel *scattering* process, our result may contribute to a substantial enlightement of the foundations of quantum-statistics for integer-spin (Bose) particles[1,3]. In this context the relevance of Compton scattering in microscopic phenomena and its relation to the establishment of Planck's distribution in an assembly of colliding particles will be considered.

Quantum Measurements in Optics, Edited by P. Tombesi and D.F. Walls, Plenum Press, New York, 1992

2. THE NON-STATIONARY FIELD-INTERFEROMETER

The "*stochastic*" or "*non stationary*" Mach-Zehnder StO-IF of our experiment is represented in Fig.1. It was excited at a wavelength λ=6328Å by a TEM_{00} single k-mode, I_L=1mW, He-NE laser (Spectra-Physics 117A). Before entering in the IF, the laser beam was sent through an amplitude/phase (Φ) modulator driven by noise-generator (A) equipped with an RC low-pass filter: $RC \approx t_c$. When activated, this set transformed the laser coherent-beam in a chaotic-superposition of $|n\rangle$ - states with a coherence-time[2]: $t_c = 5 \cdot 10^{-4}$s. The beam attenuation adopted for scanning over the experimental variable $\langle n \rangle$, the average photon-number per mode (or "*Bose-degeneracy-factor*") belonging to the input k-mode, was provided by calibrated ND-filters placed right in front of the laser. The beam was then sent to the electro-optic (EO) StO-BS of IF. The StO-BS consisted of a Pockels-cell (PC), of a Glan-Air analyzer (G) and of a z-cut quartz plate (not shown in Fig.1) inserted at one of the StO-BS outputs in order to restore equal polarization for the output-modes k_1, k_2 (labelled by i=1,2, henceforth). The PC was driven by an adjustable high-voltage generator (B) that supplied the waveform [shown in Fig.1, inset] expressed by: $v(t)=A(t) \cdot u(ft)$. $u(ft)$ represents a cw zero-mean square-wave (SW) carrier with first-harmonic frequency f=100kHz, Amplitude A-modulated by a "sawtooth" waveform A(t). This one consists of a periodic sequence of equal positive pulses each one starting at different $t=t_j \equiv (t_{j-1}+\Delta t)$ with duration Δt=200μs and expressed by $A_j(\tau)=\arcsin(\tau)$ being $\tau \equiv [(t-t_j)/\Delta t]$, $0 \leq \tau \leq 1$. $A_j(\tau)$ was syntetized at low voltage (±5V) by an EPROM device and by a digital-to-analog converter and the overall signal v(t) was linearly amplified to the maximum amplitude $|v_a(\tau)|_{max}=1kV=\underline{v}$. $\underline{v}$ was the $\lambda/4$-voltage leading, at λ, to a $\pi/4$ light-polarization rotation when applied to PC. In a different experiment, the carrier u(ft) was supplied by a narrowband gaussian-noise source, having a lorentzian spectrum with $\Delta f/f \approx 0.1$ and center-frequency $f \approx 100$kHz. The input beam polarization was such that, by deactivating B, the StO-BS became a normal, lossless Stationary-BS (StA-BS) with equal t-constant transfer-functions: $W_1(t) \equiv |w_1(t)|^2 \equiv |\underline{t}(t)|^2 = X_1 = 1/2$, $W_2(t) \equiv |w_2(t)|^2 \equiv |\underline{r}(t)|^2 = X_2 = 1/2$, $\Sigma_i X_i = 1$. By activating B, the input-beam polarization analyzed by G was driven by v(t) across the BS-symmetry condition, W_i=1/2, with the "StO-BS coherence-time" $\tau_c \approx f^{-1} << t_c$ throught a sequence of zero-mean positive and negative square-pulses with duration $\delta t = \tau_c/2$ and amplitude $|x(t)| = |\sin\Gamma(t)|$, $\Gamma(t) \equiv [\pi v_a(t)/2\underline{v}]$ being $v_a(t) \leq \underline{v}$ the voltage amplitude of the A-modulated carrier-pulses. In other words, the StO-BS transfer-functions (viz.,the *probabilities* $W_1(t)$, $W_2(t)$ of *photon-scattering* over the M=2 output i-channels) consisted of square-samples with duration δt and simultaneous amplitude: $W_1(t)=[1+x(t)]/2$, $W_2(t)=[1-x(t)]/2$, $\Sigma_i X_i(t)=1$. In order to obtain the t-display of the fringe-pattern, before i-beam interference in the output StA-BS, characterized by parameters: $|\underline{t}'|^2 = |\underline{r}'|^2 = 1/2$, one of the beams underwent a periodic delay $\Phi(t)$, $0 \leq \Phi \leq \pi$ via an EO Φ-modulator driven by a linear-sawtooth waveform at $\underline{\omega}$=10Hz. The cycle-averaged intensity of the fringe-pattern, $I(\underline{\omega}t)$, was then detected by a cooled Quantacon-photomultiplier, PM, RCA C31034A-02 inserted at one arm of the output StA-BS and having a photocurrent-integration-time $\tau_{PM} \approx 5$ms. The maximum and minimum values of the PM-photocurrent $i(t) \propto \sin(\underline{\omega}t)$ relative to each period $(2\pi/\underline{\omega})$ and for constant IF-excitation $\langle n \rangle \propto I_L$, were measured by a digital-oscilloscope (Le Croy 9400) and then sequences of 100 of such samples were numerically averaged to give the maximum and minimum I-values relative to the IF-fringe pattern. By these ones the relevant observable for 1st-order-coherence processes, the fringe-visibility (viz., the "degree of 1st-order-coherence", $g^{(1)}$) was determined[2,4,5,6]: $V \equiv (I_{max}-I_{min})/(I_{max}+I_{min})$.

3. THE TWO-CHANNEL BOSE-EINSTEIN PARTITION AND THE INTERFERENCE -FRINGE VISIBILITY

Turn now the attention to the StO-BS under excitation of a field expressed by the operators $\hat{a}^+$, $\hat{a}$ (Fig.1). Assume for instance a field's "chaotic-state" with a Bose degeneracy-factor $\langle n \rangle$, a condition obtained by activating the noise-source A. The field density-operator may be expressed in the form:

$$\hat{\rho} = \sum_{n(0-\infty)} \underline{P}_n \cdot \sum_{m,h(0-n)} | p_{nm} p_{nh} | \cdot \exp[i \tfrac{1}{2} \pi(m-h)] \cdot |m,n-m\rangle\langle n-h,h| \tag{3.1}$$

where (h,n-h) are the photon-numbers scattered over the StO-BS-output modes, $\underline{P}_n \equiv \langle n \rangle^n/(1+\langle n \rangle)^{1+n}$ and the $P_{nh} \equiv |p_{nh}|^2$ are the M=2-way BS-partition probabilities. Consider the QED theory of the beam-splitting process[5,6]. This is generally carried out, for StA-BS, by one of the two equivalent field-operator procedures:

(a) The "input-to-output": (ITO)-procedure, in which field-operators, $\hat{a}_j$, defined on the BS-input modes (j=1,2) are given as linear superposition of the output-mode (i=1,2) operators, $\hat{b}_i$: $\hat{a}_j = \Sigma_i w_i \hat{b}_i$.

(b) The "output-to-input": (OTI)-procedure, $\hat{b}_i = \Sigma_j v_j \hat{a}_j$, the inverse of (ITO). With v_j-matrix[5,6]: $V = W^{-1}$. Let us adopt here the (ITO)-procedure.

This leads to the Bernoulli partition law: $P_{nh} = W_1^{(n-h)} W_2^h \cdot {}^nC_h$ where: ${}^nC_h \equiv n!/[h!(n-h)!]$. This well known result expresses, for the StA-BS case, "classical" *absence* of photon-correlations within the BS process[2,7]. Consider now the StO-BS condition in which the photon-numbers (h,n-h) are detected in a series of measurements lasting a time Dt over which the scattering-probabilities $W_i(t) \equiv |w_i(t)|^2$ show an appreciable variation within their existence-range: $[0 \le W_i(t) \le 1]$. $W_i(t)$ may be for instance random-functions already expressed as: $W_1(t) = X_1[1+x(t)]$; $W_2(t) = X_2[1-x(t)]$. In this case the statistical distribution P_{nh} is the average over all sets $\{W_i\}$ realized in Dt. Then $[W_1^{(n-h)} W_2^h]$ in the Bernoulli expression is now replaced by:

$$S(p) \equiv (2)^{-(n+1)} \int_{-1}^{+1} [(1+x)^{(n-h)} (1-x)^h] \cdot p(x)\, dx \tag{3.2}$$

where p(x) is a distribution-density function, for instance a normalized gaussian $p_g(x)$ with width $\sigma \equiv (\sigma'/\sqrt{2})$, given here with its Fourier-transform $P(\chi)$ [Fig.2]:

$$p_q(x|\sigma) \equiv [\sqrt{\pi}\sigma' \operatorname{erf}(1/\sigma')]^{-1} \cdot \exp[-(x/\sigma')^2]; \qquad P_g(\chi) \equiv [\operatorname{erf}(1/\sigma')]^{-1} \cdot \exp[-(\pi\sigma'\chi)^2] \tag{3.3}$$

The StA-BS condition is: $p(x) = p_g(x|\sigma \to 0) = \delta(x)$. An interesting case arises for $p(x) = p_g(x|\sigma \to \infty) \equiv p' \equiv 1/2$, viz.for a *uniform* distribution. In this case S(p') is expressed by a Beta-function, $B(n-h+1, h+1) \equiv \Gamma(h+1)\Gamma(n-h+1)/\Gamma(n+2)$, and then the photon partition-law

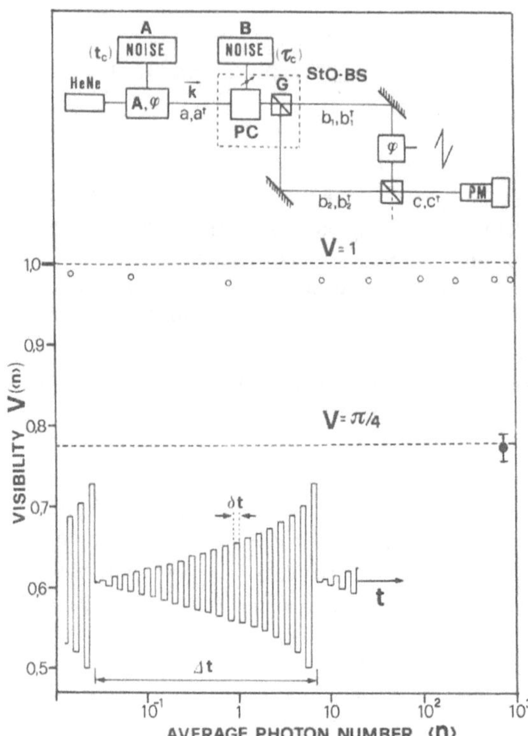

Fig.1. Fringe-*Visibility* V(⟨n⟩) vs the average photon-number per mode ⟨n⟩ for a Mach-Zehnder Interferometer (IF) excited by incoherent-light. White-dots: normal, *Stationary-IF* condition (StA-IF). Black-dot: *Stochastic-IF* (StO-IF) at $\langle n \rangle \approx 10^3$. Inset: shape of the v(t) pulse exciting StO-BS.

is given by: $P_{nh}=S(p')\cdot{}^nC_h=[n+1]^{-1}$, i.e., by the M=2 Bose-Einstein (BE) law[1,8,10]. This law is the M=2 specialization of the general BE law $(P_{nh})_{be}=n!(M-1)!/(n+M-1)!$ valid for any number of scattering-channels, M [3]. This implies that the "scrambling" effect, viz., added mutual i-beam incoherence induced by STO-BS on the input photon-stream, determines BE-correlations on the particles scattered over M=2 channels[1]. A theory of the transition from the Bernoulli-law to the BE-law for M=2 and for *any* p(x) is given in Appendix. An evidence of the transition between these regimes is expressed, for instance by the corresponding transition of the fringe-"Visibility" shown in Fig.2. Assuming the p(x)=p'=1/2 distribution, we may now express the *average-density-operator* by:

$$\langle\hat{\rho}\rangle = \sum_{n(0-\infty)} \underline{P}_n \cdot \sum_{m,h(0-n)} \underline{B}(m,h,n)\cdot \exp[i\,\tfrac{1}{2}\,\pi(m-h)] \cdot |m,n-m\rangle\langle n-h,h| \qquad (3.4)$$

where $\underline{B}\equiv B[(m+h+2)/2,\ (n+1)-(m+h)/2]x\{B[(m+1),(n+1-m)]\cdot B[(h+1),\ (n+1-h)]\}^{-1/2}$. This expression has been obtained by evaluating $|p_{nm}p_{nh}|$ for p(x)=p'=1/2. The field-intensity at the detector located at (z,t) is: $I(z,t)=Tr[\langle\hat{\rho}\rangle\cdot E^{(-)}(z,t)\ E^{(+)}(z,t)]$, where $E^{(-)}$, $E^{(+)}$ are proportional to the output-mode operators $\hat{c}^+,\hat{c}$ which are given, by OTI through the output StA-BS, in terms of the fields, $\hat{b}^+_i$, $\hat{b}_i$ to give:

$$I(z,t)\alpha\ \{|\underline{r}'|^2Tr[\langle\hat{\rho}\rangle\,\hat{b}^+_1\,\hat{b}_1]+|\underline{t}'|^2Tr[\langle\hat{\rho}\rangle\,\hat{b}^+_2\,\hat{b}_2]+2|\underline{r}'\underline{t}'|\ |Tr[\langle\hat{\rho}\rangle\,\hat{b}^+_2\,\hat{b}_1]\,|\cdot\cos\Phi(t)\} \qquad (3.5)$$

where the periodic phase $\Phi(t)$ is generated by the IF Φ-modulator. Then the general quantum expression of the if-fringe "Visibility" is:

$$V(\langle n\rangle)=2|\underline{r}'\underline{t}'|\ |Tr[\langle\hat{\rho}\rangle\,\hat{b}^+_2\,\hat{b}_1]\,|/\{|\underline{r}'|^2Tr[\langle\hat{\rho}\rangle\,\hat{b}^+_1\,\hat{b}_1]+|\underline{t}'|^2Tr[\langle\hat{\rho}\rangle\,\hat{b}^+_2\,\hat{b}_2]\} \qquad (3.6)$$

We check immediately that, for input StA-BS condition obtained by deactivating the generator B, viz., for a *normal-* IF (StA-IF), $\langle\hat{\rho}\rangle=\hat{\rho}$ and $|\underline{r}|=|\underline{t}|=|\underline{r}'|=|\underline{t}'|$, is $V(\langle n\rangle)=1$, for any input-field preparation and for any average photon-number, $\langle n\rangle$. This well known quantum result, expressing the StA-IF insensitivity to the field's state, is verified experimentally for many $\langle n\rangle$-values as shown in Fig.1 for input incoherent- and coherent-light[5]. It is worthwhile, in order to evaluate V and other observables for the StO-IF case, to give here the general expression of the quantum average of the ordered-product of the scattered fields:

$$|Tr[\langle\hat{\rho}\rangle\,\hat{b}^+_i\,\hat{b}_j]\,| = \sum_{n(0-\infty)} \underline{P}_n \sum_{h(0-n)} |{}^nC_h \cdot \mathbf{B}[a,b-a]\,| \times$$

$$\times \int_{-\infty}^{+\infty} d\chi\cdot P(\chi)\cdot\exp(-i2\pi\chi)\cdot\{{}_1F_1((b-a,b;i4\pi\chi)+{}_1F_1(a,b;i4\pi\chi)\} \qquad (3.7)$$

being ${}_1F_1(\alpha,\beta;\gamma)$ a confluent-hypergeometric function and the parameters a,b:

$$\text{for } i=j:\ a\equiv(h+1),\ b\equiv(n+2)\ ; \qquad \text{for } i\neq j:\ a\equiv(h+3/2),\ b\equiv(n+2) \qquad (3.8)$$

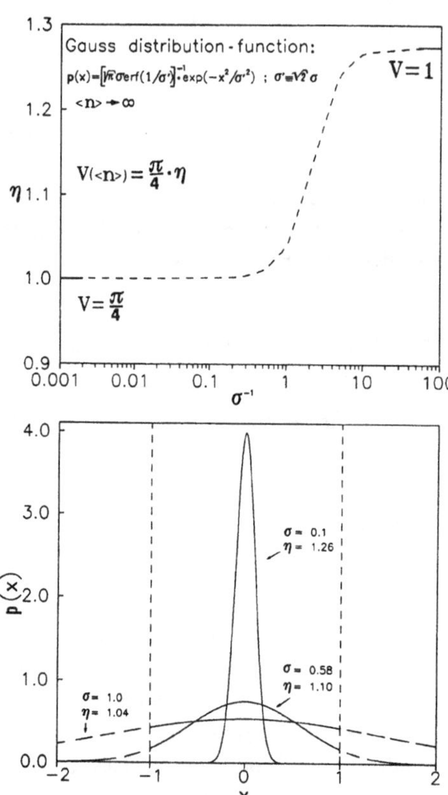

Fig.2. Transition of fringe-Visibility from the "classical" to the Bose-Einstein photon-partition by a StO-BS driven by noise with a gaussian-distribution.

These expressions result from product-field operations on the set of states appearing in [3.4][10]. For $i \neq j$ it is also found: $Tr[\langle\hat{\rho}\rangle \hat{b}^{+}_{i} \hat{b}_{j}] = -Tr[\langle\hat{\rho}\rangle \hat{b}^{+}_{j} \hat{b}_{i}]$.
A detailed discussion of the simple and significant case i=j is given in Appendix. Evaluation of the case i=1, j=2 leads to the following expression for the relevant parameter appearing within the last sum outside the integral in (3.7):

$$|{}^{n}C_{h}\cdot\mathbf{B}[a,b-a]| = \{\pi/[2^{2n}\cdot(n+1)]\}\cdot\{\Gamma[2(h+1)]\cdot\Gamma[2(n-h)]\}/\{\Gamma[h+1]\cdot\Gamma[n-h]\}^{2} \qquad (3.9)$$

We may now evaluate (3.6) by assuming the BE-case, i.e. $p(x)=1/2$, $P(\chi)=\delta(\chi)/2$ and the integral over the χ-variable in (3.7) =1. The value of the sum of the quantity (3.9) over the h-variable in (3.7) is found equal to $(\pi/8)\cdot n$ for *any* value of n. Furthermore, the sum of the traces appearing in the denominator of (3.6) is found $=\langle n\rangle$. Then, by inspection of (3.6), we easily find for $p(x)=1/2$ a value of the Visibility, $V=(\pi/4)$, *independent* of $\langle n\rangle$ and of the coherence properties of the input light. We may now evaluate the explicit value of V for *any* p(x) function and for $n\to\infty$ by replacing in (3.7) the sum over h by an integral written in terms of $z\equiv\lim_{(n=\infty)}(h/n)$ and then evaluating (3.7) in this limit. By adopting the method given in Appendix, we obtain the following expression valid for any even density-function p(x) and for large $\langle n\rangle$:

$$V=\pi/2\{[\int_{-\infty}^{+\infty} P(\chi)\cdot J_1(2\pi\chi)\cdot d\chi/\chi]/[\int_{-\infty}^{+\infty} P(\chi)\cdot \sin(2\pi\chi)\cdot d\chi/\chi]\} \qquad (3.10)$$

where J_1 is a 1st-order Bessel-function. For a gaussian $p(x)=p_g(x|\sigma)$ the integrals are easily evaluated leading to the expression:

$$V=\frac{1}{4}\,\pi\cdot\eta\,; \qquad \eta\equiv[{}_1F_1(3/2,2;\gamma^2)/{}_1F_1(1,3/2;\gamma^2)]\,; \qquad \gamma\equiv(1/\sigma') \qquad (3.11)$$

The plot of the Visibility-parameter η Vs σ is shown in Fig.2. Note there that the sharp transition between the Bernoulli to the BE-partition regimes (viz., from V=1 to $V=\pi/4$) takes place about $\sigma\approx1$, i.e., where an appreciable variation of $p_g(x)$ is established within its domain of definition $[-1\leq x\leq+1]$. By inspection in (3.11) of the asymptotic-expansion of ${}_1F_1(\alpha,\beta;\gamma)$ for the extreme values taken by σ, the corresponding, expected, values of the Visibility are found: $V(\sigma\to0)=1$ and $V(\sigma\to\infty)=\pi/4$ (10,AS ch.13). These results characterize respectively, the "classical" and the Bose-Einstein photon-partition regimes, as said. (3.11) is found to hold for any coherence propery of the input light[2]. The results of the quantum theory just given are found consistent with a Visibility theory based on *classical-electrodynamics*, viz., by assuming IF-excitation by the classical-field $E(t)\cdot\exp(-i2\pi\nu t)$ associated with the average field-energy $\langle E^2\rangle \propto I_L$. Assume the ergodic-hypotesis $\langle E^2\rangle = \{|E(t)|^2\}$ where { } brackets mean average over time T with $t_c \geq T > 0$ [11]. Consider now the behaviour of StO-BS under the nonstationary excitation described in Sect.2 leading to corresponding polarization rotations from $-\pi/4$ to $+\pi/4$. The field intensity detected by PM is: $I(t) \propto I_L\cdot|\cos[2\Gamma(t)-\pi]/4+\exp(i\Phi)\cdot\sin[2\Gamma(t)-\pi]/4|^2 = I_L\cdot\{1+\cos\Gamma(t)\cdot\cos\Phi\}$, for any realization of the phase Φ. Assume for simplicity square-wave excitation and $T=k\Delta t$, k=whole number >0. Owing to the given definition of the amplitude of the probability-samples, $|x(\tau)|=|\sin\Gamma(\tau)|$ assumed as a continuous function for $0<\tau<1$ and $(\delta t/\Delta t)\to 0$, we may write with the semantics related to [3.2]:

$$V(\{|E(t)|^2\}) \approx (1/T)\cdot\int_0^T \cos\Gamma(t)dt = \int_0^1 \cos\Gamma(\tau)d\tau = \int_{-1}^{+1}[1-x^2]^{1/2}\cdot p(x)\cdot dx \equiv V(\langle E^2\rangle) \quad (3.12)$$

where $p(x)\equiv[d\tau(x)/d|x|]/2$ is an even density-function. The shape of the A-modulating pulse $A_j(\tau)$ adopted in our work leads to $p(x)=p'\equiv 1/2$ and then to: $V(\langle E^2\rangle)=V(\{|E(t)|^2\})=\pi/4$ for any $\langle E^2\rangle \propto \langle n\rangle$. This *classical* result is characteristic of the field-interference of two beams originating from a 2-way stochastic-splitting with a *uniform* probability-distribution. This process is found to generate BE-correlations within the scattered photon-beams, in agreement with the quantum theory just given. Some preliminary results of an extended study on the IF Visibility are reported in Fig.1. There it is shown that $V\approx 1$ for a StA-IF, as expected. A single result $V\approx(\pi/4)=0.785$ for a StO-IF driven by a stochastic-noise function $p(x)=1/2$ is also given[12]. These results agree with the given theory.

4. INTERFEROMETRY, COMPLEMENTARITY AND INFORMATION THEORY

The quantum behaviour of the Stochastic-Interferometer is analyzed in simple terms by considering, with correspondence to a certain value of "Visibility", the maximum amount of information allowed by Complementarity on the "photon's path" taken within IF after beam-splitting (Wootters and Zurek, 13). According to Shannon's theory, the lacking amount of such information may be given in terms of probabilities as:

$$H=-\sum_i W_i\cdot\ln W_i\,; \qquad \sum_i W_i = 1\,. \qquad (4.1)$$

Within the pictorial semantics of Bohr Complementarity, H refers to the particle-like aspect of the photon. Since "Visibility" somewhat expresses the observer's ability to appreciate the *wave-like* content of the particle, we may introduce an approximate "lack-of-information" on this property about the condition $V=1$ by the quantity: $K=(1-V^2)/2=[W_1-W_2]^2/2 <<1$, recalling that the classical Visibility-theory gives: $V^2=4W_1 W_2$. Expression [4.1] may be now expressed in terms of K, leading to the equation: $H+K=\ln 2-O(K^2)$ where the last term is of order $K^2<<1$. Differentiation of the terms of the equation about $V\approx 1$ expresses the balance of information related to the two "complementary" aspects of the particle within IF:

$$(\text{entropy increase})_{\text{particle}} = -(\text{entropy increase})_{\text{wave}} \rightarrow (\Delta H) = -(\Delta K) \qquad (4.2)$$

In Fig.3 the range of validity of the last equation assuming W_1 as the independent variable is shown. This range coincides with the one for which the approximate equation: $(H+K)=\text{constant}$ may be considered valid. An interesting insight of the BS-induced quantum-correlation process is offered by the numerical evaluation of H in various correlation conditions. In the StA-IF case, where no interparticle quantum-correlation is induced by the beam-splitting process within a balanced IF, (4.1) leads to a "particle-like" information-entropy: $H=\ln 2=0.693$ [13]. On the other hand, in a StO-IF the Bose-Einstein correlations lead to:

$$H = -2 \cdot \int_0^1 W \cdot \ln W \, dW = 0.50 \qquad \text{(Bose-Einstein)} \qquad (4.3)$$

This shows that in a two-way scattering process (M=2) the Bose-Einstein correlation induced by the stochastic-scatterer on the particle-beam realizes a decrease of information-entropy related to the particle kinematics. This is but a manifestation of a general statistical property of the BE-condensation[3]. Of course, in our case the information gain on the "photon path" within IF is ascribable, rather trivially, to the routing effect induced by the StO-BS on the particle dynamics. As already stressed, it appears enlightening that the Bose-Einstein condensation in our StO-IF, as well as in other more significant physical systems,be connected with a simple dynamics of this sort. Note at last that the transition from the "classical" (Bernoulli) to the BE-partition process is accounted for with good approximation by (4.2). In fact for this transition we have $(\Delta H)=\ln 2-1/2=0.193$ which is close to the value $0.191=|(\Delta K)|$ obtained for the corresponding Visibility change, from V=1 to $V=(\pi/4)$ (Bose-Einstein partition).

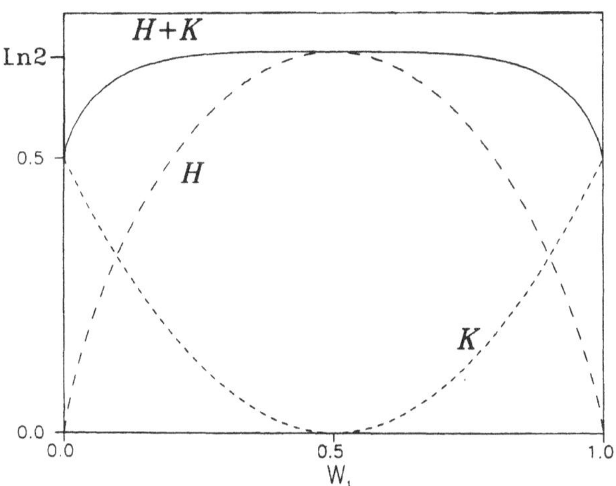

Fig.3. Information-entropy related to the photon's kinematics within a StO-IF.

5. THE STOCHASTIC INTENSITY-INTERFEROMETER

The apparatus described in the previous section can be easily adapted to the investigation of the 2nd-order-coherence, "intensity-interference" optical processes. These are important phenomena that show explicitly, by analysis via an external "correlator", of the space and/or time correlation of the detected light intensity, the "coherence-properties" of the field and, under appropriate conditions, its structure in terms of photon-particles. The basic experimental method leading to this kind of information is the well known Hanbury Brown-Twiss (HB-T) effect by which the light intensities to be correlated are measured at the output of two photomultipliers excited by the fields associated with the output-modes k_1, k_2 of a BS[2]. In work (1) we have reported the application of an original variant of the HB-T method aimed at the investigation of the statistical partition-properties of the photons belonging to a single-mode beam which is split by a StA-BS or a StO-BS. In that work we wanted to study the basic nature of the "quantum" Bose-Einstein correlations (or BE-condensation) that may determine the statistical properties of the photons within a scattering process. As already said, there the beam-splitting process is interpreted as a 2-channel, M=2, "stationary" or "stochastic" scattering process. The expressions of the M=2 stochastic-scattering probability of a photon-beam, P_{nh}, reported in Sect.2 and in the Appendix show that this one is dependent on the "size" σ of the fluctuations affecting the scattering cross-sections $W_i(x)$, $\Sigma_i W_i=1$, and on the shape of the noise density-distribution, $p(x)$. As far as the effect of $\sigma>0$ is concerned, we have seen that its value identifies a statistical conditions of large physical relevance. In fact, assuming $p(x)=p_g(x|\sigma)$, a large value of σ in StO-BS regime leads to the 2-channel BE partition-law[1], $(P_{nh})_{be}=(1+n)^{-1}$. In other words, according to our present model, the BE partition represents a *fully stochastic* process with *uniform* probability distribution. On the other hand, an extremely low perturbation level, $\sigma\approx 0$, identifies asymptotically the classical Bernoulli partition-statistics $(P_{nh})_{mb}$, the one that, when applied to each elementary inter-molecular collision-process, leads to the Maxwell-Boltzmann statistics of a "classical-gas" in *equilibrium*[3,14]. The intermediate case is found in Appendix to correspond to a partition law that is a linear superposition of a hierarchy of modified BE-distributions.

The double-photocount (HB-T)-like detection method adopted in our work consists of the measurement of the "Quantum-Noise Function" $G(1/\bar{n}) \equiv \langle\hat{g}\rangle$, where: $\hat{g} \equiv [(\hat{n}_1/\bar{n}_1) - (\hat{n}_2/\bar{n}_2)]^2$; $(1/\bar{n})\equiv[(1/\bar{n}_1)+(1/\bar{n}_2)]/4$, $\bar{n}_i\equiv\langle\hat{n}_i\rangle\equiv\langle\hat{b}_i^+ \hat{b}_i\rangle$[1,6,15]. G is related to the "degrees of 2nd-order-coherence", $g_{ij}(0)\equiv\langle\hat{b}_i^+ \hat{b}_j^+ \hat{b}_j \hat{b}_i\rangle/\langle n_i\rangle\langle n_j\rangle$ through a normal-ordering procedure. If we define $G'\equiv(g_{11}+g_{22}-2g_{12})$, we have $G'=\langle:\hat{g}:\rangle$, while $G=\langle\hat{g}\rangle$. Evaluate the ensemble-averages appearing in $G(1/\bar{n})$ by the field density-operator [3.1] expressed in terms of the "classical" and of the "quantum" distributions: $(P_{nh})_{mb}$, $(P_{nh})_{be}$. By making use of the commutation relations for Bose-operators we obtain for both cases: $G(1/\bar{n})=\underline{G}+(4/\bar{n})$. In StA-BS condition, when the "classical" law applies, is always $\underline{G}=0$. We may work out the StO-BS case by the OTI-procedure. In this way $\underline{G}$ is found to be expressed in terms of the input operator, $n\equiv(\hat{a}_k^+ \hat{a}_k)$, in the form:

$$\underline{G}=4\langle\hat{n}(\hat{n}-1)\rangle / (3\langle\hat{n}\rangle^2) \qquad \text{where:} \quad \langle\hat{n}^2\rangle=\delta\langle\hat{n}\rangle^2+\langle\hat{n}\rangle \tag{5.1}$$

is the mean-square input photon number average and $\delta=1$ or 2 for input coherent- or incoherent-fields, respectively[2]. At last, we can write in general $G(1/\bar{n})=\underline{G}+(4/\bar{n})$. Relevant values of the *interstatistics-parameter*, $\underline{G}$ for M=2 are $\underline{G}=0$, (4/3), (8/3) corresponding respectively to: "classical" partition; BE-partition for coherent input-light; BE-partition for incoherent input-light. $G(1/\bar{m})$ given in terms of the variable $(1/\bar{m})=[(1/\bar{m}_1)+(1/\bar{m}_2)]/4$, with $\bar{m}_1 \propto \bar{n}_i$ the detected photocounts, is represented in Fig.4 by the straight lines a,b,c[15]. In this respect it is easy to show by quantum theory that the value of the parameter $\underline{G}$, which represents one of the most relevant results of the present work, doesn't depend on the detector's quantum-efficiency. Then $\underline{G}$ remains *unchanged* if it is calculated by replacing in the ensemble-averages appearing in (5.1) the photon-number operators $\hat{n}_i$ by the corresponding detected photocounts, m_i. This result is valid also in the case of multichannel scattering M>>2, considered below. Furthermore, inspection of (5.1) shows that the quantum-correlations (BE condensation) caused by the scrambling effect of StO-BS on the output modes, imply additional contributions of wave-fluctuations on these modes[2,6]. In a recent work[16] the expression (5.1) of the interstatistics-parameter $\underline{G}$ has been obtained making no recourse to the field-commutation properties but by interpreting the splitting parameter given by (3.2), $^{n}C_{h}\cdot S(p)$, as a De Finetti representation theorem about indefinitely exchangeable statistical random-variables (9, Feller). In view of our present experimental test, this result is interesting as it may enlighten the inner dynamics of photon-scattering and provide a physical basis for approaching the significance of the field's commutation-properties. Note that this result is in agreement with our present statistical model.

As a significant extension of the above theory, consider a set of M≥2 scattering output-states $|\{n_i\}\rangle \equiv |n_m\rangle|n_{m-1}\rangle..|n_2\rangle|n_1\rangle$ where again photodetection is carried out over two of such states: j=1,2. The relevant operators have the form: $\hat{o}=f(\hat{n}_j+\hat{b}_j, \hat{\imath})$, being $\hat{\imath}$ the identity-operator, and then,

$$\langle\hat{o}\rangle=\mathrm{Tr}(\rho\hat{o})=\sum_{n(0-\infty)} \underline{P}n \sum_{\{n_j\}} W_i\langle\{n_i\}\rangle\, |\hat{o}|\, \{n_i\}\rangle \quad ; W_i\equiv|w_i|^2 \tag{5.2}$$

Absence of inter-particle correlation in the scattering process is expressed by the "classical" distribution for which the following expression holds[1,14]:

$$\sum_{\{n_j\}} W_i^{MB}\equiv n!\cdot \sum_{n_M(0-n)} (W_M{}^{n_M}/n_M!)\ldots \sum_{n_1(0-n_2)} (W_1{}^{n_1}/n_1!)\times\delta[n-(n_M+n_{M-1}+..n_1)]=1$$

while, for quantum-scattering of integer-spin (Bose) particles, we have[1,3,10]:

$\sum_{\{n_i\}} W_i^{BE}\equiv W^{BE}\times \sum_{\{n_i\}} \langle\hat{\imath}\rangle=1$. For both distributions we may now evaluate in terms of $\langle\hat{n}\rangle$ the relevant averages related to 2-channel photodetection: $\langle\hat{n}_1\rangle$, $\langle\hat{n}_2\rangle$, $\langle\hat{n}_1{}^2\rangle$, $\langle\hat{n}_2{}^2\rangle$, $\langle\hat{n}_1\,\hat{n}_2\rangle$, by assuming,with no loss of generality, equal output-channel probabilities: $W_i=M^{-1}$. This is obtained after some algebra by a smart redefinition of the limits of the multiple-sums at each step of a repeated elimination of the factorials. The final results are, for the two partition regimes:

Classical : $\langle\hat{n}_1\rangle = \langle\hat{n}_2\rangle = \langle\hat{n}\rangle/M$; $\langle\hat{n}_1{}^2\rangle = \langle\hat{n}_2{}^2\rangle = [\langle\hat{n}^2\rangle + (M-1)\langle\hat{n}\rangle]/M^2$;
$$\langle\hat{n}_1\,\hat{n}_2\rangle = [\langle\hat{n}^2\rangle - \langle\hat{n}\rangle]/M^2; \tag{5.3}$$

Bose-Einstein : $\langle\hat{n}_1\rangle = \langle\hat{n}_2\rangle = \langle\hat{n}\rangle/M$; $\langle\hat{n}_1{}^2\rangle = \langle\hat{n}_2{}^2\rangle = [2\langle\hat{n}^2\rangle + (M-1)\langle\hat{n}\rangle]/[M(M+1)]$;
$$\langle\hat{n}_1\,\hat{n}_2\rangle = [\langle\hat{n}^2\rangle - \langle\hat{n}\rangle]/[M(M+1)].$$

The "Scattering quantum-noise function" for M≥2 scattering-channels is then again expressed by: $G(\langle n\rangle)=\underline{G}+2M/\langle\hat{n}\rangle$. The *interstatistics-parameter* is now, for M≥2:

$$\underline{G}=0 \text{ (classical distribution)}; \qquad \underline{G}=2\delta M/(M+1) \text{ (Bose-Einstein distribution)} \tag{5.4}$$

These general formulas, which coincide with [5.1] for M=2, lead for M>>2 to the asymptotic values: $\underline{G}$=2 or 4 for coherent- or incoherent-light, respectively[1].

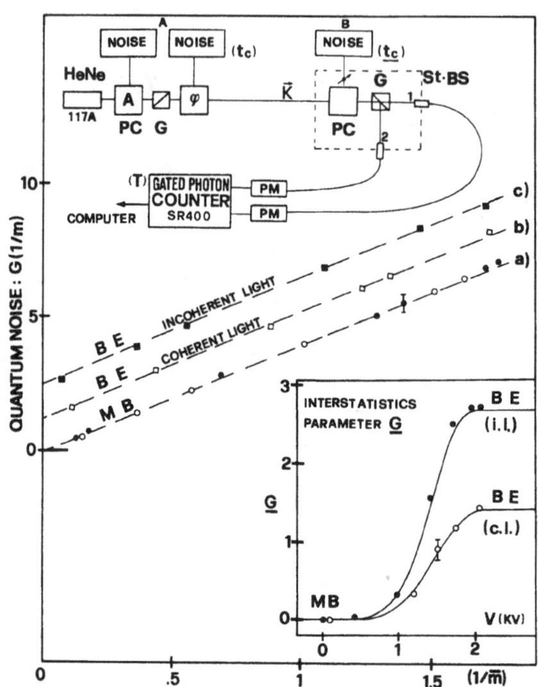

Fig.4. "*Quantum-noise-function*", G(1/m̄). Plots a, b, c, refer to classical (MB) and Bose-Einstein (BE) partition-statistics in two-channel (M=2) scattering.

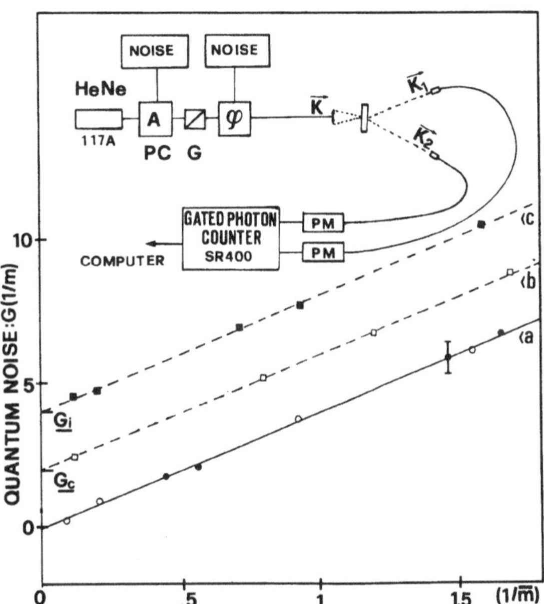

Fig.5. G(1/m̄) for a multi-channel (M>>2) scattering-experiment with polystyrene microscopic-spheres undergoing brownian-motions.

6. RESULTS OF THE INTENSITY-INTERFEROMETRY

The experimental layout of the intensity-interferometer shown in Fig.4 is derived from the one of the field-interferometer, Fig.1. The only difference is that the photons belonging to two output-beams of the IF's StO-BS are detected, via optical-fibers, by two low-noise 56DVP PM's connected to a Stanford-Research SR 400 two-channel gated photon-counter. The coherence-time of the incoherent light obtained by activating the noise-generator A was $t_c=10^{-3}$s, the StO-BS coherence-time $\tau_c=50\mu$s, the counter's gate-window $\Delta t=1\mu$s and each datum of Fig.4 corresponds to computer-evaluated averages made of $\approx 10^5$ samples. The PM noise was about 1 pulse every 100 gate-samples per channel. A continuous transition between curves (a-b) and (a-c), Fig.4, was detected by varying the value of the peak to-peak noise-voltage driving the StO-BS. The data points on the curves b), c) correspond to $\underline{v}\approx 1$kV, the PC $\lambda/4$-voltage at λ, as said. This corresponds to a variation of the scattering-probabilities $W_{1,2}$ over the full randomness range [0-1] in agreement with theory. The plots of $\underline{G}(\underline{v})$ (Fig.4, inset) show further details of the transition. The theoretical curves $\underline{G}(\underline{v})$ are computer-evaluated by the interstatistics theory given in the previous section. A full account of the experiment is reported in (1).

As a significant extension of the above scattering method to a large number of output channels, M>>2, Fig.5 shows the schematic diagram of a *Stochastic-Light-Scattering* apparatus[10]. The device A for laser light-modulation is identical to the one already described. The StO-BS of the previous experiment is here replaced by a scattering medium composed by small polystyrene spheres with diameter $\approx 0.1\mu$m undergoing brownian-motions in deionized water. The laser beam was focused on a 0.4mm thick scattering-cell by a 5cm f.l. lens and the attenuation of the beam due to scattering never exceeded 15% of the incoming light so that multiple scattering was virtually avoided. The directions of the output k_1, k_2 wavevectors were taken at 40° from the lens axis. The cell temperature T, measured by a termocouple, was varied across T=0°C by means of a salt-ice refrigerator coupled with an heating resistor. In *stationary* (StA) and *stochastic* (StO) regimes, i.e., for T<0°C and T>0°C respectively, the two PM's were excited, over output spatial single-modes, by the random and time-varying (for StO) speckle-pattern determined by the scattering dynamics[17,18]. The *scattering* coherence-time, evaluated by diffusion-theory, was $\tau_c\approx 10^{-4}$s at T=+7°C, while the other characteristic times of the experiment were $t_c=10^{-3}$s, $\Delta t=1\mu$sec. The plots of Fig.5 show that, in analogy with the BS case, the "classical" partition-statistics is realized in the StA regime, i.e., when the brownian-motion of the scattering spheres is inhibited by "freezing" the particles within the ice-crystal. By raising the value of T, in correspondence with the ice-water structural phase-transition at T≈0°C, the interstatistics-parameter was found to show a sudden jump from $\underline{G}=0$ (StA, MB-partition) to the values $\underline{G}\approx 2$ or 4 in correspondence with input coherent- or incoherent-light excitation, respectively. This is in agreement with the theoretical results reported at Section 5. We have then demonstrated once again, that the transition from the "classical" to the BE "quantum" partition-statistics is realized by a *natural* multichannel scattering process involving *microscopic* particles in virtue of a W_i-randomization mechanism satisfying rather loose statistical requirements. This appears to agree with the observed substantial p(x)-insensitivity of the "macroscopic" interstatistics processes investigated in Sect.5[10]. This also substantiates our model according to which the statistical behaviour of a gas of colliding Bose-particles and then its energy-distribution at equilibrium are due to multiple interparticle scattering events in StO regime. The *motion* of the scatterers caused either by an *external* force (a thermally-driven brownian-process, an electro-optically-driven StO-BS, a mechanically-driven moving-mirror) or by momentum-exchanges between colliding microscopic-particles (*Compton-scattering*), are expected to determine the overall statistical behaviour of the quantum-gas at equilibrium.

7. RELEVANCE OF THE COMPTON-SCATTERING: THE FREE-ELECTRON-LASER GAIN

We have just emphasized the dynamical role of the motion in establishing, at a microscopic level, the "quantum" properties of an assembly of colliding particles either in condition of thermodynamic equilibrium or of non-equilibrium, as in the case of the experiments reported in the present work. We may assume the Compton-scattering as the relevant paradigm that presides over this set of phenomena. In this respect note that at the origin of the epistemological debate over the quantum-measurement precisely the "quantum" effect of the Compton-scattering has been invoked by N.Bohr, in its debate with Einstein, to implement its first definition of the "measuring-apparatus"[19]. In facts, in order to accomplish its measuring function, no part of this apparatus should move under momentum exchange with the incoming particles. If this prescription is not fullfilled, the mechanical motion must be accounted for within the quantum theory of the overall dynamical problem. In other words a reliable measuring-apparatus is a rigid and massive "classical" device. A nice model for the Compton "quantum-mirror" consists of a relativistic electron-beam that backscatters monochromatic radiation into a field-mode. In this process the "quantum" effect consists of the stimulated-emission (StE) "gain" which is at the basis of the operation of the "free-electron-laser" (FEL, 20, 21). There a beam of electrons (e^-) with speed $\beta=(v/c)$, $\gamma=(1-\beta^2)^{-1/2}$, Compton-backscatters the Weizsäcker-Williams transformed field from a static periodic magnetic structure ("undulator") with period λ_0. At energies $E\equiv\gamma mc^2 \rangle 10$MeV the beam is highly relativistic and the "ballistic-regime" applies: $(\omega_p \lambda_0/\beta\gamma c) << \pi/2$; [$\omega_p \equiv (4\pi e\rho/\gamma m)^{1/2}$=Lorentz-invariant plasma-frequency; ρ=e^- density in laboratory-frame (l.f.)]. There the effects of Coulomb forces disappear leading to the "free-electron" (i.e., Compton) regime. A relativistic weak-field QED analysis shows the existence of two competing quantum effects: the *Compton-stimulated-emission* and the *Compton-stimulated-absorption*. Furthermore it leads to an approximate expression of the laser gain[20]:

$$\underline{G}' \propto \rho \cdot \lambda_0 \cdot \lambda^2 \cdot r_0^2 \cdot (\lambda_c/\lambda') \cdot (\delta\nu'/\nu') \tag{7.1}$$

$\underline{G}'$ is the number of excess scattered photons per meter due to the passage of a stimulating one-photon wavepacket. The classical e^- radius, r_0 is related to the Compton-wavelength, λ_c through the fine-structure constant: $r_0 \equiv (e^2/mc^2) = (2\pi\alpha\lambda_c)$. λ, λ', $(\delta\nu'/\nu')$ are respectively the wavelengths and the Doppler-shift of the back-scattered field (primed quantities are defined in electron-rest-frame: e.r.f.). The expression (7.1) is interesting for our purposes as it shows that (a): $\underline{G}'$ is proportional to $\approx(m^{-4})$; (b): $\underline{G}'$ increases with $(\lambda_c/\lambda') = \beta_e' \propto (m)^{-1}$. This last quantity expresses the speed β_e' taken by e^- in e.r.f. after Compton-collision with the photon. The same speed determines the Doppler-shift and the dynamical unbalance between the opposite effects of *stimulated-emission* and *absorption*. The conclusion is that the "quantum" behaviour of the particle, viewed as a Compton-mirror disappears quickly with the increase of its rest mass, m i.e., with its increasing inability to move in e.r.f. under momentum-transfer to the colliding-photon. Exactly for this reason the gain of an hypotetical "free-proton-laser" would be too small to be observable. Also important, the gradual replacement by increasing m of the "quantum" with the "classical" regime is found to be a rigorous result of a quantum analysis. This demonstrates once again that quantum-mechanics keeps its validity in the "classical" domain as this one generally represents an asymptotic condition determined by the value of a relevant dynamical variable of an overall quantum problem. In this example the variable is the particle-mass: this explains why the quantum-statistical behaviour is a peculiar manifestation of the *microscopic phenomena.* Note also that as far as a "mirror" or a beam-splitter, or any scatterer are concerned, there are *two* competing stimulated quantum effects: *emission* and *absorption*. These ones tend to cancel out each other within any collision process when the dynamical symmetry-breaking intrinsic of any unidirectional particle-motion is reduced by a mass increase. The cancellation is *exact* when the motion is impeded by additional constraints. Then the "classical" regime, far from being a pathological condition of theoretical physics, simply expresses the exact dynamical balance between two quantum processes having opposite effects. Note once again that these results enlighten the same Bohr's conception of the "measurement-process". If fact they give to this eminently physical process a theoretical basis instead of a mere epistemological definition. In other words and more generally, the simple FEL model is but one example of how the awkward intellectual gap existing between "quantum" and "classical" in the interpretation of a physical phenomenon can be bridged satisfactorily within the context of the existing theories.

As a final general remark inspired again by the FEL's behaviour, note that in general the dynamics of the microscopic Compton-colliding particles implies a sizable energy-spread of the corresponding De-Broglie waves. This process is well known in FEL technology as it actually limits the performance of the device. In a gas this kind of energy-spread reached in a multiple photon-atom collision process leads to the Planck's frequency distribution at thermal-equilibrium.

8. PHYSICAL INTERPRETATION AND CONCLUSIONS

The results of our scattering experiments, albeit carried out in non-equilibrium conditions, demonstrate that the statistics *is not* a fixed or intrinsic property of particles. It is in fact dependent on deterministic scattering conditions leading either to "classical" absence of particle-correlations, or to intermediate-size or, at last, to full-size (viz., BE) correlations. The consequences of that should be far reaching for fundamental physics[22]. A problem is still open however. One in fact may ask where in the present model resides the "quantum" content of quantum-statistics. To reply assume a gas of equal particles, consider the basic role plaid by the inelastic Compton-like collisions to determine the statistical properties of the gas and remind the strong quantum-dynamical effect of the (integer or half-integer) value of the *spin* of the colliding particles[23]. It is quite natural to attribute to the *particle-spin* the basic quantum-property that determines the statistical, Fermi-Dirac or Bose-Einstein properties of the gas at equilibrium. In fact, according to field theory, *bosons* and *fermions*, as properly defined by the commutation-properties of the related operators (and not necessarily by their partition properties within elementary scattering processes!) are indeed particles characterized by integer- or half-integer spins, respectively[23]. The effect of these two classes of spin-values on the dynamics of the interparticle collisions leading the gas to equilibrium determines precisely the observed quantum-statistical *equilibrium* phenomenology e.g., the Planck's distribution for integer-spin particles. All this establishes in a straigthforward way, the "*spin-statistics-connection*" within our dynamical model. Note at this point that in view of the basic relevance attributed by our model to the scattering process in determining the quantum properties of a gas, if we also define a "*gas*" as a many-body ensemble able to reach thermodynamic equilibrium through inelastic-collisions, we must conclude that the concept of "classical-gas" is self-contradictory: *every gas is a quantum-gas*. The Gibbs-paradox should be considered in this perspective[3,24]. In summary, we have shown what could be the basic mechanism of a most intriguing among physical processes: the interparticle quantum-correlations, or "exchange-interactions". We conclude with two quotations by Einstein: "The differences between Boltzmann and BE-counting express indirectly a certain hypothesis on a mutual influence of the molecules which for the time being is of quite mysterious nature...", and: "The statistical method of Herr Bose and myself is by no means beyond doubt but seems only *a posteriori* justified by its success for the case of radiation..."[25]. We believe that the epistemological perspective determined by of our present results may represent a step forward toward a realistic understanding of the basic foundations of quantum statistics. This perspective appears satisfactory: while all experimental results agree with quantum-theory, as shown, the related basic processes can be understood by a simple dynamical model with no need to appeal for any "mysterious" natural behaviour (as in Einstein's words) or for the adoption of the old and quite unnecessary "indistinguishability-principle" involving the intrinsic nature of particles. According to our model, statistics relies on the *spin-dependent dynamics* of particles. Perhaps the simplicity and the apparent richness of this approach are themselves a good indication for its validity.

APPENDIX

TRANSITION FROM THE BERNOULLI TO THE BOSE-EINSTEIN PARTITION LAWS

Making reference to the StO-BS semantics given in Sect.3, we calculate the relevant quantum probability of detecting h-photons by a detector inserted on arm 1 of IF when the input k-mode is excited by a pure $|n\rangle$ -state (Fig.1). For a StA-BS it is expressed by the well known Bernoulli-law[5]:

$$P_{nh} \equiv |\langle h|n\rangle|^2 = W_1{}^h (1-W_1)^{(n-h)} \cdot {}^nC_h \qquad ; {}^nC_h \equiv [n!/(h!(n-h)!)] \qquad (I.1)$$

while for a StO-BS driven by a signal characterized by an *even* distribution:

$$p(x)=p(-x)= \int_{-\infty}^{+\infty} P(\chi)\cos(2\pi\chi x)\, d\chi \;; \qquad P(\chi)=P(-\chi) = \int_{-\infty}^{+\infty} p(x)\cos(2\pi\chi x)\, dx \quad (I.2)$$

the same quantity, for $X_1=X_2=\frac{1}{2}$, is expressed by:

$$P_{nh}={}^nC_h \cdot 2^{-n} \int_{-\infty}^{+\infty} P(\chi)\, d\chi \int_{-1}^{+1} (1+x)^{(n-h)} (1-x)^h \cos(2\pi\chi x)\, dx \,. \qquad (I.3)$$

Consider now the integral definition of the confluent hypergeometric-function: ${}_1F_1(\alpha,\beta;\gamma)$ having a power-series expansion: ${}_1F_1(\alpha,\beta;\gamma) = \sum_{k(0-\infty)} \{(\alpha)_k \gamma^k / [(\beta)_k\, k!]\}$ where $(\alpha)_k \equiv \alpha(\alpha+1)\cdot\cdot(\alpha+k-1) = \Gamma(\alpha+k)/\Gamma(\alpha)$; $(\alpha)_0 \equiv 1$ are the Pochammer's symbols written here in terms of the gamma $\Gamma(x)$-function (10,AS). The equation (I.3) may then by given by the general, exact expression:

$$P_{nh}=(n+1)^{-1}\cdot\int_{-\infty}^{+\infty} P(\chi)\cdot\exp(-i2\pi\chi)\{{}_1F_1(a,b;i4\pi\chi)+{}_1F_1(b-a,b;i4\pi\chi)\} \qquad (I.4)$$

where $a\equiv h+1$, $b\equiv n+2$ as shown in Sect.3 for i=j. For n>>1,the Pochammer symbols in (I.4) may be replaced by simple powers $(b)_k\rightarrow(b)^k$, and then, after some simple algebra, the sum of the two ${}_1F_1$-functions within brackets {} in the integrand can be given by the approximate expression: $\{[1-(i4\pi\chi/b)]^{-a}+[1-(i4\pi\chi/b)]^{(b-a)}\}=\{\exp(-ia\phi)+\exp[i(a-b)\phi]\}$, where: $r\equiv[1+(4\pi\chi/b)^2]^{1/2}$ and $\phi=\arcsin[-4\pi\chi/(rb)]$. Then we are led to the approximate form: $\{\}\approx 2\exp(i2\pi\chi)\cos(4\pi\theta\chi)$ and finally to:

$$P_{nh}=(n+1)^{-1}\cdot 2\int_{-\infty}^{+\infty} P(\chi)\cos(2\pi\theta\chi)\,d\chi \quad ; \quad \theta\equiv(1-2z) \quad ; \quad z\equiv\lim_{(n=\infty)}(h/n) \qquad (I.5)$$

valid for n>>1 [10]. This procedure may be used for the approximate evaluation of $V(\langle n\rangle\rightarrow\infty)$ as shown in Sect.3. For a gaussian StO-BS excitation, (I.5) gives:

$$P_{nh}=(n+1)^{-1}\cdot 2[\sqrt{\pi}\,\sigma'\cdot\mathrm{erf}(1/\sigma')]^{-1}\cdot\exp[-(n-2h)/(n\sigma')]^2 \qquad (n>>1) \quad (I.6)$$

For $\sigma'\equiv(\sqrt{2}\,\sigma)\rightarrow 0$ the Bernoulli-distribution is recovered through its normal-distribution approximation for $W_i=\frac{1}{2}$ and very large $n=\sigma^{-2}$ (9,Arthurs).

REFERENCES

1. F.De Martini and S.Di Fonzo, *Europhysics Letters*, 10, 123, 1989.
2. R.Loudon,*The Quantum Theory of Light*, Oxford, 1983, chs.3,6.
3. K.Huang, *Statistical Mechanics*, Wiley, New York, 1963, Ch.9. The photon gas is a unique example of collisionless "ideal gas" if one disregards high-order QED "vacuum-polarization" effects: cfr. C.Itzykson and J.Zuber, *Quantum Field Theory*, Mc Graw Hill, N.Y., 1980, Ch.3. Note that the BE-partition: $(P_{nh})_{be}\equiv W^{BE}\equiv n!(M-1)!/(n+M-1)!$ is *independent* of the occupation-number $\{n_i\}$.
4. The gaussian-noise for sets A,B was generated by reverse-biased Motorola 2N-2369 transistors. For A, the A/Φ-modulator was PCEOD125. For B, PC was Laser metrix 1042. The IF's Φ-modulator was INRAD 621040. Condition $|v_a(t)|_{max}\leq\underline{v}$ and then $|\Gamma(t)|\leq\frac{1}{2}\pi$ was determined by sharp amplifier-saturation. The state of the IF-input field-coherence was determined by standard Hanbury Brown-Twiss StA-BS measurements: coherent- or incoherent-light leads to values of the "degree of 2nd-order-coherence": $g^{(2)}=1$ or 2, respectively[2].
5. D.F.Walls, *Am.J.Phys.* 45, 952, 1977. S.Prasad, M.O.Scully, W.Martiennsen, *Opt.Comm.* 62, 139, 1987; Z.Y.Ou, C.Hong, L.Mandel, *Opt.Comm.* 63, 118, 1987; R.Campos, B.Saleh, M.Teich, *Phys.Rev.* A 42, 4127, 1990. Quantum or classical

theories lead for the complex BS-parameters to: $\Sigma_i|w_i|^2=1$, $Re(w_1{}^*w_2)=0$, viz., for the phases of w_i: $|\phi_1-\phi_2|=\pi/2$. The IF problem requires evaluation of $\langle\hat{\rho}\rangle$ only for $|h-m|=1$ or 0.

6. H.Fearn and R.Loudon, *Opt.Comm.* 64, 485, 1987; F.De Martini and K.Strobl, *Opt. Comm.*76, 75, 1990.
7. R.Lange, J.Brendel, E.Mohler, W.Martienssen, *Europh.Lett.* 5, 619, 1988; F.De Martini in *Squeezed and Nonclassical Light*, R.Pike, P.Tombesi eds., Plenum, N.Y. 1988.
8. J.Tersoff and D.Bayer, *Phys. Rev. Lett.* 50, 553, 1983.
9. P.M.Woodward, *Probability and Information Theory*, Pergamon Press, Oxford, 1964. W.Feller, *An Introduction to Probability Theory*, Vol.2, Wiley, N.Y. 1966. A.M. Arthurs, *Probability Theory*, Dover, N.Y. 1965.
10. I.Gradshstein, I.Ryzhik, *Tables of Integrals, Series, Products*, Academic-Press, N. Y.1980. M.Abramovitz, I.Stegun, *Handbook of Mathematical Functions*, Dover, 1964. These texts will be indicated respectively by GR and AS in the paper. The BE distribution is *independent* of the occupation numbers on the scattering channels, e.g., of h for M=2. The BE-law for M>>1 has been demonstrated by a recent light-scattering experiment by: F.De Martini, R.Tommasini in *New Frontiers in QED and Quantum-Optics*, A.Barut, ed., Plenum, N.Y.1990; cf. Sect.6. For the approximate procedure leading to (I.5) cf.: L.B.W.Jolley, *Summation of Series*, Dover, N.Y.1961. Refer to (3.11): a series expansion of η for $\sigma>>0$ in terms of σ^{-2} gives: $\eta=(1 + 0.042\ \sigma^{-2} + 0.0026\sigma^{-4} + ...)$. This result and Fig.2 show that in order to obtain a 10% departure from the asymptotic value $\eta\approx1$ (i.e., from the BE-law) the value of σ should decrease from the asymptotic value=∞ to ≈0.65. This justifies the insensitivity of the BE-partition to the actual form of p(x) when this function is not too far from representing a uniform distribution. This result appears physically relevant. In fact it is possible that the apparent persistence of the Bose-Einstein phenomenology in very different physical conditions, ranging from Planck's cavity radiation to Bose-condensation in solids, has so far oriented the attention of the physicists toward a mechanism involving some *invariant* or *intrinsic* property of the "quantum" particles, as for instance their "indistinguishability". Our present work carried out, on purpose, with different microscopic and macroscopic systems, and then with different p(x)'s, shows that the size of the various accidents affecting the scattering-dynamics of different gas-ensembles in a wide variety of experimental contexts, is generally not large enough to perturb a general *equilibrium* quantum behaviour, i.e. reached after averaging all elementary effects over a large number of scattering events. The theory given in the present work supports these results.
11. The pertinence of the given stochastic-theory within our work lies on the ergodic-hypothesis prescribing equivalence between time t-averages over the field coherence-time t_c and ensemble-averages over the I(t)-samples detected within t_c and optionally A-modulated by gaussian-noise. Since within t_c the system is *stationary and gaussian*, ergodicity is valid there under broad conditions: cfr.J.Doob, *Stochastic-Processes*, Wiley, N.Y.1953, ch.11; D.Sakrison *Communication-Theory*, Wiley, 1968, ch.4. With IF-input incoherent-field the PM output results from a t-averaging of the I(t)-samples over $\tau_{PM}\approx10t_c$. The t-resolving power of our apparatus allows accurate determination of $I=I[\Phi(t)]$ and then of V, since: $\omega\tau_{PM}<<2\pi$.
12. We have reported recently [ECOOSA'90 International Conference, Rome, November 1990] experimental Visibility results showing that, for p(x)=1/2, is $V(\langle n\rangle)=(\pi/4)$ for very large $\langle n\rangle$ and $V(\langle n\rangle)\rightarrow1$ for $\langle n\rangle\rightarrow0$. This is at variance with the results of the

quantum-theory given in the present work. The reported experiment is presently under careful investigation and its results will be reported elsewhere. In the present work we report for StO-IF only *one* experimental point $V(\langle n \rangle) \approx \pi/4$ for very large $\langle n \rangle$ (Fig.1).

13. W.K.Wootters and W.H.Zurek, *Phys.Rev.* D, 19, 473, 1979; C.E.Shannon and W.Weaver, *The Mathematical Theory of Communications*, Urbana, 1975. A collection of the classic papers on Complementarity is found in: J.Wheeler and W.Zurek, *Quantum Theory and Measurement*, Princeton University, 1983. Our definition of K implies the approximation: $\ln(1+x) \approx x(1-x/2)$, $(|x| << 1)$ introduced in the calculations.
14. In the present work and in (1) we have often referred to *classical* or *quantum* partition-laws (i.e., the ones expressed by the Bernoulli or the BE-partition formulas for M=2) as Maxwell-Boltzmann (MB) or Bose-Einstein (BE) *statistics*. Actually, MB- or BE-*statistics* are usually and properly defined as representing *equilibrium* conditions of a gas. According to our model these conditions are reached through many elementary *non-equilibrium* scattering-events involving absence or presence of the BE-correlations. In our present work we are in fact dealing with such elementary *non-equilibrium* events and then the term "*statistics*" is often used here in an analogical sense.
15. For input $|\alpha\rangle$-states the statistical picture is valid owing to their n-state expansion[2]. $G(1/\bar{m})$ may be taken as the "quantum-signature" of the electromagnetic field as well as of any Bose-field. An account of this method of quantum-optics with application to gravity-waves detection is found in (6: De Martini, Strobl). The HB-T technique has been applied to high energy-physics, e.g. for study of BE-correlations in $(\pi$-$\pi)$-scattering and for measurement of cross sections and sizes of the pion-sources: cf. G.Goldhaber, W.B.Fowler, S.Goldhaber, T.Hoang, T.Kalogropoulos and W. Powell, *Phys.Rev.Lett.* 3, 181, 1959; G.Goldhaber, W.Lee and A.Pais, *Phys. Rev.* 120, 300, 1960; G.Kopylov and M.Podgoretsky *Yad.Fiz.* 15, 392, 1972; A. Giovannini, G.Mantovani and S.Ratti, *Riv.Nuovo Cimento*, 2, 1, 1979; R.Kvatadze, R.Moller and B.Lorstad, *Z.Phys.* 38C, 551, 1988; UA1-Collaboration, CERN, Geneva, *Nucl.Phys.* B, 309, 1988, and *Phys.Lett.* B, 226, 410, 1989.
16. D.Costantini and U.Garibaldi, *Phys.Lett.* A, 150, 337, 1990.
17. F.T.Arecchi, M.Giglio and U.Tartari, *Phys.Rev.* 163, 186, 1967.
18. J.Dainty, *Laser-Speckle*, Springer, N.Y., 1975; W.Martienssen and E.Spiller, *Am.J. Phys.* 32, 919, 1964.
19. N.Bohr, *Nature* 136, 65, 1935; *Phys.Rev.* 48, 696, 1935; and in P.A.Schlipp, *Albert Einstein Scientist Philosopher*, Cambridge, 1982.
20. J.Madey, *J.Appl.Phys.* 42, 1906, 1971 and in "*Free Electron Laser Handbook*", ed.by W.Colson, C.Pellegrini and A.Renieri, Elsevier, Amsterdam, 1990.
21. F.De Martini in "*Free Electron Laser Handbook*", Ref.20.
22. The most obvious one appears to be the need for rejection of the indistinguishability, [I] concept. In fact, the Bernoulli photon partition (I.1) induced by StA-BS implies *absence* of any inter-particle quantum-correlation: this rules out any *intrinsic* [I]-property of the scattered particles. In other words, at least as far as [I] is concerned, there *are no classical* or *quantum* particles. Of course the partial overlapping of the particle wave-functions within each collision process implies an [I] effect as an aspect of the Heisenberg's principle. This [I] process is not expected to play any direct role in the determination of the statistical properties of the ensemble.
23. J.Schwinger, *Particles, Sources and Fields*, ch.2, Addison-Wesley, N.Y.1989.
24. D.Dieks and V.V.Dijk, *Am. J. of Phys.* 65, 430, 1988.
25. A.Einstein, *Sitzungs.Press.Akad.Wiss.*, 261, 1924 and: 18, 1925.

WAVE FUNCTIONALS FOR THE PHOTON FIELD

E.R. Pike[*+] and S. Sarkar[*]

* Department of Physics, King's College London, UK
+ Royal Signals and Radar Establishment, Great Malvern, UK

1. INTRODUCTION

Given a Hilbert space of quantum mechanical states, in principle any basis is as good as any other if we are capable of doing exact calculations for processes. However, as is well-known, in practice it is advantageous to choose a certain basis in preference to others for a particular problem. The photon even at this basic level is still stimulating new theoretical work[1]. On the experimental front there is impressive progress in realizing some of these new calculated states[2]. On using the rules of quantum mechanics, it is of course possible to calculate accurately correlation functions which describe photodetection in these experiments. However in non-cavity situations such as parametric down-conversion we would like to understand to what extent we can think of localised photons flying through space. It is not simple to characterise such states since there is no position operator for the photon.

The search for a wave function, i.e. a complex amplitude for a particle to be at a point $\underline{x}$, is based on experience in non-relativistic quantum mechanics where we have the Schrödinger equation

$$i \frac{\partial}{\partial t} \Psi(\underline{x}, t) = H \Psi(\underline{x}, t) . \tag{1.1}$$

(Units such that $\hbar = 1$ and $c = 1$ will be used.) There is a position operator $\hat{\underline{x}}_{op}$ such that

$$\hat{\underline{x}}_{op} | \underline{x} \rangle = \underline{x} \, | \underline{x} \rangle \tag{1.2}$$

and

$$\langle \underline{x} | \underline{x}' \rangle = \delta(\underline{x} - \underline{x}') . \tag{1.3}$$

Quantum Measurements in Optics, Edited by P. Tombesi and D.F. Walls, Plenum Press, New York, 1992

The wave function $\Psi(\underline{x}\,, t)$ is the overlap amplitud

$$< \underline{x} \mid \Psi(t) > \; = \; \Psi(\underline{x}\,, t)\,. \tag{1.4}$$

For a massive relativistic particle this construction can also be followed. The Klein-Gordon equation for a spinless particle with mass m is

$$(\partial^2 + m^2)\,\phi(x) = 0\,. \tag{1.5}$$

An inner product $<|>$ structure can be introduced into the space of positive energy solutions which is most conveniently expressed in momentum space. For two solutions ϕ and Ψ

$$< \phi|\Psi> \; = \int \frac{d^3k}{k_o}\, \tilde{\phi}^*(\underline{k})\, \tilde{\Psi}(\underline{k}) \tag{1.6}$$

where

$$k_o = (\underline{k}^2 + m^2)^{1/2}$$

and

$$\tilde{\phi}(\underline{k}) = \frac{1}{(2\pi)^{3/2}} \int \frac{d^3k}{k_o}\, e^{-ik.x}\, \phi(\underline{x})\,. \tag{1.7}$$

(N.B. $k.x = k_o t - \underline{k}.\underline{x}$). This inner product defines a Hilbert space and $\hat{x}_{op}$ has to be self-adjoint with respect to $<|>$. It is given by

$$\underline{x}_{op} = i\nabla_{\underline{p}} - \frac{1}{2}\,\frac{i\,\underline{p}}{\underline{p}^2 + m^2} \tag{1.8}$$

(in the momentum space representation). The co-ordinate space representation of $|\underline{x}_o>$ is

$$\left(\frac{m}{|\underline{x} - \underline{x}_o|}\right)^{5/4} H^{(1)}_{5/4}\,(i\,m\,|\underline{x} - \underline{x}_o|)\,.$$

This sort of construction cannot be carried out for the photon as we have already noted. There is another type of Schrödinger equation for the Klein-Gordon system which, it will turn out, generalises to the case of the photon. We will discuss this in the next section.

From elementary Fourier analysis any localised photon state would require a wide spread in momentum space, i.e. a fairly broad wavepacket. However it will be instructive to review some elementary facts first about the single mode case.

2. THE HARMONIC OSCILLATOR

The simple harmonic oscillator represents a single mode. The general case of multimodes is not that different in that the Hamiltonian is a sum over modes of harmonic oscillator Hamiltonians for a particular mode. The basic Hamiltonian is

$$H = \frac{\hat{p}^2}{2m} + \frac{1}{2} K \hat{x}^2 . \tag{2.1}$$

It is customary to make the transformation to the creation and annihilation operators a^+ and a where

$$\hat{x} = \left(\frac{\omega}{2K} \right)^{1/2} \left(a^+ + a \right) \tag{2.2}$$

and

$$\hat{p} = i \left(\frac{m\omega}{2} \right)^{1/2} \left(a^+ - a \right) . \tag{2.3}$$

H then becomes

$$H = \omega \left(a^+ a + \frac{1}{2} \right) \tag{2.4}$$

where $\omega = \frac{1}{2} \left(\frac{K}{m} \right)^{1/2}$ and the well-known number states are the energy eigenstates. However it is possible to calculate the eigenstates more directly from the Schrödinger equation

$$- \frac{1}{2m} \frac{d^2\Psi}{dx^2} + \frac{1}{2} K x^2 \Psi = E \Psi . \tag{2.5}$$

The wave function Ψ is well-known[3] to be

$$\Psi(x) \propto H_n(\alpha x) \, e^{-\frac{1}{2}\alpha^2 x^2} \tag{2.6}$$

$$\text{where } \alpha^4 = mK \tag{2.7}$$

and $H_n(x)$ denotes the nth Hermite polynomial. If we have a multimode situation

$$H = \sum_n \omega_{\underline{k}_n} a^+_{\underline{k}_n} a_{\underline{k}_n} \tag{2.8}$$

then a way of generalising the previous construction is

$$\Psi\left(\ldots, x_{\underline{k}_n}, \ldots\right) \propto \prod_n H_n\left(\alpha_n x_{\underline{k}_n}\right) \tag{2.9}$$

and

$$\alpha_n = \left(m K_n\right)^{1/4} .$$

3. KLEIN-GORDON FIELD THEORY

The Klein-Gordon particle can also be discussed within the context of second quantisation. The field operator in the Schrödinger representation may be written as

$$\hat{\phi}(\underline{x}) = \int \frac{d^3k}{(2\pi)^3} \left(a(\underline{k})\, e^{-i\underline{k}\cdot\underline{x}} + a^+(\underline{k})\, e^{i\underline{k}\cdot\underline{x}} \right) . \tag{3.1}$$

This leads to

$$H = \int \frac{d^3k}{(2\pi)^3}\, \omega_{\underline{k}}\, a^+(\underline{k})\, a(\underline{k}) . \tag{3.2}$$

(Here the notation $\omega_{\underline{k}} = k_o = \sqrt{\underline{k}^2 + m^2}$ has been introduced to emphasize the similarity with the harmonic oscillator.) Our normalization is such that

$$\left[a(k), a^+(k')\right] = (2\pi)^3\, \delta^{(3)}(\underline{k} - \underline{k}') . \tag{3.3}$$

By taking Fourier transforms in (3.1) we can work in terms of $\hat{\phi}$ (and its canonically conjugate momentum $\hat{\pi}$). It is then possible to re-express (2.9) using only quantities local in co-ordinate space. For clarity of exposition however we will consider a discrete grid in configuration space, and associate a ϕ_i with each grid point i. A field operator emerges from the ϕ_i as the volume of the grid becomes infinite and the mesh site tends to zero. This limit naturally leads to functionals and functional derivatives. The natural generalization[4] of $\Psi(x)$ for the harmonic oscillator is

$$\Psi[\phi] = \langle \phi | \Psi \rangle \tag{3.4}$$

where

$$\hat{\phi}(t, \underline{x})\, |\phi\rangle = \phi(t, \underline{x})\, |\phi\rangle \tag{3.5}$$

(in terms of $\hat{\phi}_i$ on the grid $|\phi>$ is the simultaneous eigenstate of the 'coordinate' $\hat{\phi}_i$). Following elementary quantum mechanics any operator $\hat{O}$ has the representation

$$< \phi | \hat{O} | \Psi > = \int D\phi' \, O(\phi, \phi') \, \Psi[\phi'] \,. \tag{3.6}$$

For example, if $\hat{O}$ is $\hat{\phi}(x)$ then the corresponding $O(\phi, \phi')$ is $\phi(x)\delta(\phi - \phi')$. $D\phi'$, the functional measure, is best thought of as the limiting measure obtained as the grid becomes finer. Just as $\frac{id}{dx}$ is a representation of the momentum operator, $\hat{\Pi}(\underline{x})$ can be written as

$$\hat{\Pi}(x) = i\frac{\delta}{\delta\phi(x)} \delta(\phi - \phi') \,. \tag{3.7}$$

For a Hamiltonian $H(\hat{\pi}, \hat{\phi})$ the field-theoretic Schrödinger equation is

$$i\frac{\partial}{\partial t}\Psi[\phi, t] = H\left[i\frac{\delta}{\delta\phi}, \phi\right] \Psi[\phi, t] \,. \tag{3.8}$$

(This is a general equation valid even for an interacting theory, once we know the canonical variables.) The distinction between interacting and non-interacting theories lies in the difficulty in solving this equation. A generalised free Hamiltonian can be written as

$$H = \frac{1}{2}\int d^3x\, d^3y \,\{\hat{\underline{\pi}}(\underline{x}) \cdot \hat{\underline{\pi}}(\underline{y})\, \delta(\underline{x} - \underline{y}) + \hat{\underline{\phi}}(\underline{x}) \underline{\underline{K}}(\underline{x} - \underline{y}) \hat{\underline{\phi}}(\underline{y})\} \,. \tag{3.9}$$

The discrete (or grid) version of this Hamiltonian is

$$H = \sum_i \frac{p_i^2}{2} + \frac{1}{2}\sum_{<i,j>} q_i K_{ij} q_j \,. \tag{3.10}$$

K_{ij} is symmetric and it can be diagonalised on using orthogonal matrices. We write

$$q = Sq' \tag{3.11}$$

and

$$p = Sp' \tag{3.12}$$

where

$$S\,S^T = S^T S = 1. \tag{3.13}$$

(Hence q' and p' are canonical.) Moreover we require that

$$S^T_{ik} K_{kl} S_{lj} = \lambda_i S_{ij} \,. \tag{3.14}$$

(No summation convention for i is being used. For the specific scalar field Hamiltonian of interest we know implicity S and $\underset{=}{\lambda}$ since in terms of creation and annihilation operators the Hamiltonian is diagonal. However the structure of the solution in terms of functionals is more transparent when the above formalism is used. The Hamiltonian H now has the form

$$H = \sum_{i=1}^{r} \frac{1}{2} p_i'^2 + \sum_{i=1}^{r} \frac{1}{2} \lambda_i q_i'^2 \,. \tag{3.15}$$

Using (2.9) we can write the ground state $\Psi(\{q'_i\})$ as

$$\Psi(\{q'_i\}) = \prod_i H_0\left(\lambda_i^{1/4} q'_i\right) \exp\left(- q_i'^T \lambda_i^{1/2} q'_i\right) \tag{3.16}$$

up to a normalization factor. Now

$$\sum_i q_i'^T \lambda_i^{1/2} q'_i = \sum_i q_k^T S_{ki} \lambda_i^{1/2} S^T_{ij} q_j \tag{3.17a}$$

$$= q_k^T (K^{1/2})_{kj} q_j \,. \tag{3.17b}$$

On noting that $H_0(x) = 1$ for the continuum field theory we can write the ground state functional as

$$\Psi[\phi] \propto \exp\left(-\frac{1}{2}\int \phi(\underline{x})\, \Omega(\underline{x},\underline{y})\, \phi(\underline{y})\, d^3x\, d^3y\right) \tag{3.18}$$

where

$$K(\underline{x},\underline{y}) = \int \Omega(\underline{x},\underline{x}')\, \Omega(\underline{x}',\underline{y})\, d^3x' \,. \tag{3.19}$$

(The constant of proportionality can also be worked out easily. For each coordinate q'_i the n-th eigenstate has a normalization factor

$$\left(\frac{\alpha}{\pi^{1/2}\, 2^n n!}\right)^{1/2} .$$

For (3.16) the total normalization is

$$\prod_i \left(\frac{\alpha_i^{1/2}}{\pi^{1/4}} \right) = \det \left(\frac{K}{\pi^2} \right)^{1/8} = \left(\det \left(\frac{K^{1/2}}{\pi} \right) \right)^{1/4} . \tag{3.20}$$

Hence the constant of proportionality in (3.19) is

$$\left(\det \left(\frac{\Omega}{\pi} \right) \right)^{1/4} .)$$

The one and higher particle states are most simply obtained from applying the appropriate creation operator in functional form to $\Psi[\phi]$. It is straightforward to obtain this functional form on noting that

$$a_{\underline{k}} = \int d^3x \, \frac{e^{-i\underline{k}.\underline{x}}}{\sqrt{2\,\omega_{\underline{k}}}} \left(\omega_{\underline{k}} \hat{\phi}(\underline{x}) + i\hat{\pi}(\underline{x}) \right) . \tag{3.21}$$

On using (3.7) we have the representation

$$a_{\underline{k}} = \int d^3x \, \frac{e^{-i\underline{k}.\underline{x}}}{\sqrt{2\,\omega_{\underline{k}}}} \left(\omega_{\underline{k}} \phi(\underline{x}) - \frac{\delta}{\delta\phi(\underline{x})} \right) . \tag{3.22}$$

$\Psi[\phi]$ is the probability amplitude for overlap of the state Ψ with an eigenstate of the field $\hat{\phi}$ with eigenvalue ϕ. (This eigenstate is quite distinct from coherent states which are only eigenstates of the destruction part of the field. The state should not be confused with a squeezed state either since such states are eigenstates of $\mu a + \nu a^+$ with $|\mu|^2 - |\nu|^2 = 1$.) Clearly if we have a procedure for choosing a configuration ϕ which is localised in a given box, then $|\Psi[\phi]|^2$ could be used as a measure for localisability. One possible choice is ϕ constant in a box-like region of space and zero outside. This construction is not however equivalent to a system of imprimitivity. While the latter does not generalise to the case of the photon we will show that the former does.

4. THE ELECTROMAGNETIC VECTOR POTENTIAL

The difficulty of an imprimitivity construction for the photon can be seen from a classical example for the space of transverse vector functions. This same example will provide candidates for field configurations which are eigenvalues for eigenstates of the vector potential field operator.

With the 4-potential A^{μ} in mind we will consider functions $f_{\mu}(p)$ (in momentum space) which give the following representation of the Poincaré group:

$$\left[U(a,\Lambda)\, f\right]^{\mu}(p) = e^{-ia.p}\, \Lambda^{\mu}_{\nu}\, f^{\nu}(\Lambda^{-1}p) \tag{4.1}$$

where $U(a,\Lambda)$ is an element of the Poincaré group, Λ is a homogeneous Lorentz transformation and a is a translation. Moreover we impose the transversality condition

$$p^{\mu} f_{\mu}(p) = 0 \ . \tag{4.2}$$

It is natural to take functions which differ by a gauge transformation to be equivalent. We can restrict ourselves to members of the equivalence class with

$$f_0(p) = 0 \ , \tag{4.3}$$

and introduce an inner product[1]

$$<\underline{f} \mid \underline{g}> = \int d^3p\ \underline{f}^*(\underline{p}) \cdot \underline{g}(\underline{p}) \ . \tag{4.4}$$

(4.2) and (4.3) imply that

$$\underline{p} \cdot \underline{f}(p) = 0 \ . \tag{4.5}$$

These functions form a Hilbert space h (say) h is contained in a Hilbert space h^+ which has the same inner product but has functions no longer restricted by (4.5). Given any function g in h^+ we can associate a function f in h by

$$\underline{f}(p) = \underline{g}(p) - \frac{\underline{p}}{|\underline{p}|^2}\, \underline{p} \cdot \underline{g}(p) \ . \tag{4.6}$$

This mapping from h^+ to h will be denoted by P and in fact is a projection. Although it is easy to define an imprimitivity in h^+ it is not possible to do so in h. For an open set S in R^3 the imprimitivity $E(s)$ $\left(\text{in } h^+\right)$ is

$$E(s) = F^{-1} \chi(s)\, F \ . \tag{4.7}$$

Here

$$(F f)(x) = \frac{1}{(2\pi)^{3/2}} \int d^3p\, e^{ip \cdot x}\ \underline{f}(p) \tag{4.8}$$

and

$$\chi(s)\ F f(\underline{x}) = F f(\underline{x}) \text{ if } \underline{x} \in S$$

$$= 0 \text{ if } \underline{x} \notin S \ . \tag{4.9}$$

The difficulty for defining an imprimitivity in $\mathcal{h}$ is that in general

$$E(S)P \neq PE(S)\,. \tag{4.10}$$

We will now consider the photon (in the Coulomb gauge since this choice is the most physically appealing for quantum optics). Now the energy density is

$$\frac{1}{2}\left((\dot{\underline{A}}_T(x))^2 + (\underline{\nabla}\times\underline{A}_T(x))^2\right)$$

$$= \frac{1}{2}\left(\underline{\Pi}^2(x) + \left(\underline{\nabla}\times\underline{A}_T(x)\right)^2\right) \tag{4.11}$$

(since $\dot{\underline{A}}_T(x)$ and $\underline{A}_T(x)$ are canonically conjugate pairs). T denotes 'transverse' and for convenience we will not explicitly show it in the following.

$$\left(\underline{\nabla}\times\underline{A}\right)_i^2 = \partial_j\left(A_k\,\partial_j\,A_k\right) - \partial_j\left(A_k\,\partial_k\,A_j\right) - A_k\,\partial_j\,\partial_j\,A_k + A_k\,\partial_j\,\partial_k\,A_j\,. \tag{4.12}$$

The analogue of $\underline{\underline{K}}\,(x-y)$ is

$$K_{ik}\,(x-y) = \left(-\frac{\partial}{\partial x_j}\frac{\partial}{\partial x_j}\,\delta_{ik} + \frac{\partial}{\partial x_k}\frac{\partial}{\partial x_i}\right)\delta(x-y)\,. \tag{4.13}$$

The corresponding $\Omega_{ik}(x,y)$ is best found through the Fourier representation of K_{ik}:

$$K_{ik}(x,y) = \int \frac{d^3k}{(2\pi)^3}\,\underline{k}^2\left(\delta_{ij} - \frac{k_i\,k_j}{\underline{k}^2}\right)e^{i\underline{k}\cdot(\underline{x}-\underline{y})}\,. \tag{4.14}$$

Now it is easy to rewrite this as

$$K_{ik}(x,y) = \int \frac{d^3k}{(2\pi)^3}\left(\frac{\underline{k}^2\delta_{ij} - k_i\,k_j}{|\underline{k}|}\right)\times \left(\frac{\underline{k}^2\delta_{jk} - k_j\,k_k}{|\underline{k}|}\right)e^{i\underline{k}\cdot(\underline{x}-\underline{y})}\,. \tag{4.15}$$

From inspection it is now possible to see that

$$\Omega_{ij}(\underline{x},\underline{x}') = \int \frac{d^3k}{(2\pi)^3} \frac{\underline{k}^2 \delta_{ij} - k_i k_j}{|\underline{k}|} e^{i\underline{k}.(\underline{x}-\underline{x}')} . \tag{4.16}$$

The functional Schrödinger equation is

$$\int d^3k \left(-\frac{1}{2} \frac{\delta^2}{\delta A^i(x)\, \delta A^i(x)} + \frac{1}{2} \left(\underline{\nabla} \times \underline{A}(x) \right)^2 \right) \Psi[\underline{A}] = E\, \Psi[\underline{A}] . \tag{4.17}$$

This equation follows immediately from (4.11) and the representation of $\hat{\underline{\Pi}}$ as a functional derivative. Since $\underline{A}$ is transverse, the relevance of our discussion on the Hilbert space of transverse functions at the beginning of this section, should be clear. Following arguments similar to those given for the Klein–Gordon equation

$$\Psi_0[A] = \exp\left(-\frac{1}{2} \int\int d^3x d^3x' A_i(\underline{x})\, \Omega_{ij}(\underline{x},\underline{x}') A_j(\underline{x}') \right) . \tag{4.18}$$

In order to describe the many particle states it is again useful to introduce creation and annihilation operators $a_i^+(\underline{p})$, $a_i(\underline{p})$ which, instead of (3.3), satisfy

$$\left[a_i(\underline{p}), a_{i'}^+(\underline{p}') \right] = (2\pi)^3 \left(\delta_{ii'} - \frac{p_i p_{i'}}{|\underline{p}|^2} \right) \delta^{(3)}(\underline{p} - \underline{p}') . \tag{4.19}$$

In terms of canonical momenta and fields we have

$$a_j^+(\underline{p}) = \sqrt{\frac{1}{2|\underline{p}|}} \left(|\underline{p}|\, \hat{A}_j(\underline{p}) - \hat{\Pi}_j(\underline{p}) \right) \tag{4.20a}$$

and

$$a_j(\underline{p}) = \sqrt{\frac{1}{2|\underline{p}|}} \left(|\underline{p}|\, \hat{A}_j(\underline{p}) + \hat{\Pi}_j(\underline{p}) \right) . \tag{4.20b}$$

Converting to configuration space, the analogue of (3.22) is

$$a_j(\underline{p}) = \int d^3x\, e^{i\underline{p}\cdot\underline{x}} \sqrt{\frac{1}{2|\underline{p}|}} \left(|\underline{p}|\, A_j(\underline{x}) - \frac{\delta}{\delta A_j(\underline{x})} \right) . \tag{4.21}$$

Hence the one-particle wave functional (corresponding to momentum $\underline{p}$) is proportional to

$$\int d^3x \; e^{i\underline{p}\cdot\underline{x}} \sqrt{\frac{|\underline{p}|}{2}} \; A_j(\underline{x}) \, \Psi_o[A] \; .$$

Any particle wave packet can be represented by a wave functional which is a superposition of the above momentum $\underline{p}$ wave functionals.

In order to obtain a useful measure of localisation using this approach we need to choose in some sense a 'good' configuration $A_i(\underline{x})$. Intuitively an eigenstate of PE(S) seems such a good choice for localisation in a region S. Starting off with a three-vector with equal constant components inside S and zero outside we can apply the operation $\lim_{n\to\infty} (PE(S))^n$. Provided this procedure converges we will generate a suitable configuration $A_i(\underline{x})$. (The normalisation of the initial configuration can be chosen such that the average energy density agrees with that of Ψ.) In this way some of the basic ideas behind the notion of generalised imprimitivity[6, 1] can be incorporated into our approach to photon localisation through the wave functional.

REFERENCES

1. E. R. Pike and S. Sarkar in *Frontiers in Quantum Optics*, eds. E. R. Pike and S. Sarkar (Adam Hilger, Bristol 1986).

2. J. G. Rarity and P. R. Tapster in *Photons and Quantum Fluctuations*, eds. E. R. Pike and H. Walther (Adam Hilger, Bristol 1988).

   C. K. Hong and L. Mandel Phys. Rev. Lett. 59, 2044 1987.

3. L. I. Schiff *Quantum Mechanics* (McGraw-Hill, 1968).

4. R. Jackiw in *Conformal Field Theory, Anomalies and Superstrings* eds. B. E. Baaquie, C. K. Chew, C. H. Oh and K. K. Phua (World Scientific, Singapore 1988).

5. G. W. Mackey *Unitary Group Representations in Physics, Probability and Number Theory* (Benjamin, 1978).

6. W. O. Amrein Helv. Phys. Acta 42, 149 1969.

TESTS ON THE REALITY OF DE BROGLIE WAVES WITH TWO COHERENT PHOTONS

J. R. Croca

Departamento de Física
Faculdade de Ciências, Universidade de Lisboa
Campo Grande Ed. C1, 1700 Lisboa - Portugal

1. INTRODUCTION

As is well known, since the foundation of quantum mechanics, at the first quarter of this century, its interpretation has given rise to endless polemics. The reason of these debates can be rooted in the most uncommon, at least in the field of the so called exact sciences, epistemology proposed by the usual paradigm of quantum mechanics. It was traditional to believe that the facts, the results from the experiments, could be understood in the conceptual framework of spacetime. Furthermore it was supposed that, at least in principle, there would be no limits for the human possibilities in understanding Natura. Plain quantum mechanics categorically denies these assumptions. Therefore it is not surprising to observe that many people still keep trying to check its foundations. It must be stated that no one questions the main empirical predictions of the plain theory. Everybody agrees that quantum mechanics represents, perhaps one of the greatest achievements of mankind. What is under discussion is for one side whether the quantum theory is a complete theory in the sense of Bohr, meaning that the laws of nature were settled once for all. The only possible developments are those horizontal connected mainly with the practical applications of the known fundamental laws. The other point that makes people feel uneasy is the rejection of causality. If it is true that, at the quantum level, space and time are mere auxiliary concepts, without any intrinsic meaning, superfluous commodities, we have been used for too long, then the full consequences of these postulates may in fact be astonishing. It would be the greatest scientific revolution in the history of science.

For the stated reasons many experiments have been proposed and done and others are still in course with the aim of testing the interpretation of quantum theory. It may be argued that the theory have been proved correct so many times that it shall be a loss of time keeping proposing and doing new experiments. Still one knows that for each new experiment the understanding of the theory expands. Without the debate about EPR and Bell's theorem many implications of the theory would be completely overlooked.

Quantum Measurements in Optics, Edited by P. Tombesi and
D.F. Walls, Plenum Press, New York, 1992

2. ON THE NATURE OF THE WAVE FUNCTION

In this paper I want to propose a new experiment to test one aspect of the interpretation of quantum theory. It is related with the nature of the wave function ψ. According to the plain theory ψ represents a probability wave without any physical meaning, whereas for de Broglie[1] and others this same function represents a real wave evolving in space and time.

Many experiments to test the true nature of this wave have been proposed. Among them it is useful to mention those based in the so called autoreduction of the wave train and on the collapse of the wave function in a measure.

The concept of autoreduction of the wave packet was initially proposed by Koh and Sasaki[2] and was further applied, by the author[3], to the waves of de Broglie. Essentially the idea consists in supposing that when an initial matter wave packet, by the action of a beamsplitter, is divided in two, each following along a different arm of an interferometer, the visibility of the final interference pattern may eventually decrease as a function of the increasing length of the arms of the interferometer. A kind of autoreduction of the wave packet would be the responsible for the loss of visibility.

The experiments related with the collapse of the wave function by the action of a measurement are a direct consequence of the true nature of the wave function. Consider, for instance the situation where a wave packet after impinging in a beamsplitter is divided in two secondary wave packets, each following well separated trajectories. Suppose further that a detector, placed in one of the paths, is triggered. Then what conclusions can we draw from this conceptual experiment?

If the wave function is a probability wave one concludes that since the detector has revealed the particle at one path the probability of finding it at the other drops instantly to zero.

Quite different is the conclusion if one assumes that the ψ function stands for a real wave. In this case, following de Broglie, a particle is composed of an extended real wave plus a kind of singularity carrying practically all the energy of the particle. Therefore if the detector was activated at one of the paths that means, according to this model, that the singularity went through the detector while at the other path a wave without singularity, an empty wave as it is some times called, keeps going its way unaffected by a measure made at other distant place.

Experiments proposed to reveal the eventual existence of these waves can be based or in direct detection or in interferometric methods. In direct measurements, as proposed by Selleri[4], the wave without singularity is injected in a special laser gain tube where it gives origin to a zero energy-transfer stimulated emission. The interferometric processes can be with coherent or incoherent superposition[5] of the waves. The starting point of the method of incoherent overlapping was first presented by J. and M. Andrade e Silva[6] and was further developed to practical experiments by the author. An ultimate version of these experiments[7] has been recently done, in a somewhat modified version, by Mandel and his group at Rochester[8]. The results of the experiment, according to the authors, contradict the model of de Broglie and are favourable to the existence of the collapse of the wave function. Thus the experiment is a validation that the ψ function is indeed a probability wave as the plain quantum theory asserts. However, in a problem of this importance one ought to rule out any hypothesis, even if small, that may cast some shadow of doubt on the subject. On the

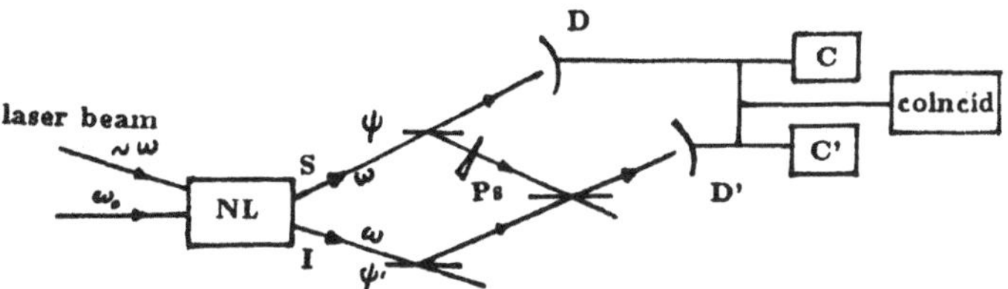

Fig. 1. Outline of the proposed experiment for detection of the empty waves of de Broglie

other hand it would prove interesting to cross confirm those results in other different and more clear experiment.

In fact more simple experiments, from the conceptual point of view, are possible. These experiments are based in coherent superposition. The first proposal was made in 1982 by Garuccio, Rapisarda and Vigier[9]. However the process to produce two nearly coherent photons at the same time do not seemed to be experimentally feasible. Now, thanks mainly to the work of Mandel and his group[10], the way to make such experiment is open. They were able to produce a source emitting two photons strongly correlated in time and mutually coherent.

3. AN EXPERIMENT WITH TWO COHERENT PHOTONS FOR THE EVENTUAL DETECTION OF THE WAVES OF DE BROGLIE

In this experiment it is convenient to start superimposing the two coherent wave packets, produced in a Mandel's source, and determine the fringe visibility of the interference pattern. Theoretically the expected fringe visibility should be one

$$I' \propto 1 + \cos\delta, \tag{1}$$

where δ is the relative phaseshift between the two waves. However, due to the misalignment of the whole apparatus, the partial monochromaticity of the two beams, imperfect overlapping of the two wave packets and other causes one should write instead

$$I \propto 1 + \alpha\cos\delta, \qquad 0 \leq \alpha \leq 1, \tag{2}$$

α being the visibility parameter which accounts for all known and unknown causes responsible for the loss of visibility. Once having the visibility parameter α experimentally determined one is ready to proceed with the test of the particle model of de Broglie.

Consider the experimental situation depicted in figure 1 where the signal photon, coming from the nonlinear crystal, is divided, at the beam splitter. If one looks for coincidences between D and D', what shall be expected? The predictions for coincidence rate depends, of course, on the theory considered.

According to the usual theory what one calls photons, as a language abuse, are no more than probability amplitudes with no physical meaning before measurement.

Therefore if detector D happens to click, meaning in common language that the photon was localized there, it turns out that the probability of being at D' is zero. That is collapse of the probability amplitude took place. In such case the expected coincidence rate shall be constant and independent of the relative phaseshift of the waves,

$$R = constant \tag{3}$$

If the realistic model of de Broglie is correct, the wave of the signal photon upon reaching the beamsplitter is divided into two waves, one directed towards D and the other to D'. The signal-photon-singularity follows one and only one of the ways. If the detector D was activated that means that the singularity went that particular way, but, and here lies the difference between the two possible interpretations, a real wave without singularity shall also arrive at D'. So there were no collapse of the wave function. In such circumstances, according to this realistic model, at D' two real photon wave packets will overlap. These two waves share a singularity, coming from I, and will guide it to the points of higher amplitude intensity, given origin as a consequence to an interference pattern. Therefore the expected coincidence R_c rate will depend on the relative phase shift between the two waves and shall be proportional to the initial distribution given by equation (2)

$$R_c \propto I \propto 1 + \alpha \cos \delta \tag{4}$$

ACKNOWLEDGMENTS

The author wants to thank Prof. Paolo Tombesi for his gentle invitation to participate at the Nato Meeting at Cortina and also Prof. John Rarity for some helpful discussion and hints concerning the present experiment.

REFERENCES

1. L. de Broglie, *The Current Interpretation of Wave Mechanics,* Elsevier, Amsterdam, 1969
2. Y. Koh and Sasaki, *in Microphysical Reality and Quantum Formalism,* eds. A van der Merwe et al., Kluwer Academic Publishing, 1969
3. J. R. Croca, *in The Concept of Probability; in The Philosophical Foundations of Quantum Formalism* eds. E. Bitsakis et al. Reidel, Dordrecht, 1988; eds. R. Nair, World Scientific Singapore, in Print
4. F. Selleri, *Found. Phys.* 12(1982)1057
5. J. R. Croca, *in Problems in Quantum Physics,* eds. L. Kostro et al. World Scientific, Singapore, 1988
6. J. and M. Andrade e Silva, *C. R. Acd. Sci. (Paris)* 20(1980)501
7. J. R. Croca, V. G. Lepore, A. Garuccio and R. N. Moreira, *Found. Phys. Lett.* 3(1990)557
8. L. J. Wang, X. Y. Zou and L. Mandel, *Phys. Rev. Lett.,* 66(1991)1111
9. A. Garuccio, V. Rapisarda and J. P. Vigier, *Phys. Lett. A,* 90(1982)17
10. Z. Y. Ou, L. J. Wang, X. Y. Zou and L. Mandel, *Phys. Rev. A,* 41(1990)566; 41(1990)1597

QUANTUM CRYPTOGRAPHY AND BELL'S THEOREM

Artur K. Ekert

Wolfson College, Oxford University
Oxford OX2 6UD (U.K.)

1. INTRODUCTION

In this article, I would like to show you a glimpse of beauty which one can easily find in the union of cryptography and quantum theory. So this is a story about quanta and ciphers. I will show how quantum mechanics protects the so-called key distribution process in cryptography. The proposed scheme is based on the well-known Bohm's version of the Einstein-Podolsky-Rosen *gedankenexperiment*[1,2]; the generalized Bell's theorem (Clauser - Horne - Shimony - Holt inequalities)[3,4] is used to test for eavesdropping. However, before I proceed any further, let me start with some historical remarks followed by some basic notions of cryptography.

2. SOME HISTORY

Origins of cryptography go back to 400 BC, when Spartans employed a device known as a *scytale* for secret communication between military commanders. The device consisted of a tapered baton around which was wrapped in a spiral strip of parchment or leather containing the message. When unwrapped, the letters of the message appeared scrambled; however, when the parchment was wrapped around another baton of the same shape, the original message reappeared. More modern cryptography and the first systematic cryptocorrespondence began around AD 1200 in the Papal state and the Italian city-states. At that time (circa 1379 - now in the Vatican archives) the first manual on ciphers appeared, written by Gabriele de Lavinde of Parma, who served Clement VII. From then onwards ciphers became a common tool of diplomatic and military communication, and of course the art of breaking ciphers - cryptoanalysis - had to follow. Cryptoanalysis often affected the course of history. In the First World War the Russians lost the Battle of Tannenberg in August 1914 because of the failure of their cryptographic communication. The entry of the United States into the First World War was partly a consequence of the breaking of one of the most famous cryptograms in history, the telegram of January 16, 1917, offering Mexico territorial gains if she would enter the war on the German side. In the Second World War, the Battle of Midway was a stunning

Quantum Measurements in Optics, Edited by P. Tombesi and
D.F. Walls, Plenum Press, New York, 1992

victory of intelligence and cryptoanalysis; but probably the most spectacular role of the cryptoanalysis of that time was revealed in the hearings held by the committee investigating the Pearl Harbor attack. After the war it was made public that the United States had been able to reconstruct the Japanese cipher machine shortly before the outbreak of the war. The machine was used for the highest level Japanese diplomatic communications, enabling the Americans to read these secret messages throughout the war. The Battle of the Pacific is often cited as one of the most important victories involving cryptoanalysis.

3. CRYPTOGRAPHY TODAY

Cryptography, despite a colourful history, only became part of mathematics and information theory this century, in the late 1940s, mainly due to the seminal papers of Claude Shannon [5]. Today, one can briefly define cryptography as a mathematical system of transforming information so that it is unintelligible and therefore useless to those who are not meant to have access to it. Originally the security of a cryptotext depended on the secrecy of the entire encrypting and decrypting procedures; however, today we use ciphers for which the algorithm for encrypting and decrypting could be revealed to anybody without compromising the security of a particular cryptogram. In such ciphers a set of specific parameters, called a *key*, is supplied together with the plaintext as an input to the encrypting algorithm, and together with the cryptogram as an input to the decrypting algorithm. The encrypting and decrypting algorithms are publicly announced; the security of the cryptogram depends entirely on the secrecy of the key, and this key, which is very important, may consist of any *randomly chosen*, sufficiently long string of bits. Once the key is established, subsequent communication involves sending cryptograms over a public channel which is vulnerable to total passive interception (e.g. public announcement in mass-media). However in order to establish the key, two users, who share no secret information initially, must at a certain stage of communication use a reliable and very secure channel. Since interception is a set of measurements performed by the eavesdropper on this channel, however difficult this might be from a technological point of view, *in principle* any classical channel can always be passively monitored, without the legitimate users being aware that any eavesdropping has taken place.

The problem of key distribution was partially solved, in the seventies, by mathematicians, who proposed public-key cryptography with two keys - public and private. Each user of the system secretly chooses a private key, which is never transmitted over any channel, and calculates from it a public key. The public key is known to everybody and everybody can send cryptograms to a given receiver, say Bob, using Bob's public key. However, once the message is scrambled one needs the private key for the decryption, which means that only Bob can read messages encrypted with his public key. In secret-key cryptography involving a community of N people any of whom may wish to communicate secretly with any other, you need to distribute $N(N-1)/2$ keys in advance, whereas in public-key cryptography you need only N keys, and even those you can distribute at any time and publicly. The secrecy of the public-key cryptosystem depends on the computing power of adversaries, and no classical public-key cryptographic system has yet been proved secure. On the contrary - one of the first public-key cryptosystems, "the knapsack code"- has been cracked by Adi Shamir of the Weizmann Institute of Science in 1982. The increasing speed of computers and new algorithms threaten further advances in public-key cryptography.

Now, suprisingly, the rescue comes from physics. Charles Bennett and Gilles Brassard have shown that quantum complementarity can be used to protect the key distribution against any passive interception, even if an adversary is equipped with unlimited computing power[6]. Their innovative work originated quantum cryptography which provides a qualitatively different solution to the secrecy of information. It is a physical law, not just a computational difficulty, that protects information, and this fact has far reaching consequences...

In the following I describe a quantum channel which distribute the key without any "element of reality" associated with the code. From a theoretical point of view the channel, by employing Bell's theorem, provides an interesting extension of Bennett's and Brassard's original idea, and from an experimental perspective offers a practical realisation by a small modification of experiments that were set up to test Bell's inequalities.

4. QUANTUM KEY DISTRIBUTION

The channel consists of a source that emits pairs of spin $\frac{1}{2}$ particles, in a singlet state. The particles fly apart along the z-axis towards the two legitimate users of the channel, say Alice and Bob, who, after the particles have separated, perform measurements on spin components along one of three directions, given by unit vectors $\vec{a}_i$ and $\vec{b}_j$ $(i,j = 1,2,3)$, respectively for Alice and Bob. For simplicity both $\vec{a}_i$ and $\vec{b}_j$ vectors lie in the x-y plane, perpendicular to the trajectory of the particles, and are characterized by azimuthal angles: $\phi_1^a = 0, \phi_2^a = \frac{1}{4}\pi, \phi_3^a = \frac{1}{2}\pi$ and $\phi_1^b = \frac{1}{4}\pi, \phi_2^b = \frac{1}{2}\pi, \phi_3^b = \frac{3}{4}\pi$. Superscripts "a" and "b" refer to Alice's and Bob's analysers respectively, and the angle is measured from the vertical x-axis. The users choose the orientation of the analysers randomly and independently for each pair of incoming particles. Each measurement, in $\frac{1}{2}\hbar$ units, can yield two results: +1 (spin up) and -1 (spin down) and can potentially reveal one bit of information.

The quantity

$$E(\vec{a}_i, \vec{b}_j) = P_{++}(\vec{a}_i, \vec{b}_j) + P_{--}(\vec{a}_i, \vec{b}_j) - P_{+-}(\vec{a}_i, \vec{b}_j) - P_{-+}(\vec{a}_i, \vec{b}_j) \tag{1}$$

is the correlation coefficient of the measurements performed by Alice along $\vec{a}_i$ and by Bob along $\vec{b}_j$. Here $P_{\pm\pm}(\vec{a}_i, \vec{b}_j)$ denotes the probability that result ± 1 has been obtained along $\vec{a}_i$ and ± 1 along $\vec{b}_j$. According to the quantum rules

$$E(\vec{a}_i, \vec{b}_j) = -\vec{a}_i \cdot \vec{b}_j. \tag{2}$$

For the two pairs of analysers of the same orientation ($\vec{a}_2$, $\vec{b}_1$ and $\vec{a}_3, \vec{b}_2$) quantum mechanics predicts total anticorrelation of the results obtained by Alice and Bob: $E(\vec{a}_2, \vec{b}_1) = E(\vec{a}_3, \vec{b}_2) = -1$.

Let us also, following Clauser, Horne, Shimony, and Holt[4], define a quantity composed of the correlation coefficients for which Alice and Bob used analysers of different orientation

$$S = E(\vec{a}_1, \vec{b}_1) - E(\vec{a}_1, \vec{b}_3) + E(\vec{a}_3, \vec{b}_1) + E(\vec{a}_3, \vec{b}_3). \tag{3}$$

Again, quantum mechanics requires

$$S = -2\sqrt{2}. \tag{4}$$

After the transmission has taken place, Alice and Bob can announce in public the orientations of the analysers they have chosen for each particular measurement and divide the measurements into two separate groups: a first group for which they used different orientation of the analysers, and a second group for which they used the same orientation of the analysers. They discard all measurements in which either or both of them failed to register a particle at all. Subsequently Alice and Bob can reveal publicly the results they obtained but within the first group of measurements only. This allows them to establish the value of S, which if the particles were not directly or indirectly "disturbed" should reproduce the result of Eq.(4). This assures the legitimate users that the results they obtained within the second second group of measurements are anticorrelated and can be converted into a secret string of bits — the key.

The eavesdropper cannot elicit any information from the particles while in transit from the source to the legitimate users, simply because there is no information encoded there! The information "comes into being" only after the legitimate users perform measurements and communicate in public afterwards. The eavesdropper may try to substitute his own prepared data for Alice and Bob to misguide them, but as he does not know which orientation of the analysers will be chosen for a given pair of particles, there is no good strategy to escape being detected. In this case his intervention will be equivalent to introducing elements of *physical reality* to the measurements of the spin components. This can easily be seen if we put appropriately modified (by the eavesdropper perfect measurement) correlation coefficients into Eq. (3). We obtain

$$S = \int \rho(\vec{n}_a, \vec{n}_b) d\vec{n}_a d\vec{n}_b [(\vec{a}_1 \cdot \vec{n}_a)(\vec{b}_1 \cdot \vec{n}_b) - (\vec{a}_1 \cdot \vec{n}_a)(\vec{b}_3 \cdot \vec{n}_b) + (\vec{a}_3 \cdot \vec{n}_a)(\vec{b}_1 \cdot \vec{n}_b) + (\vec{a}_3 \cdot \vec{n}_a)(\vec{b}_3 \cdot \vec{n}_b)], \tag{5}$$

where $\vec{n}_a$ and $\vec{n}_b$ are two unit vectors (for particles "a" and "b" respectively), oriented along the directions of the quantisation axes for which the eavesdropper acquired information about the spin component of a given particle. This information could be acquired either through a direct, "brute" measurement of the spin components or through a more subtle attack on the source, e.g. substituting a source that produces a state of two spin $\frac{1}{2}$ particles correlated with another quantum system on which the actual measurement will be performed by the eavesdropper.

One may think, for example, about delayed measurement on the third particle which is correlated with the two spin $\frac{1}{2}$ particles. By "delayed" I mean "after the orientation of the analysers has been publicly revealed by Alice and Bob". However, as we want the two particles to be in a pure, singlet state, and Alice and Bob test for it through Bell's theorem then we cannot correlate the third particle with the other two without disturbing the purity of the singlet state. There is no universal (good for all orientations $\vec{a}_i, \vec{b}_j$) state of the faked source which will pass the statistical test of the legitimate users on the subsystem of the two correlated particles "a" and "b", and provides the eavesdropper with the information about the results of Alice's and Bob's measurements[7].

In the equation above the normalised probability measure $\rho(\vec{n}_a, \vec{n}_b)$ describes the eavesdropper strategy (probability of intercepting a spin component along a given direction for a particular measurement). If only one particle (say "a") is exposed to the measurement performed by the eavesdropper along the direction $\vec{n}_a$, one may put $\vec{n}_b = -\vec{n}_a$ as a particular case in Eq. (5).

Simple calculation for a given orientation of $\vec{a}_1, \vec{a}_3$ or $\vec{b}_1, \vec{b}_3$ gives

$$S = \int \rho(\vec{n}_a, \vec{n}_b) d\vec{n}_a d\vec{n}_b [\sqrt{2}\vec{n}_a \cdot \vec{n}_b], \tag{6}$$

which implies

$$-\sqrt{2} \leq S \leq \sqrt{2} \tag{7}$$

and contradicts Eq. (4). for any strategy described by the measure $\rho(\vec{n}_a, \vec{n}_b)$. In this way it has been shown that the generalized Bell's theorem can have a practical application in cryptography, namely, it can test the safety of the key distribution. It is not a mathematical difficulty of a particular computation, but a fundamental physical law that protects the system.

5. EXPERIMENTAL REALIZATION

I have already mentioned that the proposed channel can be realized as modification of experiments that tested Bell's theorem. In particular, the celebrated Alan Aspect's experiment[8,9], in which polarized photons were used instead spin $\frac{1}{2}$ particles, would be the most obvious choice. In the experiment, every 10 nanoseconds pairs of photons were emitted in a radiative atomic cascade of calcium. Acousto-optical switches were used to change the orientation of the analysers in a time short compared with the photon transit time, and the detection efficiency was over 95%. Apart from changing the main objective of the experiment, and some details in the set-up, one will also need a software to simulate Alice, Bob, and optionally the eavesdropper.

All these changes are minor so perhaps we would not wait long for the experimental realization!

ACKNOWLEDGMENTS

The author thanks Paolo Tombesi and Elisabeth Giacobino for a wonderful week in Cortina d'Ampezzo, and David Deutsch, Peter Knight, Steve Barnett and Keith Burnett for comments and discussion.

This work was supported by Pirie-Reid Fund at Oxford University.

REFERENCES

1. A.Einstein, B.Podolsky and N.Rosen, *Phys.Rev.* 47: 777 (1935).
2. D.Bohm, *Quantum Theory* Prentice-Hall, Englewood Cliffs, N.J.,(1951).
3. J.S.Bell, *Physics (N.Y.)* 1: 195 (1965).

4. J.F.Clauser,M.A.Horne, A.Shimony, and R.A.Holt, *Phys.Rev.Lett.* 23: 880, (1969).
5. C.E.Shannon, *Bell Syst.Tech.Journ.* 28: 656 (1949).
6. C.H.Bennett and G.Brassard, *SIGACT News* 20: 78 (1989).
7. A.K.Ekert, *Ph.D. Thesis* Oxford University (in preparartion).
8. A.Aspect, P.Grangier, and G.Roger, *Phys.Rev.Lett.* 49: 91 (1982).
9. A.Aspect, J.Dalibard, and G.Roger, *Phys.Rev.Lett.* 49: 1804 (1982).

IS COMPLEMENTARITY COMPATIBLE WITH QUANTUM OPTICS?

Augusto Garuccio and Vito Luigi Lepore

Dipartimento di Fisica, Università di Bari
INFN, Sezione di Bari
Via G. Amendola, 173
I-70126 Bari, Italy

1. INTRODUCTION

Complementarity, viz., the fact that causality excludes spatio-temporal description and vice-versa has been placed by Bohr at the heart of the Copenhagen interpretation of quantum mechanics. However, notwithstanding the impressive success of the quantum-mechanical formalism, the foundational problems are far from being settled[1]. Although complementarity involves many aspects of quantum mechanics, the most well-known consequence of it is the impossibility of observing interference fringes and simultaneously the route of the corpuscle in Young's double-slit experiment[2]. In the following, we shall present a two-port interferometer in which it is possible to detect of the path of the corpuscle even though the photon follows both the branches of the interferometer, and, therefore one might expects to have interference on the basis of classical optics. We shall also analyse the proposed experiment within the quantum-optical theory and show that on detecting the path of the corpuscle one destroys interference. Therefore, at least for the considered experimental situation, Bohr's complementarity is compatible with the second quantization of the electromagnetic field.

2. TWO IMPORTANT EXPERIMENTS ON PAIRS OF PHOTONS

In the process of parametric down-conversion photons from an incident laser beam with wavevector $\mathbf{k}_0$, and angular frequency ω_0, interact with a nonlinear crystal, and split into two lower-frequency signal and idler photons of frequencies ω_1 and ω_2, and wavevectors $\mathbf{k}_1$ and $\mathbf{k}_2$, such that

Quantum Measurements in Optics, Edited by P. Tombesi and
D.F. Walls, Plenum Press, New York, 1992

$$\omega_0 = \omega_1 + \omega_2$$

$$\mathbf{k}_0 = \mathbf{k}_1 + \mathbf{k}_2$$

The photons of the pair are produced simultaneously[3,4] in their respective modes. In the degenerate case both signal and idler photons are characterized by the same frequency

$$\omega_1 = \omega_2 = \frac{1}{2}\omega_0$$

On directing the photons on the opposite sides of a 50%:50% beam splitter as is shown in Fig. 1, one observes[5,6] that the coincidence counting rate between D_1 and D_2 vanishes.

From the theoretical point of view we shall assume that the photons can be described by monochromatic fields; such a description is an over-simplification, however, it contains the main physical features. Denoting by $\hat{a}_i$ and $\hat{a}_s$ the annihilation operators along the idler and signal modes, the state produced is

$$|\psi\rangle = \hat{a}_s^{\dagger}\hat{a}_i^{\dagger}|vac\rangle \tag{1}$$

The joint probability of detection of photons at both detectors, D_1 and D_2, is given[7] by

$$P_{12} = K_{12} \langle\psi|\hat{E}_1^{(-)}\hat{E}_2^{(-)}\hat{E}_2^{(+)}\hat{E}_1^{(+)}|\psi\rangle \tag{2}$$

where K_{12} is a constant characteristic of the detection process, $|\psi\rangle$ is the state of the photons. For the fields we assume the single-mode treatment

$$\hat{E}_1^{(+)}(\mathbf{r}_1,t) = i\left(\frac{\hbar\omega}{2\varepsilon_0 V}\right)^{1/2}\hat{b}_1\, e^{i(\mathbf{k}_1\cdot\mathbf{r}_1 - \omega t)}$$

$$\hat{E}_2^{(+)}(\mathbf{r}_2,t) = i\left(\frac{\hbar\omega}{2\varepsilon_0 V}\right)^{1/2}\hat{b}_2\, e^{i(\mathbf{k}_2\cdot\mathbf{r}_2 - \omega t)} \tag{3}$$

with $\hat{b}_1$ and $\hat{b}_2$ denoting the corresponding annihilation operators of the outgoing modes at the detectors.

According to quantum optics[8], a beam splitter is described by the complex reflection and transmission coefficients r and t, which satisfy the relations

$$|r|^2 + |t|^2 = 1$$
$$rt^* + r^*t = 0 \tag{4}$$

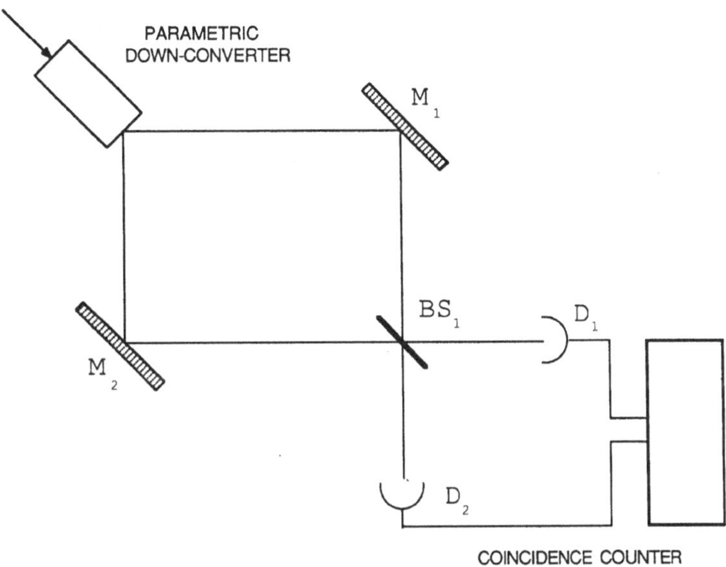

Fig. 1 - *Two photons sent from opposite sides emerge both along the same path.*

The outgoing annihilation operators $\hat{b}_1$ and $\hat{b}_2$ can be written in terms of the incoming annihilation operators at the beam splitter BS_1 $\hat{a}_i$ and $\hat{a}_s$ (see Fig. 1)

$$\hat{b}_1 = r\hat{a}_i + t\hat{a}_s \tag{5}$$

$$\hat{b}_2 = t\hat{a}_i + r\hat{a}_s \tag{6}$$

On making use of the expressions (2), (3), (5) and (6) and assuming that the state of the photons is given by (1), one obtains

$$P_{12} = K_{12} \left(\frac{\hbar\omega}{2\varepsilon_0 V}\right)^2 \left(|r|^2 - |t|^2\right)^2 \quad (7)$$

For a 50%:50% beam splitter $|r|^2=|t|^2= 1/2$, the joint detection probability P_{12} vanishes. The lack of coincidence is considered as an evidence of the fact that both the corpuscles go along the same path[8].

However, a simple corpuscular picture of the previous experiment is misleading. Indeed, a second experiment[9,10] has shown that, on directing the outgoing beams on a second beam splitter as in Fig. 2, interference fringes appear.

The photons produced by the parametric down-converter following different routes, after the reflection on the mirrors M_1 and M_2, pass through the beam splitter BS_1. The position of BS_1 is chosen so that the conditions of the previous experiment are satisfied. The state produced is allowed to interfere in the Mach-Zehnder formed by the beam splitters BS_1 and BS_2 by the mirrors M_3 and M_4. The joint detection probability at the detectors D_1 and D_2 depends on the optical lengths z_1, z_2, z_3 and z_4 of the interferometer, therefore, interference displays. The dependence upon the optical length difference is predicted by quantum optics. The outgoing annihilation operators $\hat{b}_1$ as a function of the incoming annihilation operators $\hat{a}_i$ and $\hat{a}_s$ is

$$\hat{b}_1 = \left[r^2 e^{ik(z_1 + z_2)} + t^2 e^{ik(z_3 + z_4)} \right] \hat{a}_i + \left[rt\, e^{ik(z_1 + z_2)} + rt\, e^{ik(z_3 + z_4)} \right] \hat{a}_s \quad (8)$$

Similarly, for the outgoing operator $\hat{b}_2$ one has

$$\hat{b}_2 = \left[rt\, e^{ik(z_1 + z_2)} + rt\, e^{ik(z_3 + z_4)} \right] \hat{a}_i + \left[t^2 e^{ik(z_1 + z_2)} + r^2 e^{ik(z_3 + z_4)} \right] \hat{a}_s \quad (9)$$

Inserting (8) and (9) in the expression (2), and assuming that the state of the photons is given by (1), one obtains

$$P_{12} = K_{12} \left(\frac{\hbar\omega}{2\varepsilon_0 V}\right)^2 \left[\left(|r|^2 - |t|^2\right)^2 - 4|r|^2\, |t|^2 \cos k\Delta z \right]^2 \quad (10)$$

where $\Delta z = (z_3 + z_4) - (z_1 + z_2)$ is the optical length difference between the two paths of the interferometer. For a 50%:50% beam splitter $|r|^2=|t|^2= 1/2$,

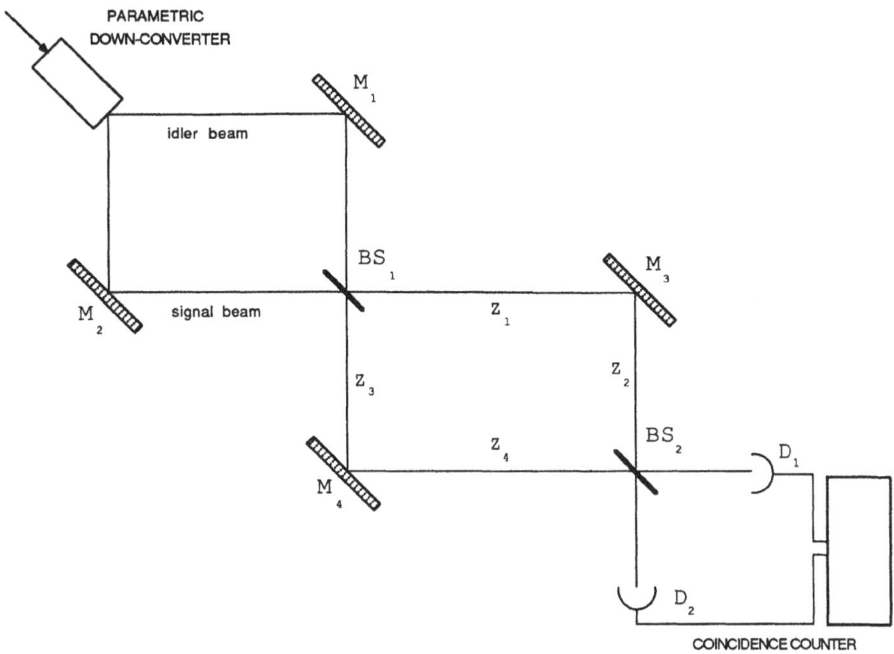

Fig. 2 - *Two photons sent from opposite sides in a the two-port Mach-Zehnder interferometer produce interference.*

the joint detection probability is

$$P_{12} = K_{12} \left(\frac{\hbar\omega}{2\varepsilon_0 V} \right)^2 \frac{1}{2} \left(1 + \cos 2k\Delta z \right) \tag{11}$$

The dependence upon the optical lengths clearly displays interference.

3. PARTICLE TRAJECTORY AND INTERFERENCE

From the preceding section we can draw the conclusion that for a Mach-Zehnder interferometer, when two photons are simultaneously directed on the opposite sides of the entrance beam splitter they both follow the same route (first experiment) and produce interference (second experiment). The situation is very similar to the standard Mach-Zehnder interferometer with single-photon states[11]: An interference pattern is built up on collecting

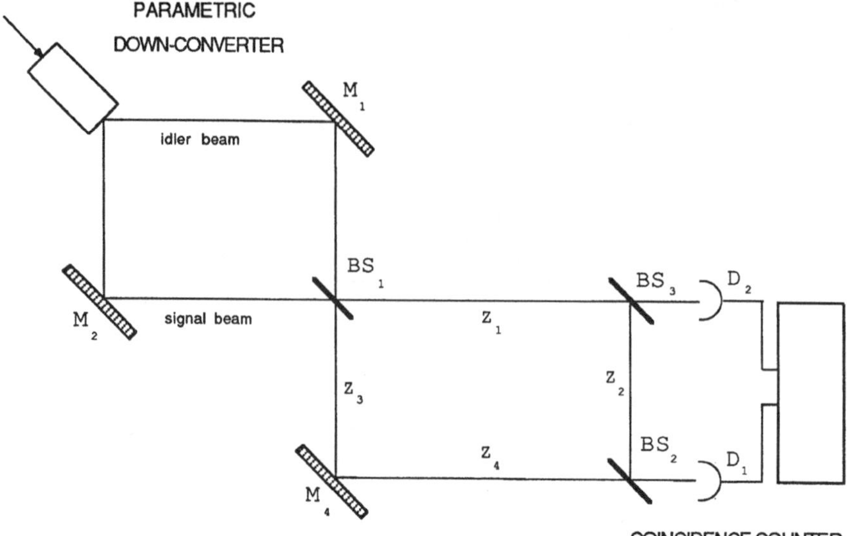

Fig. 3 - *The interferometer of Fig. 2 is modified substituting the mirror* M_3 *with the beam splitter* BS_3 *and moving the photomultiplier* D_2 *in front of the new output port. If a photon is revealed in* D_2 *its twin is on the route* BS_1-BS_3-BS_2. *Does it display interference on passing through the beam splitter* BS_2*?*

repeated events with a detector placed at one of the output port of the interferometer; on the other hand, if two detectors are placed along the routes of the interferometer coincidences disappears. However, in the two input interferometer it is possible to reveal the path of both the corpuscles on detecting one of them.

Now, let us consider the Mach-Zehnder interferometer, modified as is shown in Fig. 3. Two photons simultaneously directed on the opposite sides of the beam splitter BS_1 emerge both along the same path. Consider the case in which one of the photons is detected in D_2; then one can say that the twin photon followed the route BS_1-BS_3-BS_2. If the probability of revealing a photon in D_1 and a photon in D_2 depends upon the optical length difference between the two arms of the interferometer, one observes simultaneously the trajectory and the interference. To carry out the quantum optical predictions for the experiment under discussion we need the expression for the outgoing annihilation operator $\hat{b}_1$

$$\hat{b}_1 = \left[r^3 e^{ik(z_1 + z_2)} + t^2 e^{ik(z_3 + z_4)} \right] \hat{a}_i +$$

$$\left[r^2 t\, e^{ik(z_1 + z_2)} + rt\, e^{ik(z_3 + z_4)} \right] \hat{a}_s \tag{12}$$

and for the outgoing annihilation operator $\hat{b}_2$

$$\hat{b}_2 = rt\, e^{ikz_1} \hat{a}_i + t^2 e^{ikz_1} \hat{a}_s \tag{13}$$

Inserting eqns. (12) and (13) in (3) and (2), one easily obtains (ϑ_r is the argument of r)

$$P_{12} = K_{12} \left(\frac{\hbar\omega}{2\varepsilon_0 V}\right)^2 |t|^4 \Big[4|r|^4 +$$

$$\left(|r|^2 - |t|^2 \right) \left(|r|^2 - |t|^2 + 4|r|^3 \cos (k\Delta z - \vartheta_r) \right) \Big] \tag{14}$$

which for a 50%:50% beam splitter becomes

$$P_{12} = K_{12} \left(\frac{\hbar\omega}{2\varepsilon_0 V}\right)^2 \frac{1}{4} \tag{15}$$

The joint-detection probability (15) does not show any dependence on the difference of optical lengths of the interferometer, i.e., according to quantum optics on detecting the trajectories, one destroys interference. This is exactly what complementarity claims.

Thus, on the one hand, we can deduce from the experimental evidence that[5,6] both the photons follow either the path BS_1-M_3-BS_2 or the path BS_1-M_4-BS_2 in the interferometer of Fig. 2; however, on measuring the coincidence rate between the two output ports one observes interference[9,10]. On the other hand, quantum-optical formalism predicts that the interference pattern is wiped out if the trajectories of photons are displayed.

The proposed experiment is, thus, a test for both complementarity and quantum optics, which can be carried out with the available technology, since similar apparata have recently been set up[5,6,9,10].

REFERENCES

1. F. Selleri, *Quantum Paradoxes and Physical Reality*, Kluwer, Dordrecht (1990).

2. M. Jammer, *The Conceptual Development of Quantum Mechanics*, McGraw-Hill, New York (1966), p. 345.
3. D.C. Burnham and D.L. Weinberg, *Phys. Rev. Lett.* 25:84 (1970).
4. S. Friberg, C.K. Hong, and L. Mandel *Phys. Rev. Lett.* 54:2011 (1985).
5. C.K. Hong, Z.Y. Ou, and L. Mandel *Phys. Rev. Lett.* 59:2044 (1987).
6. J.G. Rarity and P.R. Tapster, *Phys. Rev.* A41:5139 (1990).
7. R.J. Glauber, *Phys. Rev.* 130:2529 (1963); R.J. Glauber, *Phys. Rev.* 131:2766 (1963).
8. H. Fearn and R. Loudon, *Opt. Commun.* 64:485 (1987).
9. Z.Y. Ou, X.Y. Zou, L.J. Wang, and L. Mandel, *Phys. Rev.* A42:2957 (1990).
10. J.G. Rarity, P.R. Tapster, E. Jakeman, T. Larchuk, R.A. Campos, M.C. Teich, and B.E.A. Saleh, *Phys. Rev. Lett.* 65:1348 (1990).
11. P. Grangier, G. Roger, and A. Aspect, *Europhys. Lett.* 1:173 (1986).

Advance Research Workshop on Quantum Measurements in Optics
January 21 - 25, Cortina d'Ampezzo, Italy

INDEX

Zeitfracht Medien GmbH
Ferdinand-Jühlke-Straße 7
99095 Erfurt, Deutschland
produktsicherheit@kolibri360.de